ADVANCED
VIBRATION
ANALYSIS

MECHANICAL ENGINEERING
A Series of Textbooks and Reference Books

Founding Editor

L. L. Faulkner

*Columbus Division, Battelle Memorial Institute
and Department of Mechanical Engineering
The Ohio State University
Columbus, Ohio*

1. *Spring Designer's Handbook*, Harold Carlson
2. *Computer-Aided Graphics and Design*, Daniel L. Ryan
3. *Lubrication Fundamentals*, J. George Wills
4. *Solar Engineering for Domestic Buildings*, William A. Himmelman
5. *Applied Engineering Mechanics: Statics and Dynamics*, G. Boothroyd and C. Poli
6. *Centrifugal Pump Clinic*, Igor J. Karassik
7. *Computer-Aided Kinetics for Machine Design*, Daniel L. Ryan
8. *Plastics Products Design Handbook, Part A: Materials and Components; Part B: Processes and Design for Processes*, edited by Edward Miller
9. *Turbomachinery: Basic Theory and Applications*, Earl Logan, Jr.
10. *Vibrations of Shells and Plates*, Werner Soedel
11. *Flat and Corrugated Diaphragm Design Handbook*, Mario Di Giovanni
12. *Practical Stress Analysis in Engineering Design*, Alexander Blake
13. *An Introduction to the Design and Behavior of Bolted Joints*, John H. Bickford
14. *Optimal Engineering Design: Principles and Applications*, James N. Siddall
15. *Spring Manufacturing Handbook*, Harold Carlson
16. *Industrial Noise Control: Fundamentals and Applications*, edited by Lewis H. Bell
17. *Gears and Their Vibration: A Basic Approach to Understanding Gear Noise*, J. Derek Smith
18. *Chains for Power Transmission and Material Handling: Design and Applications Handbook*, American Chain Association
19. *Corrosion and Corrosion Protection Handbook*, edited by Philip A. Schweitzer
20. *Gear Drive Systems: Design and Application*, Peter Lynwander
21. *Controlling In-Plant Airborne Contaminants: Systems Design and Calculations*, John D. Constance
22. *CAD/CAM Systems Planning and Implementation*, Charles S. Knox
23. *Probabilistic Engineering Design: Principles and Applications*, James N. Siddall
24. *Traction Drives: Selection and Application*, Frederick W. Heilich III and Eugene E. Shube

25. *Finite Element Methods: An Introduction,* Ronald L. Huston and Chris E. Passerello

26. *Mechanical Fastening of Plastics: An Engineering Handbook,* Brayton Lincoln, Kenneth J. Gomes, and James F. Braden

27. *Lubrication in Practice: Second Edition,* edited by W. S. Robertson

28. *Principles of Automated Drafting,* Daniel L. Ryan

29. *Practical Seal Design,* edited by Leonard J. Martini

30. *Engineering Documentation for CAD/CAM Applications,* Charles S. Knox

31. *Design Dimensioning with Computer Graphics Applications,* Jerome C. Lange

32. *Mechanism Analysis: Simplified Graphical and Analytical Techniques,* Lyndon O. Barton

33. *CAD/CAM Systems: Justification, Implementation, Productivity Measurement,* Edward J. Preston, George W. Crawford, and Mark E. Coticchia

34. *Steam Plant Calculations Manual,* V. Ganapathy

35. *Design Assurance for Engineers and Managers,* John A. Burgess

36. *Heat Transfer Fluids and Systems for Process and Energy Applications,* Jasbir Singh

37. *Potential Flows: Computer Graphic Solutions,* Robert H. Kirchhoff

38. *Computer-Aided Graphics and Design: Second Edition,* Daniel L. Ryan

39. *Electronically Controlled Proportional Valves: Selection and Application,* Michael J. Tonyan, edited by Tobi Goldoftas

40. *Pressure Gauge Handbook,* AMETEK, U.S. Gauge Division, edited by Philip W. Harland

41. *Fabric Filtration for Combustion Sources: Fundamentals and Basic Technology,* R. P. Donovan

42. *Design of Mechanical Joints,* Alexander Blake

43. *CAD/CAM Dictionary,* Edward J. Preston, George W. Crawford, and Mark E. Coticchia

44. *Machinery Adhesives for Locking, Retaining, and Sealing,* Girard S. Haviland

45. *Couplings and Joints: Design, Selection, and Application,* Jon R. Mancuso

46. *Shaft Alignment Handbook,* John Piotrowski

47. *BASIC Programs for Steam Plant Engineers: Boilers, Combustion, Fluid Flow, and Heat Transfer,* V. Ganapathy

48. *Solving Mechanical Design Problems with Computer Graphics,* Jerome C. Lange

49. *Plastics Gearing: Selection and Application,* Clifford E. Adams

50. *Clutches and Brakes: Design and Selection,* William C. Orthwein

51. *Transducers in Mechanical and Electronic Design,* Harry L. Trietley

52. *Metallurgical Applications of Shock-Wave and High-Strain-Rate Phenomena,* edited by Lawrence E. Murr, Karl P. Staudhammer, and Marc A. Meyers

53. *Magnesium Products Design,* Robert S. Busk

54. *How to Integrate CAD/CAM Systems: Management and Technology,* William D. Engelke

55. *Cam Design and Manufacture: Second Edition; with cam design software for the IBM PC and compatibles, disk included,* Preben W. Jensen

56.   Solid-State AC Motor Controls: Selection and Application, Sylvester Campbell
57.   Fundamentals of Robotics, David D. Ardayfio
58.   Belt Selection and Application for Engineers, edited by Wallace D. Erickson
59.   Developing Three-Dimensional CAD Software with the IBM PC, C. Stan Wei
60.   Organizing Data for CIM Applications, Charles S. Knox, with contributions by Thomas C. Boos, Ross S. Culverhouse, and Paul F. Muchnicki
61.   Computer-Aided Simulation in Railway Dynamics, by Rao V. Dukkipati and Joseph R. Amyot
62.   Fiber-Reinforced Composites: Materials, Manufacturing, and Design, P. K. Mallick
63.   Photoelectric Sensors and Controls: Selection and Application, Scott M. Juds
64.   Finite Element Analysis with Personal Computers, Edward R. Champion, Jr. and J. Michael Ensminger
65.   Ultrasonics: Fundamentals, Technology, Applications: Second Edition, Revised and Expanded, Dale Ensminger
66.   Applied Finite Element Modeling: Practical Problem Solving for Engineers, Jeffrey M. Steele
67.   Measurement and Instrumentation in Engineering: Principles and Basic Laboratory Experiments, Francis S. Tse and Ivan E. Morse
68.   Centrifugal Pump Clinic: Second Edition, Revised and Expanded, Igor J. Karassik
69.   Practical Stress Analysis in Engineering Design: Second Edition, Revised and Expanded, Alexander Blake
70.   An Introduction to the Design and Behavior of Bolted Joints: Second Edition, Revised and Expanded, John H. Bickford
71.   High Vacuum Technology: A Practical Guide, Marsbed H. Hablanian
72.   Pressure Sensors: Selection and Application, Duane Tandeske
73.   Zinc Handbook: Properties, Processing, and Use in Design, Frank Porter
74.   Thermal Fatigue of Metals, Andrzej Weronski and Tadeusz Hejwowski
75.   Classical and Modern Mechanisms for Engineers and Inventors, Preben W. Jensen
76.   Handbook of Electronic Package Design, edited by Michael Pecht
77.   Shock-Wave and High-Strain-Rate Phenomena in Materials, edited by Marc A. Meyers, Lawrence E. Murr, and Karl P. Staudhammer
78.   Industrial Refrigeration: Principles, Design and Applications, P. C. Koelet
79.   Applied Combustion, Eugene L. Keating
80.   Engine Oils and Automotive Lubrication, edited by Wilfried J. Bartz
81.   Mechanism Analysis: Simplified and Graphical Techniques, Second Edition, Revised and Expanded, Lyndon O. Barton
82.   Fundamental Fluid Mechanics for the Practicing Engineer, James W. Murdock
83.   Fiber-Reinforced Composites: Materials, Manufacturing, and Design, Second Edition, Revised and Expanded, P. K. Mallick
84.   Numerical Methods for Engineering Applications, Edward R. Champion, Jr.
85.   Turbomachinery: Basic Theory and Applications, Second Edition, Revised and Expanded, Earl Logan, Jr.
86.   Vibrations of Shells and Plates: Second Edition, Revised and Expanded, Werner Soedel

87. *Steam Plant Calculations Manual: Second Edition, Revised and Expanded*, V. Ganapathy

88. *Industrial Noise Control: Fundamentals and Applications, Second Edition, Revised and Expanded*, Lewis H. Bell and Douglas H. Bell

89. *Finite Elements: Their Design and Performance*, Richard H. MacNeal

90. *Mechanical Properties of Polymers and Composites: Second Edition, Revised and Expanded*, Lawrence E. Nielsen and Robert F. Landel

91. *Mechanical Wear Prediction and Prevention*, Raymond G. Bayer

92. *Mechanical Power Transmission Components*, edited by David W. South and Jon R. Mancuso

93. *Handbook of Turbomachinery*, edited by Earl Logan, Jr.

94. *Engineering Documentation Control Practices and Procedures*, Ray E. Monahan

95. *Refractory Linings Thermomechanical Design and Applications*, Charles A. Schacht

96. *Geometric Dimensioning and Tolerancing: Applications and Techniques for Use in Design, Manufacturing, and Inspection*, James D. Meadows

97. *An Introduction to the Design and Behavior of Bolted Joints: Third Edition, Revised and Expanded*, John H. Bickford

98. *Shaft Alignment Handbook: Second Edition, Revised and Expanded*, John Piotrowski

99. *Computer-Aided Design of Polymer-Matrix Composite Structures*, edited by Suong Van Hoa

100. *Friction Science and Technology*, Peter J. Blau

101. *Introduction to Plastics and Composites: Mechanical Properties and Engineering Applications*, Edward Miller

102. *Practical Fracture Mechanics in Design*, Alexander Blake

103. *Pump Characteristics and Applications*, Michael W. Volk

104. *Optical Principles and Technology for Engineers*, James E. Stewart

105. *Optimizing the Shape of Mechanical Elements and Structures*, A. A. Seireg and Jorge Rodriguez

106. *Kinematics and Dynamics of Machinery*, Vladimír Stejskal and Michael Valásek

107. *Shaft Seals for Dynamic Applications*, Les Horve

108. *Reliability-Based Mechanical Design*, edited by Thomas A. Cruse

109. *Mechanical Fastening, Joining, and Assembly*, James A. Speck

110. *Turbomachinery Fluid Dynamics and Heat Transfer*, edited by Chunill Hah

111. *High-Vacuum Technology: A Practical Guide, Second Edition, Revised and Expanded*, Marsbed H. Hablanian

112. *Geometric Dimensioning and Tolerancing: Workbook and Answerbook*, James D. Meadows

113. *Handbook of Materials Selection for Engineering Applications*, edited by G. T. Murray

114. *Handbook of Thermoplastic Piping System Design*, Thomas Sixsmith and Reinhard Hanselka

115. *Practical Guide to Finite Elements: A Solid Mechanics Approach*, Steven M. Lepi

116. *Applied Computational Fluid Dynamics*, edited by Vijay K. Garg

117. *Fluid Sealing Technology*, Heinz K. Muller and Bernard S. Nau
118. *Friction and Lubrication in Mechanical Design*, A. A. Seireg
119. *Influence Functions and Matrices*, Yuri A. Melnikov
120. *Mechanical Analysis of Electronic Packaging Systems*, Stephen A. McKeown
121. *Couplings and Joints: Design, Selection, and Application, Second Edition, Revised and Expanded*, Jon R. Mancuso
122. *Thermodynamics: Processes and Applications*, Earl Logan, Jr.
123. *Gear Noise and Vibration*, J. Derek Smith
124. *Practical Fluid Mechanics for Engineering Applications*, John J. Bloomer
125. *Handbook of Hydraulic Fluid Technology*, edited by George E. Totten
126. *Heat Exchanger Design Handbook*, T. Kuppan
127. *Designing for Product Sound Quality*, Richard H. Lyon
128. *Probability Applications in Mechanical Design*, Franklin E. Fisher and Joy R. Fisher
129. *Nickel Alloys*, edited by Ulrich Heubner
130. *Rotating Machinery Vibration: Problem Analysis and Troubleshooting*, Maurice L. Adams, Jr.
131. *Formulas for Dynamic Analysis*, Ronald L. Huston and C. Q. Liu
132. *Handbook of Machinery Dynamics*, Lynn L. Faulkner and Earl Logan, Jr.
133. *Rapid Prototyping Technology: Selection and Application*, Kenneth G. Cooper
134. *Reciprocating Machinery Dynamics: Design and Analysis*, Abdulla S. Rangwala
135. *Maintenance Excellence: Optimizing Equipment Life-Cycle Decisions*, edited by John D. Campbell and Andrew K. S. Jardine
136. *Practical Guide to Industrial Boiler Systems*, Ralph L. Vandagriff
137. *Lubrication Fundamentals: Second Edition, Revised and Expanded*, D. M. Pirro and A. A. Wessol
138. *Mechanical Life Cycle Handbook: Good Environmental Design and Manufacturing*, edited by Mahendra S. Hundal
139. *Micromachining of Engineering Materials*, edited by Joseph McGeough
140. *Control Strategies for Dynamic Systems: Design and Implementation*, John H. Lumkes, Jr.
141. *Practical Guide to Pressure Vessel Manufacturing*, Sunil Pullarcot
142. *Nondestructive Evaluation: Theory, Techniques, and Applications*, edited by Peter J. Shull
143. *Diesel Engine Engineering: Thermodynamics, Dynamics, Design, and Control*, Andrei Makartchouk
144. *Handbook of Machine Tool Analysis*, Ioan D. Marinescu, Constantin Ispas, and Dan Boboc
145. *Implementing Concurrent Engineering in Small Companies*, Susan Carlson Skalak
146. *Practical Guide to the Packaging of Electronics: Thermal and Mechanical Design and Analysis*, Ali Jamnia
147. *Bearing Design in Machinery: Engineering Tribology and Lubrication*, Avraham Harnoy
148. *Mechanical Reliability Improvement: Probability and Statistics for Experimental Testing*, R. E. Little

149. *Industrial Boilers and Heat Recovery Steam Generators: Design, Applications, and Calculations*, V. Ganapathy

150. *The CAD Guidebook: A Basic Manual for Understanding and Improving Computer-Aided Design*, Stephen J. Schoonmaker

151. *Industrial Noise Control and Acoustics*, Randall F. Barron

152. *Mechanical Properties of Engineered Materials*, Wolé Soboyejo

153. *Reliability Verification, Testing, and Analysis in Engineering Design*, Gary S. Wasserman

154. *Fundamental Mechanics of Fluids: Third Edition*, I. G. Currie

155. *Intermediate Heat Transfer*, Kau-Fui Vincent Wong

156. *HVAC Water Chillers and Cooling Towers: Fundamentals, Application, and Operation*, Herbert W. Stanford III

157. *Gear Noise and Vibration: Second Edition, Revised and Expanded*, J. Derek Smith

158. *Handbook of Turbomachinery: Second Edition, Revised and Expanded*, edited by Earl Logan, Jr. and Ramendra Roy

159. *Piping and Pipeline Engineering: Design, Construction, Maintenance, Integrity, and Repair*, George A. Antaki

160. *Turbomachinery: Design and Theory*, Rama S. R. Gorla and Aijaz Ahmed Khan

161. *Target Costing: Market-Driven Product Design*, M. Bradford Clifton, Henry M. B. Bird, Robert E. Albano, and Wesley P. Townsend

162. *Fluidized Bed Combustion*, Simeon N. Oka

163. *Theory of Dimensioning: An Introduction to Parameterizing Geometric Models*, Vijay Srinivasan

164. *Handbook of Mechanical Alloy Design*, edited by George E. Totten, Lin Xie, and Kiyoshi Funatani

165. *Structural Analysis of Polymeric Composite Materials*, Mark E. Tuttle

166. *Modeling and Simulation for Material Selection and Mechanical Design*, edited by George E. Totten, Lin Xie, and Kiyoshi Funatani

167. *Handbook of Pneumatic Conveying Engineering*, David Mills, Mark G. Jones, and Vijay K. Agarwal

168. *Clutches and Brakes: Design and Selection, Second Edition*, William C. Orthwein

169. *Fundamentals of Fluid Film Lubrication: Second Edition*, Bernard J. Hamrock, Steven R. Schmid, and Bo O. Jacobson

170. *Handbook of Lead-Free Solder Technology for Microelectronic Assemblies*, edited by Karl J. Puttlitz and Kathleen A. Stalter

171. *Vehicle Stability*, Dean Karnopp

172. *Mechanical Wear Fundamentals and Testing: Second Edition, Revised and Expanded*, Raymond G. Bayer

173. *Liquid Pipeline Hydraulics*, E. Shashi Menon

174. *Solid Fuels Combustion and Gasification*, Marcio L. de Souza-Santos

175. *Mechanical Tolerance Stackup and Analysis*, Bryan R. Fischer

176. *Engineering Design for Wear*, Raymond G. Bayer

177. *Vibrations of Shells and Plates: Third Edition, Revised and Expanded*, Werner Soedel

178. *Refractories Handbook*, edited by Charles A. Schacht

179. *Practical Engineering Failure Analysis*, Hani M. Tawancy, Anwar Ul-Hamid, and Nureddin M. Abbas

180. *Mechanical Alloying and Milling*, C. Suryanarayana

181. *Mechanical Vibration: Analysis, Uncertainties, and Control, Second Edition, Revised and Expanded*, Haym Benaroya

182. *Design of Automatic Machinery*, Stephen J. Derby

183. *Practical Fracture Mechanics in Design: Second Edition, Revised and Expanded*, Arun Shukla

184. *Practical Guide to Designed Experiments*, Paul D. Funkenbusch

185. *Gigacycle Fatigue in Mechanical Practive*, Claude Bathias and Paul C. Paris

186. *Selection of Engineering Materials and Adhesives*, Lawrence W. Fisher

187. *Boundary Methods: Elements, Contours, and Nodes*, Subrata Mukherjee and Yu Xie Mukherjee

188. *Rotordynamics*, Agnieszka (Agnes) Muszńyska

189. *Pump Characteristics and Applications: Second Edition*, Michael W. Volk

190. *Reliability Engineering: Probability Models and Maintenance Methods*, Joel A. Nachlas

191. *Industrial Heating: Principles, Techniques, Materials, Applications, and Design*, Yeshvant V. Deshmukh

192. *Micro Electro Mechanical System Design*, James J. Allen

193. *Probability Models in Engineering and Science*, Haym Benaroya and Seon Han

194. *Damage Mechanics*, George Z. Voyiadjis and Peter I. Kattan

195. *Standard Handbook of Chains: Chains for Power Transmission and Material Handling, Second Edition*, American Chain Association and John L. Wright, Technical Consultant

196. *Standards for Engineering Design and Manufacturing*, Wasim Ahmed Khan and Abdul Raouf S.I.

197. *Maintenance, Replacement, and Reliability: Theory and Applications*, Andrew K. S. Jardine and Albert H. C. Tsang

198. *Finite Element Method: Applications in Solids, Structures, and Heat Transfer*, Michael R. Gosz

199. *Microengineering, MEMS, and Interfacing: A Practical Guide*, Danny Banks

200. *Fundamentals of Natural Gas Processing*, Arthur J. Kidnay and William Parrish

201. *Optimal Control of Induction Heating Processes*, Edgar Rapoport and Yulia Pleshivtseva

202. *Practical Plant Failure Analysis: A Guide to Understanding Machinery Deterioration and Improving Equipment Reliability*, Neville W. Sachs, P.E.

203. *Shaft Alignment Handbook, Third Edition*, John Piotrowski

204. *Advanced Vibration Analysis*, Graham S. Kelly

ADVANCED VIBRATION ANALYSIS

S. GRAHAM KELLY

CRC Press
Taylor & Francis Group
Boca Raton London New York

CRC Press is an imprint of the
Taylor & Francis Group, an **informa** business
A TAYLOR & FRANCIS BOOK

CRC Press
Taylor & Francis Group
6000 Broken Sound Parkway NW, Suite 300
Boca Raton, FL 33487-2742

First issued in paperback 2019

ISBN-13: 978-0-8493-3419-1 (hbk)
ISBN-13: 978-0-367-38965-9 (pbk)

Library of Congress Cataloging-in-Publication Data

Kelly, S. Graham.
   Advanced vibration analysis / S. Graham Kelly.
     p. cm. -- (Dekker mechanical engineering)
   Includes bibliographical references and index.
   ISBN 0-8493-3419-5
   1. Vibration--Mathematical models. I. Title. II. Series.

TA355.K45 2006
620.1'1248--dc22   2006049912

Visit the Taylor & Francis Web site at
http://www.taylorandfrancis.com

and the CRC Press Web site at
http://www.crcpress.com

Dedication

To

Seala and Graham

Preface

Advanced Vibration Analysis is a result of three graduate classes that I have regularly taught at The University of Akron: Vibrations of Discrete Systems, Vibrations of Continuous Systems, and Engineering Analysis. The latter teaches a framework for the development of mathematical solutions for models of engineering systems, relying on knowledge of the physics of the system to guide the mathematical solution. The foundation upon which this framework is built is linear algebra, focusing on vector spaces and linear operators whose domain is a subspace of a larger vector space. While teaching the vibrations courses, it became clear that the analysis of vibration of a multi-degree-of-freedom system or a continuous system is best performed in this framework. This book ties all three courses together.

The objectives of this book are to develop a general mathematical framework for the analysis of a model of a physical system undergoing vibration and to illustrate how the physics of a problem is used to develop a more specific framework for the analysis of that problem. Such an analysis includes the determination of an exact solution for a linear problem and approximate solutions for problems in which an exact solution is difficult to obtain.

A general theory is developed that is applicable to both discrete and continuous systems. Presentation of the theory includes proofs of important results, especially proofs that are themselves instructive for a comprehensive understanding of the result. The application of the theory to discrete systems is straightforward, and its understanding requires little addition to what is developed in this book. A thorough understanding of the application of the theory to continuous systems requires additional discussion regarding convergence of sequences and series in infinite dimensional vector spaces. A basic discussion of the required theory is presented, but proofs are lengthy and not contained within this book.

The use of physics in the development of the mathematical framework is evident in the formation of concepts such as energy inner products. To this end, the book begins with a discussion of the physics of dynamic systems comprised of particles, rigid bodies, and deformable bodies. Components of mechanical systems are classified as storing kinetic energy, storing potential energy, dissipating energy, or being a source of energy. Chapter 1 concludes with a discussion of the physics and mathematics for the analysis of a system with a single-degree-of-freedom.

The development of mathematical models using energy methods is presented in Chapter 2. The chapter begins with a brief introduction to the calculus of variations. It shows how Newton's second law is used to develop an energy formulation for the analysis of a system, leading to Hamilton's Principle and Extended Hamilton's Principle. Calculus of variations applied to these principles leads to a method for the formulation of the mathematical model for both discrete and continuous systems, conservative and non-conservative. Application is made to a variety of problems.

The mathematical foundation for the framework is presented in Chapter 3. Vector spaces and concepts of basis and dimension are defined. Examples of finite- and infinite-dimensional vector spaces are provided. Operations between vectors called inner products are defined as well as a measure of the length of a vector called a norm. Linear operators and their domain and range are defined, and examples relevant to vibration analysis are provided. Inner products and norms are used to develop concepts of self-adjoint and positive-definite operators. Finally, energy inner products are developed for positive-definite and self-adjoint operators. The mathematical foundation developed in Chapter 3 is used throughout the remainder of the book to develop the framework for analyzing vibrations.

Chapter 4 illustrates the development and analysis of the linear operators used in vibrations problems. The variational methods developed in Chapter 2 are used to formulate the linear operators, while the mathematics of Chapter 3 is used to analyze the operators. Operators for a variety of systems, discrete and continuous, are developed and analyzed.

The framework developed in Chapter 1 through Chapter 4 leads to the formulation of the differential equations governing the response of a conservative linear system in terms of self-adjoint linear operators, the inertia operator, and the stiffness operator. The domains on which these operators exist are defined by the physics of the problem. For a continuous system, the domain is defined by boundary conditions formulated using the variational methods of Chapter 2. The inertia operator is shown to be positive-definite and the stiffness operator at least positive-definite.

Chapter 5 focuses on the free response of linear conservative systems. The chapter begins with a discussion of the use of a normal mode solution to formulate an eigenvalue–eigenvector problem to determine the natural frequencies and mode shapes for any system. General properties of eigenvalue problems for self-adjoint and positive-definite operators are discussed including eigenvector orthogonality and an expansion theorem. The results are applied to the operators developed in Chapter 4 with many examples given.

Chapter 6 focuses on the free response of non-self-adjoint systems. These include non-conservative systems due to the presence of viscous damping and gyroscopic systems. Eigenvalue problems for non-self-adjoint systems are discussed including the use of eigenvectors of the adjoint operator and the bi-orthogonality theorem. Examples of discrete and continuous systems with viscous damping and discrete gyroscopic systems are presented.

Three methods to determine the forced response are considered in Chapter 7. The method of undetermined coefficients is used to determine the forced response to systems with harmonic excitation. The Laplace transform method is applied for discrete and continuous systems. The method is used to illustrate the waveform nature of the response for continuous systems. The expansion theorem is used to develop modal analysis for conservative and non-conservative systems.

Approximate methods of solution for continuous systems are presented in Chapter 8. A functional, Rayleigh's quotient, is defined for all elements of the domain of the stiffness operator. The functional is stationary only for mode shapes of the free response. An approximate solution is sought from a finite-dimensional subspace of the domain. The Rayleigh–Ritz method uses Rayleigh's quotient to develop the best approximation for the exact solution from the subspace, using an energy inner product generated norm. An assumed modes method is developed from the Rayleigh–Ritz method. The finite-element method is shown to be an application of the Rayleigh–Ritz method when a class of functions called admissible functions, piecewise defined, are used as a basis for the subspace. The assumed modes method is used to develop a rubric for application of the finite-element method.

The use of the mathematical foundation and the application of the physics to build a framework for the modeling and development of the response is emphasized throughout the book, including in Chapter 8. The framework includes the identification of a vector space for the formulation of the problem, the use of variational methods to develop the differential equations, the definition of the stiffness and inertia operators and their domains, and the formulation of an appropriate eigenvalue problem to determine the free response. This framework is applied to a variety of discrete and continuous systems.

The presence of the framework becomes more important as the complexity of the system increases. Applications to MEMS problems and nanoscale problems for which continuous modeling is acceptable are complex and often involve multiple deformable bodies. Several examples are included.

An engineering book written at this level presents either new research or new pedagogical methods but always builds upon the work of previous authors. In this regard, gratitude is owed to those who have written distinguished works in the field. There are many good books in advanced vibrations, but those of Meirovitch are exemplary. It is hard to improve upon Meirovitch's work, especially in the development of variational principles for the derivation of the differential equations. Indeed, this book does follow

Meirovitch's ideas in the development of such principles but with some deviation. Meirovitch also published a general mathematical theory. This text builds the foundation for theory, formalizes it, extends it, and uses it in a consistent fashion including an application to contemporary research using linear vibrations.

The author is indebted to many for assistance in the development and publication of this book. The greatest acknowledgment of appreciation is to Jong Beom Suh (J.B.) who aided in the solution of many of the examples and the development of graphical representations of results as well as helping me identify errors and misprints. However, it is inevitable that such errors and misprints remain and for these I take sole responsibility. Some of the examples are a result of my work with other students. For this work I acknowledge Sathish Martin, Hari Parthasatari, Shirish Srinivas, and Jeff Slisk. I also appreciated the suggestions and patience of the graduate students who have taken my classes at The University of Akron.

Special acknowledgments are due to my wife Seala Fletcher-Kelly and my son Samuel G. Kelly IV, not only for their support, but for their contributions (especially my wife's) in preparing all figures and line drawings.

I also offer my appreciation to those at Taylor & Francis who believed in this project and helped guide it toward completion. This declaration of gratitude is specifically directed to, but not limited to, B.J. Clark who encouraged the development of this project; Michael Slaughter, acquiring editor; Marsha Pronin, project coordinator; Elizabeth Spangenberger, editorial assistant; and Khrysti Nazzaro and Joette Lynch, project editors.

Graham Kelly
Akron, Ohio

Table of Contents

Chapter 1   Introduction and Vibration of Single-Degree-of-Freedom
Systems 1

1.1 Introduction 1
    1.1.1 Degrees of Freedom and Generalized Coordinates 1
    1.1.2 Scope of Study 7
1.2 Newton's Second Law, Angular Momentum, and Kinetic Energy 8
    1.2.1 Particles 8
    1.2.2 Systems of Particles 9
    1.2.3 Rigid Bodies 13
1.3 Components of Vibrating Systems 17
    1.3.1 Inertia Elements 17
    1.3.2 Stiffness Elements 22
    1.3.3 Energy Dissipation 30
    1.3.4 External Energy Sources 34
1.4 Modeling of One-Degree-of-Freedom Systems 38
    1.4.1 Introduction and Assumptions 38
    1.4.2 Static Spring Forces 39
    1.4.3 Derivation of Differential Equations 42
    1.4.4 Model Systems 48
    1.4.5 One-Degree-of-Freedom Models of Continuous Systems 49
1.5 Qualitative Aspects of One-Degree-of-Freedom Systems 56
1.6 Free Vibrations of Linear Single-Degree-of-Freedom Systems 63
1.7 Response of a Single-Degree-of-Freedom System
Due to Harmonic Excitation 70
    1.7.1 General Theory 70
    1.7.2 Frequency-Squared Excitation 73
    1.7.3 Motion Input 75
    1.7.4 General Periodic Input 80
1.8 Transient Response of a Single-Degree-of-Freedom System 82

Chapter 2   Derivation of Differential Equations
Using Variational Methods 87

2.1 Functionals 87
2.2 Variations 91

2.3 Euler–Lagrange Equation 93
2.4 Hamilton's Principle 100
2.5 Lagrange's Equations for Conservative Discrete Systems 104
2.6 Lagrange's Equations for Non-Conservative Discrete Systems 112
2.7 Linear Discrete Systems 122
    2.7.1 Quadratic Forms 122
    2.7.2 Differential Equations for Linear Systems 125
    2.7.3 Linearization of Differential Equations 127
2.8 Gyroscopic Systems 130
2.9 Continuous Systems 136
2.10 Bars, Strings, and Shafts 138
2.11 Euler–Bernoulli Beams 150
2.12 Timoshenko Beams 166
2.13 Membranes 170

Chapter 3   Linear Algebra 173

3.1 Introduction 173
3.2 Three-Dimensional Space 174
3.3 Vector Spaces 177
3.4 Linear Independence 182
3.5 Basis and Dimension 185
3.6 Inner Products 189
3.7 Norms 193
3.8 Gram–Schmidt Orthonormalization Method 197
3.9 Orthogonal Expansions 202
3.10 Linear Operators 206
3.11 Adjoint Operators 212
3.12 Positive Definite Operators 219
3.13 Energy Inner Products 222

Chapter 4   Operators Used in Vibration Problems 225

4.1 Summary of Basic Theory 225
4.2 Differential Equations for Discrete Systems 227
4.3 Stiffness Matrix 227
4.4 Mass Matrix 233
4.5 Flexibility Matrix 234
4.6 $M^{-1}K$ and AM 240
4.7 Formulation of Partial Differential Equations for Continuous
    Systems 242
4.8 Second-Order Problems 245
4.9 Euler–Bernoulli Beam 253

4.10 Timoshenko Beams 262
4.11 Systems with Multiple Deformable Bodies.................................... 266
4.12 Continuous Systems with Attached Inertia Elements 272
4.13 Combined Continuous and Discrete Systems 278
4.14 Membranes.. 283

Chapter 5   Free Vibrations of Conservative Systems 287

5.1 Normal Mode Solution.. 287
5.2 Properties of Eigenvalues and Eigenvectors 292
    5.2.1 Eigenvalues of Self-Adjoint Operators 292
    5.2.2 Positive Definite Operators.. 297
    5.2.3 Expansion Theorem.. 298
    5.2.4 Summary.. 302
5.3 Rayleigh's Quotient 303
5.4 Solvability Conditions 306
5.5 Free Response Using the Normal Mode Solution.......................... 309
    5.5.1 General Free Response... 309
    5.5.2 Principal Coordinates 314
5.6 Discrete Systems.. 316
    5.6.1 The Matrix Eigenvalue Problem 317
5.7 Natural Frequency Calculations Using Flexibility Matrix 326
5.8 Matrix Iteration 330
5.9 Continuous Systems... 341
5.10 Second-Order Problems (Wave Equation) 342
5.11 Euler–Bernoulli Beams 360
5.12 Repeated Structures 375
5.13 Timoshenko Beams 398
5.14 Combined Continuous and Discrete Systems 409
5.15 Membranes... 414
5.16 Green's Functions 430

Chapter 6   Non-Self-Adjoint Systems 437

6.1 Non-Self-Adjoint Operators.. 437
6.2 Discrete Systems with Proportional Damping............................... 441
6.3 Discrete Systems with General Damping....................................... 446
6.4 Discrete Gyroscopic Systems 452
6.5 Continuous Systems with Viscous Damping 458

Chapter 7   Forced Response 465

7.1 Response of Discrete Systems for Harmonic Excitations 465

7.1.1 General Theory 465
7.1.2 Vibration Absorbers 470
7.2 Harmonic Excitation of Continuous Systems 480
7.3 Laplace Transform Solutions 490
7.3.1 Discrete Systems 491
7.3.2 Continuous Systems 497
7.4 Modal Analysis for Undamped Discrete Systems 501
7.5 Modal Analysis for Undamped Continuous Systems 504
7.6 Discrete Systems with Damping 516
7.6.1 Proportional Damping 516
7.6.2 General Viscous Damping 517

Chapter 8   Rayleigh–Ritz and Finite-Element Methods 525

8.1 Fourier Best Approximation Theorem 525
8.2 Rayleigh–Ritz Method 528
8.3 Galerkin Method 531
8.4 Rayleigh–Ritz Method for Natural Frequencies and Mode
    Shapes 532
8.5 Rayleigh–Ritz Methods for Forced Response 551
8.6 Admissible Functions 556
8.7 Assumed Modes Method 560
8.8 Finite-Element Method 570
8.9 Assumed Modes Development of Finite-Element Method 575
8.10 Bar Element 577
8.11 Beam Element 584

Chapter 9   Exercises 595

9.1 Chapter 1 595
9.2 Chapter 2 602
9.3 Chapter 3 611
9.4 Chapter 4 614
9.5 Chapter 5 617
9.6 Chapter 6 620
9.7 Chapter 7 622
9.8 Chapter 8 625

References 627

Index 629

1 Introduction and Vibrations of Single-Degree-of-Freedom Systems

1.1 INTRODUCTION

1.1.1 DEGREES OF FREEDOM AND GENERALIZED COORDINATES

Vibrations are the oscillations of a system about an equilibrium position. This study concerns the analysis of vibrations. More accurately it concerns the development of a framework for the analysis of vibrations of linear systems. The framework serves as a guide to the analysis of complex systems.

Vibrations are time dependent. Thus, time is an independent variable in the analysis of vibrations of all systems. Vibration analysis involves tracking the motion, as a function of time, of particles in a system. A coordinate is a time dependent variable that is used to track the motion of a particle. Two particles are kinematically independent if there is no geometric or kinematic relationship, which constrains the motion of one particle relative to the other.

Let x_1 and x_2 represent the displacements of the particles in the system of Figure 1.1. The particles of Figure 1.1a are connected by a rigid rod and they have the displacement such that $x_1 = x_2$, and thus these particles are not kinematically independent; instead they are kinematically dependent. The particles of Figure 1.1b are connected by a flexible component, a spring. The motions of the particles are related kinetically, as they are subject to the same force from the spring, but they are kinematically independent.

The number of degrees of freedom necessary to analyze a system is equal to the minimum number of kinematically independent coordinates necessary to specify the motion of every particle in the system. Any such set of kinematically independent coordinates is called a set of generalized coordinates. The number of degrees of freedom is unique, whereas the choice of generalized coordinates is not unique.

1

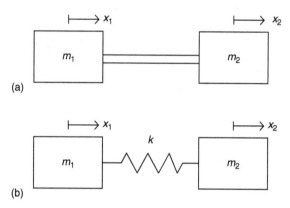

(a)

(b)

FIGURE 1.1 (a) The particles are not kinematically independent, as they are connected by a rigid rod and $x_1 = x_2$. (b) Since the particles are connected by an elastic spring x_1 is kinematically independent of x_2.

This study encompasses the motion of particles, rigid bodies, and deformable bodies. A particle is a body whose mass can be assumed to be concentrated at a single point. The mass of a rigid body is distributed about a mass center, but the relative position between two particles on the rigid body is constant. A deformable body has a distribution of mass throughout the body, but particles may move relative to one another as motion occurs.

A particle whose motion is unconstrained has three degrees of freedom, as it may move independently in three spatial directions. If the particle's motion is constrained, then it has less than three degrees of freedom. The particle of Figure 1.2a is constrained to move in the plane of the paper, but can move anywhere within the plane. Thus, this particle has two degrees of freedom.

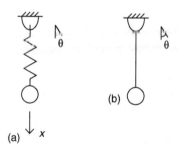

(b)

(a)

FIGURE 1.2 (a) Particle is constrained to move in the plane of the paper, but x and θ are independent. (b) The particle must move along a circular path in the plane.

A suitable choice for the set of generalized coordinates includes x, the downward displacement of the particle, measured from the system's equilibrium position, and θ, the counterclockwise angular displacement of the particle, measured from the vertical position. The particle of Figure 1.2b is constrained to move in the plane of the paper and in a circular path centered at O. Since there are two constraints on the motion of the particle, it has one degree of freedom. The counterclockwise angular displacement θ is a suitable choice for the generalized coordinate.

A rigid body whose motion is unconstrained has six degrees of freedom. Its mass center may move independently in three directions while the body may independently rotate about three axes. The gyroscope of Figure 1.3 rotates independently about three axes, but its mass center is fixed in space. Thus, it has three degrees of freedom. The set of generalized coordinates illustrated in Figure 1.3 are called Eulerian angles. The angle ϕ measures the precession, the angle θ measures the nutation, and the angle ψ measures the spin.

A rigid body undergoing planar motion is one whose mass center is constrained to move within a plane and rotation occurs only about an axis perpendicular to the plane of motion of the mass center. The mass center is said to translate within the plane. Motion of a rigid body undergoing planar

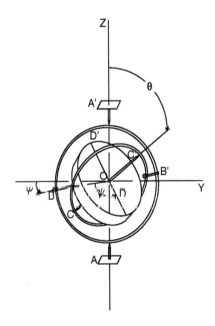

FIGURE 1.3 A gyroscope has three degrees of freedom.

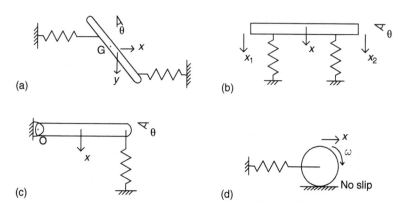

FIGURE 1.4 A rigid body undergoing planar motion with (a) three degrees of freedom, (b) two degrees of freedom, (c) one degree of freedom, and (d) one degree of freedom when disk rolls without slip.

motion is a superposition of the translation and the rotation. If the motion of the mass center is unconstrained, then the system has three degrees of freedom. The bar of Figure 1.4a has three degrees of freedom, as it is unconstrained to move in the plane. A choice for the set of generalized coordinates may be θ, the clockwise angular displacement of the bar, as well as x and y, the horizontal and vertical displacements of the mass center of the bar. The bar of Figure 1.4b is constrained such that its mass center may not move horizontally, thus it has two degrees of freedom. One choice of a set of generalized coordinates is x, the vertical displacement of the mass center, and θ, the clockwise angular displacement of the bar, both measured from an equilibrium position. An alternate choice of a set of generalized coordinates is x_1 and x_2 the vertical displacements of the ends of the bar. The bar of Figure 1.4c is constrained such that its mass center moves in a circular path centered at O. Choices of the generalized coordinate include x, the vertical displacement of the mass center or θ, the clockwise angular displacement of the bar, both measured from the system's equilibrium position. Note that x and θ are not independent in that $x = (L/2)\sin(\theta)$. The disk of Figure 1.4d, which rolls without slip, is a single degree of freedom system in that since it rolls without slip, the angular velocity ω is kinematically related to the velocity of the mass center of the disk $\bar{v}$ by $\bar{v} = r\omega$.

Systems are often composed of more than one particle of rigid body, each of which has some number of degrees of freedom. However, constraints between the bodies reduce the number of degrees of freedom needed to analyze the system. The three particles of the system of Figure 1.5a are

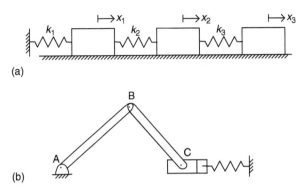

(a)

(b)

FIGURE 1.5 Systems composed of multiple bodies with constraints with (a) three degrees of freedom, and (b) one degree of freedom.

constrained to move only in the horizontal direction. There are no kinematic constraints on the relative motion of the particles; the system is a three-degree-of-freedom system. The slider crank-mechanism of Figure 1.5b is composed of three bodies, bar AB, the crank, bar BC, the connecting rod, and the collar at C, the slider. Bar AB has one degree of freedom, as it must rotate in a circular path centered at A. The slider, if considered by itself, has one degree of freedom, in that it is constrained to move in the horizontal direction. The connecting rod, if analyzed independent of the other bodies, has three degrees of freedom. Since the connecting rod is pinned to the crank at B, both its horizontal and vertical displacements are constrained to be the same as those of the end of the crank at B. The rod has a third constraint inbecause its end at C is constrained to move horizontally. The slider is also subject to a constraint, in that it must have the same displacement as the end of the rod. If each component were treated independently of the others, the system would have five degrees of freedom. The four constraints reduces the system to a single-degree-of-freedom system.

The relative position of particles changes during the motion of a deformable body, such as the beam of Figure 1.6. An assumption often made in the analysis of the deformation of a beam is that plane sections remain

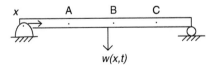

FIGURE 1.6 The beam, a deformable body, is a continuous system.

plane. Every particle in a cross-section whose normal is along the axis of the beam has the same displacement. The neutral axis, a plane in which the normal stress is zero, is usually used as a reference for the displacement. If x is a spatial variable measured along the neutral axis of the beam when it is in equilibrium, then $w(x,t)$ is the transverse displacement of a particle on the neural axis, a distance x from the left end of the beam. The displacement of particle A is kinematically independent of the displacement of particle B, particle C, and any other particle along the neutral axis for some other value of x. Thus, there are an infinite number of kinematically independent particles and the system has an infinite number of degrees of freedom. Instead of using an infinite number of generalized coordinates, the single variable x is used as a generalized spatial variable.

The plate of Figure 1.7 is another example of a system with an infinite number of degrees of freedom. However, for the plate, the motion of particles A, B, and C are all kinematically independent. Two spatial variables, x and y, are necessary to completely describe the motion of all particles.

A system with a finite number of degrees of freedom is called a discrete system, while a system with an infinite number of degrees of freedom is called a continuous system or distributed parameter system. The beam is a one-dimensional continuous system while the plate is a two-dimensional continuous system.

The system of Figure 1.8 (a) is composed of a particle connected to a beam through a spring. The beam, by itself, is of course a continuous system. The particle, by itself, is a discrete system. The system is then a combined discrete–continuous system. It may be said to have infinity plus one degree of freedom.

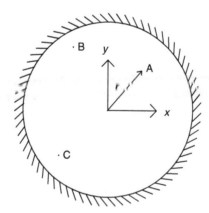

FIGURE 1.7 A plate is a two-dimensional continuous system.

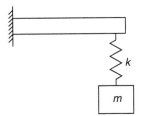

FIGURE 1.8 Combined continuous-discrete system.

The system of Figure 1.9 is composed of two beams, each a continuous system, connected through a layer of springs. The system is a two-degree-of-freedom continuous system.

1.1.2 SCOPE OF STUDY

This study is titled *Advanced Vibration Analysis*. The first adjective, "Advanced," implies that the study is beyond that of a first study of vibrations, which would generally include a study of the free and forced response of single-degree-of-freedom systems, and perhaps an introduction to multi-degree-of-freedom systems. Chapter 1 is a review of one-degree-of-freedom systems designed to review concepts used in this more comprehensive study.

The term "Vibration" implies that this study concerns systems undergoing oscillatory motion about a stable equilibrium position. The study does not address stability or control of dynamic systems.

The term "Analysis" implies that the study is focused on the analysis, not the design, of such systems. There are two parts to the analysis: derivation of the governing equations and determination of the response. Frameworks for both parts of the analysis are presented. A framework for deriving the governing equations is presented in Chapter 2. Chapter 3 presents the mathematical background for developing the framework for

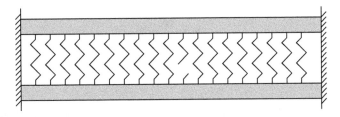

FIGURE 1.9 Two-degree-of-freedom continuous system.

determination of responses. The framework is developed partially in Chapter 4 by introducing linear operators used in the analysis. Each subsequent chapter further develops this framework while concentrating on a specific aspect of the analysis. Approximate solutions using Rayleigh–Ritz and finite element methods are presented in Chapter 8.

1.2 NEWTON'S SECOND LAW, ANGULAR MOMENTUM, AND KINETIC ENERGY

1.2.1 PARTICLES

Only a few basic laws of nature exist. These are laws that apply to any system, regardless of the geometry of the system or the materials from which it is made. These include conservation of mass, conservation of energy, the Second and Third Laws of Thermodynamics, and of course, Newton's Second Law. Newton's Second Law, developed for a particle, is

$$\mathbf{F} = \frac{d\mathbf{L}}{dt} \tag{1.1}$$

where $\mathbf{F}$ is the resultant of all external forces acting on the particle and

$$\mathbf{L} = m\mathbf{v} \tag{1.2}$$

is the particle's linear momentum. When the mass of the system is constant Equation 1.1 becomes

$$\mathbf{F} = m\mathbf{a} \tag{1.3}$$

Newton's second law is a basic law of nature. It cannot be derived from any law more fundamental and can be proved only by observation.

Let $\mathbf{r}$ be the position vector of the particle measured from a fixed origin. Then by definition the velocity and acceleration of the particle are

$$\mathbf{v} = \frac{d\mathbf{r}}{dt} = \dot{\mathbf{r}} \tag{1.4}$$

$$\mathbf{a} = \frac{d\mathbf{v}}{dt} = \ddot{\mathbf{r}} \tag{1.5}$$

Equation 1.4 and Equation 1.5 employ the dot notation, in which a dot above a variable indicates differentiation of the variable with respect to time.

Newton's second law for a particle of fixed mass is

$$\mathbf{F} = m\ddot{\mathbf{r}} \tag{1.6}$$

The angular momentum of a particle about a point is defined as the moment of linear momentum about the point. Thus, the angular momentum about O is defined as

$$\mathbf{H}_O = \mathbf{r} \times m\dot{\mathbf{r}} \tag{1.7}$$

The moment of a force about a point is equal to

$$\mathbf{M} = \mathbf{r} \times \mathbf{F} \tag{1.8}$$

where $\mathbf{r}$ is a vector from the point to the force. Taking the moment of Equation 1.6 about the origin leads to

$$\mathbf{M}_O = \mathbf{r} \times \mathbf{F} = \mathbf{r} \times m\ddot{\mathbf{r}} \tag{1.9}$$

Noting that $\dot{\mathbf{H}}_O = (d/dt)(\mathbf{r} \times m\dot{\mathbf{r}}) = m(\dot{\mathbf{r}} \times \dot{\mathbf{r}} + \mathbf{r} \times \ddot{\mathbf{r}})$ and that $\dot{\mathbf{r}} \times \dot{\mathbf{r}} = \mathbf{0}$, Equation 1.9 becomes

$$\dot{\mathbf{H}}_O = \mathbf{M}_O \tag{1.10}$$

Equation 1.10 states that the rate of change of angular momentum about the origin of the coordinate system is equal to the resultant moment of the external forces about the origin. Nothing in the derivation of Equation 1.10 prevented any fixed point form being chosen as the origin, so Equation 1.10 can be generalized as the rate of change of angular momentum of a particle about any point is equal to the resultant moment of the external forces about that point.

The kinetic energy of a particle is defined as

$$T = \frac{1}{2}m\mathbf{v} \cdot \mathbf{v} = \frac{1}{2}m\dot{\mathbf{r}} \cdot \dot{\mathbf{r}} \tag{1.11}$$

1.2.2 SYSTEMS OF PARTICLES

Consider a system of n particles. Let $\mathbf{r}_i$ be the position vector of the ith particle whose mass is m_i. If $\tilde{\mathbf{F}}_i$ represents the resultant of all external forces acting on the particle, then application of Newton's second law to the particle leads to

$$\tilde{\mathbf{F}}_i = m_i \ddot{\mathbf{r}}_i \tag{1.12}$$

The free-body diagram of a particle in the system illustrates forces between the particle and all other particles in the system. Let $\mathbf{f}_{ij}$ be the force acting on particle i from particle j. From Newton's third law the force $\mathbf{f}_{ji}$ acting on particle j, from particle i, is $-\mathbf{f}_{ij}$. Using these definitions

$$\tilde{\mathbf{F}}_i = \mathbf{F}_i + \sum_{\substack{j=1 \\ j \neq 1}}^{n} \mathbf{f}_{ij} \tag{1.13}$$

where $\mathbf{F}_i$ is the resultant of all forces external to the system acting on particle i.

Adding all of the equations of the form of Equation 1.12 to each other leads to

$$\sum_{i=1}^{n} \tilde{\mathbf{F}}_i = \sum_{i=1}^{n} m_i \ddot{\mathbf{r}}_i$$

$$\sum_{i=1}^{n} \left(\mathbf{F}_i + \sum_{\substack{j=1 \\ j \neq i}}^{n} \mathbf{f}_{ij} \right) = \sum_{i=1}^{n} m_i \ddot{\mathbf{r}}_i \tag{1.14}$$

$$\sum_{i=1}^{n} \mathbf{F}_i + \sum_{i=1}^{n} \sum_{\substack{j=1 \\ j \neq i}}^{n} \mathbf{f}_{ij} = \sum_{i=1}^{n} m_i \ddot{\mathbf{r}}_i$$

Since for each $\mathbf{f}_{ij}$, $\mathbf{f}_{ji} = -\mathbf{f}_{ij}$ the double summation in Equation 1.14 is identically equal to zero. The resulting summation of forces is the resultant of all external forces acting on the system of particles. Then Equation 1.14 reduces to

$$\sum \mathbf{F} = \sum_{i=1}^{n} m_i \ddot{\mathbf{r}}_i \tag{1.15}$$

The mass center of the system of particles is located by the vector $\mathbf{\bar{r}}$ defined by

$$\bar{\mathbf{r}} = \frac{\displaystyle\sum_{i=1}^{n} m_i \mathbf{r}_i}{\displaystyle\sum_{i=1}^{n} m_i} = \frac{1}{m} \sum_{i=1}^{n} m_i \mathbf{r}_i \tag{1.16}$$

where m is the total mass of the system of particles. Using Equation 1.16 in Equation 1.15 gives

$$\sum \mathbf{F} = m\ddot{\bar{\mathbf{r}}} = m\bar{\mathbf{a}} \tag{1.17}$$

Equation 1.17 states that the resultant of the external forces acting on a system of particles is equal to the total mass of the system, times the acceleration of its mass center. Note that the mass center may not correspond to any particle in the system, and may only be defined by its mathematical definition of Equation 1.16. Then $\bar{\mathbf{r}}$ is simply a mathematically defined function of time and $\bar{\mathbf{a}} = \ddot{\bar{\mathbf{r}}}$.

The angular momentum of a system of particles about a point O is

$$\mathbf{H}_O = \sum_{i=1}^{n} \mathbf{r}_i \times m_i \dot{\mathbf{r}}_i \tag{1.18}$$

Following the same steps as above it is shown that

$$\dot{\mathbf{H}}_O = \sum_{i=1}^{n} m_i \mathbf{r}_i \times \ddot{\mathbf{r}}_i \tag{1.19}$$

Substitution of Equation 1.12 into Equation 1.19 leads to

$$\dot{\mathbf{H}}_O = \sum_{i=1}^{n} \mathbf{r}_i \times \tilde{\mathbf{F}}_i \tag{1.20}$$

When Equation (1.13) is used Equation 1.20 contains terms of the form

$$\mathbf{r}_i \times \mathbf{f}_{ij} + \mathbf{r}_j \times \mathbf{f}_{ji} = \mathbf{r}_i \times \mathbf{f}_{ij} + \mathbf{r}_j \times (-\mathbf{f}_{ij}) = (\mathbf{r}_i - \mathbf{r}_j) \times \mathbf{f}_{ij} \tag{1.21}$$

The vectors involved in this cross product are illustrated in Figure 1.10. The vectors $\mathbf{r}_i$, $\mathbf{r}_j$, and $\mathbf{f}_{ij}$ are coplanar. A vector proportional to $\mathbf{f}_{ij}$ forms a triangle with $\mathbf{r}_i$ and $\mathbf{r}_j$. Using the triangle rule for vector addition, as illustrated in Figure 1.10, the vector $\mathbf{r}_i - \mathbf{r}_j$, is parallel to $\mathbf{f}_{ij}$, and thus their cross product is zero. Since all of the terms of the form of Equation 1.21 are zero, Equation 1.20 reduces to

$$\dot{\mathbf{H}}_O = \sum_{i=1}^{n} \mathbf{r}_i \times \mathbf{F}_i = \mathbf{M}_O \tag{1.22}$$

Equation 1.22 implies that the time rate of change of angular momentum about a point is equal to the resultant of the external moments about that point.

The total kinetic energy of a system of particles is the sum of the kinetic energies of the individual particles

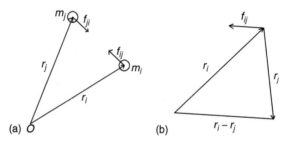

FIGURE 1.10 (a) Illustration of internal force between two particles. The vectors $\mathbf{r_i}$, $\mathbf{r_j}$ and $\mathbf{f_{ij}}$ are coplanar and $\mathbf{f_{ji}} = -\mathbf{f_{ij}}$. (b) Triangle rule for vector addition shows that $\mathbf{r_i} - \mathbf{r_j}$ is parallel to $\mathbf{f_{ij}}$.

$$T = \sum_{i=1}^{n} T_i = \frac{1}{2} \sum_{i=1}^{n} m_i \dot{\mathbf{r}}_i \cdot \dot{\mathbf{r}}_i \tag{1.23}$$

As illustrated in Figure 1.11,

$$\mathbf{r}_i = \bar{\mathbf{r}} + \mathbf{r}_{i/G} \tag{1.24}$$

where $\mathbf{r}_{i/G}$ is the position vector of a particle located at point i, relative to the mass center G. Substitution of Equation 1.24 into Equation 1.23 leads to

$$T = \frac{1}{2} \sum_{i=1}^{n} m_i (\dot{\bar{\mathbf{r}}} + \dot{\mathbf{r}}_{i/G}) \cdot (\dot{\bar{\mathbf{r}}} + \dot{\mathbf{r}}_{i/G})$$

$$= \frac{1}{2} \left[\sum_{i=1}^{n} m_i \dot{\bar{\mathbf{r}}} \cdot \dot{\bar{\mathbf{r}}} + 2\dot{\bar{\mathbf{r}}} \sum_{i=1}^{n} m_i \dot{\mathbf{r}}_{i/G} + \sum_{i=1}^{n} m_i \dot{\mathbf{r}}_{i/G} \cdot \dot{\mathbf{r}}_{i/G} \right] \tag{1.25}$$

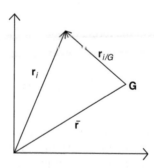

FIGURE 1.11 Illustration of relative position vector.

The term $\sum_{i=1}^{n} m_i \dot{\mathbf{r}}_{i/G}$ is equal to the total mass times the velocity of the mass center of the system relative to G. However, since G is the mass center, this sum must be zero. Thus,

$$T = \frac{1}{2} \left[\sum_{i=1}^{n} m_i \dot{\mathbf{r}} \cdot \dot{\mathbf{r}} + \sum_{i=1}^{n} m_i \dot{\mathbf{r}}_{i/G} \cdot \dot{\mathbf{r}}_{i/G} \right] \tag{1.26}$$

The first term in equation is the kinetic energy of the mass center of the system, while the second term is the kinetic energy due to motion of particles relative to the mass center.

1.2.3 RIGID BODIES

A rigid body is a system in which the mass has a continuous distribution about the mass center, and as motion occurs, each particle maintains its position relative to the mass center. If a coordinate system, $x'y'z'$ is attached to the body with its origin at the mass center, then for any particle the relative position, vector between the particle and the mass center is a constant when viewed from the $x'y'z'$ reference frame.

The derivations of the equations of motion for a rigid body are similar to those for a system of particles. The rigid body is composed of a number of particles of mass Δm_i. Then Equation 1.15 becomes

$$\sum \mathbf{F} = \sum_{i=1}^{n} \Delta m_i \ddot{\mathbf{r}}_i \tag{1.27}$$

Taking the limit as the number of particles grows large and each Δm_i approaches zero leads to

$$\sum \mathbf{F} - \lim_{n \to \infty} \sum_{i=1}^{n} \Delta m_i \ddot{\mathbf{r}}_i = \int_m \ddot{\mathbf{r}} \, dm \tag{1.28}$$

The mass center of a rigid body is defined by

$$\bar{\mathbf{r}} = \frac{1}{m} \int_m \mathbf{r} \, dm \tag{1.29}$$

The acceleration of the mass center is

$$\bar{\mathbf{a}} = \ddot{\bar{\mathbf{r}}} = \frac{1}{m} \int_m \ddot{\mathbf{r}} \, dm \tag{1.30}$$

Thus, Equation 1.27 becomes

$$\sum \mathbf{F} = m\bar{\mathbf{a}} \tag{1.31}$$

The same method as above is used to show that Equation 1.22 applies for a rigid body; the rate of change of angular momentum about a point is equal to the resultant of the moments taken about that point. However, the expression for the rate of change of angular momentum may be clarified. Beginning with Equation 1.18 and performing the limiting process to transform the equation from a system of particles to a rigid body leads to

$$\mathbf{H_O} = \lim_{n \to \infty} \sum_{i=1}^{n} \mathbf{r}_i \times \Delta m_i \dot{\mathbf{r}}_i = \int_m \mathbf{r} \times \dot{\mathbf{r}} \, dm \tag{1.32}$$

The relative velocity equation is used to determine the velocity of any particle in the rigid body in terms of the velocity of the mass center

$$\dot{\mathbf{r}} = \mathbf{v} = \bar{\mathbf{v}} + \omega \times \mathbf{r}_G \tag{1.33}$$

where $\mathbf{r}_G$ is a position vector from the mass center to a point in the body, and ω is the angular velocity vector. In addition $\mathbf{r} = \bar{\mathbf{r}} + \mathbf{r}_G$.

Substitution of Equation 1.33 into Equation 1.32 leads to

$$\mathbf{H_O} = \int_m (\bar{\mathbf{r}} + \mathbf{r}_G) \times (\bar{\mathbf{v}} + \omega \times \mathbf{r}_G) dm$$

$$= \bar{\mathbf{r}} \times m\bar{\mathbf{v}} + \bar{\mathbf{r}} \times \left[\omega \times \int_m \mathbf{r}_G \, dm \right] + \int_m \mathbf{r}_G dm \times \bar{\mathbf{v}} \tag{1.34}$$

$$+ \int_m \mathbf{r}_G \times (\omega \times \mathbf{r}_G) \, dm$$

By definition, $\int_m \mathbf{r}_G dm = 0$ and Equation 1.34 becomes

$$\mathbf{H_O} = \bar{\mathbf{r}} \times m\bar{\mathbf{v}} + \int_m \mathbf{r}_G \times (\omega \times \mathbf{r}_G)dm \qquad (1.35)$$

Note that if the point O is chosen to be the mass center, then $\bar{\mathbf{r}} = \mathbf{0}$ and Equation 1.35 is written as

$$\mathbf{H_G} = \int_m \mathbf{r}_G \times (\omega \times \mathbf{r}_G)dm \qquad (1.36)$$

If the angular momentum about the mass center is known, then the angular momentum about any point is determined as

$$\mathbf{H_O} = \mathbf{H_G} + \bar{\mathbf{r}} \times m\bar{\mathbf{v}} \qquad (1.37)$$

For simplification, first consider the case when the rigid body is undergoing planar motion. In this case

$$\omega = \omega\mathbf{k} \qquad (1.38)$$

$$\mathbf{r}_G = x\mathbf{i} + y\mathbf{j} \qquad (1.39)$$

and

$$\mathbf{H_G} = \omega \int_m (x^2 + y^2)\, dm\mathbf{k} = \bar{I}\omega\mathbf{k} \qquad (1.40)$$

where $\bar{I} = \int_m (x^2 + y^2)dm$ is the body's centroidal moment of inertia about the z axis. The time rate of change of angular momentum is

$$\dot{\mathbf{H}}_\mathbf{O} = \bar{I}\alpha \qquad (1.41)$$

where $\alpha - d\omega/dt$ is the body's angular acceleration. Thus, the angular momentum equation for a rigid body undergoing planar motion is

$$\sum \mathbf{M}_G = \bar{I}\alpha\mathbf{k} \qquad (1.42)$$

When moments are taken about a point other than the mass center, Equation 1.37 and Equation 1.42 are combined to give

$$\sum \mathbf{M}_O = \bar{I}\alpha\mathbf{k} + \bar{\mathbf{r}} \times m\bar{\mathbf{a}} \qquad (1.43)$$

For a general three-dimensional motion of a rigid body, with an angular velocity defined by

$$\omega = \omega_x \mathbf{i} + \omega_y \mathbf{j} + \omega_z \mathbf{k} \tag{1.44}$$

Equation 1.36 becomes

$$\mathbf{H}_G = H_{G,x}\mathbf{i} + H_{G,y}\mathbf{j} + H_{G,z}\mathbf{k} \tag{1.45}$$

where

$$H_{G,x} = \omega_x I_{xx} - \omega_y I_{xy} - \omega_z I_{xz} \tag{1.46}$$

$$H_{G,y} = \omega_y I_{yy} - \omega_x I_{yx} - \omega_z I_{yz} \tag{1.47}$$

$$H_{G,x} = \omega_z I_{zz} - \omega_x I_{zx} - \omega_y I_{zy} \tag{1.48}$$

The time rate of change of angular momentum about the mass center is determined from

$$\dot{\mathbf{H}}_G = \frac{\mathrm{d}}{\mathrm{d}t}\left[\int_m \mathbf{r}_G \times (\omega \times \mathbf{r}_G)\mathrm{d}m\right] \tag{1.49}$$

The moment equations for general three-dimensional motion assuming x–y–z are a set of principal axes ($I_{xy} = I_{yz} = I_{zx} = 0$) become

$$\sum M_x = I_{xx}\dot{\omega}_x - (I_{yy} - I_{zz})\omega_y\omega_z \tag{1.50}$$

$$\sum M_y = I_{yy}\dot{\omega}_y - (I_{zz} - I_{xx})\omega_z\omega_x \tag{1.51}$$

$$\sum M_z = I_{zz}\dot{\omega}_z - (I_{xx} - I_{yy})\omega_x\omega_y \tag{1.52}$$

The kinetic energy of a rigid body is determined by taking the limit of Equation 1.26 as the number of particles grows large and each m_i grows smaller:

$$T = \frac{1}{2} \lim_{n \to \infty} \left[\sum_{i=1}^{n} m_i \dot{\bar{r}} \cdot \dot{\bar{r}} + 2\dot{\bar{r}} \sum_{i=1}^{n} m_i \dot{r}_{i/G} + \sum_{i=1}^{n} m_i \dot{r}_{i/G} \cdot \dot{r}_{i/G} \right]$$

$$= \frac{1}{2} m \bar{v} \cdot \bar{v} + \frac{1}{2} \int_m \dot{r}_G \cdot \dot{r}_G \, dm \qquad (1.53)$$

Since $\dot{r}_G$ is the velocity of a particle relative to the mass center $\dot{r}_G = \omega \times r_G$, the kinetic energy for a rigid body becomes

$$T = \frac{1}{2} m \bar{v} \cdot \bar{v} + \frac{1}{2} \int_m (\omega \times r_G) \cdot (\omega \times r_G) \, dm$$

$$= \frac{1}{2} m \bar{v}^2 + \frac{1}{2} (\bar{I}_x \omega_x^2 + \bar{I}_y \omega_y^2 + \bar{I}_z \omega_z^2 - 2\bar{I}_{xy} \omega_x \omega_y - 2\bar{I}_{yz} \omega_y \omega_z$$

$$- 2\bar{I}_{zx} \omega_z \omega_x) \qquad (1.54)$$

1.3 COMPONENTS OF VIBRATING SYSTEMS

The basic components of a mechanical system are described in terms of energy principles:

- Inertia components are system components with finite mass. During motion of the system, the inertia elements store kinetic energy, the energy associated with the motion of individual particles.
- Stiffness elements are components which store potential energy.
- Damping elements are elements which dissipate energy which is usually transferred to heat.
- External components do work on the system or work is done by the system on the external components.

The total energy in the system is the sum of the kinetic energy and the potential energy. External components may add energy to the system while damping elements dissipate energy.

1.3.1 INERTIA ELEMENTS

An inertia element is a body with finite mass. As motion occurs, the inertia element stores and releases kinetic energy. Kinetic energy is stored as the body's speed increases and is released as its speed decreases. The kinetic energy of a body is a function of the body's inertia properties as well as the time rate of change of its position vector. In general the kinetic energy may be a function of the system's generalized coordinates and their time derivatives,

$$T = T(x_1, x_2, \ldots, x_n, \dot{x}_1, \dot{x}_2, \ldots, \dot{x}_n) \tag{1.55a}$$

The following reviews the equations for the kinetic energy of inertia elements which are derived in Section 1.2. The kinetic energy of a particle of mass m moving with a velocity v is

$$T = \frac{1}{2}mv^2 \tag{1.55b}$$

A rigid body has translational kinetic energy and rotational kinetic energy. The total kinetic energy of a rigid body is

$$T = \frac{1}{2}m\bar{v}^2 + \frac{1}{2}(\bar{I}_x\omega_x^2 + \bar{I}_y\omega_y^2 + \bar{I}_z\omega_z^2 - 2\bar{I}_{xy}\omega_x\omega_y - 2\bar{I}_{yz}\omega_y\omega_z$$
$$- 2\bar{I}_{zx}\omega_z\omega) \tag{1.56}$$

where $\bar{v}$ is the velocity of the body's mass center, $\omega = \omega_x\mathbf{i} + \omega_y\mathbf{j} + \omega_z\mathbf{k}$ is its angular velocity vector, and the $\bar{I}$ terms are components of the body's inertia tensor about axes through the center of mass. If the x–y–z axes are principal axes, then the product of inertia terms are all zero.

Planar motion occurs when the path of the mass center is in a plane and rotation occurs only about an axis perpendicular to that plane. For a rigid body undergoing planar motion, Equation 1.56 reduces to

$$T = \frac{1}{2}m\bar{v}^2 + \frac{1}{2}\bar{I}\omega^2 \tag{1.57}$$

where $\bar{I}$ is moment of inertia about an axis through the center of mass, perpendicular to the plane in which the mass center moves, and ω is the angular velocity about the axis. If the body is rotating about a fixed axis through a point O, application of the parallel axis theorem leads to an alternate form of Equation 1.57 as

$$T = \frac{1}{2}I_O\omega^2 \tag{1.58}$$

Example 1.1. The rigid body of Figure 1.4b is a uniform bar of length L and mass m. Its mass moment of inertia about its centroidal axis is $\bar{I} = (1/12)mL^2$. Determine the kinetic energy of the bar at an arbitrary instant, written in terms of the chosen generalized coordinates using: (a) x and θ as generalized coordinates, and (b) x_1 and x_2 as generalized coordinates.

Solution: The kinetic energy of a rigid body undergoing planar motion is given by Equation 1.57. Note that Equation 1.58 does not apply, as there is no fixed axis of rotation for the bar. (a) Since x is the displacement of the mass center, the velocity of the mass center is $\dot{x}$. Since θ is the angular displacement of the bar, its angular velocity is $\omega = \dot{\theta}$. The kinetic energy of the bar at an arbitrary instant is determined using Equation 1.57 as

$$T = \frac{1}{2}m\dot{x}^2 + \frac{1}{2}\left(\frac{1}{12}mL^2\right)\dot{\theta}^2 \tag{a}$$

(b) Referring to the geometry of bar at an arbitrary instant, as illustrated in Figure 1.12, it is noted that the displacement of the mass center is

$$\bar{x} = \frac{1}{2}(x_1 + x_2) \tag{b}$$

and its angular displacement is

$$\sin\theta = \frac{1}{L}(x_2 - x_1) \tag{c}$$

Differentiating Equation c with respect to time

$$\dot{\theta}\cos\theta = \frac{1}{L}(\dot{x}_2 - \dot{x}_1) \tag{d}$$

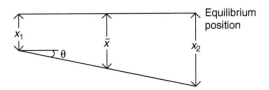

FIGURE 1.12 Displacement of bar of Example 1.1 at an arbitrary instant.

Geometry shows that

$$\cos\theta = \frac{1}{L}\sqrt{L^2 - (x_2 - x_1)^2} \tag{e}$$

The angular velocity is determined using Equation d and Equation e as

$$\dot{\theta} = \frac{\dot{x}_2 - \dot{x}_1}{\sqrt{L^2 - (x_2 - x_1)^2}} \tag{f}$$

Equation b and Equation f are used in Equation 1.57, leading to

$$T = \frac{1}{2}m(\dot{x}_2 - \dot{x}_1)^2 + \frac{1}{2}\left(\frac{1}{12}mL^2\right)\left[\frac{\dot{x}_2 - \dot{x}_1}{\sqrt{L^2 - (x_2 - x_1)^2}}\right]^2 \tag{g}$$

Example 1.2. The slender bar of Figure 1.13 is of mass m and length L ($\bar{I} = (1/12)mL^2$) and is pinned at point O, which has a prescribed horizontal displacement $x(t)$ and a prescribed vertical displacement $y(t)$. Determine the kinetic energy of the bar at an arbitrary instant using θ, the counterclockwise angular displacement of the bar, measured from the bar's vertical position, as the chosen generalized coordinate.

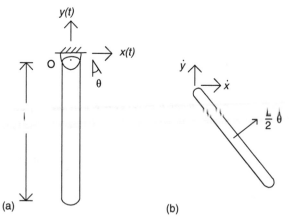

(a)   (b)

FIGURE 1.13 (a) Slend bar has pinned at O, which has a prescribed velocity of $\dot{x}\vec{i} + \dot{y}\vec{j}$. (b) Illustration of use of relative velocity equation.

Solution: The velocity of point O, the pin support is

$$\mathbf{v}_O = \dot{x}\mathbf{i} + \dot{y}\mathbf{j} \tag{a}$$

The velocity of the mass center of the bar, $\bar{\mathbf{v}}$, is determined using the relative velocity equation

$$\bar{\mathbf{v}} = \mathbf{v}_O + \mathbf{v}_{G/O} = \mathbf{v}_O + \omega \times \mathbf{r}_{G/O}$$

$$= \dot{x}\mathbf{i} + \dot{y}\mathbf{j} + \dot{\theta}\mathbf{k} \times \left(\frac{L}{2}\sin\theta\mathbf{i} - \frac{L}{2}\cos\theta\mathbf{j}\right)$$

$$= \left(\dot{x} + \frac{L}{2}\dot{\theta}\sin\theta\right)\mathbf{i} + \left(\dot{y} + \frac{L}{2}\dot{\theta}\cos\theta\right)\mathbf{j} \tag{b}$$

The kinetic energy at an arbitrary instant is determined using Equation 1.57 as

$$T = \frac{1}{2}m\left[\left(\dot{x} + \frac{L}{2}\dot{\theta}\sin\theta\right)^2 + \left(\dot{y} + \frac{L}{2}\dot{\theta}\cos\theta\right)^2\right] + \frac{1}{2}\left(\frac{1}{12}mL^2\right)\dot{\theta}^2$$

$$= \frac{1}{2}m\left(\dot{x}^2 + \dot{y}^2 + L\dot{x}\dot{\theta}\cos\theta + L\dot{y}\dot{\theta}\sin\theta + \frac{1}{3}L^2\dot{\theta}^2\right) \tag{c}$$

The kinetic energy of a deformable body is determined by calculating the kinetic energy of a differential element of the body and then integrating over the entire body. Consider the bar of Figure 1.14 as it undergoes longitudinal vibrations. Let $u(x,t)$ be the time dependent displacement of a particle in a cross-section a distance x from its left end. It is assumed that all particles in the cross-section have the same longitudinal displacement. Figure 1.14b shows

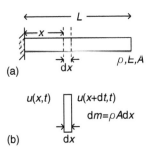

(a)

(b)

FIGURE 1.14 (a) $u(x,t)$ is the displacement function for longitudinal motion of bar. (b) Differential element of mass dm. Displacement of left face is $u(x,t)$, while displacement of right face is $u(x+dx,t)$.

a differential element of thickness dx, whose left face is a distance x from the left end of the bar. The mass of the element is

$$dm = \rho A dx \qquad (1.59)$$

where ρ is the mass density of the bar and A is its cross-section area. The kinetic energy of this differential element is

$$dT = \frac{1}{2} dm \left(\frac{\partial u}{\partial t}\right)^2 = \frac{1}{2}\rho A \left(\frac{\partial u}{\partial t}\right)^2 dx \qquad (1.60)$$

The total kinetic energy of the bar of length L at any instant is

$$T = \int dT = \frac{1}{2} \int_0^L \rho A \left(\frac{\partial u}{\partial t}\right)^2 dx \qquad (1.61)$$

1.3.2 STIFFNESS ELEMENTS

The work done by force $\mathbf{F}$ as the particle where it is attached travels along a path that begins at a position vector $\mathbf{r}_1$ is:

$$W = \int_{\mathbf{r}} \mathbf{F} \cdot d\mathbf{r} \qquad (1.62)$$

The force is conservative if the work done by the force is independent of path. That is, the work done by a force as the particle to where it is attached moves between two points, is the same regardless of the path traveled between these points. In such a case, the work is a function of only the force, the particle's initial position and the particle's final position.

A potential energy function $V(\mathbf{r})$, a function of the position vector of the particle is defined for a conservative force such that the work done by the force as the particle to where it is attached travels between $\mathbf{r}_1$ and $\mathbf{r}_2$ is

$$W = V(\mathbf{r}_1) - V(\mathbf{r}_2) \qquad (1.63)$$

Using Cartesian coordinates with x, y and z as the final position

$$W = V(x_1, y_1, z_1) - V(x, y, z) \qquad (1.64)$$

Equation 1.64 implies

$$dW = -dV = -\frac{\partial V}{\partial x}dx - \frac{\partial V}{\partial y}dy - \frac{\partial V}{\partial z}dz = -\nabla V \cdot d\mathbf{r} \qquad (1.65)$$

Using the definition of work, Equation 1.62,

$$dW = \mathbf{F} \cdot d\mathbf{r} \qquad (1.66)$$

Equation 1.65 and Equation 1.66 are equal for all possible differential displacement vectors $d\mathbf{r}$ if, and only if,

$$\mathbf{F} = -\nabla V \qquad (1.67)$$

The force due to gravity, written using a right-handed Cartesian coordinate system is

$$F = -mg\mathbf{j} \qquad (1.68)$$

Applying Equation 1.67 to Equation 1.68

$$-mg\mathbf{j} = -\frac{\partial V}{\partial x}\mathbf{i} - \frac{\partial V}{\partial y}\mathbf{j} - \frac{\partial V}{\partial z}\mathbf{k} \qquad (1.69)$$

Equating like components on both sides of Equation 1.69 leads to

$$\frac{\partial V}{\partial x} = 0 \qquad (1.70)$$

$$\frac{\partial V}{\partial y} - mg \qquad (1.71)$$

$$\frac{\partial V}{\partial z} = 0 \qquad (1.72)$$

Integration of Equation 1.71 gives

$$V = mgy + f(x, z) \qquad (1.73)$$

Equation 1.70 implies that V is independent of x, while Equation 1.72 implies that V is independent of z. Thus, the arbitrary function in Equation 1.73 is a constant. Work is the difference between the potential energies at two spatial positions. Thus, when the work is calculated the constant is eliminated. Hence, the constant is truly arbitrary and may be chosen conveniently. Usually the constant is chosen such that the potential energy due to gravity is zero at a convenient reference position. The plane in which the potential energy due to gravity is zero is called the datum plane or simply datum. The potential energy is positive when the particle is above the datum and negative when below the datum.

The particle of the system of Figure 1.15 is displaced a distance x from equilibrium. The force developed in the linear spring of stiffness k is

$$F = -kx\mathbf{i} \tag{1.74}$$

The work done by the spring force is independent of path, and thus, a conservative force. Application of Equation 1.67 leads to

$$\frac{\partial V}{\partial x} = kx \tag{1.75}$$

$$\frac{\partial V}{\partial y} = 0 \tag{1.76}$$

$$\frac{\partial V}{\partial z} = 0 \tag{1.77}$$

Integration of Equation 1.75 gives

$$V = \frac{1}{2}kx^2 + f(y, z) \tag{1.78}$$

It is inferred from Equation 1.76 and Equation 1.77 that $f(y,z) = C$, a constant. Since the work is the difference in potential energies, the constant may

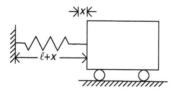

FIGURE 1.15 The force developed in a linear spring is conservative.

arbitrarily be chosen as zero leading to

$$V = \frac{1}{2}kx^2 \tag{1.79}$$

A spring naturally exists in an unstretched state. In this state the spring is not subject to an applied force and has a unstretched length ℓ. When a force is applied to the spring it changes in length according to a constitutive equation which for a linear spring is

$$F = kx \tag{1.80}$$

where x is the change in length of the spring from its unstretched length. Equation 1.79 may be used to calculate the potential energy in the spring when x is taken as the change in length of the spring from its unstretched length.

Example 1.3. Determine the potential energy in the system of Figure 1.16 in terms of the generalized coordinate x, the downward displacement of the block of mass m_1. In the position shown, the length of the spring is $\ell - \delta$.

Solution: The total potential energy is the sum of the potential energies of the two blocks and the spring. Choose the datum for potential energy calculations as the position of the blocks of Figure 1.16 in the position shown. If the block of mass m_1 is displaced a distance x downward, then the block of mass m_2 is displaced $2x$ upward. Thus, the total potential energy of the system is

$$V = -m_1 gx + m_2 g(2x) + \frac{1}{2}k(2x - \delta)^2$$

Potential energy is a stored energy, as it represents the potential of the system to do work. Elements which store potential energy are called stiffness

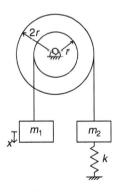

FIGURE 1.16 System of Example 1.3.

elements. Springs and gravity forces are two examples of stiffness elements in a discrete system.

Example 1.4. Determine the total potential energy of the three-degree-of-freedom system of Figure 1.17 at an arbitrary instant in terms of the generalized coordinates x_1, x_2, and x_3.

Solution: The total potential energy is the sum of the potential energies in each spring.

$$V = V_1 + V_2 + V_3 \tag{a}$$

The potential energy stored in a spring is calculated using Equation 1.79, where x is the change in length of the spring from its unstretched length. At an arbitrary instant, the change in length of the leftmost spring is x_1, thus

$$V_1 = \frac{1}{2}k_1x_1^2 \tag{b}$$

The right end of the spring between the two leftmost particles has displaced a distance x_2 at an arbitrary instant, while the displacement of the left end of the spring is x_1. The change in length of the spring is x_2-x_1. The spring has increased in length from its unstretched length when $x_2>x_1$ and has decreased in length when $x_2<x_1$. The potential energy stored in this spring is

$$V_2 = \frac{1}{2}k_2(x_2 - x_1)^2 \tag{c}$$

Through a similar analysis, the potential energy stored in the third spring is

$$V_3 = \frac{1}{2}k_3(x_3 - x_2)^2 \tag{d}$$

The total energy stored in the springs is obtained by combining Equation b through Equation d, yielding

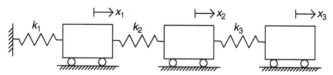

FIGURE 1.17 System of Example 1.4.

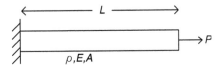

FIGURE 1.18 As the load is applied, the bar changes in length. As long as the yield stress is not exceeded when the force is removed, the bar assumes its original length.

$$V = \frac{1}{2}k_1 x_1^2 + \frac{1}{2}k_2(x_2 - x_1)^2 + \frac{1}{2}k_3(x_3 - x_2)^2 \qquad \text{(e)}$$

Potential energy in the form of strain energy is stored in deformable systems. Consider the deformable system of Figure 1.18. A bar of length L and cross-sectional area A is made of a material of elastic modulus E. A static load P is applied to the end of the bar. As the bar is loaded, the normal stress σ and the normal strain ε follow the linear stress strain curve of Figure 1.19, which in the linear range follows Hooke's Law

$$\sigma = E\varepsilon \qquad (1.81)$$

When the maximum stress is less than the yield stress as the force is removed, the stress–strain curve traces the same path down such that the bar is in its original state once the force is removed. This suggests that the system is conservative and a potential energy function exists. As the force is applied, a form of potential energy, called strain energy, is developed in the bar. As the force is removed, the strain energy is released.

The strain energy per volume is the area under the stress strain curve,

$$e = \frac{1}{2}\sigma\varepsilon \qquad (1.82)$$

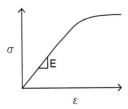

FIGURE 1.19 Illustration of Hooke's Law.

Application of Hooke's Law to Equation 1.82 leads to

$$e = \frac{1}{2}E\varepsilon^2 \tag{1.83}$$

The normal strain is defined as change in length per unit length. Thus, if x is a coordinate measured along the axis of the bar from its left end and $u(x,t)$ is the displacement of the particle from its initial position, the normal strain is

$$\varepsilon = \frac{\partial u}{\partial x} \tag{1.84}$$

and the strain energy per unit volume is

$$e = \frac{1}{2}E\left(\frac{\partial u}{\partial x}\right)^2 \tag{1.85}$$

A differential segment of thickness dx has a volume of $A\,dx$ and a total strain energy of

$$dV = \frac{1}{2}E\left(\frac{\partial u}{\partial x}\right)^2 (A\,dx) \tag{1.86}$$

The total strain energy in the bar is

$$V = \int dV = \frac{1}{2}\int_0^L AE\left(\frac{\partial u}{\partial x}\right)^2 dx \tag{1.87}$$

Example 1.5. The beam of Figure 1.20 is made of an elastic material of elastic modulus E and mass density ρ. It is of length L, with a cross-section area A and area moment of inertia I. Let x be a coordinate measured along the neutral axis of the beam from its left end and $w(x,t)$ the transverse deflection of the beam. Assuming the normal stress due to bending is less than the yield stress, and the usual assumptions of beam theory (small radius of curvature, plane sections remain plane) apply, determine an expression for the total strain energy of the beam in a form similar to that of Equation 1.87.

Solution: The strain energy per unit volume is given by Equation 1.85. Application of Hooke's Law written as $\varepsilon = (\sigma/E)$ gives

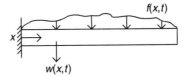

FIGURE 1.20 Application of an external load leads to strain energy in the beam.

$$e = \frac{\sigma^2}{2E} \tag{a}$$

Let y be a coordinate measured from the neutral axis in the cross-section of the beam, as shown in Figure 1.21. From elementary beam theory, the normal stress at a point in the cross-section due to bending is

$$\sigma = \frac{M(x)y}{I} \tag{b}$$

where $M(x)$ is the bending moment in that cross-section. Assuming small deflections, the bending moment is related to the transverse displacement by

$$M = EI\frac{\partial^2 w}{\partial x^2} \tag{c}$$

Substitution of Equation b and Equation c into Equation a leads to

$$e = \frac{M^2 y^2}{2EI^2} = \frac{Ey^2}{2}\left(\frac{\partial^2 w}{\partial x^2}\right)^2 \tag{d}$$

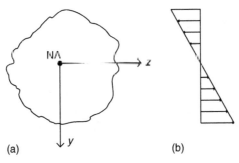

(a)   (b)

FIGURE 1.21 (a) Cross section of beam in which y is measured positive downward from the beam's neutral axis. (b) The normal stress due to bending is linear across the cross section if the elastic limit is not exceeded.

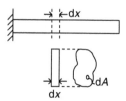

FIGURE 1.22 Differential plug of thickness dx and area dA.

The strain energy of a differential plug of thickness dx and area dA, as illustrated in Figure 1.22, is

$$dV = e\, dA\, dx = \frac{Ey^2}{2}\left(\frac{\partial^2 w}{\partial x^2}\right)^2 dA\, dx \tag{e}$$

The total strain energy in the beam is

$$V = \int_0^L \int_A \frac{Ey^2}{2}\left(\frac{\partial^2 w}{\partial x^2}\right)^2 dA\, dx = \int_0^L \frac{E}{2}\left(\frac{\partial^2 w}{\partial x^2}\right)^2 \left(\int_A y^2 dA\right) dx \tag{f}$$

By definition $I = \int_A y^2 dA$ and Equation f becomes

$$V = \frac{1}{2}\int_0^L EI\left(\frac{\partial^2 w}{\partial x^2}\right)^2 dx \tag{g}$$

1.3.3 ENERGY DISSIPATION

Energy is dissipated from mechanical systems through damping mechanisms. Many forms of damping are modeled by viscous damping. Viscous dampers, by themselves, lead to linear terms in the governing differential equations, whereas other forms of damping lead to nonlinear terms. A viscous damper, illustrated in Figure 1.23a, develops a force proportional to the velocity of the particle to which it is attached. The force developed in the viscous damper of Figure 1.23b is

$$F = c\dot{x} \tag{1.88}$$

The viscous damping force opposes the direction of motion.

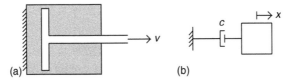

FIGURE 1.23 (a) Schematic of a viscous damper. As piston moves in cylinder of viscous liquid, a force opposing direction of motion and proportional to velocity is developed. (b) Representation of viscous damper in mechanical system, where c is the viscous damping coefficient.

The work done by the viscous damping force of Figure 1.23b is determined using Equation 1.62

$$W = \int_{x_1}^{x_2} (-c\dot{x}\mathbf{i}) \cdot (dx\mathbf{i}) = -\int_{x_1}^{x_2} c\dot{x}\,dx = -\int_{t_1}^{t_2} c\dot{x}^2 dt \qquad (1.89)$$

The work done by the viscous damping force is equal to the negative of the energy dissipated. Power is the time rate of change of energy. Thus, the power dissipated by the viscous damper is

$$P = -\frac{dW}{dt} = c\dot{x}^2 \qquad (1.90)$$

Equation 1.89 illustrates that the work done by a viscous damping force is always negative, leading to energy dissipation.

The viscous damper in the system of Figure 1.24 is installed between two particles. In such a case, the force developed in the viscous damper is the product of the damping coefficient and the relative velocity between the two points, and is in opposition to the direction of the relative velocity. The force developed in the viscous damper in the system of Figure 1.24 is

$$F = c(\dot{x}_2 - \dot{x}_1) \qquad (1.91)$$

The forces applied to each of the blocks from the viscous dampers are equal and opposite as illustrated in Figure 1.25. The total work done by the viscous

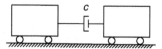

FIGURE 1.24 The force developed in the viscous damper is proportional to the relative velocity of the two particles and opposes the direction of the relative velocity.

damping force is

$$
W = -\int_{x_{2,1}}^{x_{2,2}} c(\dot{x}_2 - \dot{x}_1)dx_2 + \int_{x_{1,1}}^{x_{1,2}} c(\dot{x}_2 - \dot{x}_1)dx_1
$$

$$
= -\int_{t_1}^{t_2} c(\dot{x}_2 - \dot{x}_1)\dot{x}_2 dt + \int_{t_1}^{t_2} c(\dot{x}_2 - \dot{x}_1)\dot{x}_1 dt \qquad (1.92)
$$

$$
= -\int_{t_1}^{t_2} c(\dot{x}_2 - \dot{x}_1)^2 dt
$$

The power dissipated by this viscous damper is

$$
P = -\frac{dW}{dt} = c(\dot{x}_2 - \dot{x}_1)^2 \qquad (1.93)
$$

Coulomb damping is the dissipation of energy that occurs when two bodies slide relative to one another. Dry sliding friction is a simple example of Coulomb damping. Coulomb's friction law states that the tangential component of the reaction force, the friction force, between the particle of Figure 1.26 and the surface on which it slides is

$$
F = \mu N \qquad (1.94)
$$

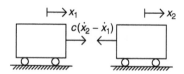

FIGURE 1.25 Illustration of direction of forces developed in viscous damper. If $\dot{x}_2 > \dot{x}_1$, then relative velocity is positive and damping force is in direction indicated.

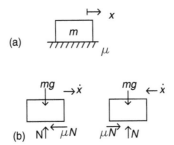

(a)

(b)

FIGURE 1.26 (a) Block slides on surface with a coefficient of friction between block and surface. (b) Free-body diagrams when $\dot{x} < 0$ and $\dot{x} > 0$.

where N is the normal component of the reaction force and μ is the coefficient of friction. The friction force always opposes the direction of motion. It is possible to take the direction of the friction force into account in Equation 1.94 by writing

$$F = -\mu N \frac{|\dot{x}|}{\dot{x}} \tag{1.95}$$

Other examples of Coulomb damping include rope friction, axle friction, belt friction, and pulley friction. Unfortunately, Coulomb damping leads to a nonlinear term in the differential equation, modeling a mechanical system.

The previous discussion of strain energy per unit volume assumes ideal material behavior. In a real material, energy is dissipated due to dislocations moving along slip planes and the breaking of molecular bonds. When a real material is loaded and unloaded, a stress–strain diagram similar to that of Figure 1.27 occurs. Recalling that the strain energy per unit volume is the area under the stress–strain diagram, the energy stored during the loading process is given by Equation 1.70, but the energy recovered during the unloading process

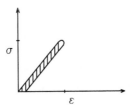

FIGURE 1.27 Stress–strain diagram of a real material during loading and unloading. Energy lost during this cycle is shaded area between loading and unloading curves.

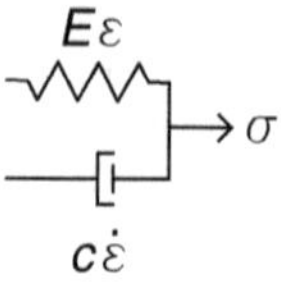

FIGURE 1.28 Schematic representation of Kelvin–Voight model of viscoelastic material.

is $e - e_1$, where e_1 is the area between the loading and unloading curves on the stress–strain diagram. The area e_1 is the energy lost per unit volume during this process, called hysteresis. The resulting loss of energy is called hysteretic damping.

The energy lost during each cycle of motion during hysteretic damping is dependent upon many factors including geometry and manufacturing methods used in fabrication of the object. Thus, hysteretic damping cannot be quantified using analytical methods. Empirical results are used instead. Empirical evidence does suggest that hysteretic damping can be modeled using an equivalent viscous damping coefficient in some situations.

The behavior of a viscoelastic material is governed by a constitutive equation in which the stress is a function of strain and strain rate. The Kelvin–Voight model of Figure 1.28 is a common model for a viscoelastic material where the stress is a linear combination of strain and strain rate:

$$\sigma = E\varepsilon + c\frac{d\varepsilon}{dt} \tag{1.96}$$

The Kelvin–Voight model is comparable to a model of a spring in parallel with a viscous damper.

1.3.4 External Energy Sources

Energy is provided to a mechanical system through externally supplied force and motion input. An external force supplies input to the system of Figure 1.29a. A force may be provided through a force actuator or some device which specifies a time dependent force applied to a given point. The work done by the force is calculated using Equation 1.62 and is dependent upon the motion of the particle to which the force is applied. The power transmitted to the system by application of the force is

$$P = \frac{dW}{dt} = Fv \tag{1.97}$$

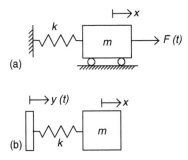

FIGURE 1.29 Illustration of (a) force input, and (b) motion input.

where v is the velocity of the point to which the force is applied. Power may also be transmitted to the system by application of a torque or moment.

Motion input is illustrated in Figure 1.29b. The motion of a particle in the mechanical system is prescribed. The motion of this particle affects the motion of the mechanical system. Motion input may be provided by a motion actuator or a kinematic device such as a cam and follower system. The energy delivered to the system through the motion input in the system of Figure 1.29b is

$$V = \frac{1}{2}k[y(t)]^2 \qquad (1.98)$$

External input to a system is classified as periodic or transient. A periodic input is an input $F(t)$ such that $F(t+T) = F(t)$ for some T, called the period of the input. A single-frequency harmonic input is a periodic input of the form

$$F(t) = F_0\sin \omega t \qquad (1.99)$$

where F_0 is the amplitude of the input and ω is its frequency. The period of the input is

$$T = \frac{2\pi}{\omega} \qquad (1.100)$$

Each of the periodic inputs of Figure 1.30 have a Fourier series representation, an infinite series expansion which converges pointwise to $F(t)$. The fundamental frequency of a periodic input is

$$\omega_1 = \frac{2\pi}{T} \qquad (1.101)$$

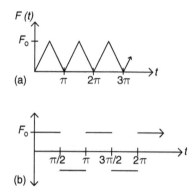

FIGURE 1.30 Periodic inputs (a) triangular or sawtooth wave, and (b) square wave.

The long-term response of a system due to a periodic input is called the steady-state response.

An example of a transient input is a force or motion input applied over a finite duration and then removed. Figure 1.31 presents examples of transient input. An input that suddenly changes the motion of the system is another type of transient input. An impulsive force is a very large force applied over a very short time. An impulsive force is often modeled by

$$F(t) = I\delta(t) \tag{1.102}$$

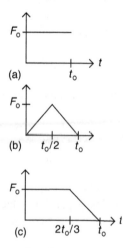

FIGURE 1.31 Examples of transient input (a) truncated step input, (b) triangular pulse, and (c) step input with linear decay.

where I is the magnitude of the impulse and

$$\delta(t) = \begin{cases} 0 & t \neq 0 \\ \infty & t = 0 \end{cases} \tag{1.103}$$

is called the Dirac delta function, or the unit impulse function. The unit impulse function is the mathematical model of an impulsive force, which results in application of an impulse of magnitude one (a unit impulse) to the system.

The principle of impulse and momentum may be used to examine the effect of application of a unit impulse. If a particle of mass m is at rest when subject to an impulse of magnitude I, then the principle of impulse and momentum is used to determine that immediately after application of the impulse, the particle has a velocity of I/m. The velocity is apparently discontinuous as a result of application of the impulse. In reality the force is applied over a finite time; the model of the impulse is only used for mathematical convenience. The velocity remains continuous but has a large gradient. The response of a system due to application of a unit impulse is called its impulsive response.

The step response of a system is the response due to a sudden change in the applied force, from one constant value to another. The model of a unit step is illustrated in Figure 1.32. The sudden change in force leads to a large acceleration.

A special type of motion input occurs when the system itself is rotating about an axis and motion occurs relative to the rotating axis. Inertia forces are generated due to the Coriolis acceleration. These forces are called circulatory forces, or gyroscopic forces. Such a system is called a gyroscopic system.

External forces applied to continuous systems may be functions of the independent spatial variable as illustrated in Figure 1.33. The force per unit length is described by $F(x,t)$. A force applied to a single point in a continuous system is modeled using the Dirac delta function. The mathematical model for a concentrated force applied to a beam a distance a from the left support is

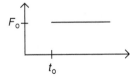

FIGURE 1.32 Step input at $t = t_o$ is modeled as $F_o\, u(t - t_o)$, where $u(t)$ is the unit step function.

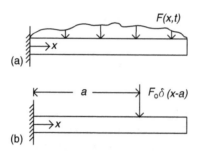

FIGURE 1.33 (a) Distributed load applied to beam. (b) Concentrated load is modeled by Dirac delta function.

$$F(x, t) = F_0(t)\delta(x - a) \tag{1.104}$$

The displacement, slope, and moment are continuous in a beam with a concentrated load. However, the shear force is discontinuous at $x = a$.

1.4 MODELING OF ONE-DEGREE-OF-FREEDOM SYSTEMS

1.4.1 INTRODUCTION AND ASSUMPTIONS

Two methods can be developed to derive a mathematical model governing the motion of a one-degree-of-freedom system. The first, presented here, is the application of basic conservation laws to a free-body diagram of the system. The second is an energy method. The energy method for one-degree-of-freedom systems is a special case of Lagrange's equations, which are derived in Chapter 2.

The governing differential equation for a one-degree-of-freedom system is derived by drawing a free-body diagram of the system at an arbitrary instant. Forces are labeled in terms of the generalized coordinate, and drawn consistent with the chosen positive direction of the generalized coordinate. The appropriate forms of the conservation laws, Newton's second law and the angular momentum equation, are applied. With only one generalized coordinate needed to completely describe the motion, scalar forms of the conservation laws are often sufficient. For example, the component forms of Newton's second law applied to a particle are

$$\sum F_x = m\ddot{x} \tag{1.105}$$

$$\sum F_y = m\ddot{y} \tag{1.106}$$

$$\sum F_z = m\ddot{z} \qquad (1.107)$$

Mathematical modeling of a mechanical system requires effective use of assumptions. Basic assumptions used include:

- All forms of energy dissipation other than viscous damping are neglected
- Gravity is the only body force
- Earth is an inertial reference frame
- Relativistic effects are ignored
- Springs and viscous dampers are massless
- Springs are linear

Other assumptions such as small displacements are made in some problems.

1.4.2 STATIC SPRING FORCES

A particle of mass m slides on a frictionless surface. It is attached to a spring of stiffness in parallel with a viscous damper of damping coefficient c and acted on by a time dependent force $F(t)$. Let x, the displacement of the particle from its equilibrium position, be the chosen generalized coordinate for this one-degree-of-freedom system. The free-body diagram of the system at an arbitrary instant is illustrated in Figure 1.34. Application of Newton's second law to the particle in its component form gives

$$\sum F = ma_x$$
$$F(t) - kx - c\dot{x} = m\ddot{x}$$
$$m\ddot{x} + c\dot{x} + kx = F(t) \qquad (1.108)$$

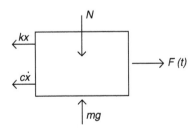

FIGURE 1.34 Free-body diagram of particle sliding on a frictionless surface.

Equation 1.108 is the differential equation governing the motion of a mass-spring-viscous damper system.

The particle of the system of Figure 1.35 is suspended from a support through a spring in parallel with a viscous damper. The particle is subject to a time-dependent force, $F(t)$. Let x, the downward displacement of the block measured from the system's equilibrium position, be the chosen generalized coordinate.

When the system of Figure 1.35 is in its equilibrium position, in the absence of any external force, the gravity force is balanced by a force in the spring $F_s = k\Delta_s$ where Δ_s is the static deflection of the spring, defined as the change in length of the spring from its unstretched length when the system is in a stable equilibrium position. Setting the resultant force acting on the free-body diagram of Figure 1.36 to zero leads to

$$mg - k\Delta_s = 0 \tag{1.109}$$

Equation 1.109 is the system's static equilibrium condition.

A free-body diagram of the system of Figure 1.35 at an arbitrary instant is illustrated in Figure 1.37. Application of Newton's second law for a particle in the direction of motion of the particle leads to

$$F(t) - k(x + \Delta_s) + mg - c\dot{x} = m\ddot{x}$$
$$m\ddot{x} + c\dot{x} + kx = F(t) + mg - k\Delta_s \tag{1.110}$$

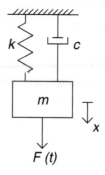

FIGURE 1.35 The generalized coordinate used to derive the differential equation for this system is measured from the system's equilibrium position. The gravity force and static spring force cancel with one another.

FIGURE 1.36 Free-body diagram of system of Figure 1.35 in its static equilibrium position.

Use of the equilibrium condition, Equation 1.109 in Equation 1.110 leads to

$$m\ddot{x} + c\dot{x} + kx = F(t) \tag{1.111}$$

Equation 1.108 and Equation 1.111 are identical. While the gravity force and the static spring force are drawn and labeled on the free-body diagram, they cancel with one another from the differential equation when the equilibrium condition is applied.

The reason for this occurrence is explained by looking at the system's potential energies. The potential energy of the system of Figure 1.34 at an arbitrary instant is

$$V = \frac{1}{2}kx^2 \tag{1.112}$$

Using the equilibrium position of the system as a datum the potential energy of the system of Figure 1.35 at an arbitrary instant is

$$V = \frac{1}{2}k(x + \Delta_s)^2 - mgx = \frac{1}{2}kx^2 + (k\Delta_s - mg)x + \frac{1}{2}k\Delta_s^2 \tag{1.113}$$

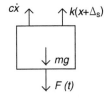

FIGURE 1.37 Free-body diagram of system of Figure 1.35 at an arbitrary instant.

Use of the static equilibrium condition, Equation 1.109, in Equation 1.113 leads to

$$V = \frac{1}{2}kx^2 + \frac{1}{2}k\Delta_s^2 \tag{1.114}$$

The term $(1/2)k\Delta_s^2$, which appears in Equation 1.114, is the initial potential energy in the system when the equilibrium position is used as a datum. The system response is dependent on the work done by all forces, which is the difference in potential energies. For both systems, the difference in potential energy between an arbitrary position and the static equilibrium position is $(1/2)kx^2$. Thus, since the systems have the same mass and viscous dampers, the two systems should be dynamically equivalent.

This is a general principle that can be proven using energy methods. The static forces in springs, and the gravity force or other forces for which static spring forces are necessary to maintain a stable equilibrium condition, cancel with each other in the governing differential equation, provided the system is linear and the generalized coordinates are measured from the system's equilibrium position. This result is used in the remainder of this chapter and throughout the book. Static spring forces, and forces which make static spring forces necessary, cancel with one another from the differential equation governing the motion of the system when the equilibrium condition is imposed. These forces will not be drawn on free-body diagrams of linear systems.

1.4.3 DERIVATION OF DIFFERENTIAL EQUATIONS

Differential equations governing the motion of a one-degree-of-freedom are derived through application of basic conservation laws to free-body diagrams of the system drawn at an arbitrary instant. Constitutive relations, such as those for forces developed in springs and viscous dampers, are applied to represent such forces in terms of the generalized coordinate. Principles of kinematics are often necessary to write kinematic properties in terms of the generalized coordinate.

Application of the equations of motion for rigid bodies requires an expression for the acceleration of the mass center of the body. The relative acceleration equation

$$\mathbf{a}_B = \mathbf{a}_A + \boldsymbol{\alpha} \times \mathbf{r}_{B/A} + \boldsymbol{\omega} \times (\boldsymbol{\omega} \times \mathbf{r}_{B/A}) \tag{1.115}$$

is used to determine the acceleration of particle B on a rigid body, in terms of the acceleration of particle A, on the same rigid body.

Example 1.6. The compound pendulum of Figure 1.38 consists of a slender bar of length L and mass m pinned at O. Let θ be the counterclockwise angular displacement of the bar, measured from the vertical equilibrium position. Derive a differential equation governing the motion of the bar when disturbed from equilibrium, using θ as the chosen generalized coordinate.

Solution: A free-body diagram of the bar, drawn at an arbitrary instant, is shown in Figure 1.39. The moment equation for a rigid body undergoing planar motion taken about point O is Equation 1.43, which when written for this system becomes

$$\sum \mathbf{M_O} = \bar{I}\alpha\mathbf{k} + \bar{\mathbf{r}} \times m\bar{\mathbf{a}} \tag{a}$$

The following are noted for this system:

$$\sum \mathbf{M_O} = -mg\frac{L}{3}\sin\theta\mathbf{k} \tag{b}$$

$$\bar{I} = \frac{1}{12}mL^2 \tag{c}$$

$$\alpha = \ddot{\theta} \tag{d}$$

$$\bar{\mathbf{r}} = \mathbf{r}_{G/O} = \frac{L}{3}(\sin\theta\mathbf{i} - \cos\theta\mathbf{j}) \tag{e}$$

$$\mathbf{a}_O = \mathbf{0}$$

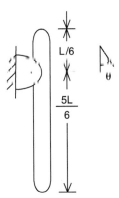

FIGURE 1.38 Compound pendulum of Example 1.6.

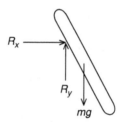

FIGURE 1.39 Free-body diagram of compound pendulum of Example 1.6 drawn at an arbitrary instant.

$$\bar{a} = a_O + (\ddot{\theta}\mathbf{k}) \times \left[\frac{L}{3}(\sin\theta\mathbf{i} - \cos\theta\mathbf{j})\right] + (\dot{\theta}\mathbf{k})$$

$$\times \left[(\dot{\theta}\mathbf{k}) \times \frac{L}{3}(\sin\theta\mathbf{i} - \cos\theta\mathbf{j})\right]$$

$$= \frac{L}{3}\ddot{\theta}(\cos\theta\mathbf{i} + \sin\theta\mathbf{j}) + \frac{L}{3}\dot{\theta}^2(-\sin\theta\mathbf{i} + \cos\theta\mathbf{j}) \qquad (f)$$

Equation b through Equation f are used in Equation a, leading to

$$-mg\frac{L}{3}\sin\theta = \frac{1}{12}mL^2\ddot{\theta} + m\frac{L^2}{9}\ddot{\theta}$$

$$\frac{7}{36}mL^2\ddot{\theta} + mg\frac{L}{3}\sin\theta = 0 \qquad (g)$$

$$\ddot{\theta} + \frac{12g}{7L}\sin\theta = 0$$

Example 1.7. A thin disk rolls on a surface as shown in Figure 1.40. Let x, the displacement of the mass center, be the chosen generalized coordinate. Assuming no slip, derive the differential equation governing the motion of the system.

Solution: A free-body diagram of the system at an arbitrary instant is drawn in Figure 1.41. No slip is not the same as no friction. In the case of no slip, the velocity of the point on the disk that is instantaneously in contact with the surface is zero. The friction force does no work. Kinematics is used to show that the angular acceleration, α, of the disk is related to the acceleration of the mass center $\bar{a}$ by

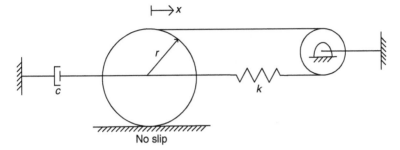

FIGURE 1.40 System of Example 1.7.

$$\bar{a} = r\alpha \qquad \text{(a)}$$

In order to eliminate the friction force from the resulting equation, moments are taken about C, the point of contact between the disk and the surface

$$\sum \mathbf{M_C} = \bar{I}\alpha\mathbf{k} + \mathbf{r}_{C/G} \times m\bar{\mathbf{a}} \qquad \text{(b)}$$

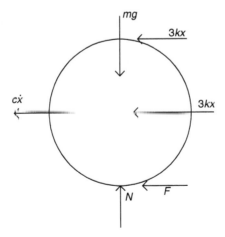

FIGURE 1.41 Free-body diagram of system of Example 1.7, drawn at an arbitrary instant.

Assuming no slip the following are used:

$$\sum \mathbf{M}_C = [-(3kx)(r)-(3kx)(2r)-c\dot{x}(r)+M(t)]\mathbf{k}$$
$$= [M(t)-9krx-cr\dot{x}]\mathbf{k} \tag{c}$$

$$\bar{I} = \frac{1}{2}mr^2 \tag{d}$$

$$\bar{\mathbf{a}} = \ddot{x}\mathbf{i} \tag{e}$$

$$\mathbf{r}_{C/G} = -r\mathbf{j} \tag{f}$$

$$\alpha = \frac{\ddot{x}}{r} \tag{g}$$

Substituting Equation c through Equation g into Equation b leads to

$$M(t)-9krx-cr\dot{x} = \left(\frac{1}{2}mr^2\right)\left(\frac{\ddot{x}}{r}\right)+r(m\ddot{x})$$

$$\frac{3}{2}mr\ddot{x}+cr\dot{x}+9krx = M(t) \tag{h}$$

Example 1.8. The pulley in the system of Figure 1.42 has a centroidal mass moment of inertia I_p. Let x be the displacement of the cart, measured to the right from the system's equilibrium position. Determine the differential equation governing the motion of the system, using x as the generalized coordinate.

Solution: Free-body diagrams of the pulley and the cart, drawn at an arbitrary instant, are shown in Figure 1.43. Let θ be the angular displacement of the

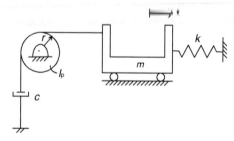

FIGURE 1.42 System of Example 1.8.

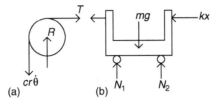

FIGURE 1.43 Free-body diagrams of component of system of Example 1.8 drawn at an arbitrary instant. (a) Pulley and (b) Cart.

pulley, measured counterclockwise from the systems' equilibrium position. Assuming no slip between the cable and the pulley

$$\theta = \frac{x}{r} \tag{a}$$

Summing moments about O using the free-body diagram of Figure 1.43a, noting that O is the mass center of the pulley, leads to

$$\sum \mathbf{M_O} = I_p \alpha_p \tag{b}$$

The following are used in Equation b:

$$\sum \mathbf{M_O} = [(T)(r) - cr\dot{\theta}]\mathbf{k} \tag{c}$$

$$\alpha_p = \frac{\ddot{x}}{r} \tag{d}$$

$$\dot{\theta} = \frac{\dot{x}}{r} \tag{e}$$

Substitution of Equation c through Equation e into Equation b, leads to

$$Tr - (cr)\left(\frac{\dot{x}}{r}\right) = I_p\left(\frac{\ddot{x}}{r}\right)$$

$$T = c\dot{x} + \frac{I_p}{r^2}\ddot{x} \tag{f}$$

The appropriate equation for summing moments on the free-body diagram of the cart is

$$\sum F_x = ma_x \tag{g}$$

$$F(t) - T - kx = m\ddot{x}$$

$$\left(m + \frac{I_p}{r^2}\right)\ddot{x} + c\dot{x} + kx = F(t) \tag{h}$$

1.4.4 MODEL SYSTEMS

The differential equations, Equation 1.108, Equation h of Example 1.7, and Equation d of Example 1.8, are linear second-order ordinary differential equations with constant coefficients. Furthermore, they are all of the same form

$$a_2\ddot{x} + a_1\dot{x} + a_0x = F(t) \tag{1.116}$$

Since the equations for a linear system are all of the same form, their solutions will be of the same form and a single model equation may be used to describe the motion of this class of problems. The simple mass-spring-viscous damping system of Figure 1.34 serves as a model for all linear one-degree-of-freedom systems. Equation 1.108 serves as a model equation for all linear one-degree-of-freedom systems.

Equation 1.108 uses a linear displacement for the generalized coordinate. Often, as in Example 1.6, the generalized coordinate is chosen as an angular displacement. In this case, the model system is that of a thin disk attached to a torsional spring and torsional viscosus damper in parallel, and subject to an external moment. This model system is illustrated in Figure 1.44. The model

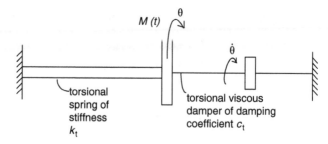

FIGURE 1.44 Model system for system, with angular coordinate used as generalized coordinate.

equation is

$$I\ddot{\theta} + c_t\dot{\theta} + k_t\theta = M(t) \tag{1.117}$$

1.4.5 ONE-DEGREE-OF-FREEDOM MODELS OF CONTINUOUS SYSTEMS

Distributed parameter systems are modeled using partial differential equations, whose solutions often require considerable effort. Under certain conditions, a one-degree-of-freedom model for a system that should be modeled as a continuous system, leads to good estimates for vibration properties of the system.

Consider a particle of mass m attached to the end of an elastic bar. When the particle is displaced a distance x from equilibrium, the bar is stretched a distance x. Such a change in length can only result when a normal stress is developed in the cross section of the bar. The normal stress has a resultant axial force, F, which is constant over the length of the bar. The normal stress also results in strain energy being stored in the bar. A free-body diagram of the particle is drawn illustrating the axial force acting on the particle. As the particle moves the axial force changes with time.

The motion of the particle at the end of the bar is similar to that of a particle attached to a spring. The bar, indeed, behaves like a spring. A linear spring is one in which the force is proportional to the change in length of the spring. Application of principles of elementary Mechanics of Materials leads to the result that when a static load F is applied to a uniform elastic bar of cross section area A and elastic modulus E, the bar has an increase in length Δ calculated by

$$\Delta = \frac{FL}{AE} \tag{1.118}$$

Inversion of Equation 1.118 leads to the force required for the bar to increase in length by Δ is

$$F = \left(\frac{AE}{L}\right)\Delta \tag{1.119}$$

Since the force is proportional to the change in length, this suggests that Equation 1.119 is rewritten as $F = k\Delta$ where k, the equivalent stiffness of the bar, is

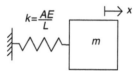

FIGURE 1.45 Model one-degree-of-freedom for particle attached to end of elastic bar.

$$k = \frac{AE}{L} \qquad (1.120)$$

A one-degree-of-freedom model of the motion of a particle at the end of an elastic bar is illustrated in Figure 1.45. It is shown in Chapter 2 that a continuous system model for the motion of this system is governed by the wave equation with a special type of boundary condition. The approximation of Figure 1.45 does not allow for wave phenomena in the bar. The accuracy of this type of approximation is considered in Chapter 5 and Chapter 6.

When the thin disk in the system of Figure 1.46 is rotated through an angle θ, a shear stress develops in the elastic shaft. The shear stress distribution, which is linear across the cross section of the shaft, has a resultant torque. Principles of Mechanics of Materials are used to show that the angular displacement of the end of the shaft in terms of the resultant torque is

$$\theta = \frac{TL}{JG} \qquad (1.121)$$

where G is the shear modulus of the uniform shaft, which has a cross section polar area moment of inertia J. Equation 1.121 is rearranged to solve for the torque required to rotate the end of the shaft through an angle θ as

$$T = \left(\frac{JG}{L}\right)\theta \qquad (1.122)$$

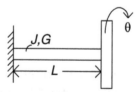

FIGURE 1.46 Torsional system consisting of thin disk at end of elastic shaft.

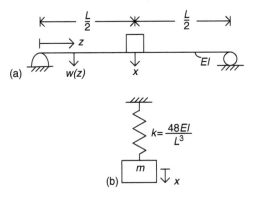

(a)

(b)

FIGURE 1.47 (a) Machine attached at midspan of simply supported beam. (b) One-degree-of-freedom model.

Equation 1.122 shows that under the application of a static torque, the angular displacement of the end of the shaft is proportional to the torque. This is a characteristic of a torsional spring, in which application of a torque leads to angular displacement of the end of the spring. The relation between torque and angular displacement for a torsional spring is

$$T = k_t \theta \tag{1.123}$$

where k_t is called the spring's torsional stiffness. Comparing Equation 1.123 with Equation 1.122, it is apparent that the torsional stiffness of the shaft is

$$k_t = \frac{JG}{L} \tag{1.124}$$

A one-degree-of-freedom model for a machine attached to the midspan of a simply supported beam is illustrated in Figure 1.47. When a concentrated load P, is applied to the midspan of the beam, the beam has a deflection curve $w(z)$. The deflection at the midspan is

$$w\left(\frac{L}{2}\right) = \frac{PL^3}{48EI} \tag{1.125}$$

The concentrated load required to cause a specific midspan deflection is

$$P = \left(\frac{48EI}{L^3}\right) w\left(\frac{L}{2}\right) \tag{1.126}$$

The equivalent stiffness for the model of Figure 1.47b is

$$k = \frac{48EI}{L^3} \tag{1.127}$$

The equivalent stiffness for a one-degree-of-freedom model of the vibrations of a machine placed along the span of a beam is dependent upon the beam's supports, as well as the location of the machine. If $w(z)$ is the deflected shape of the beam for a concentrated load P applied at $z = a$, then the equivalent stiffness for a one-degree-of-freedom model is

$$k = \frac{P}{w(a)} \tag{1.128}$$

Equation 1.128 is applied to show that the equivalent stiffness for the system of Figure 1.48, a machine attached to the end of a cantilever beam, is

$$k = \frac{3EI}{L^3} \tag{1.129}$$

Combinations of springs may be replaced by a single spring of an equivalent stiffness, such that the same mass attached to the single spring has the same motion as the mass attached to the combination of springs.

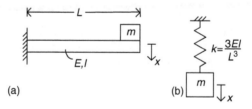

(a)   (b)

FIGURE 1.48 (a) Machine attached at end of cantilever beam. (b) One-degree-of-freedom model of system.

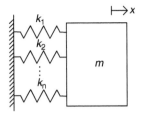

FIGURE 1.49 Springs in parallel have same displacement and resultant is sum of spring forces.

Parallel springs, illustrated in Figure 1.49, are characterized by each spring having the same displacement and that the resultant force from the parallel combination is the sum of the forces in each spring. The equivalent stiffness of a parallel combination of n springs is

$$k_{eq} = \sum_{i=1}^{n} k_i \tag{1.130}$$

Series springs, illustrated in Figure 1.50, are characterized by each spring having the same force, and the total change in length of the combination is the sum of the changes in length of the individual springs. The equivalent stiffness of a series spring combination is

$$k_{eq} = \frac{1}{\displaystyle\sum_{i=1}^{n} \frac{1}{k_i}} \tag{1.131}$$

Example 1.9. Determine the equivalent stiffness that should be used for a one-degree-of-freedom model for each system in Figure 1.51.

Solution: (a) The midspan of the beam and the discrete spring have the same displacement when motion occurs, thus these springs are in parallel. The equivalent stiffness for a one-degree-of-freedom model is obtained using Equation 1.126 and Equation 1.130 as

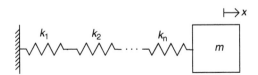

FIGURE 1.50 Springs in series each have the same force, but total displacement is sum of individual spring displacements.

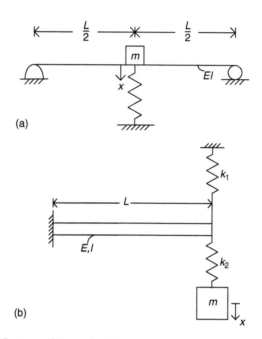

FIGURE 1.51 Systems of Example 1.9.

$$k_{eq} = k + \frac{48EI}{L^3} \tag{a}$$

(b) The displacement of the end of the beam is the same as the change in length of the discrete spring of stiffness k_1. The equivalent stiffness of this combination is

$$k_{eq1} = k_1 + \frac{3EI}{L^3} \tag{b}$$

The displacement of the particle is the displacement of the parallel combination, plus the change in length of the discrete spring of stiffness k_2. Thus, the parallel combination is in series with the discrete spring with an equivalent stiffness of

$$k_{eq} = \frac{1}{\frac{1}{k_2} + \frac{1}{k_{eq1}}} \tag{c}$$

The above approximations ignore the inertia of the elastic member; they assume the elastic members are massless. The importance of inertia can be assessed by comparing the kinetic energy of the particle to the kinetic energy of the bar, shaft, or beam. Consider again the thin disk attached to the shaft of Figure 1.46 at a time when the angular velocity of the shaft is ω. The total kinetic energy of the system at this instant is

$$T = \frac{1}{2}I_D\omega^2 + T_s \tag{1.132}$$

where I_D is the mass moment of inertia of the disk about the axis of rotation, and T_s is the kinetic energy of the shaft. Define z as a coordinate measured along the axis of the shaft. After static application of a torque the angular displacement is linear with z,

$$\theta(z,t) = \theta(L,t)\frac{z}{L} \tag{1.133}$$

Differentiation of Equation 1.133 with respect to time leads to

$$\frac{\partial\theta}{\partial t}(z,t) = \left[\frac{d\theta}{dt}(L,t)\right]\frac{z}{L} = \omega\frac{z}{L} \tag{1.134}$$

The mass moment of inertia of the differential element of the shaft, illustrated in Figure 1.52, is $dI = \rho J dz$ where ρ is the mass density of the shaft and J is the polar area moment of inertia of the cross section of the shaft. The kinetic energy of this element is

$$dT_s = \frac{1}{2}dI\left(\frac{\partial\theta}{\partial t}(z,t)\right)^2 = \frac{1}{2}(\rho J dz)\left(\omega\frac{z}{L}\right)^2 \tag{1.135}$$

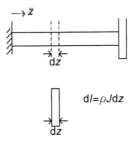

FIGURE 1.52 Differential element of shaft.

The shaft's total kinetic energy is

$$T_s = \int dT_s = \int_0^L \frac{1}{2} \frac{\rho J \omega^2}{L^2} z^2 dz = \frac{1}{2} \left(\frac{\rho J \omega^2}{L^2} \right) \left(\frac{z^3}{3} \right)_{z=0}^{z=L} = \frac{1}{2} \left(\frac{\rho J L}{3} \right) \omega^2$$

$$= \frac{1}{2} \left(\frac{I_s}{3} \right) \omega^2 \tag{1.136}$$

where I_s is the total mass moment of inertia of the shaft.

Substitution of Equation 1.136 into Equation 1.132 leads to

$$T = \frac{1}{2} \left(I_D + \frac{1}{3} I_s \right) \omega^2 \tag{1.137}$$

Equation 1.137 is used to assess the importance of the inertia effect of the shaft on the vibrations of the system. If $I_s \ll 3I_D$, then the inertia of the shaft is negligible and may be neglected. If $I_s < 3I_D$, the inertia of the shaft may be important and can be approximated by virtually adding a disk of mass moment of inertia $I_s/3$ to the system at the end of the shaft. If $I_s \geq I_D$, the inertia of the shaft is important and a continuous system model should be used for accurate modeling.

A similar analysis for longitudinal motion of a bar is applied with similar results. The inertia effects of a longitudinal bar are approximated by adding a particle of mass $1/3m_b$ to the end of the bar. A similar result applies for a helical coil spring.

The deflection of a beam due to a concentrated load is a piecewise-defined fourth-order polynomial in z. The displacement is continuous at $z = a$, but the shear force which is proportional to the third derivative of displacement has a jump discontinuity at $z = a$. The coefficients in the polynomial depend upon the beam supports as well as a. While the added mass concept may be used to improve the accuracy of a one-degree-of-freedom model of a machine mounted along the span of a beam, a general formula, such as the 1/3 formula for bars, shafts, and springs, does not exist for beams.

1.5 QUALITATIVE ASPECTS OF ONE-DEGREE-OF-FREEDOM SYSTEMS

Differential equations governing the motion of one-degree-of-freedom dynamic systems considered in this study are of the form

$$\ddot{\theta} + f(t,\theta,\dot{\theta}) = F(t) \qquad (1.138)$$

where $F(t)$ is from an external input, and $f(t,\theta,\dot{\theta})$ is determined during the derivation of the governing equation. The system is linear if $f(t, \theta, \dot{\theta})$ is a linear function of θ and $\dot{\theta}$. The differential equations derived in Examples 1.7 and Example 1.8 are linear. A system is nonlinear if $f(t,\theta,\dot{\theta})$ is a nonlinear function of either θ or $\dot{\theta}$. The differential equation derived in Example 1.6 is nonlinear. In most cases $f(t,\theta,\dot{\theta})$ is explicitly independent of t. If $f(t,\theta,\dot{\theta})$ is explicitly dependent upon t, the system is self-excited.

The equilibrium points of a system are those values of, θ such that $f(t,\theta,\dot{\theta}) = 0$. A stable equilibrium point is one for which when the system is slightly perturbed from this position, it will eventually return to this equilibrium position. An unstable equilibrium position occurs when the system will move farther away from this position and will eventually approach a stable equilibrium position, if one exists. A neutrally stable equilibrium position occurs when the system neither returns to the original equilibrium position, nor moves unbounded away from the perturbed state.

Let $\theta = \theta_0$ correspond to an equilibrium position. The stability of an equilibrium position is determined by letting $\theta = \theta_0 + \varepsilon\theta_1$, where ε is a small dimensionless parameter, and determining the fate of θ_1. Substituting into Equation 1.138 with $F(t) = 0$ leads to

$$\varepsilon\ddot{\theta}_1 + f(t,\varepsilon\dot{\theta}_1,\theta_0 + \varepsilon\theta_1) = 0 \qquad (1.139)$$

Using a Taylor series expansion in Equation 1.139, and keeping only first order terms in ε, leads to

$$\varepsilon\ddot{\theta}_1 + f(t,0,0) + \varepsilon\dot{\theta}_1 \frac{\partial f}{\partial \dot{\theta}}(t,0,0) + \varepsilon\theta_1 \frac{\partial f}{\partial \theta}(t,0,0) = 0 \qquad (1.140)$$

The linearized differential equation for θ_1 becomes

$$\ddot{\theta}_1 + \frac{\partial f}{\partial \dot{\theta}}(t,0,0)\dot{\theta}_1 + \frac{\partial f}{\partial \theta}(t,0,0)\theta_1 - 0 \qquad (1.141)$$

If the solution of Equation 1.141 grows without bound, then the system is unstable. If the solution of Equation 1.141 eventually approaches zero, then the system is stable. If the solution is bounded, but does not approach zero, the system is neutrally stable.

If f is not explicitly dependent upon t, that is $f = f(\dot{\theta},\theta)$, a solution to Equation 1.141 is sought as

$$\theta_1 = Ae^{\alpha t} \tag{1.142}$$

Substitution of Equation 1.142 into Equation 1.141 leads to

$$\alpha^2 + \frac{\partial f}{\partial \dot{\theta}}(0,0)\alpha + \frac{\partial f}{\partial \theta}(0,0) = 0 \tag{1.143}$$

The solutions of Equation 1.143 are

$$\alpha = \frac{1}{2}\left[-\frac{\partial f}{\partial \dot{\theta}}(0,0) \pm \sqrt{\left(\frac{\partial f}{\partial \dot{\theta}}(0,0)\right)^2 - 4\frac{\partial f}{\partial \theta}(0,0)} \right] \tag{1.144}$$

The stability of the system depends upon the real parts of both values of α. The system is unstable if either value of α has a positive real part. The system is neutrally stable if the real part of one value of α is zero, and the other has a real part that is either zero negative. The system is stable if the real parts of both values of α are negative. Study of Equation 1.144 leads to the following conclusions:

- The system is unstable if $\partial f/\partial \theta(0,0) < 0$.
- The system is neutrally stable if $\partial f/\partial \theta(0,0) = 0$.
- The system is unstable if $(\partial f/\partial \dot{\theta})(0,0) < 0$. This case corresponds to a negative damping.

Example 1.10. The differential equation governing the motion of a one-degree-of-system is

$$\ddot{\theta} + \omega_n^2 \sin \theta = 0 \tag{a}$$

Determine the equilibrium points for the system and their stability.

Solution: The equilibrium points are the zeros of

$$f(\theta) = \sin \theta \tag{b}$$

which are

$$\theta_0 = 0, \pm \pi, \pm 2\pi, \dots \tag{c}$$

The stability of the equilibrium points are determined by the sign of

$$\frac{df}{d\theta} = \cos\theta \tag{d}$$

Noting that $\cos(2n\pi) = 1$ and, $\cos[(2n-1)\pi] = -1$ it is shown that the equilibrium points $\theta_0 = 0, \pm 2\pi, \pm 4\pi,\ldots$ are neutrally stable and the equilibrium points $\theta_0 = \pm\pi, \pm 3\pi, \pm 5\pi\ldots$ are unstable.

The differential equation for Example 1.10 is that of a simple or compound pendulum. The results imply that the pendulum is stable when it is hanging downward from its support and unstable when it is suspended upward from its support (the inverted pendulum).

The phase plane is a plot of $\dot\theta$ on the vertical axis versus θ on the horizontal axis. Equilibrium points are classified by the behavior of the trajectories in the phase plane in the vicinity of the equilibrium point. If both values of α are real and they are of the same sign, the equilibrium point is called a node. If both are positive, the node is unstable, and if both are negative, the node is stable. If the values of α are of opposite signs, the equilibrium point is called a saddle point and is unstable. If the values of α are complex conjugates, the equilibrium point is called a focus and is stable. If the values of α are purely imaginary, the equilibrium point is called a center and is neutrally stable.

Sketches of phase plane plots in the vicinity of equilibrium points are presented in Figure 1.53.

Example 1.11. The differential equation for the response of a system with a quadratic spring is

$$\ddot{x} + x + \lambda x^2 = 0 \tag{a}$$

Determine the equilibrium points for the system, their stability, and their type.

Solution: The equilibrium points and their nature are determined from

$$f(x) = x + \lambda x^2 \tag{b}$$

The equilibrium points are the zeroes of $f(x)$ which are

$$x_0 = 0, -\frac{1}{\lambda} \tag{c}$$

The stability of the equilibrium points is determined by examining

For $x = 0$, $df/dx = 1$ and the values of α calculated using Equation 1.144, are $\alpha = \pm i$. Since the values of α are purely imaginary, the system is neutrally stable and the equilibrium point $x = 0$ is a center.

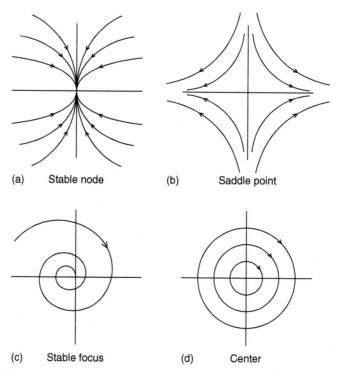

(a)   Stable node   (b)   Saddle point

(c)   Stable focus   (d)   Center

FIGURE 1.53 Phase planes in the vicinity of equilibrium points. (a) A stable node occurs when values of α are of same sign. (b) A saddle point is always unstable. (c) A focus is stable for positive damping. (d) A center is neutrally stable.

For $x = -1/\lambda$, $df/dx = -1$ and the values of α calculated using Equation 1.131, are $\alpha = \pm 1$. Since both values of α are real, but with different signs, the equilibrium point $x = -1/\lambda$ is an unstable saddle point.

Nonlinear differential equations are difficult to solve. Since exact solutions do not currently exist for most nonlinear differential equations, numerical or approximate solutions must be used. One method of approximation is to linearize the differential equation about an equilibrium position. This is essentially the method employed in obtaining Equation 1.141. Such an equation is called a first-order approximation.

An alternate method of linearization is to make an *a priori* linearization assumption, an assumption that when used, the resulting differential equation is linear. A common example of an a priori linearizing assumption is a small angle or small displacement assumption. Consider Example 1.6 and the derivation of the differential equation governing the motion of a compound

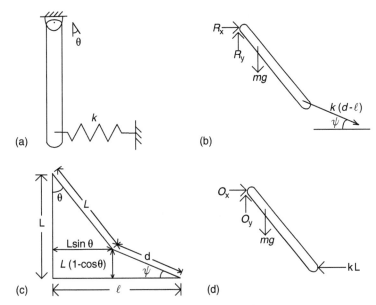

FIGURE 1.54 (a) System of Example 1.12. (b) Free-body diagram at arbitrary instant. (c) Illustration of geometry of system. (d) Free-body diagram using small angle assumption.

pendulum. An *a priori* linearizing assumption is that the angle through which the pendulum rotates is small enough such that $\sin \theta \approx \theta$. When this assumption is used, the differential equation derived is

$$\frac{7}{36} mL^2 \ddot{\theta} + \frac{mgL}{3} \theta = 0 \tag{1.145}$$

Example 1.12. Consider the system of Figure 1.54a. Let θ, the counter-clockwise angular displacement of the bar from the system's stable vertical equilibrium position, be the chosen generalized coordinate. (a) Derive the nonlinear differential equation governing the motion of the system. (b) Linearize the differential equation by assuming a perturbation about the equilibrium position. (c) Derive the differential equation using the small angle assumption a *priori*.

Solution: (a) The free-body diagram of the system at an arbitrary instant is shown in Figure 1.54(b). Let ℓ be the unstretched length of the spring and d the length of the spring at the instant the free-body diagram is drawn. The magnitude of the spring force is $k(d-\ell)$. At this instant the spring makes an angle ψ with the horizontal. Summing moments about the pin support leads to

$$\sum M_O = I_o\alpha - mg\left(\frac{L}{2}\sin\theta\right) - k(d-\ell)(\cos\psi)(L\sin\theta) + k(d-\ell)$$

$$\times(\sin\psi)(L\cos\theta)$$

$$= \frac{1}{3}mL^2\ddot{\theta} \tag{a}$$

The diagram of Figure 1.54c illustrates the geometry used to determine the instantaneous length of the spring

$$d = \sqrt{(\ell - L\sin\theta)^2 + L(1-\cos\theta)^2} \tag{b}$$

as well as

$$\sin\psi = \frac{L(1-\cos\theta)}{d} \tag{c}$$

$$\cos\psi = \frac{\ell - L\sin\theta}{d} \tag{d}$$

Equation a is rearranged to

$$\frac{1}{3}mL^2\ddot{\theta} + mg\frac{L}{2}\sin\theta + k(\ell - L\sin\theta)(L\sin\theta) - kL(1-\cos\theta)L\cos\theta$$

$$- \frac{k\ell(\ell - L\sin\theta)(L\sin\theta) - \ell L^2(1-\cos\theta)\cos\theta}{\sqrt{(\ell - L\sin\theta)^2 + L^2(1-\cos\theta)^2}} = 0 \tag{e}$$

$$\frac{1}{3}mL^2\ddot{\theta} + mg\frac{L}{2}\sin\theta + kL(\ell\sin\theta - L\cos\theta + 2L\cos^2\theta - L)$$

$$\left(1 - \frac{\ell}{\sqrt{\ell^2 - 2\ell L\sin\theta - 2L^2\cos\theta + 2L^2}}\right) = 0$$

(b) Assuming small θ, and perturbing about $\theta = 0$ such that $\theta = \varepsilon\theta_1$ leads to

$$\sin(\varepsilon\theta_1) \approx \varepsilon\theta_1 \tag{f}$$

$$\cos(\varepsilon\theta) = 1 - \frac{1}{2}\varepsilon^2\theta^2 \tag{g}$$

$$1 - \cos(\varepsilon\theta) = \frac{1}{2}\varepsilon^2\theta^2 \tag{h}$$

Substituting Equation f through Equation h into Equation e leads to

$$\frac{1}{3}mL^2(\varepsilon\ddot{\theta}) + \frac{mgL}{2}(\varepsilon\theta) + k(\ell - L\varepsilon\theta)(L\varepsilon\theta) - kL\left(\frac{1}{2}\varepsilon^2\theta^2\right)L(1 - \frac{1}{2}\varepsilon^2\theta^2)$$

$$-\frac{k(\ell - L\varepsilon\theta)(L\varepsilon\theta) - kL\left(\frac{1}{2}\varepsilon^2\theta^2\right)L(1 - \frac{1}{2}\varepsilon^2\theta^2)}{\sqrt{(\ell - L\varepsilon\theta)^2 + L^2\left(\frac{1}{2}\varepsilon^2\theta\right)^2}} = 0 \tag{i}$$

Keeping only terms of $O(\varepsilon)$ in Equation i leads to

$$\frac{1}{3}mL^2\ddot{\theta} + \frac{mgL}{2}\theta + k\ell L\theta\left(\frac{L}{\ell}\right) = 0$$

$$\frac{1}{3}mL^2\ddot{\theta} + \left(\frac{mgL}{2} + kL^2\right)\theta = 0 \tag{j}$$

(c) The free-body diagram of Figure 1.54d is drawn using the small angle assumption. Summing moments about O, using Equation f = Equation h, leads to

$$\sum M_O = I_O\alpha - mg\frac{L}{2}\theta - kL^2\theta = \frac{1}{3}mL^2\ddot{\theta} \tag{k}$$

It is noted that Equation k is identical to Equation j.

1.6 FREE VIBRATIONS OF LINEAR SINGLE-DEGREE-OF-FREEDOM SYSTEMS

The differential equation governing the free vibrations of the model system of Figure 1.55 is

$$m\ddot{x} + c\dot{x} + kx = 0 \tag{1.146}$$

Dividing by m leads to

$$\ddot{x} + 2\zeta\omega_n\dot{x} + \omega_n^2 x = 0 \tag{1.147}$$

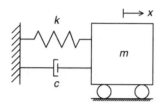

FIGURE 1.55 Single-degree-of-freedom mass-spring-viscous damper system.

where

$$\omega_n = \sqrt{\frac{k}{m}} \tag{1.148}$$

is the system's natural frequency and

$$\zeta = \frac{c}{2m\omega_n} \tag{1.149}$$

is called the damping ratio. Equation 1.147 is called the standard form of the differential equation, governing the free vibrations of a single-degree-of-freedom linear system.

The free response of a single-degree-of-freedom system is characterized by two parameters, the natural frequency and the damping ratio. The natural frequency, as defined in Equation 1.148, has dimensions of T^{-1}, indicating that that natural frequency is related to the time scale at which the free response occurs. Typical units for the natural frequency are rad/s. The damping ratio is dimensionless and is a measure of the damping force relative to the inertia force.

The general solution of Equation 1.146 is obtained by assuming

$$x(t) = Ae^{\alpha t} \tag{1.150}$$

Substitution of Equation 1.150 in Equation 1.147 leads to a quadratic equation in α with solutions

$$\alpha = -\zeta\omega_n \pm \omega_n\sqrt{\zeta^2 - 1} \tag{1.151}$$

The mathematical form of the general solution depends upon the numerical value of ζ. The solution is in terms of constants of integration, which are determined by application of initial conditions of the form

$$x(0) = x_0 \qquad \dot{x}(0) = \dot{x}_0 \qquad (1.152)$$

The mathematical form of the free response depends upon the value of the damping ratio. Four cases are considered: (1) $\zeta = 0$, in which case the system is undamped; (2) $0 < \zeta < 1$, in which case the system is underdamped; (3) $\zeta = 1$, in which case the system is critically damped; and (4), $\zeta > 0$ in which case the system is overdamped.

1. When $\zeta = 0$, the system is undamped. The values of α determined from Equation (1.151) are $\pm i\omega_n$. The free vibration response is

$$x(t) = \hat{C}_1 e^{-i\omega_n t} + \hat{C}_2 e^{i\omega_n t} \qquad (1.153)$$

Euler's identity is used to replace the exponential functions with complex exponents by a linear combination of trigonometric functions as

$$x(t) = C_1 \cos(\omega_n t) + C_2 \sin(\omega_n t) \qquad (1.154)$$

Trigonometric identities are used to rewrite Equation 1.154 in the form

$$x(t) = A \sin(\omega_n t + \phi) \qquad (1.155)$$

where the amplitude A is

$$A = \sqrt{x_0^2 + \left(\frac{\dot{x}_0}{\omega_n}\right)^2} \qquad (1.156)$$

and the phase angle is

$$\phi = \tan^{-1}\left(\frac{\omega_n x_0}{\dot{x}_0}\right) \qquad (1.157)$$

The free response of an undamped system, illustrated in Figure 1.56, is cyclic and periodic.

$$T = \frac{2\pi}{\omega_n} \qquad (1.158)$$

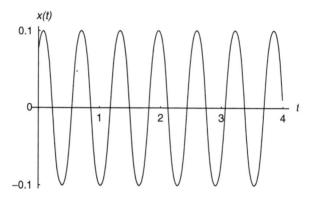

FIGURE 1.56 Free response of an undamped system with $\omega_n = 10$ rad/s.

A cycle of motion occurs between the times when the system reaches its peak responses. The time required to execute one cycle, called the period, is the same for each cycle for a linear system.

The reciprocal of the period is the frequency of motion. If time is measured in sec., then the frequency is measured in cycles/sec. or hertz (Hz). The natural frequency, ω_n, is the frequency, measured in rad/sec.

Since there are no dissipative forces in an undamped system, energy is conserved. The total energy, the kinetic energy plus the potential energy, in the system at all times is

$$E = \frac{1}{2}m\dot{x}_0^2 + \frac{1}{2}kx_0^2 = \frac{1}{2}kA^2 \tag{1.159}$$

In the absence of dissipative forces, free vibrations are sustained at the natural frequency and with the constant total energy of Equation 1.159.

2. When $0 < \zeta < 1$ the system is said to be underdamped. The values of α obtained from Equation 1.151 are $\alpha = -\zeta\omega_n \pm i\omega_n\sqrt{1-\zeta^2}$. The free response of the system is

$$x(t) = Ae^{-\zeta\omega_n t}\sin(\omega_d t + \phi_d) \tag{1.160}$$

where the damped natural frequency is

$$\omega_d = \omega_n\sqrt{1-\zeta^2} \tag{1.161}$$

the initial amplitude is

$$A = \sqrt{x_0^2 + \left(\frac{\dot{x}_0 + \zeta\omega_n x_0}{\omega_d}\right)^2} \tag{1.162}$$

and the phase angle is

$$\phi_d = \tan^{-1}\left(\frac{x_0\omega_d}{\dot{x}_0 + \zeta\omega_n x_0}\right) \tag{1.163}$$

The free response of an underdamped system, illustrated in Figure 1.57, is cyclic of period

$$T_d = \frac{2\pi}{\omega_d} \tag{1.164}$$

Since the damping force is nonconservative and its work is negative, then in the absence of any external excitation, the system's total energy continually decreases from its initial value of $(1/2)kx_0^2 + (1/2)m\dot{x}_0^2$. The damping force provides more resistance to motion when the damping ratio is larger. Thus, it takes longer to execute one cycle of motion. If the system is underdamped, the damping force is not large enough to dissipate the total energy within one cycle. It can be shown that the energy dissipated during the nth cycle is

$$\Delta E_n = \left(1 - e^{-\frac{4\pi\zeta}{\sqrt{1-\zeta^2}}}\right)E_n \tag{1.165}$$

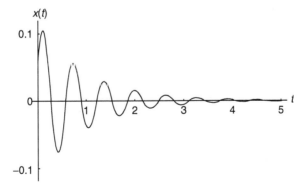

FIGURE 1.57 Free response of an underdamped system with $\omega_n = 10$ rad/s and $\zeta = 0.1$.

where E_n is the total energy present at the beginning of the cycle. Equation 1.165 shows that a constant fraction of energy is dissipated over each cycle. The total energy and the amplitude decrease exponentially. The ratio of amplitudes on successive cycles is a constant

$$\frac{A_{n+1}}{A_n} = e^{-\frac{2\pi\zeta}{\sqrt{1-\zeta^2}}} \tag{1.166}$$

The amplitudes on successive cycles form a geometric sequence.

3. When $\zeta = 1$, the free vibrations are said to be critically damped. Only one value of α is obtained from Equation 1.151, $\alpha = -\omega_n$. The free vibration response of a critically damped system is

$$x(t) = e^{-\omega_n t}[x_0 + (\dot{x}_0 + \omega_n x_0)t] \tag{1.167}$$

Equation 1.167 is plotted for various initial conditions in Figure 1.58. When $\zeta = 1$, the damping force is just large enough such that all of the initial energy is dissipated within one cycle. Indeed, a critically damped system does not execute a full cycle of motion; its response decays exponentially. When the initial conditions are of opposite signs and the initial kinetic energy is greater than the initial potential energy, the critically damped system will overshoot equilibrium and approach equilibrium from the opposite side.

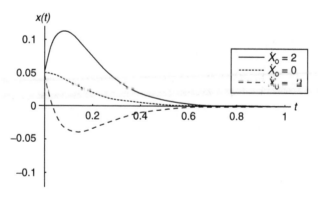

FIGURE 1.58 Free response of a critically damped system with $\omega_n = 10$ rad/s and $x_0 = 0.05$.

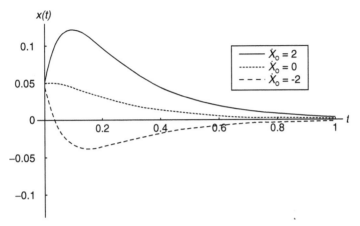

FIGURE 1.59 Free response of an overdamped system with $\omega_n = 10$ rad/s, $\zeta = 1.15$, and $x_o = 0.05$.

4. When $\zeta > 1$, the free vibrations are said to be overdamped. The values of α obtained from Equation 1.151 are $\alpha = -\zeta\omega_n \pm \omega_n\sqrt{\zeta^2 - 1}$. The free response for an overdamepd system is

$$x(t) = \frac{e^{-\zeta\omega_n t}}{2\sqrt{\zeta^2 - 1}}\left[\frac{\dot{x}_0}{\omega_n} + x_0\left(\zeta + \sqrt{\zeta^2 - 1}\right)\right]e^{\omega_n\sqrt{\zeta^2 - 1}t}$$

$$+ \frac{e^{-\zeta\omega_n t}}{2\sqrt{\zeta^2 - 1}}\left[-\frac{\dot{x}_0}{\omega_n} + x_0\left(-\zeta + \sqrt{\zeta^2 - 1}\right)\right]e^{-\omega_n\sqrt{\zeta^2 - 1}t} \qquad (1.168)$$

which is illustrated in Figure 1.59. Overdamped free vibrations decay exponentially, as the damping force is more than sufficient to dissipate the initial energy within one cycle.

The nature of the free response of a system with viscous damping depends upon the value of the damping ratio, ς. If the system is undamped, the free response is periodic with a frequency ω_n. Motion continues indefinitely with constant amplitude. When the system is underdamped the motion is cyclic, but not periodic. The larger the damping ratio, the smaller the frequency and the larger the period of the cyclic motion. When the system is underdamped, a constant fraction of energy is dissipated on each cycle. The amplitude decays exponentially. When the damping ratio is one, the system is critically damped. The motion is not cyclic, but decays exponentially. When the system is overdamped, motion decays exponentially. The rate at which the response decays decreases as the damping ratio increases.

1.7 RESPONSE OF A SINGLE-DEGREE-OF-FREEDOM SYSTEM DUE TO HARMONIC EXCITATION

1.7.1 GENERAL THEORY

The differential equation governing the response of a single-degree-of-freedom system subject to a single-frequency harmonic excitation can be written in a standard form of

$$\ddot{x} + 2\zeta\omega_n\dot{x} + \omega_n^2 x = \frac{F_0}{m}\sin{(\omega t)} \qquad (1.169)$$

where F_0 is the amplitude of the excitation and ω is the frequency of the excitation, which is independent of the natural frequency. For $\zeta > 0$, the homogeneous solution of Equation 1.169 decays exponentially and, after some time, is negligible in comparison to the particular solution. A particular solution of Equation 1.169 is obtained using the method of undetermined coefficients by assuming

$$x(t) = A\cos{(\omega t)} + B\sin{(\omega t)} \qquad (1.170)$$

The resulting solution, also called the steady-state response, is written as

$$x(t) = X\sin(\omega t - \phi) \qquad (1.171)$$

The steady-state amplitude, X, is

$$X = \frac{F_0}{m\sqrt{(\omega_n^2 - \omega^2)^2 + (2\zeta\omega\omega_n)^2}} \qquad (1.172)$$

The phase difference between the excitation and the response is

$$\phi = \tan^{-1}\left(\frac{2\zeta\omega\omega_n}{\omega_n^2 - \omega^2}\right) \qquad (1.173)$$

A nondimensional form of the above equations is obtained through introduction of the dimensionless frequency ratio

$$r = \frac{\omega}{\omega_n} \tag{1.174}$$

Introduction of Equation 1.174 into Equation 1.172 and Equation 1.173, leads to

$$\frac{m\omega_n^2 X}{F_0} = \frac{1}{\sqrt{(1-r^2)^2 + (2\zeta r)^2}} \tag{1.175}$$

$$\phi = \tan^{-1}\left(\frac{2\zeta r}{1-r^2}\right) \tag{1.176}$$

The frequency response of a system is the variation of steady-state amplitude and phase with frequency. The frequency response of a one-degree-of-freedom system due to a single frequency input is given by Equation 1.175 and Equation 1.176, which are graphically illustrated in Figure 1.60 and Figure 1.61. Equation 1.175 and Equation 1.176, and Figure 1.60 and Figure 1.61, lead to the following:

- The right-hand side of Equation 1.175 is the ratio of the maximum force developed in the spring during the steady-state response to the maximum value of the excitation force. It is also called the magnification factor.

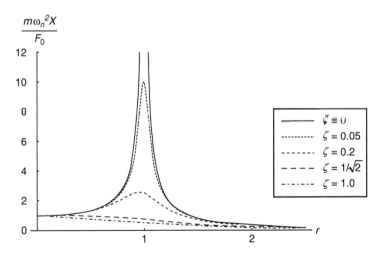

FIGURE 1.60 Frequency response curves for a single-degree-of-freedom system.

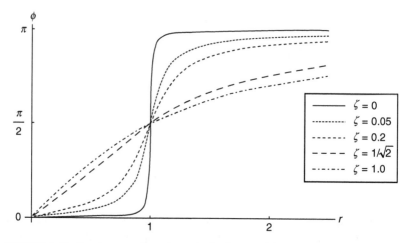

FIGURE 1.61 Steady-state phase for a single-degree-of-freedom system.

- The magnification factor is near one for small r and approaches zero as $1/r^2$ for large r.
- For $\zeta < 1/\sqrt{2}$, the magnification factor increases with increasing r, until it reaches a maximum and then decreases.
- When a maximum exists, it occurs for $r = \sqrt{1 - 2\zeta^2}$ and is equal to $1/(2\zeta\sqrt{1 - \zeta^2})$.
- For a given value of r, the magnification factor is smaller for larger ζ.
- For $\zeta > 1/\sqrt{2}$, the magnification factor decreases as r increases.
- For $r = 0$, $\phi = 0$; for $r = 1$, $\phi = \pi/2$.
- $\phi \to \pi$ for large r.
- $\zeta = 0$ is a special case and must be examined separately.

For an undamped system, $\zeta = 0$, Equation 1.175 and Equation 1.176 reduce to

$$\frac{m\omega_n^2 X}{F_0} = \frac{1}{|1 - r^2|} \tag{1.177}$$

$$\phi = \begin{cases} 0 & r < 1 \\ \pi & r > 1 \end{cases} \tag{1.178}$$

A steady-state amplitude cannot be calculated using Equation 1.177 for an undamped system when $r = 1$. In this case, the non-homogeneous term in

Equation 1.169 is equal to a homogeneous solution. The correct particular solution is determined as

$$x(t) = -\frac{F_0}{2m\omega_n} t \cos(\omega_n t) \tag{1.179}$$

Equation 1.179 suggests that the response grows without bound, a condition called resonance, as illustrated in Figure 1.62. However, the solution is valid only within the limits of the linearity assumptions used in deriving the differential equation. Eventually these assumptions will become invalid, leading to inelastic behavior or catastrophic failure.

Resonance occurs in an undamped system when the frequency of excitation coincides with the natural frequency. Given an initial energy, free vibrations of an undamped system are sustained at the natural frequency at the initial energy level. The excitation force does work on the system. When the excitation frequency is different than the natural frequency, the work done by the external force is necessary to sustain the response at the excitation frequency. However, when the excitation frequency coincides with the natural frequency, the work done by the external force is not needed to sustain the motion at the natural frequency. The work leads to a continual increase in the total energy in the system and an increase in amplitude.

1.7.2 Frequency-Squared Excitation

A case of interest is the frequency-squared excitation in which the excitation amplitude is proportional to the square of the frequency (frequency squared

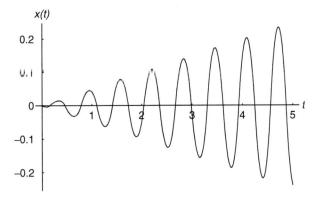

FIGURE 1.62 Resonance response of a single-degree-of-freedom system is characterized by an amplitude that grows indefinitely.

excitation)

$$F_0 = K\omega^2 \tag{1.180}$$

A machine on an elastic foundation with an unbalanced rotating component, as illustrated in Figure 1.63, is an example on a single-degree-of-freedom system subject to a frequency-squared excitation. In this case the constant of proportionality is

$$K = m_0 e \tag{1.181}$$

where m_0 is the mass of the rotating component and e is the distance from the axis of rotation to the center of mass of the rotating component.

The steady-state amplitude of the response due to a frequency-squared excitation is obtained using Equation 1.175. Substituting Equation 1.180 into Equation 1.175 and rearranging, leads to

$$\frac{mX}{K} = \frac{r^2}{\sqrt{(1 - r^2)^2 + (2\zeta r)^2}} \tag{1.182}$$

which is illustrated in Figure 1.64, Equation 1.182, and Figure 1.64 are used to determine the following regarding the steady-state amplitude for a one-degree-of-freedom system subject to a frequency-squared excitation:

- The steady-state amplitude is small for small r
- As $r \to \infty$, $mX/K \to 1$
- For $\zeta < 1/\sqrt{2}$, mX/K increases as r increases until it reaches a maximum and then decreases, approaching 1 as r gets large

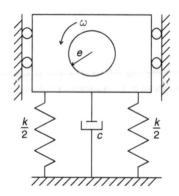

FIGURE 1.63 Machine with rotating unbalance is an example of a system with a frequency squared excitation.

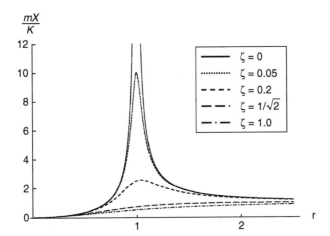

FIGURE 1.64 Frequency response for a system in which excitation amplitude is proportional to square of excitation frequency.

- When a maximum exists it occurs for a value of $r = 1/\sqrt{1-2\zeta^2}$ and is equal to $1/(2\zeta\sqrt{1-\zeta^2})$
- For $\zeta < 1/\sqrt{2}$, mX/K increases as r increases, never reaching a maximum, and asymptotically approaching 1 for large r

1.7.3 MOTION INPUT

Consider the system of Figure 1.65, in which the system mass is connected, through a spring in parallel with a viscous damper, to a support, which has a prescribed time-dependent displacement, $y(t)$. Let x be the absolute displacement of the mass, measured from the system's equilibrium position.

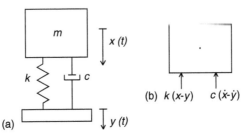

FIGURE 1.65 (a) Machine whose undetermined displacement is $x(t)$ is connected to a base with prescribed displacement $y(t)$. (b) Free-body diagram at an arbitrary instant.

The differential equation governing the motion of the mass is derived as

$$m\ddot{x} + c\dot{x} + kx = ky + c\dot{y} \tag{1.183}$$

An alternate choice for the generalized coordinate for this single-degree-of-freedom system is the displacement of the mass relative to its base

$$z(t) = x(t) - y(t) \tag{1.184}$$

The differential equation whose solution leads to $z(t)$ is

$$m\ddot{z} + c\dot{z} + kz = -m\ddot{y} \tag{1.185}$$

which is written in standard form as

$$\ddot{z} + 2\zeta\omega_n\dot{z} + \omega_n^2 z = -\ddot{y} \tag{1.186}$$

Consider a harmonic base displacement such that

$$y(t) = Y \sin(\omega t) \tag{1.187}$$

in which case Equation 1.186 becomes

$$\ddot{z} + 2\zeta\omega_n\dot{z} + \omega_n^2 z = \omega^2 Y \sin(\omega t) \tag{1.188}$$

Thus, the harmonic base motion provides a frequency-squared excitation when solving for the relative displacement such that $K = mY$. The steady-state response for the relative displacement is

$$z(t) = Z \sin(\omega t - \phi) \tag{1.189}$$

where ϕ is given by Equation 1.176 and

$$Z = \frac{r^2}{\sqrt{(1-r^2)^2 + (2\zeta r)^2}} Y \tag{1.190}$$

The absolute displacement of the mass is

$$x(t) = z(t) + y(t) = Z \sin(\omega t - \phi) + Y \sin(\omega t) \tag{1.191}$$

which after some algebra leads to

$$x(t) = X \sin(\omega t - \lambda) \tag{1.192}$$

where

$$X = \sqrt{\frac{1 + (2\zeta r)^2}{(1 - r^2)^2 + (2\zeta r)^2}} Y \tag{1.193}$$

$$\lambda = \tan^{-1}\left(\frac{2\zeta r^3}{1 + (4\zeta^2 - 1)r^2}\right) \tag{1.194}$$

Equation 1.193 is illustrated in Figure 1.66. The following is deduced from Equation 1.193 and Figure 1.66

- The ratio X/Y is called the transmissibility ratio.
- The transmissibility ratio is near one for small r.
- As r gets large, the transmissibility ratio approaches zero as $2\zeta/r$.
- For all values of ζ, the transmissibility ratio is equal to one when $r = \sqrt{2}$.
- The transmissibility ratio is less than one for all ζ when $r < \sqrt{2}$. This region on Figure 1.66 is called the amplification range.
- The transmissibility ratio is greater than one for all ζ when $r > \sqrt{2}$. This region of Figure 1.66 is called the isolation range.
- Within the amplification range for a specific r, the transmissibility ratio is smaller for larger ζ.

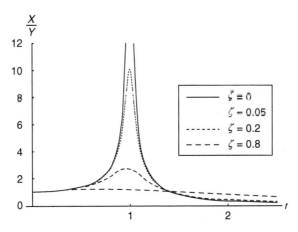

FIGURE 1.66 Frequency response due to harmonic motion input.

- Within the range of isolation for a specific r, the transmissibility ratio is smaller for smaller ζ.

Example 1.13. A simplified single-degree-of-freedom model of a vehicle suspension system is illustrated in Figure 1.67. The vehicle travels with a constant horizontal velocity v over a road of sinusoidal contour.

1. Determine the amplitude response of the vehicle in terms of the velocity v.
2. Determine the amplitude of the force developed in the spring in terms of the velocity.
3. The comfort of the passengers in the vehicle is determined by the amplitude of the absolute acceleration. Determine and plot the amplitude of absolute acceleration as a function of the velocity for damping ratios of 0.1, 0.4, and 0.7.

Solution: Let ξ be the horizontal distance traveled, at a time t, by the vehicle from a reference location, which coincides with a point where $y(\xi) = 0$. The vertical displacement of the wheel is

$$y(t) = d \sin\left(\frac{2\pi}{\ell}\xi\right) \tag{a}$$

For a constant velocity, $\xi = vt$, thus

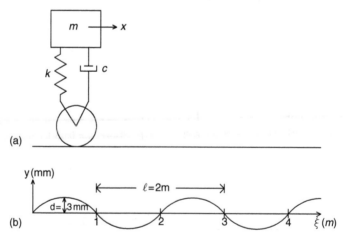

FIGURE 1.67 (a) Simplified one-degree-of-freedom model of suspension system. (b) Idealization of road contour.

$$y(t) = d \sin\left(\frac{2\pi v}{\ell} t\right)$$ (b)

The vehicle, and its suspension system, is modeled by the system of Figure 1.65, with a harmonic excitation of amplitude d and frequency $\omega = 2\pi v/\omega = 2\pi v/\ell$. Thus:

1. Using Equation 1.193, the amplitude of absolute acceleration of the vehicle is:

$$X = d\sqrt{\frac{1 + (2\zeta r)^2}{(1 - r^2)^2 + (2\zeta r)^2}}$$ (c)

where

$$r = \frac{\omega}{\omega_n} = \frac{2\pi v}{\ell \omega_n}$$ (d)

$$\omega_n = \sqrt{\frac{k}{m}}$$ (e)

$$\zeta = \frac{c}{2m\omega_n}$$ (f)

2. The force developed in the spring is:

$$F_s = k(y - x) = kz(t)$$ (g)

Using Equation 1.182, the amplitude of the spring force is:

$$F = \frac{r^2}{\sqrt{(1 - r^2)^2 + (2\zeta r)^2}} kd$$ (h)

3. The steady-state absolute acceleration of the vehicle is.

$$\ddot{x}(t) = -\omega^2 X \sin(\omega t - \lambda)$$ (i)

and the amplitude of the steady-state acceleration is

$$R = \omega^2 X = \omega^2 d \sqrt{\frac{1 + (2\zeta r)^2}{(1 - r^2)^2 + (2\zeta r)^2}}$$ (j)

which can be rearranged to

$$\frac{R}{d\omega_n^2} = r^2 \sqrt{\frac{1 + (2\zeta r)^2}{(1 - r^2)^2 + (2\zeta r)^2}} \qquad (k)$$

This non-dimensional representation of the steady-state amplitude is plotted in Figure 1.68 for several values of ζ.

1.7.4 GENERAL PERIODIC INPUT

A general periodic excitation of period T has a Fourier series representation as

$$F_s(t) = \frac{a_0}{2} + \sum_{i=1}^{\infty} [a_i \cos(\omega_i t) + b_i \sin(\omega_i t)] \qquad (1.195)$$

where the Fourier coefficients are

$$a_i = \frac{2}{T} \int_0^T F(t) \cos(\omega_i t)\, dt \quad i = 0, 1, 2... \qquad (1.196)$$

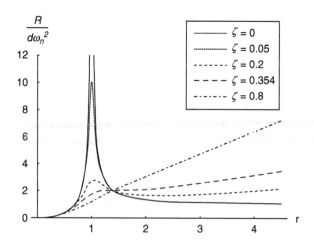

FIGURE 1.68 Frequency response of a single-degree-of-freedom system due to motion input in which amplitude is proportional to square of frequency.

$$b_i = \frac{2}{T} \int_0^T F(t) \sin(\omega_i t)\, dt \quad i = 1,2\dots \tag{1.197}$$

and the component frequencies are

$$\omega_i = \frac{2\pi i}{T} \tag{1.198}$$

An alternate representation of the Fourier series is

$$F_s(t) = \frac{a_0}{2} + \sum_{i=1}^{\infty} c_i \sin(\omega_i t + \kappa_i) \tag{1.199}$$

where

$$c_i = \sqrt{a_i^2 + b_i^2} \tag{1.200}$$

$$\kappa_i = \tan^{-1}\left(\frac{a_i}{b_i}\right) \tag{1.201}$$

The Fourier series representation for $F(t)$, Equation 1.195 or Equation 1.199, converges pointwise to $F(t)$ at all t, $0 \le t \le T$, where $F(t)$ is continuous. If $F(t)$ is piecewise continuous at t, then the Fourier series representation converges to the average value of $F(t)$ as approached from the right and the left. The Fourier series representation for $F(t)$ is periodic of period T,

$$F_s(t + T) = F_s(t) \tag{1.202}$$

If $F(t)$ is an even function ($F(-t) = F(t)$, $0 \le t \le T$), then the Fourier series representation is an even function and $b_i = 0$ $i = 1,2,\dots$. If $F(t)$ is an odd function ($F(-t) = -F(t)$, $0 \le t \le T$), then the Fourier series representation is an odd function and $a_i = 0$ $i = 0,1,2,\dots$.

The steady-state response of a single-degree-of-freedom system subject to a force input with a Fourier series representation of Equation 1.199, is obtained by substituting Equation 1.199 into the standard form of the governing differential equation leading to

$$\ddot{x} + 2\zeta\omega_n\dot{x} + \omega_n^2 x = \frac{1}{m}\left[\frac{a_0}{2} + \sum_{i=1}^{\infty} c_i \sin(\omega_i t + \kappa_i)\right] \tag{1.203}$$

The response of a single-degree-of-freedom system due to a single frequency excitation is given by Equation 1.171 through Equation 1.176. Since Equation 1.203 is a linear differential equation, and the Fourier series representation converges, the steady-state solution of Equation 1.203 is a linear superposition of solutions of the form of Equation 1.171. The result is

$$x(t) = \frac{a_0}{2m\omega_n^2} + \sum_{i=1}^{\infty} X_i \sin(\omega_i t + \kappa_i - \phi_i) \qquad (1.204)$$

where

$$X_i = \frac{c_i}{m\omega_n^2 \sqrt{(1 - r_i^2)^2 + (2\zeta r_i)^2}} \qquad (1.205)$$

and

$$\phi_i = \tan^{-1}\left(\frac{2\zeta r_i}{1 - r_i^2}\right) \qquad (1.206)$$

with

$$r_i = \frac{\omega_i}{\omega_n} \qquad (1.207)$$

1.8 TRANSIENT RESPONSE OF A SINGLE-DEGREE-OF-FREEDOM SYSTEM

The Laplace transform method is used to determine the response of a single-degree-of-freedom system due to an arbitrary excitation. A mass-spring-viscous damper system is subject to a time dependent force $F(t)$. Define

$$X(s) = \mathcal{L}\{x(t)\} = \int_0^{\infty} x(t)e^{-st} dt \qquad (1.208)$$

$$f(s) = \mathcal{L}\{F(t)\} \qquad (1.209)$$

as the Laplace transforms of the generalized coordinate and externally applied force, respectively.

The differential equation governing the response of the particle whose displacement is described by $x(t)$ is

$$\ddot{x} + 2\zeta\omega_n\dot{x} + \omega_n^2 x = \frac{1}{m}f(t) \tag{1.210}$$

Taking the Laplace transform of Equation 1.210, and using the properties of linearity of the transform, and transform of first and second derivatives, gives

$$s^2 X(s) - sx(0) - \dot{x}(0) + 2\zeta\omega_n[sX(s) - x(0)] + \omega_n^2 X(s) = \frac{1}{m}F(s) \tag{1.211}$$

Assuming $x(0) = \dot{x}(0) = 0$, Equation 1.211 is rearranged to

$$X(s) = \frac{F(s)}{m(s^2 + 2\zeta\omega_n s + \omega_n^2)} \tag{1.212}$$

The convolution theorem is used to invert Equation 1.212, leading to

$$x(t) = f(t) * h(t) = \int_0^t f(\tau)h(t - \tau)\,d\tau \tag{1.213}$$

The function $h(t)$ is the response of the system due to the application of a unit impulse. The forms of $h(t)$ for ranges of ζ are given in Table 1.1. If the free response of the system is underdamped, Equation 1.213 can be written as

$$x(t) = \frac{1}{m\omega_d}\int_0^t f(\tau)e^{-\zeta\omega_n(t-\tau)}\sin[\omega_d(t - \tau)]d\tau \tag{1.214}$$

TABLE 1.1
Impulsive Response of Single-Degree-of-Freedom System of Mass m and Natural Frequency ω_n

Damping Ratio, ζ	Impulsive Response, $h(t)$	Comments
$\zeta = 0$	$\frac{1}{m\omega_n}\sin(\omega_n t)$	
$0 < \zeta < 1$	$\frac{1}{m\omega_d}e^{-\zeta\omega_n t}\sin(\omega_d t)$	$\omega_d = \omega_n\sqrt{1 - \zeta^2}$
$\zeta = 1$	$\frac{1}{m}te^{-\omega_n t}$	
$\zeta > 1$	$\frac{1}{m(r_1 - r_2)}(e^{r_2 t} - e^{r_1 t})$	$r_1 = -\omega_n\left(\zeta + \omega_n\sqrt{\zeta^2 - 1}\right)$
		$r_2 = -\omega_n\left(\zeta - \omega_n\sqrt{\zeta^2 - 1}\right)$

Equation 1.214 is called the convolution integral solution for $x(t)$, and is the solution of Equation 1.210 under the condition that the initial velocity and displacement are zero.

Example 1.14. Determine the response of a single-degree-of-freedom system when subject to the excitation of Figure 1.69, when the free response of the system is underdamped. Use (a) the Laplace transform method, and (b) direct application of the convolution integral in both cases.

Solution: The input of Figure 1.69 is written in a concise mathematical form as

$$F(t) = F_0[u(t) - u(t - t_0)] \tag{a}$$

The Laplace transform of Equation a is

$$F(s) = \frac{1 - e^{-st_0}}{s} \tag{b}$$

Substitution of Equation b into Equation 1.212, with $\zeta = 0$, leads to

$$X(s) = \frac{F_0}{m} \left[\frac{1 - e^{-st_0}}{s(s^2 + \omega_n^2)} \right] \tag{c}$$

A partial fraction decomposition of the right-hand side of Equation c leads to

$$X(s) = \frac{F_0}{m\omega_n^2} \left(\frac{1}{s} - \frac{s}{s^2 + \omega_n^2} \right)(1 - e^{-st_0}) \tag{d}$$

The second shifting theorem is used to invert Equation d, leading to

$$x(t) = \frac{F_0}{m\omega_n^2} \{[1 - \cos(\omega_n t)]u(t) - [1 - \cos(\omega_n(t - t_0))]u(t - t_0)\} \tag{e}$$

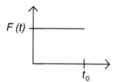

F (t)

t_0

FIGURE 1.69 Input to system of Example 1.14.

(b) Choose $t < t_0$. Evaluation of the convolution integral gives

$$x(t) = \frac{1}{m\omega_n} \int_0^t F_0 \sin[\omega_n(t-\tau)]d\tau = \frac{F_0}{m\omega_n^2} \cos[\omega_n(t-\tau)]\big|_{\tau=0}^{\tau=t}$$

$$= \frac{F_0}{m\omega_n^2}[1 - \cos(\omega_n t)] \tag{f}$$

Choose $t > t_0$. Evaluation of the convolution integral gives

$$x(t) = \frac{1}{m\omega_n} \int_0^{t_0} F_0 \sin[\omega_n(t-\tau)]d\tau = \frac{F_0}{m\omega_n^2} \cos[\omega_n(t-\tau)]\big|_{\tau=0}^{\tau=t_0}$$

$$= \frac{F_0}{m\omega_n^2}\{\cos[\omega_n(t-t_0)] - \cos(\omega_n t)\} \tag{g}$$

Equation f and Equation g are summarized by

$$x(t) = \frac{F_0}{m\omega_n^2} \begin{cases} 1 - \cos(\omega_n t) & t < t_0 \\ \cos[\omega_n(t-t_0)] - \cos(\omega_n t) & t > t_0 \end{cases} \tag{h}$$

Equation e and Equation h are identical.

2 Derivation of Differential Equations Using Variational Methods

Variational methods provide a consistent method for deriving differential equations governing the motion of discrete and continuous systems. The method involves application of the principles of calculus of variations to dynamic systems. Newton's Law is used to derive energy principles to which the calculus of variations is applied. Thus, the methods presented are often referred to as "energy methods."

In order to understand why Hamilton's Principle and Lagrange's equations work, the chapter begins with a survey of the basic principles of calculus of variations. Newton's Second Law is used to establish variational principles for dynamic systems. The governing equations for discrete systems are obtained using Lagrange's equations. Conservative systems are considered first, then the method is extended to include the effects of nonconservative forces. Extended Hamilton's Principle is used to develop the differential equations and requisite boundary conditions for continuous systems.

Variational principles are used in later chapters to develop methods of approximation of solutions of the derived equations. The Rayleigh–Ritz and finite element methods of Chapter 8 are true variational methods.

2.1 FUNCTIONALS

A function of a real continuous variable is a mapping of the domain of the variable into its range. The domain of the function is a dense subset of the real number line. The range of a function is determined by the definition of the function. For example if $f(x) = \sin x$ the domain of $f(x)$ is the entire real number line, $\Re$. The range of $f(x)$ is all $y, -1 \leq y \leq 1$. If $f(x) = 1/(x-2)$ then the domain of $f(x)$ is all real x, except $x = 2$. The range of $f(x)$ is all real y, except $y = 0$.

Functions are often required to satisfy certain conditions. For example, the solution to a differential equation may be required to satisfy certain boundary conditions. The elastic curve, $w(x)$ of a fixed—fixed beam of length L is required to satisfy $w(0) = 0$, $dw/dx(0) = 0$, $w(L) = 0$ and $dw/dx(L) = 0$. In finding a solution to the differential equation, only functions satisfying these conditions are sought.

A functional is a mapping of a set of functions into a set of real numbers, such that the mapping of each function in the set is a unique real number. In other words, a functional is a function of a set of functions which results in a real number. The domain of a functional is a set of functions and the range is a set of real numbers.

Consider some common functionals:

- $df/dx(2)$ is a functional whose domain is all functions differentiable at $x = 2$.
- $\int_0^{2\pi} f(x)\sin(nx)dx$ is a functional whose domain is all functions that are piecewise continuous on $[0,2\pi]$.
- $\left[\int_0^1 [f(x)]^2 w(x)dx\right]^{1/2}$ is a functional considered in Chapter 3, which measures in some fashion the magnitude of $f(x)$. The domain of the functional is the set of all piecewise continuous functions defined on $[0,1]$.
- $\max_{0 \le x \le 1} |f(x)|$ is a functional, which is also a measure of the magnitude of $f(x)$. Its domain is the set of all piecewise continuous functions on $[0,1]$.
- $V = \frac{1}{2}\int_0^L EA\left(\frac{\partial u}{\partial x}\right)^2 dx$ is the potential energy of an elastic bar of length L, elastic modulus E and cross sectional area A. The domain of V is all longitudinal displacements $u(x,t)$ that satisfy boundary conditions specified for the bar. For example, if the bar is fixed at $x = 0$ and at $x = L$ then $u(0,t) = 0$ and $u(L,t) = 0$.

Of particular importance to this study are functionals of the form

$$I = \int_a^b F(x,y,y')dx \qquad (2.1)$$

where x is an independent variable, $y(x)$ is a continuously differentiable function on $[a,b]$ that satisfies specified conditions at $x = a$ and $x = b$,

$y' = dy/dx$, and F is a defined function of these variables. The potential energy of the elastic bar is an example of such a functional as is the kinetic of the bar.

Two classic functionals are presented and are used to illustrate calculus of variations before applying the principles to dynamic systems. These problems are illustrated in Figure 2.1. The first is the simple geometric problem to calculate the arc length of a curve $y(x)$, which passes through the points (x_1, y_1) and (x_2, y_2). The arc length of a curve between two points is

$$I = \int_{x_1}^{x_2} \sqrt{1 + y'^2}\, dx \tag{2.2}$$

The arc length is a functional of the form of Equation 2.1 with $F = \sqrt{1 + y'^2}$. The calculus of variations problem associated with the

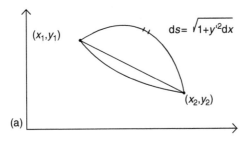

(a)

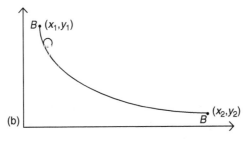

(b)

FIGURE 2.1 Two classic calculus of variations problems. (a) Is a straight line the shortest distance between two points? (b) The brachistochrone problem: Along what path should a particle travel without friction such that it travels between A and B in the shortest possible time?

functional of Equation 2.2 is to determine the curve $y(x)$, which minimizes the arc length between the two points.

The second classic calculus of variations problem considered is Bernoulli's brachistochrone problem, which is to find the path between two specified points along which a particle will travel without friction, such that the time the particle takes to travel between the two points is a minimum. The formulation of the brachistochrone problem follows.

Let $y(x)$ represent a path between the points (x_1,y_1) and (x_2,y_2), with $x_2>x_1$ and $y_2<y_1$ such that a particle starts at (x_1,y_1) with a velocity of zero and travels along the path. Let v represent the speed of the particle at any instant. By definition,

$$v = \frac{ds}{dt} \tag{2.3}$$

where s is a coordinate measured along the curve. Equation 2.3 is rearranged such that the time it takes to traverse the path of length ds is

$$dt = \frac{ds}{v} \tag{2.4}$$

The arc length is related to $y(x)$ by

$$ds = \sqrt{1 + y'^2}dx \tag{2.5}$$

The total time required to traverse the path is

$$t = \int_{x_1}^{x_2} \frac{\sqrt{1 + y'^2}}{v} dx \tag{2.6}$$

The velocity is a function of position along the path. Since there is no friction, the only force that does work is the gravity force, which is conservative. Thus, energy is conserved along the path. Taking the datum for potential energy calculations as the initial position of the particle, conservation of energy is expressed between the initial position and an arbitrary position along the path of the particle by

$$0 = \frac{1}{2}mv^2 + mg(y-y_1) \tag{2.7}$$

Equation 2.7 is solved for v, leading to

$$v = \sqrt{2g(y_1 - y)} \tag{2.8}$$

Substituting Equation 2.8 into Equation 2.6 leads to

$$t = \frac{1}{\sqrt{2g}} \int_{x_1}^{x_2} \sqrt{\frac{1 + y'^2}{y_1 - y}} \, dx \tag{2.9}$$

The time required for the particle to travel between the two points along the path described by $y(x)$ is determined using the functional of Equation 2.9. The associated calculus of variations problem is to determine the $y(x)$ for which t is a minimum.

2.2 VARIATIONS

If $y(x)$ is a continuous function of x the differential dy is a change in y due to a change in the dependent variable x,

$$dy = \lim_{\Delta x \to 0} [y(x + \Delta x) - y(x)] \tag{2.10}$$

It might also be written that

$$dy = y(x + dx) - y(x) \tag{2.11}$$

Let $y(x)$ be a curve between two points (x_1, y_1) and (x_2, y_2). Let ε be a small parameter and let $\eta(x)$ be any curve such that $\eta(x_1) = 0$ and $\eta(x_2) = 0$. A family of curves is then defined by

$$\tilde{y}(x) = y(x) + \varepsilon \eta(x) \tag{2.12}$$

The change from $y(x)$ to $\tilde{y}(x)$ is called a variation in y and is denoted as δy. That is

$$\delta y(x) = \varepsilon \eta(x) \tag{2.13}$$

A variation is a change in y due to a change in a path taken between two points. The concept of variation is illustrated in Figure 2.2. The functions $y(x)$ and $\tilde{y}(x)$ begin and end at the same points. There are infinitely many variations of the path represented by δy.

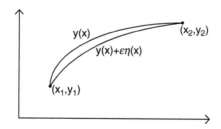

FIGURE 2.2 $Y(x)$ is the path from (x_1,y_1) to (x_2,y_2). A variation ∂y is defined by $\varepsilon\eta(x)$ where $\eta(x_1) = \eta(x_2) = 0$.

The concepts of differential and variations are compared in Figure 2.3. The horizontal axis ranges from some value x to $x+dx$. The function $y(x)$ between x and $x+dx$, is illustrated by the curve from point A to point B. The differential dy is the distance from A' to B. A variation of y is illustrated by the curve between C and D. The variation in y at x, δy, is the distance from A to C. The distance from C' to D is the change in the variation due to the change in x, $d(y+\delta y)$ whereas the distance from B to D is the variation between y and dy or $\delta(y+dy)$.

The distance from A' to D on Figure 2.3 can be calculated as either $A'B+BD$ or $A'C'+C'D$. Since both distances are equal,

$$dy + \delta(y + dy) = \delta y + d(y + \delta dy) \qquad (2.14)$$

Both the differential and variation are linear in that $d(y+\delta dy) = dy+d(\delta y)$ and $\delta(y+dy) = \delta y+\delta(dy)$. Using these expansions in Equation 2.14 leads to

$$dy + \delta y + \delta(dy) = \delta y + dy + d(\delta y) \qquad (2.15)$$

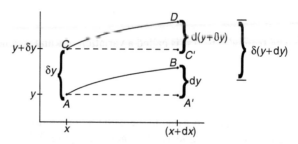

FIGURE 2.3 Illustration showing concepts of differential and variation that are used to show $d(\delta y) = \delta(dy)$.

Simplification of Equation 2.15 leads to

$$\delta(dy) = d(\delta y) \tag{2.16}$$

Equation 2.16 shows that the differential and variation operations are commutative. This can be extended to show that

$$\frac{d}{dx}(\delta y) = \delta\left(\frac{dy}{dx}\right) \tag{2.17}$$

which implies that the order of variation and differentiation can be interchanged.

The variation of a functional of the form of Equation 2.1, also called the Gateaux variation, is defined as

$$\delta I = \lim_{\varepsilon \to 0} \left[\frac{I(y + \varepsilon\eta, y' + \varepsilon\eta') - I(y, y')}{\varepsilon} \right] \tag{2.18}$$

The evaluation of I for any y in the domain of I leads to a real value. This real value is a function of the parameter ε. Thus, the definition of the Gateaux variation, Equation 2.18, is equivalent to the definition of the differentiation of I with respect to ε and evaluated for $\varepsilon = 0$:

$$\delta I = \frac{dI}{d\varepsilon}(\varepsilon = 0) \tag{2.19}$$

2.3 EULER–LAGRANGE EQUATION

The brachistochrone problem and the arc length problem involve the minimization of a functional of the form of Equation 2.1. Let $y(x)$ be the function that minimizes the functional and let $\delta y = \varepsilon\eta$ be a variation of $y(x)$. Then Equation 2.1 becomes

$$I(y + \delta y, y' + \delta y') = \int_{x_1}^{N_j} F(x, y + \varepsilon\eta, y' + \varepsilon\eta')dx \tag{2.20}$$

The functional I is a function of the parameter ε:

$$I(y + \delta y, y' + \delta y') = I(\varepsilon) \tag{2.21}$$

Thus, an extremum of the functional is obtained by setting $dI/d\varepsilon = 0$. By hypothesis, the function $y(x)$ leads to an extremum, which implies that an extremum occurs for $\varepsilon = 0$. Thus, the functional is rendered stationary when

$$\frac{dI}{d\varepsilon}(\varepsilon = 0) = 0 \qquad (2.22)$$

Equation 2.22 shows that evaluation of $dI/d\varepsilon$ at $\varepsilon = 0$ is equal to the variation of I. Thus, the functional is stationary

$$\delta I = 0 \qquad (2.23)$$

For functionals of the form of Equation 2.1, Equation 2.23 leads to

$$\delta \int_{x_1}^{x_2} F(x,y,y')dx = 0 \qquad (2.24)$$

The order of variation and integration in Equation 2.24 may be interchanged. This can be proved by using Equation 2.16. Then the condition for I to be stationary becomes

$$\int_{x_1}^{x_2} \delta F(x,y,y')dx = 0 \qquad (2.25)$$

Noting that F itself is a functional,

$$\delta F(x,y,y') = F(x,y + \delta y, y' + \delta y') - F(x,y,y') \qquad (2.26)$$

Using a Taylor series expansion keeping only through linear terms, Equation 2.26 becomes

$$\delta F(x,y,y') = F(x,y,y') + \frac{\partial F}{\partial y}(x,y,y')\delta y + \frac{\partial F}{\partial y'}(x,y,y')\delta y' - F(x,y,y')$$

$$= \frac{\partial F}{\partial y}(x,y,y')\delta y + \frac{\partial F}{\partial y'}(x,y,y')\delta y' \qquad (2.27)$$

Substitution of Equation 2.27 into Equation 2.25 leads to

$$\int_{x_1}^{x_2} \left[\frac{\partial F}{\partial y}(x,y,y')\delta y + \frac{\partial F}{\partial y'}(x,y,y')\delta y' \right] dx = 0 \qquad (2.28)$$

Noting from Equation 2.16 that $\delta y' = (d(\delta y)/dx)$ application of integration by parts leads to

$$\int_{x_1}^{x_2} \frac{\partial F}{\partial y'} \delta y' \, dx = \left[\left(\frac{\partial F}{\partial y'} \right) \delta y \right]_{x=x_1}^{x=x_2} - \int_{x_1}^{x_2} \frac{d}{dx} \left(\frac{\partial F}{\partial y'} \right) \delta y \, dx \qquad (2.29)$$

By definition, the variation satisfies $\delta y(x_1) = 0$ and $\delta y(x_2) = 0$. Use of these conditions in Equation 2.29, and subsequent substitution in Equation 2.28, leads to

$$\int_{x_1}^{x_2} \left[\frac{\partial F}{\partial y} - \frac{d}{dx} \left(\frac{\partial F}{\partial y'} \right) \right] \delta y \, dx = 0 \qquad (2.30)$$

A modified form of the DuBois–Reymond Lemma states the following: Let $h(x)$ be a function defined such that $h(x_1) = h(x_2) = 0$, then if

$$\int_{x_1}^{x_2} f(x)h(x) \, dx = 0 \qquad (2.31)$$

for all $h(x)$ that satisfy the stated conditions then $f(x) = 0$ for all x, $x_1 \leq x \leq x_2$.

The variation δy satisfies the conditions on $h(x)$ in the statement of the DuBois–Reymond Lemma. Thus, its application to Equation 2.30 requires

$$\frac{\partial F}{\partial y} - \frac{d}{dx} \left(\frac{\partial F}{\partial y'} \right) = 0 \qquad (2.32)$$

Equation 2.32 is called the Euler–Lagrange equation. It is used to derive differential equations which must be satisfied by a function that extremizes a functional of the form of Equation 2.1.

Example 2.1. Use the Euler Lagrange equation to determine the function $y(x)$ which provides the shortest path between the points (x_1, y_1) and (x_2, y_2).

Solution: The curve which provides the shortest distance between two points is the function $y(x)$, which minimizes the functional

$$I = \int_{x_1}^{x_2} \sqrt{1 + y'^2} \, dx \qquad (a)$$

for which

$$F(x,y,y') = \sqrt{1 + y'^2}$$ (b)

Noting that

$$\frac{\partial F}{\partial y} = 0$$ (c)

$$\frac{\partial F}{\partial y'} = \frac{y'}{\sqrt{1 + y'^2}}$$ (d)

application of the Euler–Lagrange equation leads to

$$-\frac{d}{dx}\left(\frac{y'}{\sqrt{1 + y'^2}}\right) = 0$$ (e)

Integration of Equation e with respect to x leads to

$$\frac{y'}{\sqrt{1 + y'^2}} = C_1$$ (f)

where C_1 is an arbitrary constant. Equation f is rearranged and solved for y' leading to

$$y' = \sqrt{\frac{C_1^2}{1 - C_1^2}} = C_2$$ (g)

Integration of Equation g leads to

$$y(x) = C_2 x + C_3$$ (h)

Equation h proves what is well known; a straight line is the shortest route between two points. The constants are evaluated by requiring the line pass through the points (x_1, y_1) and (x_2, y_2), which leads to

$$y(x) = \frac{(y_2 - y_1)x + y_1 x_2 - y_2 x_1}{x_2 - x_1}$$ (i)

Example 2.2. Solve the brachistochrone problem.

Solution: The functional for the brachistochrone problem is

$$I = \int_{x_1}^{x_2} \frac{\sqrt{1+y'^2}}{\sqrt{2g(y_1-y)}} \, dx \tag{a}$$

with

$$F(x,y,y') = \frac{\sqrt{1+y'^2}}{\sqrt{2g(y_1-y)}} \tag{b}$$

The necessary derivatives for application of the Euler–Lagrange equation to Equation b are

$$\frac{\partial F}{\partial y'} = \frac{y'}{\sqrt{2g(y_1-y)(1+y'^2)}} \tag{c}$$

$$\frac{\partial F}{\partial y} = \frac{\sqrt{1+y'^2}}{2\sqrt{2g}(y_1-y)^{3/2}} \tag{d}$$

Substituting Equation c and Equation d into the Euler–Lagrange equation and simplifying leads to

$$(1+y'^2) - 2y''(y_1-y) = 0 \tag{e}$$

Equation e is a nonlinear second-order differential equation. Since the equation does not explicitly involve x, the independent variable, its order may be reduced by making the substitutions

$$w = y' \tag{f}$$

$$y'' = w\frac{dw}{dy} \tag{g}$$

resulting in

$$(1+w^2) - 2w\frac{dw}{dy}(y_1-y) = 0 \tag{h}$$

Equation h is a separable first-order differential equation with w as the dependent variable and y as the independent variable. It may be rearranged and integrated, leading to

$$\int \frac{w}{(1+w^2)}\,dw = \int \frac{dy}{y_1 - y} \qquad \text{(i)}$$

Integration of Equation i and substitution of Equation f leads to

$$C = (y_1 - y)(1 + y'^2) \qquad \text{(j)}$$

Equation j may be solved for y' and then integrated. After a little algebra, a parametric representation of the solution is obtained as

$$x = C_1 + C_2(\theta - \sin \theta) \qquad \text{(k)}$$

$$y = y_1 + C_2(1 - \cos \theta) \qquad \text{(l)}$$

where C_1 and C_2 are chosen such that the curve passes through the appropriate points. Equation k and Equation l describe a cycloid, which is illustrated in Figure 2.4.

A functional that depends upon more than one dependent variable is of the form

$$I = \int_{x_1}^{x_2} F(x, y_1, y_2, \ldots, y_n, y'_1, y'_2, \ldots y'_n)\,dx \qquad (2.33)$$

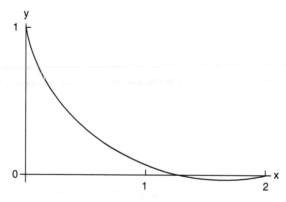

FIGURE 2.4 The solution to the brachistochrone problem is a cycloid.

The condition for extremizing the functional is

$$\delta I = 0$$

$$\int_{x_1}^{x_2} \delta F \, dx = 0 \tag{2.34}$$

where

$$\delta F = F(x, y_1 + \delta y_1, y_2 + \delta y_2, \ldots, y_n + \delta y_n, y'_1 + \delta y'_1, y'_2 + \delta y'_2, \ldots y'_n$$

$$+ \delta y'_n) - F(x, y_1, y_2, \ldots, y_n, y'_1, y'_2, \ldots y'_n) \tag{2.35}$$

where $\delta y_k = \varepsilon \eta_k(x)$ is the variation of the kth function, where $\eta_k(x)$ is independent of $\eta_\ell(x)$ for all $\ell = 1, 2, \ldots, n$ but $\ell \neq k$. A Taylor series expansion is used to expand δF as

$$\delta F = \sum_{i=1}^{n} \frac{\partial F}{\partial y_i} \delta y_i + \sum_{i=1}^{n} \frac{\partial F}{\partial y'_i} \delta y'_i \tag{2.36}$$

Substituting Equation 2.36 into Equation 2.34 and applying integration by parts to the terms contained in the second summation leads to

$$\int_{x_1}^{x_2} \sum_{i=1}^{n} \left\{ \left[\frac{\partial F}{\partial y_i} - \frac{d}{dx} \left(\frac{\partial F}{\partial y'_i} \right) \right] \delta y_i \right\} dx = 0$$

$$\sum_{i=1}^{n} \int_{x_1}^{x_2} \left\{ \left[\frac{\partial F}{\partial y_i} - \frac{d}{dx} \left(\frac{\partial F}{\partial y'_i} \right) \right] \delta y_i \right\} dx = 0 \tag{2.37}$$

Since the δy_i are independent of one another, Equation 2.37 can be satisfied only if each integral is identically zero (a subsequent result from the DuBois–Reymond Lemma). This leads to a set of Euler–Lagrange equations

$$\frac{\partial F}{\partial y_i} - \frac{d}{dx} \left(\frac{\partial F}{\partial y'_i} \right) = 0 \quad i = 1, 2, \ldots, n \tag{2.38}$$

2.4 HAMILTON'S PRINCIPLE

Let $v_1, v_2, ..., v_n$ be a set of generalized coordinates chosen for a n-degree-of-freedom system. The motion (displacement, velocity, and acceleration) of any particle in the system can be described in terms of its chosen generalized coordinates. Consider a particle of mass m and let $\mathbf{r}$ be the position vector, measured from the origin of the coordinate system, for this particle:

$$\mathbf{r} = x(t)\mathbf{i} + y(t)\mathbf{j} + z(t)\mathbf{k} \tag{2.39}$$

Since the motion of the particle can be described using the generalized coordinates

$$x = x(v_1, v_2, ..., v_n) \tag{2.40a}$$

$$y = y(v_1, v_2, ..., v_n) \tag{2.40b}$$

$$z = z(v_1, v_2, ..., v_n) \tag{2.40c}$$

$$\mathbf{r} = \mathbf{r}(v_1, v_2, ..., v_n) \tag{2.41}$$

Newton's Second Law, written for this particle, is of the form of Equation 1.14 repeated below,

$$\mathbf{F} + \sum \mathbf{f} = m\ddot{\mathbf{r}} \tag{2.42}$$

where $\mathbf{F}$ is the resultant of all forces external to the system acting on the particle and $\sum \mathbf{f}$ refers to the forces acting on this particle from other particles in the system.

The principle of work and energy is derived by taking the dot product of both sides of Equation 2.42, with the differential displacement vector $d\mathbf{r}$, and integrating between two positions along the path of motion

$$\int_{\mathbf{r}_1}^{\mathbf{r}_2} \mathbf{F} \cdot d\mathbf{r} + \sum \int_{\mathbf{r}_1}^{\mathbf{r}_2} \mathbf{f} \cdot d\Delta\mathbf{r} = \int_{\mathbf{r}_1}^{\mathbf{r}_2} m\ddot{\mathbf{r}} \cdot d\mathbf{r} \tag{? 43}$$

where $\Delta\mathbf{r}$ represents the vector between two particles. The work done by the external forces is defined as

$$W_{1 \to 2} = \int_{\mathbf{r}_1}^{\mathbf{r}_2} \mathbf{F} \cdot d\mathbf{r} \tag{2.44}$$

The term on the right-hand side of Equation 2.43 is rewritten as

$$\int_{\mathbf{r}_1}^{\mathbf{r}_2} m\ddot{\mathbf{r}}\cdot d\mathbf{r} = \int_{t_1}^{t_2} m\ddot{\mathbf{r}}\cdot\frac{d\mathbf{r}}{dt}dt = \int_{t_1}^{t_2} m\frac{1}{2}\frac{d}{dt}(\dot{r}\cdot\dot{r})dt = \frac{1}{2}m\dot{\mathbf{r}}\cdot\dot{r}|_{t_1}^{t_2} \tag{2.45}$$

The kinetic energy of the particle is

$$T = \frac{1}{2}m\dot{\mathbf{r}}\cdot\dot{r} \tag{2.46}$$

Equation 2.43 now becomes

$$W_{1\to 2} + \sum\int_{\mathbf{r}_1}^{\mathbf{r}_2} \mathbf{f}\cdot d\Delta\mathbf{r} = T_2 - T_1 \tag{2.47}$$

A similar equation may be written for every particle in the system. When all equations are added, since the force acting particle A from particle B is equal and opposite to the force acting on particle B from particle A, the total work done by the internal forces is zero. The total kinetic energy is the sum of the kinetic energies of the individual particles, and the total work is the sum of the work done by all external forces. The general principle of work and energy becomes

$$W_{1\to 2} = T_2 - T_1 \tag{2.48}$$

It is clear from the above that both the work and kinetic energy at any instant are functionals of the generalized coordinates. Recalling that the equations of motion for a rigid body $(\sum \mathbf{F} = m\mathbf{a}$ and $\sum \mathbf{M_O} - \dot{\mathbf{H}}_O)$ are derived by taking the limit of the sum of the individual equations, obtained for individual particles in a system of particles, it is clear that Equation 2.48 is also applicable for systems which include rigid bodies.

Each of the generalized coordinates are functions of time, and as such, variations in each of the generalized coordinates can be defined between times t_1 and t_2. The variations are $\delta v_1, \delta v_2, \ldots \delta v_n$, and satisfy

$$\delta v_k(t_1) = \delta v_k(t_2) = 0 \quad k = 1,2,\ldots n \tag{2.49}$$

The variation in the position vector is

$$\delta\mathbf{r} = \mathbf{r}(v_1 + \delta v_1, v_2 + \delta v_2, \ldots, v_n + \delta v_n) - \mathbf{r}(v_1, v_2, \ldots, v_n)$$
$$= \frac{\partial\mathbf{r}}{\partial v_1}\delta v_1 + \frac{\partial\mathbf{r}}{\partial v_2}\delta v_2 + \ldots + \frac{\partial\mathbf{r}}{\partial v_n}\delta v_n \tag{2.50}$$

Taking the dot product of Equation 2.42 with the variation in position vector leads to

$$\mathbf{F}\cdot\delta\mathbf{r} + (\sum \mathbf{f}\cdot\delta\Delta r) = m\ddot{\mathbf{r}}\cdot\delta\mathbf{r} \tag{2.51}$$

The variation in the work, using the definition of Equation 2.44, is

$$\delta W = \mathbf{F}\cdot\delta\mathbf{r} \tag{2.52}$$

The work defined in Equation 2.52 is called the virtual work. Consider

$$\frac{d}{dt}(\dot{\mathbf{r}}\cdot\delta\mathbf{r}) = \ddot{\mathbf{r}}\cdot\delta\mathbf{r} + \dot{\mathbf{r}}\cdot\frac{d}{dt}(\delta\mathbf{r}) \tag{2.53}$$

It has been shown, Equation 2.16, that the order of variation and differentiation may be interchanged. Equation 2.53 is rearranged as

$$\ddot{\mathbf{r}}\cdot\delta\mathbf{r} = \frac{d}{dt}(\dot{\mathbf{r}}\cdot\delta\mathbf{r}) - \dot{\mathbf{r}}\cdot\delta\dot{\mathbf{r}} \tag{2.54}$$

It is further noted that

$$\delta(\dot{\mathbf{r}}\cdot\dot{\mathbf{r}}) = \delta\dot{\mathbf{r}}\cdot\dot{\mathbf{r}} + \dot{\mathbf{r}}\cdot\delta\dot{\mathbf{r}} = 2\dot{\mathbf{r}}\cdot\delta\dot{\mathbf{r}} \tag{2.55}$$

The use of Equation 2.55 in Equation 2.54, and subsequently Equation 2.51, leads to

$$\delta W + \left(\sum \mathbf{f}\right)\cdot\delta\Delta r = m\frac{d}{dt}(\dot{\mathbf{r}}\cdot\delta\mathbf{r}) - \delta\left(\frac{1}{2}m\dot{\mathbf{r}}\cdot\dot{\mathbf{r}}\right) \tag{2.56}$$

Noting that the variation in kinetic energy for this particle is

$$\delta T = \delta\left(\frac{1}{2}m\dot{\mathbf{r}}\cdot\dot{\mathbf{r}}\right) \tag{2.57}$$

Equation 2.56 becomes

$$\delta T + \delta W + \left(\sum \mathbf{f} \cdot \delta \Delta \mathbf{r} \right) = m \frac{\mathrm{d}}{\mathrm{d}t} (\dot{\mathbf{r}} \cdot \delta \mathbf{r}) \tag{2.58}$$

Equation 2.58 is written for a single particle. A similar equation can be written for each particle in the system. The result of adding all equations together is

$$\delta T + \delta W = \sum m_i \frac{\mathrm{d}}{\mathrm{d}t} (\dot{\mathbf{r}}_i \cdot \delta \mathbf{r}_i) \tag{2.59}$$

The total virtual work of the forces between particles is zero.

External forces acting on the system may be conservative (spring forces, gravity) or nonconservative (viscous damping, externally applied loads). The work done by conservative forces can be determined from a potential energy functional, $V(v_1, v_2, ... v_n)$, such that $\delta W = -\delta V$. The total virtual work is

$$\delta W = \delta W_{\mathrm{nc}} - \delta V \tag{2.60}$$

where δW_{nc} is the virtual work of all nonconservative forces. Equation 2.59 thus becomes

$$\delta T - \delta V + \delta W_{\mathrm{nc}} = m \frac{\mathrm{d}}{\mathrm{d}t} (\dot{\mathbf{r}} \cdot \delta \mathbf{r}) \tag{2.61}$$

Integrating Equation 2.61, with respect to t between t_1 and t_2, leads to

$$\int_{t_1}^{t_2} (\delta T - \delta V + \delta W_{\mathrm{nc}}) \mathrm{d}t = \int_{t_1}^{t_2} m \frac{\mathrm{d}}{\mathrm{d}t} (\dot{\mathbf{r}} \cdot \delta \mathbf{r}) \mathrm{d}t = m \dot{\mathbf{r}} \cdot \delta \mathbf{r} |_{t_1}^{t_2} \tag{2.62}$$

The variations in generalized coordinates are defined such that they are identically zero at t_1 and t_2. Thus, the right-hand side of Equation 2.62 is zero, leading to

$$\int_{t_1}^{t_2} (\delta T - \delta V + \delta W_{\mathrm{nc}}) \, \mathrm{d}t = 0 \tag{2.63}$$

The lagrangian functional is defined as

$$L = T - V \tag{2.64}$$

With this definition Equation 2.63 becomes

$$\int_{t_1}^{t_2} (\delta L + \delta W_{nc})\, dt = 0 \tag{2.65}$$

Equation 2.65 is a statement of Extended Hamilton's Principle. If all forces are conservative $\delta W_{nc} = 0$ and Equation 2.65 reduces to

$$\int_{t_1}^{t_2} \delta L\, dt = 0 \tag{2.66}$$

Except in unusual cases, not applicable in this study, the order of variation and integration may be interchanged leading to

$$\delta \int_{t_1}^{t_2} L\, dt = 0 \tag{2.67}$$

which is referred to as Hamilton's Principle.

2.5 LAGRANGE'S EQUATIONS FOR CONSERVATIVE DISCRETE SYSTEMS

It is shown in Section 2.3 that setting the variation of a functional to zero is equivalent to minimizing the functional over all possible functions in the domain of the functional. Thus, Hamilton's Principle implies that for a conservative system the time dependent response of the generalized coordinates must render the functional $\int_{t_1}^{t_2} L\, dt$ stationary.

The Lagrangian is a function of n variables dependent upon time, the generalized coordinates. The Euler–Lagrange equation, Equation 2.48 provides sufficient conditions to render the functional stationary. To this end

$$\frac{\partial L}{\partial v_k} - \frac{d}{dt}\left(\frac{\partial L}{\partial \dot{v}_k}\right) = 0 \quad k = 1,2,\ldots,n \tag{2.68}$$

Use of Equation 2.68 for each value of k leads to n independent differential equations whose solution provides the time-dependent response of the system.

Lagrange's equations are used to derive the governing differential equations for linear and nonlinear systems.

Example 2.3. Use Lagrange's equation to derive the differential equation governing the motion of the system of Figure 2.5. Use θ as the chosen generalized coordinate. Linearize the resulting equation.

Solution: The kinetic energy of the system is derived in Example 1.2 as

$$T = \frac{1}{2}m\left(\dot{x}^2 + \dot{y}^2 + L\dot{x}\dot{\theta}\cos\theta + L\dot{y}\dot{\theta}\sin\theta + \frac{1}{3}L^2\dot{\theta}^2\right) \tag{a}$$

The potential energy of the system is calculated using the equilibrium position of the mass center as the datum as

$$V = mg\left[y + \frac{L}{2}(1 - \cos\theta)\right] \tag{b}$$

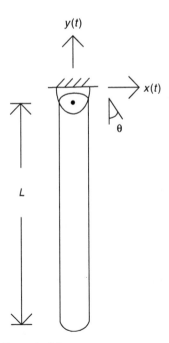

FIGURE 2.5 System of Example 2.3.

The lagrangian for this system is

$$L = T - V$$

$$= \frac{1}{2}m\left(\dot{x}^2 + \dot{y}^2 + L\dot{x}\dot{\theta}\cos\theta + L\dot{y}\dot{\theta}\sin\theta + \frac{1}{3}L^2\dot{\theta}^2\right) \tag{c}$$

$$- mg\left[y + \frac{L}{2}(1 - \cos\theta)\right]$$

Lagrange's equation for this conservative single-degree-of-freedom system is

$$\frac{d}{dt}\left(\frac{\partial L}{\partial\dot{\theta}}\right) - \frac{\partial L}{\partial\theta} = 0 \tag{d}$$

Use of Equation c in Equation d leads to

$$\frac{d}{dt}\left[\frac{1}{2}m\left(L\dot{x}\cos\theta + L\dot{y}\sin\theta + \frac{2}{3}L^2\dot{\theta}\right)\right]$$

$$- \left[\frac{1}{2}m(-L\dot{x}\dot{\theta}\sin\theta + L\dot{y}\dot{\theta}\cos\theta) - \frac{mgL}{2}\sin\theta\right] = 0 \tag{e}$$

Expanding the time derivative and cleaning up the algebra leads to

$$\frac{1}{2}m\left(L\ddot{x}\cos\theta - L\dot{x}\dot{\theta}\sin\theta + L\ddot{y}\sin\theta + L\dot{y}\dot{\theta}\cos\theta + \frac{2}{3}L^2\ddot{\theta}\right)$$

$$+ \frac{1}{2}mL\dot{x}\dot{\theta}\sin\theta - \frac{1}{2}mL\dot{y}\dot{\theta}\cos\theta + \frac{mgL}{2}\sin\theta = 0 \tag{f}$$

$$\frac{1}{3}mL^2\ddot{\theta} + \frac{1}{2}mL\ddot{x}\cos\theta + \frac{1}{2}mL\ddot{y}\sin\theta + \frac{1}{2}mgL\sin\theta = 0 \tag{g}$$

When θ is assumed small $\sin\theta \approx \theta$ and $\cos\theta \approx 1$, which when used in Equation g leads to

$$\frac{1}{3}mL^2\ddot{\theta} + \frac{1}{2}mL(\ddot{y} + g)\theta = -\frac{1}{2}mL\ddot{x} \tag{h}$$

When the small angle assumption is used, the horizontal motion acts as an external excitation, while the vertical motion of the base acts as a self-excitation.

Example 2.4. Use Lagrange's equations to derive the equations of motion for the system of Figure 2.6 using x_1, the displacement of the mass center of the cart and x_2, the absolute displacement of the mass center of the disk as the chosen generalized coordinates. Assume the disk rolls without slip.

Solution: The kinetic energy of the cart is

$$T_C = \frac{1}{2}m\dot{x}_1^2 \tag{a}$$

Since the disk rolls without slip, the velocity of the point in contact between the disk and the cart is $\dot{x}_1$. Thus, the angular velocity of the disk is

$$\omega_D = \frac{1}{r}(\dot{x}_2 - \dot{x}_1) \tag{b}$$

Thus, the kinetic energy of the disk is

$$T_D = \frac{1}{2}m_D\dot{x}_2^2 + \frac{1}{2}\left(\frac{1}{2}m_D r^2\right)\left(\frac{\dot{x}_2 - \dot{x}_1}{r}\right)^2 \tag{c}$$

The total kinetic energy of the system is

$$T = T_C + T_D = \frac{1}{2}m\dot{x}_1^2 + \frac{1}{2}m_D\dot{x}_2^2 + \frac{1}{2}\left(\frac{1}{2}m_D\right)(\dot{x}_2 - \dot{x}_1)^2 \tag{d}$$

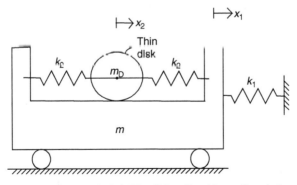

FIGURE 2.6 System of Example 2.4. The disk rolls without slip relative to the cart.

The potential energy of the system is

$$V = \frac{1}{2}k_1 x_1^2 + \frac{1}{2}2k_2(x_1 - x_2)^2 \tag{e}$$

The system's Lagrangian is

$$L = T - V$$

$$= \frac{1}{2}m\dot{x}_1^2 + \frac{1}{2}m_D\dot{x}_2^2 + \frac{1}{2}\left(\frac{1}{2}m_D\right)(\dot{x}_2 - \dot{x}_1)^2$$

$$- \left[\frac{1}{2}k_1 x_1^2 + \frac{1}{2}2k_2(x_1 - x_2)^2\right] \tag{f}$$

Application of Lagrange's equation for x_1 leads to

$$\frac{d}{dt}\left(\frac{\partial L}{\partial \dot{x}_1}\right) - \frac{\partial L}{\partial x_1} = 0$$

$$\frac{d}{dt}\left[m\dot{x}_1 - \frac{1}{2}m_D(\dot{x}_2 - \dot{x}_1)\right] + [k_1 x_1 + 2k_2(x_1 - x_2)] = 0 \tag{g}$$

$$\left(m + \frac{1}{2}m_D\right)\ddot{x}_1 - \frac{1}{2}m_D\ddot{x}_2 + (k_1 + 2k_2)x_1 - 2k_2 x_2 = 0$$

Application of Lagrange's equation for x_2 leads to

$$\frac{d}{dt}\left(\frac{\partial L}{\partial \dot{x}_2}\right) - \frac{\partial L}{\partial x_2} = 0$$

$$\frac{d}{dt}\left[m_D\dot{x}_2 + \frac{1}{2}m_D(\dot{x}_2 - \dot{x}_1)\right] + [-2k_2(x_1 - x_2)] = 0 \tag{h}$$

$$-\frac{1}{2}m_D\ddot{x}_1 + \frac{3}{2}m_D\ddot{x}_2 - 2k_2 x_1 + 2k_2 x_2 = 0$$

Equation g and Equation h are summarized in matrix form as

$$\begin{bmatrix} m + \frac{1}{2}m_D & -\frac{1}{2}m_D \\ -\frac{1}{2}m_D & \frac{3}{2}m_D \end{bmatrix}\begin{bmatrix} \ddot{x}_1 \\ \ddot{x}_2 \end{bmatrix} + \begin{bmatrix} k_1 + 2k_2 & -2k_2 \\ -2k_2 & 2k_2 \end{bmatrix}\begin{bmatrix} x_1 \\ x_2 \end{bmatrix} = \begin{bmatrix} 0 \\ 0 \end{bmatrix} \tag{i}$$

Example 2.5. Use Lagrange's equations to derive the differential equations governing the motion of the three-degree-of-freedom system of Figure 2.7. Use x_1, x_2 and θ as the chosen generalized coordinates.

Solution: The kinetic energy of the system at an arbitrary instant is

$$T = \frac{1}{2}m_1\dot{x}_1^2 + \frac{1}{2}m_2\dot{x}_2^2 + \frac{1}{2}I\dot{\theta}^2 \tag{a}$$

The potential energy of the system at an arbitrary instant is

$$V = \frac{1}{2}kx_1^2 + \frac{1}{2}k(x_1 - 2r\theta)^2 + \frac{1}{2}2k(r\theta - x_2)^2 \tag{b}$$

The Lagrangian for the system is

$$\begin{aligned}
L &= T - V \\
&= \frac{1}{2}m_1\dot{x}_1^2 + \frac{1}{2}m_2\dot{x}_2^2 + \frac{1}{2}I\dot{\theta}^2 \\
&\quad - \left[\frac{1}{2}kx_1^2 + \frac{1}{2}k(x_1 - 2r\theta)^2 + \frac{1}{2}2k(r\theta - x_2)^2\right]
\end{aligned} \tag{c}$$

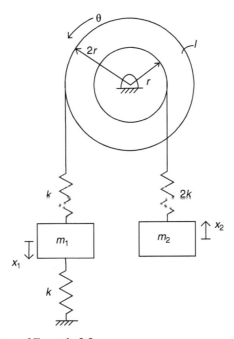

FIGURE 2.7 System of Example 2.5.

Application of Lagrange's equations for x_1 leads to

$$\frac{d}{dt}\left(\frac{\partial L}{\partial \dot{x}_1}\right) - \frac{\partial L}{\partial x_1} = 0$$

$$\frac{d}{dt}(m_1\dot{x}_1) + [kx_1 + k(x_1 - 2r\theta)] = 0 \qquad (d)$$

$$m\ddot{x}_1 + 2kx_1 - 2kr\theta = 0$$

Application of Lagrange's equation for x_2 leads to

$$\frac{d}{dt}\left(\frac{\partial L}{\partial \dot{x}_2}\right) - \frac{\partial L}{\partial x_2} = 0$$

$$\frac{d}{dt}(m_2\dot{x}_2) + [2k(-1)(r\theta - x_2)] \qquad (e)$$

$$m_2\ddot{x}_2 + 2kx_2 - 2kr\theta = 0$$

Application of Lagrange's equation for θ leads to

$$\frac{d}{dt}\left(\frac{\partial L}{\partial \dot{\theta}}\right) - \frac{\partial L}{\partial \theta} = 0$$

$$\frac{d}{dt}(I\dot{\theta}) + [k(-2r)(x_1 - 2r\theta) + r(2k)(r\theta - x_2)] = 0 \qquad (f)$$

$$I\dot{\theta} - 2krx_1 - 2krx_2 + 6kr^2\theta = 0$$

Equation d through Equation f are summarized in matrix form as

$$\begin{bmatrix} m_1 & 0 & 0 \\ 0 & m_2 & 0 \\ 0 & 0 & I \end{bmatrix} \begin{bmatrix} \ddot{x}_1 \\ \ddot{x}_2 \\ \ddot{\theta} \end{bmatrix} + \begin{bmatrix} 2k & 0 & -2kr \\ 0 & 2k & -2kr \\ -2kr & -2kr & 6kr^2 \end{bmatrix} \begin{bmatrix} x_1 \\ x_2 \\ \theta \end{bmatrix} = \begin{bmatrix} 0 \\ 0 \\ 0 \end{bmatrix} \qquad (g)$$

Example 2.6. Use Lagrange's equations to derive the differential equations governing the motion of the system of Figure 2.8. Use θ_1 and θ_2 as the chosen generalized coordinates. The spring is unstretched in the position shown.

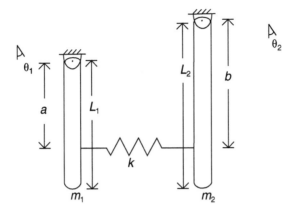

FIGURE 2.8 System of Example 2.6.

Solution: The kinetic energy of the system at an arbitrary instant is

$$T = \frac{1}{2}m_1\left(\frac{L_1}{2}\dot{\theta}_1\right)^2 + \frac{1}{2}\left(\frac{1}{12}m_1L_1^2\right)\dot{\theta}_1^2 + \frac{1}{2}m_2\left(\frac{L_2}{2}\dot{\theta}_2\right)^2$$

$$+ \frac{1}{2}\left(\frac{1}{12}m_2L_2^2\right)\dot{\theta}_2^2$$

$$= \frac{1}{2}\left(\frac{1}{3}m_1L_1^2\dot{\theta}_1^2 + \frac{1}{3}m_2L_2^2\dot{\theta}_2^2\right) \tag{a}$$

The unstretched length of the spring is ℓ. Geometry shows that the length of the spring at an arbitrary instant is

$$\tilde{\ell} = [(\ell - a\sin\theta_1)^2 + b^2(1-\cos\theta_2)^2]^{1/2} \tag{b}$$

The change in length of the spring is $\ell - \tilde{\ell}$. Taking the datum for potential energy calculations as the horizontal plane through the mass center of each bar, the potential energy of the system at an arbitrary instant is

$$V = \frac{1}{2}k\{[(\ell - a\sin\theta_1)^2 + b^2(1-\cos\theta_2)^2]^{1/2} - \ell\}^2$$

$$+ m_1g\frac{L_1}{2}(1-\cos\theta_1) + m_2g\frac{L_2}{2}(1-\cos\theta_2) \tag{c}$$

Noting that $L = T - V$, application of Lagrange's equation for θ_1 leads to

$$\frac{d}{dt}\left(\frac{\partial L}{\partial \dot{\theta}_1}\right) - \frac{\partial L}{\partial \theta_1} = 0$$

$$\frac{d}{dt}\left(\frac{1}{3}m_1 L_1^2 \dot{\theta}_1\right) + k\{[(\ell - a\sin\theta_1)^2 + b^2(1-\cos\theta_2)^2]^{1/2} - \ell\} \qquad \text{(d)}$$

$$\times \frac{(\ell - a\sin\theta_1)(-a\cos\theta_1)}{[(\ell - a\sin\theta_1)^2 + b^2(1-\cos\theta_2)^2]^{1/2}} + m_1 g \frac{L_1}{2}\sin\theta_1 = 0$$

Application of Lagrange's equation for θ_2 leads to

$$\frac{d}{dt}\left(\frac{1}{3}m_2 L_2^2 \dot{\theta}_2\right) + k\{[(\ell - a\sin\theta_1)^2 + b^2(1-\cos\theta_2)^2]^{1/2} - \ell\}$$

$$\qquad\qquad\qquad\qquad\qquad\qquad\qquad\qquad\qquad\qquad\qquad \text{(e)}$$

$$\times \frac{b^2(1-\cos\theta_2)\sin\theta_2}{[(\ell - a\sin\theta_1)^2 + b^2(1-\cos\theta_2)^2]^{1/2}} + m_2 g \frac{L_2}{2}\sin\theta_2 = 0$$

Equation c and Equation d can be simplified.

2.6 LAGRANGE'S EQUATIONS FOR NON-CONSERVATIVE DISCRETE SYSTEMS

When nonconservative forces exist, it is necessary to apply Extended Hamilton's Principle

$$\int_{t_1}^{t_2} (\delta L + \delta W_{nc})dt = 0 \qquad\qquad (2.69)$$

The virtual work of the nonconservative forces, when calculated, cannot depend upon the variations of the time derivatives of the generalized coordinates. It can only depend upon the variations of the generalized coordinates, and is thus of the form

$$\delta W_{nc} = \sum_{i=1}^{n} \frac{\partial W_{nc}}{\partial v_i} \delta v_i \qquad\qquad (2.70)$$

Equation 2.70 can be written as

$$\delta W_{nc} = \sum_{i=1}^{n} Q_i \delta v_i \qquad (2.71)$$

where Q_i, $i = 1, 2, ..., n$ are called the generalized forces. The generalized forces are determined through determination of the virtual work of the nonconservative forces and writing it in the form of Equation 2.71.

Substituting Equation 2.71 into Equation 2.69 leads to

$$\int_{t_1}^{t_2} \left(\delta L + \sum_{i=1}^{n} Q_i \delta v_i \right) dt = 0 \qquad (2.72)$$

An analysis similar to that leading to Equation 2.38 is used to rewrite Equation 2.72 as

$$\sum_{i=1}^{n} \int_{t_1}^{t_2} \left[\frac{\partial L}{\partial v_i} - \frac{d}{dt} \left(\frac{\partial L}{\partial \dot{v}_i} \right) + Q_i \right] \delta v_i dt = 0 \qquad (2.73)$$

An extension of the DuBois–Reymond Lemma is used to argue that the term in the square brackets in each integral must be identically zero in order to satisfy Equation 2.73 for all possible sets of variations. Thus,

$$\frac{\partial L}{\partial v_i} - \frac{d}{dt} \left(\frac{\partial L}{\partial \dot{v}_i} \right) + Q_i = 0 \quad 1, 2, ..., n \qquad (2.74)$$

Equation 2.74 can be rearranged as

$$\frac{d}{dt} \left(\frac{\partial L}{\partial \dot{v}_i} \right) - \frac{\partial L}{\partial v_i} = Q_i \quad i = 1, 2, ..., n \qquad (2.75)$$

Example 2.7. Use Lagrange's equations to derive the equations governing the motion of the bar of Figure 2.9. Use x_1, x_2, and x_3 as the chosen generalized coordinates. Assume small displacements as an *a priori* linearizing assumption.

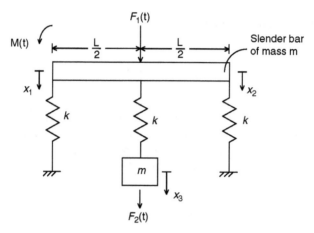

FIGURE 2.9 Three-degree-of-freedom system of Example 2.7.

Solution: Figure 2.10 illustrates the geometry of the system when the small displacement assumption is used. The displacement of the mass center of the bar is

$$\bar{x} = \frac{1}{2}(x_1 + x_2) \qquad \text{(a)}$$

and the clockwise angular displacement of the bar is

$$\theta = \frac{1}{L}(x_2 - x_1) \qquad \text{(b)}$$

The kinetic energy of the system at an arbitrary instant is

$$T = \frac{1}{2}m\left(\frac{\dot{x}_1 + \dot{x}_2}{2}\right)^2 + \frac{1}{2}\left(\frac{1}{12}mL^2\right)\left(\frac{\dot{x}_2 - \dot{x}_1}{L}\right)^2 + \frac{1}{2}m\dot{x}_3^2 \qquad \text{(c)}$$

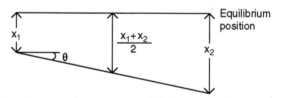

FIGURE 2.10 Geometry of system of Example 2.9, assuming small displacements.

The potential energy at an arbitrary instant is

$$V = \frac{1}{2}kx_1^2 + \frac{1}{2}kx_2^2 + \frac{1}{2}k\left[x_3 - \left(\frac{x_1 + x_2}{2}\right)\right]^2 \tag{d}$$

The virtual work of a force is the product of the force and the variation in the displacement of the particle where the force acts. The virtual work of a moment is the product of the moment, with the variation of the angular displacement of the body to which the moment is applied.

$$\delta W_{nc} = F_1(t)\delta\left(\frac{x_1 + x_2}{2}\right) - M(t)\delta\left(\frac{x_2 - x_1}{L}\right) + F_2(t)\delta x_3$$

$$= \left[\frac{1}{2}F_1(t) + \frac{1}{L}M(t)\right]\delta x_1 + \left[\frac{1}{2}F_1(t) - \frac{1}{L}M(t)\right]\delta x_2 + F_2(t)\delta x_3 \tag{e}$$

The generalized forces are obtained from Equation 2.71 as

$$Q_1 = \frac{1}{2}F_1(t) + \frac{1}{L}M(t) \tag{f}$$

$$Q_2 = \frac{1}{2}F_1(t) - \frac{1}{L}M(t) \tag{g}$$

$$Q_3 = F_2(t) \tag{h}$$

Application of Lagrange's equations to this system for x_1 leads to

$$\frac{d}{dt}\left(\frac{\partial L}{\partial \dot{x}_1}\right) - \frac{\partial L}{\partial x_1} = Q_1$$

$$\frac{d}{dt}\left[m\left(\frac{\dot{x}_1 + \dot{x}_2}{2}\right)\left(\frac{1}{2}\right) + \left(\frac{1}{12}mL^2\right)\left(\frac{\dot{x}_2 - \dot{x}_1}{L}\right)\left(-\frac{1}{L}\right)\right]$$

$$+ kx_1 + k\left[x_3 - \left(\frac{x_1 + x_2}{2}\right)\right]\left(-\frac{1}{2}\right) = \frac{1}{2}F_1(t) + \frac{1}{L}M(t) \tag{i}$$

$$\frac{1}{3}m\ddot{x}_1 + \frac{1}{6}m\ddot{x}_2 + \frac{5}{4}kx_1 + \frac{1}{4}kx_2 - \frac{1}{2}kx_3 = \frac{1}{2}F_1(t) + \frac{1}{L}M(t)$$

Application of Lagrange's equations to this system for x_2 leads to

$$\frac{\mathrm{d}}{\mathrm{d}t}\left(\frac{\partial L}{\partial \dot{x}_2}\right) - \frac{\partial L}{\partial x_2} = Q_2$$

$$\frac{\mathrm{d}}{\mathrm{d}t}\left[m\left(\frac{\dot{x}_1 + \dot{x}_2}{2}\right)\left(\frac{1}{2}\right) + \left(\frac{1}{12}mL^2\right)\left(\frac{\dot{x}_2 - \dot{x}_1}{L}\right)\left(\frac{1}{L}\right)\right]$$

$$+ kx_2 + k\left[x_3 - \left(\frac{x_1 + x_2}{2}\right)\right]\left(-\frac{1}{2}\right) = \frac{1}{2}F_1(t) - \frac{1}{L}M(t) \tag{j}$$

$$\frac{1}{6}m\ddot{x}_1 + \frac{1}{3}m\ddot{x}_2 + \frac{1}{4}kx_1 + \frac{5}{4}kx_2 - \frac{1}{2}kx_3 = \frac{1}{2}F_1(t) - \frac{1}{L}M(t)$$

Application of Lagrange's equations to this system for x_3 leads to

$$\frac{\mathrm{d}}{\mathrm{d}t}\left(\frac{\partial L}{\partial \dot{x}_3}\right) - \frac{\partial L}{\partial x_3} = Q_3$$

$$\frac{\mathrm{d}}{\mathrm{d}t}(m\dot{x}_3) + k\left[x_3 - \left(\frac{x_1 + x_2}{2}\right)\right] = F_2(t) \tag{k}$$

$$m\ddot{x}_3 - \frac{1}{2}kx_1 - \frac{1}{2}kx_2 + kx_3 = F_2(t)$$

Equation i through Equation k are written in matrix form as

$$\begin{bmatrix} \frac{1}{3}m & \frac{1}{6}m & 0 \\ \frac{1}{6}m & \frac{1}{3}m & 0 \\ 0 & 0 & m \end{bmatrix}\begin{bmatrix} \ddot{x}_1 \\ \ddot{x}_2 \\ \ddot{x}_3 \end{bmatrix} + \begin{bmatrix} \frac{5}{4}k & \frac{1}{4}k & -\frac{1}{2}k \\ \frac{1}{4}k & \frac{5}{4}k & -\frac{1}{2}k \\ -\frac{1}{2}k & -\frac{1}{2}k & k \end{bmatrix}\begin{bmatrix} x_1 \\ x_2 \\ x_3 \end{bmatrix}$$

$$= \begin{bmatrix} \frac{1}{2}F_1(t) + \frac{1}{L}M(t) \\ \frac{1}{2}F_1(t) - \frac{1}{L}M(t) \\ F_2(t) \end{bmatrix} \tag{l}$$

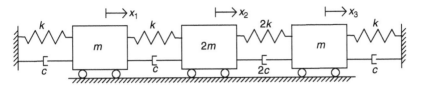

FIGURE 2.11 System of Example 2.10.

Example 2.10. Use Lagrange's equations to derive the differential equations governing the motion of the system of Figure 2.11

Solution: The kinetic energy of the system at an arbitrary instant is

$$T = \frac{1}{2}m\dot{x}_1^2 + \frac{1}{2}2m\dot{x}_2^2 + \frac{1}{2}m\dot{x}_3^2 \qquad (a)$$

The potential energy of the system at an arbitrary instant is

$$V = \frac{1}{2}kx_1^2 + \frac{1}{2}k(x_2 - x_1)^2 + \frac{1}{2}2k(x_3 - x_2)^2 + \frac{1}{2}kx_3^2 \qquad (b)$$

Assume virtual displacements δx_1, δx_2, and δx_3. The force applied to the block of mass $2m$ from the viscous damper placed between the two leftmost blocks at an arbitrary instant is $c(\dot{x}_2 - \dot{x}_1)$ acting away from the block. The virtual displacement of the block is δx_2 in the opposite direction. Thus, the virtual work due to this force is

$$\delta W_{1,2} = -c(\dot{x}_2 - \dot{x}_1)\delta x_2 \qquad (c)$$

The same force acts on the leftmost block, whose virtual displacement is δx_1 in the same direction as the force. The virtual work done by this force is

$$\delta W_{2,1} = c(\dot{x}_2 - \dot{x}_1)\delta x_1 \qquad (d)$$

The total virtual work is the work done by the viscous damping forces due to the variations in the generalized coordinates and is

$$\delta W_{nc} = -c\dot{x}_1\delta x_1 + c(\dot{x}_2 - \dot{x}_1)\delta x_1 - c(\dot{x}_2 - \dot{x}_1)\delta x_2 + 2c(\dot{x}_3 - \dot{x}_2)\delta x_2$$

$$- 2c(\dot{x}_3 - \dot{x}_2)\delta x_3 - c\dot{x}_3\delta x_3$$

$$= (-2c\dot{x}_1 + c\dot{x}_2)\delta x_1 + (c\dot{x}_1 - 3c\dot{x}_2 + 2c\dot{x}_3)\delta x_2 + (2c\dot{x}_2 - 3c\dot{x}_3)\delta x_3 \qquad (e)$$

The generalized forces are obtained from the virtual work as

$$Q_1 = -2c\dot{x}_1 + c\dot{x}_2 \tag{f}$$

$$Q_2 = c\dot{x}_1 - 3c\dot{x}_2 + 2c\dot{x}_3 \tag{g}$$

$$Q_3 = 2c\dot{x}_2 - 3c\dot{x}_3 \tag{h}$$

Application of Lagrange's equation for x_1 leads to

$$\frac{d}{dt}\left(\frac{\partial L}{\partial \dot{x}_1}\right) - \frac{\partial L}{\partial x_1} = Q_1$$

$$\frac{d}{dt}(m\dot{x}_1) + kx_1 + k(x_2 - x_1)(-1) = -2c\dot{x}_1 + c\dot{x}_2 \tag{i}$$

$$m\ddot{x}_1 + 2c\dot{x}_1 - c\dot{x}_2 + 2kx_1 - kx_2 = 0$$

Application of Lagrange's equation for x_2 leads to

$$\frac{d}{dt}\left(\frac{\partial L}{\partial \dot{x}_2}\right) - \frac{\partial L}{\partial x_2} = Q_2$$

$$\frac{d}{dt}(2m\dot{x}_2) + k(x_2 - x_1) + 2k(x_3 - x_2)(-1) = c\dot{x}_1 - 3c\dot{x}_2 + 2c\dot{x}_3 \tag{j}$$

$$2m\ddot{x}_2 - 2c\dot{x}_1 + 3c\dot{x}_2 - 2c\dot{x}_3 - kx_1 + 3kx_2 - 2kx_3 = 0$$

Application of Lagrange's equation for x_3 leads to

$$\frac{d}{dt}\left(\frac{\partial L}{\partial \dot{x}_3}\right) - \frac{\partial L}{\partial x_3} = Q_3$$

$$\frac{d}{dt}(m\dot{x}_3) + 2k(x_3 - x_2) + kx_3 = 2c\dot{x}_2 - 3c\dot{x}_3 \tag{k}$$

$$m\ddot{x}_3 - c\dot{x}_2 + 3c\dot{x}_3 - 2kx_2 + 3kx_3 = 0$$

Equation i through Equation k are summarized in matrix form as

$$
\begin{bmatrix} m & 0 & 0 \\ 0 & 2m & 0 \\ 0 & 0 & m \end{bmatrix} \begin{bmatrix} \ddot{x}_1 \\ \ddot{x}_2 \\ \ddot{x}_3 \end{bmatrix} + \begin{bmatrix} 2c & -c & 0 \\ -c & 3c & -2c \\ 0 & -2c & 3c \end{bmatrix} \begin{bmatrix} \dot{x}_1 \\ \dot{x}_2 \\ \dot{x}_3 \end{bmatrix}
$$

$$
+ \begin{bmatrix} 2k & -k & 0 \\ -k & 3k & -2k \\ 0 & -2k & 3k \end{bmatrix} \begin{bmatrix} x_1 \\ x_2 \\ x_3 \end{bmatrix} = \begin{bmatrix} 0 \\ 0 \\ 0 \end{bmatrix} \tag{1}
$$

The power dissipated by a viscous damper with one end fixed is calculated using Equation 1.90, while the power dissipated by a viscous damper, which is placed between two particles with nonzero velocities, is calculated as in Equation 1.93. The power dissipation created by the viscous dampers in Example 2.10 is

$$
P = c\dot{x}_1^2 + c(\dot{x}_2 - \dot{x}_1)^2 + 2c(\dot{x}_3 - \dot{x}_2)^2 + c\dot{x}_3^2 \tag{2.76}
$$

Note that

$$
\frac{\partial P}{\partial \dot{x}_1} = 2cx_1 + 2c(\dot{x}_2 - \dot{x}_1)(-1) = 4c\dot{x}_1 - 2c\dot{x}_2 = -2Q_1 \tag{2.77a}
$$

$$
\frac{\partial P}{\partial \dot{x}_2} = 2c(\dot{x}_2 - \dot{x}_1) + 4c(\dot{x}_3 - \dot{x}_2)(-1)
$$

$$
= -2c\dot{x}_1 + 6c\dot{x}_2 - 4c\dot{x}_3 = -2Q_2 \tag{2.77b}
$$

$$
\frac{\partial P}{\partial \dot{x}_3} = 4c(\dot{x}_3 - \dot{x}_2) + 2c\dot{x}_3 = 4c\dot{x}_2 + 6c\dot{x}_3 = -2Q_3 \tag{2.77c}
$$

Equation 2.76 and Equation 2.77 are generalized as

$$
Q_i = -\frac{1}{2} \frac{\partial P}{\partial \dot{x}_i} \tag{2.78}
$$

Rayleigh's dissipation function is defined as

$$
\mathfrak{I} = -\frac{1}{2} P \tag{2.79}
$$

where P is the power dissipated by the viscous dampers written in terms of the generalized coordinates. Using Rayleigh's dissipation function, the generalized force corresponding to viscous damping is

$$Q_i = \frac{\partial \mathfrak{I}}{\partial \dot{v}_i} \tag{2.80}$$

The virtual work of the nonconservative forces can be broken up into the virtual work done by viscous damping forces and the virtual work of other nonconservative forces:

$$\delta W_{nc} = \delta W_v + \delta W_o = Q_{v,i}\delta v_i + Q_{o,i}\delta v_i = \left(\frac{\partial \mathfrak{I}}{\partial \dot{v}_i} + Q_{o,i}\right)\delta v_i \tag{2.81}$$

Substituting for the generalized forces into Equation 2.75 leads to

$$\frac{d}{dt}\left(\frac{\partial L}{\partial \dot{v}_i}\right) - \frac{\partial L}{\partial v_i} = \frac{\partial \mathfrak{I}}{\partial \dot{v}_i} + Q_{o,i}$$

$$\tag{2.82}$$

$$\frac{d}{dt}\left(\frac{\partial L}{\partial \dot{v}_i}\right) - \frac{\partial L}{\partial v_i} - \frac{\partial \mathfrak{I}}{\partial \dot{v}_i} = Q_{o,i}$$

Equation 2.82 is a form of Lagrange's equations when Rayleigh's dissipation function is used to include the effect of viscous damping.

Example 2.11. Derive the differential equations governing the motion of the system of Figure 2.12 using x and θ as the chosen generalized coordinates.

Solution: The kinetic energy of the system at an arbitrary instant is

$$T = \frac{1}{2}m\dot{x}^2 + \frac{1}{2}I\dot{\theta}^2 \tag{a}$$

The potential energy of the system at an arbitrary instant is

$$V = \frac{1}{2}k(x - r_1\theta)^2 \tag{b}$$

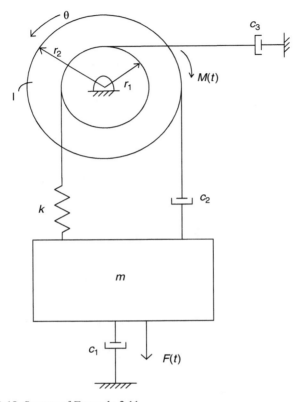

FIGURE 2.12 System of Example 2.11.

Rayleigh's dissipation function is the negative of one half of the power dissipated by the viscous dampers:

$$\mathfrak{I} = -\frac{1}{2}[c_1\dot{x}^2 + c_2(\dot{x} - r_2\dot{\theta})^2 + c_3(r_1\dot{\theta})^2] \tag{c}$$

Assuming variations δx and $\delta\theta$, the virtual work of the external forces is

$$\delta W_o = F(t)\delta x - M(t)\delta\theta \tag{d}$$

The generalized forces are obtained from Equation d as

$$Q_1 = F(t) \tag{e}$$

$$Q_2 = -M(t) \tag{f}$$

Application of Lagrange's equation, Equation 2.82, for x leads to

$$\frac{\mathrm{d}}{\mathrm{d}t}\left(\frac{\partial L}{\partial \dot{x}}\right) - \frac{\partial L}{\partial x} - \frac{\partial \mathfrak{I}}{\partial \dot{x}} = Q_1$$

$$\frac{\mathrm{d}}{\mathrm{d}t}(m\dot{x}) + k(x - r_1\theta) + [c_1\dot{x} + c_2(\dot{x} - r_2\dot{\theta})] = F(t) \tag{g}$$

$$m\ddot{x} + (c_1 + c_2)\dot{x} - c_2 r_2\dot{\theta} + kx - kr_1\theta = F(t)$$

Application of Lagrange's equation for the generalized coordinate θ leads to

$$\frac{\mathrm{d}}{\mathrm{d}t}\left(\frac{\partial L}{\partial \dot{\theta}}\right) - \frac{\partial L}{\partial \theta} - \frac{\partial \mathfrak{I}}{\partial \dot{\theta}} = Q_2$$

$$\frac{\mathrm{d}}{\mathrm{d}t}(I\dot{\theta}) + k(x - r_1\theta)(-r_1) + c_2(\dot{x} - r_2\dot{\theta})(-r_2) + c_3 r_1^2\dot{\theta} = -M(t) \tag{h}$$

$$I\ddot{\theta} - c_2 r_2\dot{x} + (c_2 r_2^2 + c_3 r_1^2)\dot{\theta} - kr_1 x + kr_1^2\theta = -M(t)$$

The matrix formulation of Equation g and Equation h is

$$\begin{bmatrix} m & 0 \\ 0 & I \end{bmatrix}\begin{bmatrix} \ddot{x} \\ \ddot{\theta} \end{bmatrix} + \begin{bmatrix} c_1 + c_2 & -c_2 r_2 \\ -c_2 r_2 & c_2 r_2^2 + c_3 r_1^2 \end{bmatrix}\begin{bmatrix} \dot{x} \\ \dot{\theta} \end{bmatrix} + \begin{bmatrix} k & -kr_1 \\ -kr_1 & kr_1^2 \end{bmatrix}\begin{bmatrix} x \\ \theta \end{bmatrix}$$

$$= \begin{bmatrix} F(t) \\ -M(t) \end{bmatrix} \tag{i}$$

2.7 LINEAR DISCRETE SYSTEMS

2.7.1 Quadratic Forms

The kinetic energy of a system of p particles is

$$T = \frac{1}{2}\sum_{i=1}^{p} m_i \dot{\mathbf{r}}_i r \cdot \mathbf{r}_i \tag{2.83}$$

The position vector for each particle is a function of the chosen generalized coordinates

$$\mathbf{r}_i = \mathbf{r}_i(v_1, v_2, \ldots, v_n) \tag{2.84}$$

If the system is linear and nongyroscopic, its position vector is a linear function of the generalized coordinates

$$\mathbf{r_i} = \left(\sum_{j=1}^{n} b_{x,j,i} v_j\right)\mathbf{i} + \left(\sum_{j=1}^{n} b_{y,j,i} v_j\right)\mathbf{j} + \left(\sum_{j=1}^{n} b_{z,j,i} v_j\right)\mathbf{k} \tag{2.85}$$

where the b coefficients are determined through geometry. Using Equation 2.85 the particle's velocity vector is

$$\dot{\mathbf{r}}_i = \left(\sum_{j=1}^{n} b_{x,j,i} \dot{v}_j\right)\mathbf{i} + \left(\sum_{j=1}^{n} b_{y,j,i} \dot{v}_j\right)\mathbf{j} + \left(\sum_{j=1}^{n} b_{z,j,i} \dot{v}_j\right)\mathbf{k} \tag{2.86}$$

The particle's kinetic energy is calculated as

$$
\begin{aligned}
T_i &= \frac{1}{2} m_i \left[\left(\sum_{j=1}^{n} b_{x,j,i} \dot{v}_j\right)\left(\sum_{k=1}^{n} b_{x,k,i} \dot{v}_k\right) + \left(\sum_{j=1}^{n} b_{y,j,i} \dot{v}_j\right)\left(\sum_{k=1}^{n} b_{y,k,i} \dot{v}_k\right) \right. \\
&\quad \left. + \left(\sum_{j=1}^{n} b_{z,j,i} \dot{v}_j\right)\left(\sum_{k=1}^{n} b_{x,k,i} \dot{v}_k\right) \right] \\
&= \frac{1}{2} \sum_{j=1}^{n} \sum_{k=1}^{n} (a_{j,k})_i \dot{v}_j \dot{v}_k
\end{aligned}
$$

$$\tag{2.87}$$

where

$$(a_{j,k})_i = m_i(b_{x,j,i} b_{x,k,i} + b_{y,j,i} b_{y,k,i} + b_{z,j,i} b_{z,k,i}) \tag{2.88}$$

The total kinetic energy of the system is

$$T = \frac{1}{2} \sum_{j=1}^{n} \sum_{k=1}^{n} m_{j,k} \dot{v}_j \dot{v}_k \tag{2.89}$$

where

$$m_{j,k} = \sum_{i=1}^{p} (a_{j,k})_i = \sum_{i=1}^{p} m_i(b_{x,j,i}b_{x,k,i} + b_{y,j,i}b_{y,k,i} + b_{z,j,i}b_{z,k,i}) \qquad (2.90)$$

The form of Equation 2.90 is called the quadratic form of kinetic energy. The kinetic energy of all linear systems may be written in a quadratic form. The coefficients $m_{i,j}$ are called inertia coefficients. By definition:

$$m_{k,j} = \sum_{i=1}^{n} (a_{k,j})_i = \sum_{i=1}^{n} m_i(b_{x,k,i}b_{x,j,i} + b_{y,k,i}b_{y,j,i} + b_{z,k,i}b_{z,j,i}) \qquad (2.91)$$

Comparison of Equation 2.90 and Equation 2.91 shows that

$$m_{j,k} = m_{k,j} \qquad (2.92)$$

Equation 2.92 shows that the inertia coefficients are symmetric.

Consider a system in which the only sources of potential energy are from p discrete linear springs. The potential energy of the system is

$$V = \frac{1}{2} \sum_{i=1}^{p} k_i \Delta_i^2 \qquad (2.93)$$

where Δ_i is the change in length of the ith spring. The change in length of each spring is a function of the chosen generalized coordinates. For a linear system

$$\Delta_i = \sum_{j=1}^{n} d_{j,i} v_j \qquad (2.94)$$

where the d coefficients are determined from geometry. Substitution of Equation 2.94 into Equation 2.93 leads to a quadratic form of potential energy for a linear system as

$$V = \frac{1}{2} \sum_{k=1}^{n} \sum_{j=1}^{n} k_{j,k} v_j v_k \qquad (2.95)$$

where

$$k_{j,k} = \sum_{i=1}^{p} k_i d_{j,i} d_{k,i} \qquad (2.96)$$

are called stiffness coefficients. It is easy to show that the stiffness coefficients satisfy

$$k_{j,k} = k_{k,j} \qquad (2.97)$$

An analysis similar to those above is used to show that, for a linear system, Rayleigh's dissipation function has a quadratic form, which is written as

$$\mathfrak{I} = -\frac{1}{2} \sum_{j=1}^{n} \sum_{k=1}^{n} c_{j,k} \dot{v}_j \dot{v}_k \tag{2.98}$$

where the $c_{j,k}$ are called the damping coefficients, which satisfy

$$c_{j,k} = c_{k,j} \tag{2.99}$$

2.7.2 Differential Equations for Linear Systems

Application of Lagrange's equations, Equation 2.82, to a linear nongyroscopic system, using Equation 2.89, Equation 2.95, and Equation 2.98 for the kinetic energy; potential energy; and Rayleigh's dissipation function, gives

$$\frac{d}{dt} \left[\frac{\partial}{\partial \dot{v}_i} \left(\frac{1}{2} \sum_{j=1}^{n} \sum_{k=1}^{n} m_{j,k} \dot{v}_j \dot{v}_k \right) \right] + \frac{\partial}{\partial v_i} \left(\frac{1}{2} \sum_{k=1}^{n} \sum_{j=1}^{n} k_{j,k} v_j v_k \right)$$

$$- \frac{\partial}{\partial \dot{v}_i} \left(-\frac{1}{2} \sum_{j=1}^{n} \sum_{k=1}^{n} c_{j,k} \dot{v}_j \dot{v}_k \right) = Q_i \quad i = 1,2,\ldots,n \tag{2.100}$$

Simplifying the derivative of the potential energy term leads to

$$\frac{\partial}{\partial v_i} \left(\frac{1}{2} \sum_{k=1}^{n} \sum_{j=1}^{n} k_{j,k} v_j v_k \right) = \frac{1}{2} \sum_{k=1}^{n} \sum_{j=1}^{n} k_{j,k} \frac{\partial}{\partial v_i} (v_j v_k)$$

$$= \frac{1}{2} \sum_{k=1}^{n} \sum_{j=1}^{n} k_{j,k} \left[v_j \frac{\partial}{\partial v_i} (v_k) + v_k \frac{\partial}{\partial v_i} (v_j) \right]$$

$$= \frac{1}{2} \sum_{k=1}^{n} \sum_{j=1}^{n} k_{j,k} [v_j \delta_{i,k} + v_k \delta_{i,j}]$$

$$= \frac{1}{2} \left[\sum_{k=1}^{n} v_k \left(\sum_{j=1}^{n} k_{j,k} \delta_{i,j} \right) + \sum_{j=1}^{n} v_j \left(\sum_{k=1}^{n} k_{j,k} \delta_{i,k} \right) \right]$$

$$= \frac{1}{2} \left[\sum_{k=1}^{n} v_k k_{i,k} + \sum_{j=1}^{n} v_j k_{j,i} \right]$$

$$= \frac{1}{2} \left[\sum_{k=1}^{n} v_k k_{i,k} + \sum_{k=1}^{n} v_k k_{k,i} \right]$$

$$\tag{2.101}$$

Symmetry of the stiffness coefficients is used in Equation 2.101, leading to

$$\frac{\partial}{\partial v_i}\left(\frac{1}{2}\sum_{k=1}^{n}\sum_{j=1}^{n}k_{j,k}v_jv_k\right) = \sum_{k=1}^{n}k_{i,k}v_k \quad i = 1,2,\ldots,n \qquad (2.102)$$

Similar reductions for the kinetic energy and dissipation function terms allow Equation 2.100 to be written as

$$\sum_{k=1}^{n}(m_{i,k}\ddot{v}_k + c_{i,k}\dot{v}_k + k_{i,k}v_k) = Q_i \quad i = 1,2,\ldots,n \qquad (2.103)$$

The set of n equations represented by Equation 2.103 can be rewritten in a matrix form as

$$
\begin{bmatrix}
m_{1,1} & m_{1,2} & m_{1,3} & \cdots & m_{1,n} \\
m_{2,1} & m_{2,2} & m_{2,3} & \cdots & m_{2,n} \\
m_{3,1} & m_{3,2} & m_{3,3} & \cdots & m_{3,n} \\
\vdots & \vdots & \vdots & \ddots & \vdots \\
m_{n,1} & m_{n,2} & m_{n,3} & \cdots & m_{n,n}
\end{bmatrix}
\begin{bmatrix}
\ddot{v}_1 \\
\ddot{v}_2 \\
\ddot{v}_3 \\
\vdots \\
\ddot{v}_n
\end{bmatrix}
$$

$$
+
\begin{bmatrix}
c_{1,1} & c_{1,2} & c_{1,3} & \cdots & c_{1,n} \\
c_{2,1} & c_{2,2} & c_{2,3} & \cdots & c_{2,n} \\
c_{3,1} & c_{3,2} & c_{3,3} & \cdots & c_{3,n} \\
\vdots & \vdots & \vdots & \ddots & \vdots \\
c_{n,1} & c_{n,2} & c_{n,3} & \cdots & c_{n,n}
\end{bmatrix}
\begin{bmatrix}
\dot{v}_1 \\
\dot{v}_2 \\
\dot{v}_3 \\
\vdots \\
\dot{v}_n
\end{bmatrix} \qquad (2.104)
$$

$$
+
\begin{bmatrix}
k_{1,1} & k_{1,2} & k_{1,3} & \cdots & k_{1,n} \\
k_{2,1} & k_{2,2} & k_{2,3} & \cdots & k_{2,n} \\
k_{3,1} & k_{3,2} & k_{3,3} & \cdots & k_{3,n} \\
\vdots & \vdots & \vdots & \ddots & \vdots \\
k_{n,1} & k_{n,2} & k_{n,3} & \cdots & k_{n,n}
\end{bmatrix}
\begin{bmatrix}
v_1 \\
v_2 \\
v_3 \\
\vdots \\
v_n
\end{bmatrix}
=
\begin{bmatrix}
Q_1 \\
Q_2 \\
Q_3 \\
\vdots \\
Q_n
\end{bmatrix}
$$

The set of n coupled linear differential equations formulated in Equation 2.104 is of the general form

$$\mathbf{M}\ddot{v} + \mathbf{C}\dot{v} + \mathbf{K}v = \mathbf{Q} \qquad (2.105)$$

where $\mathbf{M}$ is the symmetric $n \times n$ mass matrix, $\mathbf{C}$ is the symmetric $n \times n$ damping matrix, $\mathbf{K}$ is the symmetric $n \times n$ stiffness matrix, $\mathbf{Q}$ is the $n \times 1$ generalized force vector, $\mathbf{v}$ is the $n \times 1$ displacement vector (the vector of generalized coordinates), $\dot{\mathbf{v}}$ is the $n \times 1$ velocity vector (the vector of generalized velocities), and $\ddot{\mathbf{v}}$ is the $n \times 1$ acceleration vector (vector of generalized accelerations).

Equation 2.105 is the standard form of the differential equations governing the motion of a linear n-degree-of-freedom system. Subsequent theory is based upon use of these equations. It is important to note that $\mathbf{M}$, $\mathbf{C}$, and $\mathbf{K}$ are symmetric.

The nature of the coupling of a system using a set of generalized coordinates depends upon the mass and stiffness matrices. If the mass matrix is not a diagonal matrix, the system is said to be dynamically coupled. If the stiffness matrix is not a diagonal, matrix the system is said to be statically coupled.

2.7.3 LINEARIZATION OF DIFFERENTIAL EQUATIONS

In general, the kinetic energy is a function of the generalized coordinates x_1, $x_2, \ldots x_n$, and the generalized velocities and the potential energy is a function of the generalized coordinates.

$$T = T(x_1, x_2, \ldots, x_n, \dot{x}_1, \dot{x}_2, \ldots, \dot{x}_n) \tag{2.106}$$

$$V = V(x_1, x_2, \ldots, x_n) \tag{2.107}$$

Consider an equilibrium configuration of the system defined by a set of generalized coordinates. For simplicity, assume this equilibrium position corresponds to $x_1 = x_2 = \ldots = x_n = 0$. The generalized coordinates are then displacements from equilibrium. The velocities in the equilibrium position are, of course, all zero. Let E refer to this equilibrium state. Considering the kinetic energy in this light, a Taylor series expansion may be employed as

$$
\begin{aligned}
T = {} & T(x_1, x_2, \ldots, x_n, \dot{x}_1, \dot{x}_2, \ldots, \dot{x}_n) \\
= {} & T(E) + \sum_{i=1}^{n} \left[\frac{\partial T}{\partial x_i}(E) \right] x_i + \sum_{i=1}^{n} \left[\frac{\partial T}{\partial \dot{x}_i}(E) \right] \dot{x}_i \\
& + \frac{1}{2} \sum_{j=1}^{n} \sum_{i=1}^{n} \left[\frac{\partial^2 T}{\partial x_i \partial x_j}(E) \right] x_i x_j + \frac{1}{2} \sum_{j=1}^{n} \sum_{i=1}^{n} \left[\frac{\partial^2 T}{\partial x_i \partial \dot{x}_j}(E) \right] x_i \dot{x}_j \\
& + \frac{1}{2} \sum_{j=1}^{n} \sum_{i=1}^{n} \left[\frac{\partial^2 T}{\partial \dot{x}_i \partial \dot{x}_j}(E) \right] \dot{x}_i \dot{x}_j + \ldots
\end{aligned}
\tag{2.108}
$$

Application of Lagrange's equations requires evaluation of

$$\frac{\mathrm{d}}{\mathrm{d}t}\left(\frac{\partial T}{\partial \dot{x}_k}\right) - \frac{\partial T}{\partial x_k} = -\frac{\partial T}{\partial x_k}(E) + \sum_{i=1}^{n} m_{i,k}\ddot{x}_i + \sum_{i=1}^{n} \ell_{i,k}\dot{x}_i$$

$$- \sum_{i=1}^{n} n_{i,k}\dot{x}_i - \sum_{i=1}^{n} p_{i,k}x_i \tag{2.109}$$

where

$$m_{i,k} = \frac{\partial^2 T}{\partial \dot{x}_i \partial \dot{x}_k}(E) \tag{2.110}$$

$$\ell_{i,k} = \frac{\partial^2 T}{\partial \dot{x}_i \partial x_k}(E) \tag{2.111}$$

$$n_{i,k} = \frac{\partial^2 T}{\partial x_i \partial \dot{x}_k}(E) = \ell_{i,k} \tag{2.112}$$

$$p_{i,k} = \frac{\partial^2 T}{\partial x_i \partial x_k}(E) \tag{2.113}$$

Equation 2.109 can be rewritten as

$$\frac{\mathrm{d}}{\mathrm{d}t}\left(\frac{\partial T}{\partial \dot{x}_k}\right) - \frac{\partial T}{\partial x_k} = -\frac{\partial T}{\partial x_k}(E) + \sum_{i=1}^{n} m_{i,k}\ddot{x}_i + \sum_{i=1}^{n} s_{i,k}\dot{x}_i - \sum_{i=1}^{n} p_{i,k}x_i \tag{2.114}$$

where

$$s_{i,k} = \ell_{i,k} - n_{i,k} \tag{2.115}$$

Equation 2.114 is the general form for the use of the kinetic energy functional in Lagrange's equations for a linear system. When linearization of the dissipation function and potential energy are performed in a similar manner, and equations are written and summarized in a matrix form, the general result is

$$\mathbf{M}\ddot{x} + \mathbf{S}\dot{x} - \mathbf{P}x + \mathbf{C}\dot{x} + \mathbf{K}x = -\mathbf{J} \tag{2.116}$$

The vector $\mathbf{J}$ is the vector whose components are

$$J_k = \frac{\partial T}{\partial x_k}(E) \qquad (2.117)$$

It is rare to include the vector $\mathbf{J}$ in the formulation of the differential equations. The components of $\mathbf{J}$, $\mathbf{M}$, $\mathbf{S}$, and $\mathbf{P}$ can be determined directly from writing the kinetic energy functional in the form of

$$T = T(x_1, x_2, \ldots, x_n, \dot{x}_1, \dot{x}_2, \ldots, \dot{x}_n)$$

$$= T(E) + \sum_{i=1}^{n} J_i x_i + \sum_{i=1}^{n} \left[\frac{\partial T}{\partial \dot{x}_i}(E)\right] \dot{x}_i - \frac{1}{2} \sum_{j=1}^{n} \sum_{i=1}^{n} p_{i,j} x_i x_j$$

$$+ \frac{1}{2} \sum_{j=1}^{n} \sum_{i=1}^{n} s_{i,j} x_i \dot{x}_j + \frac{1}{2} \sum_{j=1}^{n} \sum_{i=1}^{n} m_{i,j} \dot{x}_i \dot{x}_j + \ldots \qquad (2.118)$$

The mass matrix $\mathbf{M}$ stiffness matrix $\mathbf{K}$, and viscous damping matrix $\mathbf{C}$, are as previously defined. Note that the symmetry of these matrices are guaranteed. For example,

$$m_{i,k} = \frac{\partial^2 T}{\partial \dot{x}_i \partial \dot{x}_k} \qquad (2.119)$$

$$m_{k,i} = \frac{\partial^2 T}{\partial \dot{x}_k \partial \dot{x}_i} \qquad (2.120)$$

When the order of differentiation is reversed the evaluation remains the same.

The matrix $\mathbf{S}$ exists even in the absence of viscous damping due to gyroscopic forces. Since the order of differentiation does not matter, it is clear that $\ell_{i,k} = \ell_{k,i}$ and $n_{i,k} = n_{k,i}$. Thus,

$$s_{i,k} = \ell_{i,k} - \ell_{k,i} \qquad (2.121)$$

$$s_{k,i} = \ell_{k,i} - \ell_{i,k} \qquad (2.122)$$

Equation 2.121 and Equation 2.122 show that $s_{i,k} = -s_{k,i}$. The matrix $\mathbf{S}$ is a skew-symmetric matrix, a matrix with all zeroes along the main diagonal and the corresponding off-diagonal elements are negatives of one another. The matrix $\mathbf{S}$ occurs in gyroscopic problems.

The matrix $\mathbf{P}$ is symmetric and is an addition to the stiffness matrix from gyroscopic terms.

2.8 GYROSCOPIC SYSTEMS

The rigid body of Figure 2.13 is rotating about a fixed axis with an angular velocity ω. Let X–Y–Z represent an inertial set of axes while x–y–z represent a set of axes rotating with the body. The origins of the two sets of axes are coincident. If r represents the position vector of a particle in the x–y–z reference frame to a point on the rigid body, then r is constant.

$$\mathbf{r} = x\mathbf{i} + y\mathbf{j} \tag{2.123}$$

If the angular velocity of the body is constant, then the angle between the x and X axes is $\theta = \omega t$, and the position vector written in the X–Y–Z system is

$$\mathbf{R} = \cos(\omega t)\mathbf{I} + \sin(\omega t)\mathbf{J} \tag{2.124}$$

Assume the axis of rotation is the z axis. The velocity of a point on the rigid body is

$$\mathbf{v} = \omega\mathbf{k}\mathbf{r} \tag{2.125}$$

Let P be a particle on the rigid body that is at a point on the body whose position coincides with the point whose position vector is $\mathbf{r}$. Particle P moves relative to the rigid body with a velocity, $\mathbf{v}_{rel} = \dot{\mathbf{r}}$ such that the velocity of the

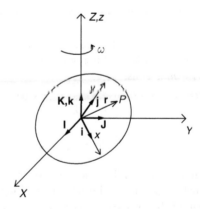

FIGURE 2.13 X–Y–Z axes as fixed in space, while x–y–z axes rotate with rigid body.

particle is

$$\mathbf{v}_p = \omega\mathbf{k} \times \mathbf{r} + \mathbf{v}_{rel} \tag{2.126}$$

When differentiated with respect to time, the acceleration of the particle is obtained as

$$\mathbf{a}_p = \alpha\mathbf{k}r + \omega\mathbf{k} \times (\omega\mathbf{k} \times \mathbf{r}) + \mathbf{a}_{rel} + 2\omega\mathbf{k} \times \mathbf{v}_{rel} \tag{2.127}$$

where $\mathbf{a}_{rel} = \ddot{\mathbf{r}}$ and $2\omega\mathbf{k} \times \mathbf{v}_{rel}$ is called the Coriolis acceleration. The velocity and acceleration components of the particle are illustrated in Figure 2.14.

The Coriolis acceleration is a result of the motion of a particle relative to a rotating reference frame (or rotating body). Systems in which the Coriolis acceleration is present are called gyroscopic systems. Gyroscopic systems are inherently linear, but the kinetic energy functional is not quadratic for gyroscopic systems, as illustrated in the following example.

Example 2.12. The arm of Figure 2.15 rotates with a constant angular velocity ω about the fixed Z axis. A mass-spring-viscous damper system rotates with the arm. Use Lagranage's equation to derive the differential equation for $z(t)$, the displacement of the mass along the arm, measured from an equilibrium position.

Solution: The rotating x–y–z system is defined such that the arm is along the x axis. Let ℓ be the length of the spring when the system is in an equilibrium position. The position vector of the particle in the rotating system is

$$\mathbf{r} = (\ell + u)\mathbf{i} \tag{a}$$

The velocity of the particle relative to the arm is

$$\mathbf{v}_{\iota \cup \mathsf{J}} = \dot{u}\mathbf{i} \tag{b}$$

Thus, the velocity of the particle is

$$\mathbf{v} = \omega\mathbf{k} \times [(\ell + u)\mathbf{i}] + \dot{u}\mathbf{i} = \dot{u}\mathbf{i} + \omega(\ell + u)\mathbf{j} \tag{c}$$

The kinetic energy of the system is

$$T = \frac{1}{2}m[\dot{u}^2 + \omega^2(\ell + u)^2] \tag{d}$$

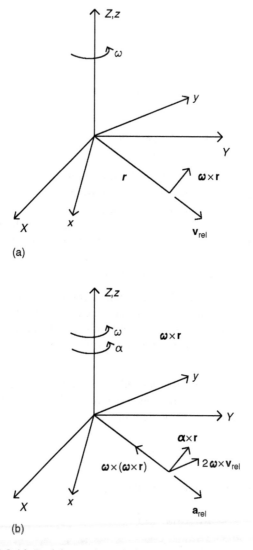

FIGURE 2.14 Particle moving relative to rotating body. (a) Velocity components. (b) Acceleration components.

The potential energy of the system at an arbitrary instant is

$$V = \frac{1}{2}k(u - \ell_e)^2 \qquad (e)$$

where ℓ_e is the length of the spring when it is unstretched.

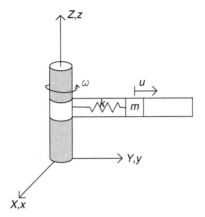

FIGURE 2.15 As arm rotates about z axis, particle moves relative to arm. System of Example 2.12.

Application of Lagrange's equation leads to

$$\frac{d}{dt}\left(\frac{\partial L}{\partial \dot{u}}\right) - \frac{\partial L}{\partial u} = 0$$

$$\frac{d}{dt}(m\dot{u}) - m\omega^2(\ell + u) + k(u - \ell_e) = 0 \qquad \text{(f)}$$

$$m\ddot{u} + (k - m\omega^2)u - (k\ell_e + m\omega^2\ell) = 0$$

The equilibrium position for this system is one in which the inertia forces due to rotation of the arm are balanced by the spring force. When the particle is not moving relative to the arm, it is still subject to an inertia force. The term $\omega\mathbf{k}\times(\omega\mathbf{k}\times\mathbf{r})$ in Equation 2.127 along the arm toward the axis of rotation. The particle is subject to an inertia force of $m\ell\omega^2$ acting away from the axis of rotation. In order to maintain the particle in this position, a force is developed in the spring to balance the inertia force. The spring is compressed from its equilibrium length by $\ell_e - \ell$. Summing forces of the free-body diagram of Figure 2.16 leads to

$$\ell_e = -\frac{m\omega^2\ell}{k} \qquad \text{(g)}$$

Substitution of Equation g into Equation f leads to

$$m\ddot{u} + (k - m\omega^2)u = 0 \qquad \text{(h)}$$

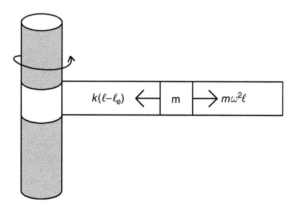

FIGURE 2.16 Free-body diagram of particle in equilibrium.

The term $-m\omega^2 u$ in the differential equation, Equation g, is destabilizing. If $(k - m\omega^2) < 0$ the system is unstable.

Example 2.12 is a one-degree-of-freedom system with a gyroscopic effect, which reduces the stiffness of the system. This is the one-degree-of-freedom equivalent of the **P** matrix of Equation 2.116.

Example 2.13. The turntable of Figure 2.17 rotates at a constant angular velocity ω. A particle is mounted on the turntable, connected to the side of the turntable by linear springs. The springs are both unstretched when the particle is at the center of the turntable. Assuming x and y are small, derive the differential equations governing their motion relative to the turntable.

Solution: The position vector from the center of the turntable in the x–y–z reference frame is

$$\mathbf{r} = x\mathbf{i} + y\mathbf{j} \qquad \text{(a)}$$

The velocity of the particle at an arbitrary instant is

$$\mathbf{v} = \omega\mathbf{k} \times (x\mathbf{i} + y\mathbf{j}) + \dot{x}\mathbf{i} + \dot{y}\mathbf{j} = (\dot{x} - \omega y)\mathbf{i} + (\dot{y} + \omega x)\mathbf{j} \qquad \text{(b)}$$

The kinetic energy of the particle at an arbitrary instant is

$$T = \frac{1}{2}m[(\dot{x} - \omega y)^2 + (\dot{y} + \omega x)^2]$$

$$= \frac{1}{2}m(\dot{x}^2 + \dot{y}^2 - 2\omega\dot{x}y + 2\omega\dot{y}x + \omega^2 x^2 + \omega^2 y^2) \qquad \text{(c)}$$

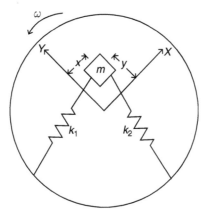

FIGURE 2.17 Particle of mass m moves relative to turntable, creating gyroscopic effects. System of Example 2.13.

The kinetic energy is written in the form of Equation 2.118 for a two-degree-of-freedom system. The coefficients of **M**, **S**, and **P** can be obtained by comparing the quadratic forms appearing in Equation j, with Equation 2.110 through Equation 2.113. The mass matrix is obtained as

$$\mathbf{M} = \begin{bmatrix} m & 0 \\ 0 & m \end{bmatrix} \tag{d}$$

The gyroscopic matrix is obtained as

$$\mathbf{S} = \begin{bmatrix} 0 & -2m\omega \\ 2m\omega & 0 \end{bmatrix} \tag{e}$$

Recalling that the added stiffness matrix is multiplied by a negative sign in the formulation of the differential equations for a gyroscopic system

$$\mathbf{P} = \begin{bmatrix} m\omega^2 & 0 \\ 0 & m\omega^2 \end{bmatrix} \tag{f}$$

The change in length of the spring of stiffness k_1 is

$$d_1 = \ell - \sqrt{(\ell + x)^2 + y^2}$$
$$= \ell\left\{ 1 - \left[\left(1 + \frac{x}{\ell}\right)^2 + \left(\frac{y}{\ell}\right)^2 \right]^{1/2} \right\} \tag{g}$$

Assuming small x and y, and using a binomial expansion to expand the square root leads to

$$d_1 = \ell \left\{ 1 - \left[1 + \frac{1}{2}\left(2\frac{x}{\ell} \right) \right] + \dots \right\} = x \tag{h}$$

A similar analysis is used to show that the change in length of the spring of stiffness k_2 is approximately y. Using these approximations, the potential energy of the system is

$$V = \frac{1}{2}k_1 x^2 + \frac{1}{2}k_2 y^2 \tag{i}$$

The stiffness matrix is obtained from Equation i as

$$\mathbf{K} = \begin{bmatrix} k_1 & 0 \\ 0 & k_2 \end{bmatrix} \tag{j}$$

The differential equations governing the motion of this linear system are

$$\begin{bmatrix} m & 0 \\ 0 & m \end{bmatrix} \begin{bmatrix} \ddot{x} \\ \ddot{y} \end{bmatrix} + \begin{bmatrix} 0 & -2m\omega \\ 2m\omega & 0 \end{bmatrix} \begin{bmatrix} \dot{x} \\ \dot{y} \end{bmatrix} + \begin{bmatrix} k_1 - m\omega^2 & 0 \\ 0 & k_2 - m\omega^2 \end{bmatrix} \begin{bmatrix} x \\ y \end{bmatrix}$$

$$= \begin{bmatrix} 0 \\ 0 \end{bmatrix} \tag{k}$$

2.9 CONTINUOUS SYSTEMS

The response of a continuous system is a function of one or more spatial coordinates, as well as time. In general, let $w(\mathbf{r},t)$ represent the response of a continuous system, where $\mathbf{r}$ is a position vector, written in terms of spatial coordinates, and is defined in a region R. The system is one dimensional if $\mathbf{r}$ is a function of one spatial variable, in which case R is a finite segment of the positive real axis. The system is two dimensional if $\mathbf{r}$ is a function of two variables. In this case, R is a surface which can be described by a unit vector normal to the surface, and a curve that bounds the surface. If $\mathbf{r}$ is a function of three spatial coordinates, the system is three dimensional and the region R is a volume in space bounded by a closed surface.

The kinetic energy and potential energy of a continuous system are written as functionals of the form

$$T = \int_R f\left(\mathbf{r}, w(\mathbf{r},t), \frac{\partial w}{\partial t}\right) dR \tag{2.128}$$

$$V = \int_R g(\mathbf{r}, w(\mathbf{r},t), \nabla w) dR \tag{2.129}$$

Hamilton's Principle is used to derive a mathematical problem that $w(\mathbf{r},t)$ must solve.

For a conservative system,

$$\int_{t_1}^{t_2} (\delta T - \delta V) dt = 0 \tag{2.130}$$

For a system with nonconservative forces whose virtual work is δW_{nc}, Extended Hamilton's Principle is written as

$$\int_{t_1}^{t_2} (\delta T - \delta V + \delta W_{\mathrm{nc}}) dt = 0 \tag{2.131}$$

The variation in $w(\mathbf{r},t)$ is of the form $\delta w = \varepsilon \eta(\mathbf{r},t)$ and must be chosen such that $\eta(\mathbf{r},t_1) = \eta(\mathbf{r},t_2) = 0$.

Consider a one-dimensional system in which x, $0 \leq x \leq L$ is the spatial variable. The kinetic and potential energies are of the form

$$T = \frac{1}{2} \int_0^L f\left(x, w(x,t), \frac{\partial w}{\partial t}\right) dx \tag{2.132}$$

$$V = \frac{1}{2} \int_0^L g\left(x, w(x,t), \frac{\partial w}{\partial x}\right) dx \tag{2.133}$$

Applications of Hamilton's Principle and Extended Hamilton's Principle are considered in subsequent sections to derive differential equations governing the motion of various continuous systems.

2.10 BARS, STRINGS, AND SHAFTS

The process of applying Hamilton's Principle toward the application of the partial differential equations governing the motion of bars, strings, and shafts is the same for each. Table 2.1 shows the potential energy, kinetic energy, resulting differential equations, and potential boundary conditions for each. The application to bars is illustrated in this section.

Example 2.14. Use Hamilton's Principle to formulate the problem to be solved for the longitudinal displacement $u(x,t)$ of the elastic bar of Figure 2.18. The bar is fixed at $x = 0$ and free at $x = L$.

Solution: The kinetic energy functional for the elastic bar is

$$T = \frac{1}{2} \int_0^L \rho A \left(\frac{\partial u}{\partial t} \right)^2 dx \tag{a}$$

The potential energy functional for the elastic bar is determined as Equation 1.87, as

$$V = \frac{1}{2} \int_0^L EA \left(\frac{\partial u}{\partial x} \right)^2 dx \tag{b}$$

Application of Hamilton's Principle leads to

$$\frac{1}{2} \int_{t_1}^{t_2} \left[\delta \int_0^L \rho A \left(\frac{\partial u}{\partial t} \right)^2 dx - \delta \int_0^L EA \left(\frac{\partial u}{\partial x} \right)^2 dx \right] dt = 0 \tag{c}$$

Each term is considered separately. Recalling that the order of variation and integration may be interchanged, and that the order of differentiation and variation may also be interchanged, the first term on the left-hand side of

TABLE 2.1
Kinetic Energy, Potential Energy, and Boundary Conditions for Second-Order Systems

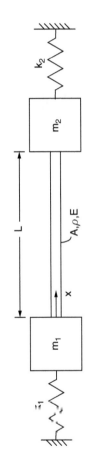

I. Bar $u(x,t)$ = longitudinal displacement

Parameters

$A(x)$ = area

$\rho(x)$ = mass density

$E(x)$ = elastic modulus

L = length

k_1 = stiffness of spring attached at $x = 0$

k_2 = stiffness of spring attached at $x = L$

m_1 = mass of particles attached at $x = 0$

m_2 = mass of particles attached at $x = L$

Geometric boundary conditions

Fixed end at $x = 0$ $u(0,t) = 0$

Fixed end at $x = L$ $u(0,t) = 0$

Kinetic energy

$$T = \frac{1}{2} \int_0^L \rho A \left(\frac{\partial u}{\partial t} \right)^2 \, dx + \frac{1}{2} m_1 \left[\frac{\partial u}{\partial t}(0,t) \right]^2 + \frac{1}{2} m_2 \left[\frac{\partial u}{\partial t}(L,t) \right]^2$$

Potential energy

$$V = \frac{1}{2} \int_0^L EA \left(\frac{\partial u}{\partial x} \right)^2 \, dx + \frac{1}{2} k_1 \left[\frac{\partial u}{\partial x}(0,t) \right]^2 + \frac{1}{2} k_2 \left[\frac{\partial u}{\partial x}(L,t) \right]^2$$

Natural boundary conditions

$$E(0)A(0)\frac{\partial u}{\partial x}(0,t) - k_1 u(0,t) = m_1 \frac{\partial^2 u}{\partial t^2}(0,t)$$

$$-E(L)A(L)\frac{\partial u}{\partial x}(L,t) - k_2 u(L,t) = m_2 \frac{\partial^2 u}{\partial t^2}(L,t)$$

(continued)

TABLE 2.1 (Continued)

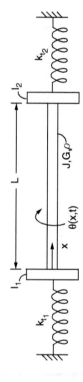

II. Shaft $\theta(x,t)$ = angular displacement

Parameters

$J(x)$ = polar area moment of inertia

$\rho(x)$ = mass density

$G(x)$ = shear modulus

L = length

k_{t1} = torsional stiffness of spring attached at $x = 0$

k_{t2} = torsional stiffness of spring attached at $x = L$

I_1 = polar mass moment of inertia of disk attached at $x = 0$

I_2 = polar mass moment of inertia of disk attached at $x = L$

Geometric boundary conditions

Fixed end at $x = 0$ $\theta(0,t) = 0$

Fixed end at $x = L$ $\theta(L,t) = 0$

Kinetic energy

$$T = \frac{1}{2}\int_0^L \rho J\left(\frac{\partial\theta}{\partial t}\right)^2 dx + \frac{1}{2}I_1\left[\frac{\partial\theta}{\partial t}(0,t)\right]^2 + \frac{1}{2}I_2\left[\frac{\partial\theta}{\partial t}(L,t)\right]^2$$

Potential energy

$$V = \frac{1}{2}\int_0^L JG\left(\frac{\partial\theta}{\partial x}\right)^2 dx + \frac{1}{2}k_{t1}\left[\frac{\partial\theta}{\partial x}(0,t)\right]^2 + \frac{1}{2}k_{t2}\left[\frac{\partial\theta}{\partial x}(L,t)\right]^2$$

Natural boundary conditions

$$J(0)G(0)\frac{\partial\theta}{\partial x}(0,t) - k_{t1}\theta(0,t) = I_1\frac{\partial^2\theta}{\partial t^2}(0,t)$$

$$-J(L)G(L)\frac{\partial\theta}{\partial x}(L,t) - k_{t2}\theta(L,t) = I_2\frac{\partial^2\theta}{\partial t^2}(L,t)$$

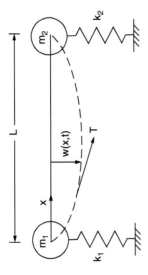

III. String $w(x,t)$ = transverse displacement

Parameters

$T(x)$ = tension

$\mu(x)$ = mass per length

L = length

k_1 = stiffness of spring attached at $x = 0$

k_2 = stiffness of spring attached at $x = L$

m_1 = mass of particle attached at $x = 0$

m_2 = mass of particle attached at $x = L$

Geometric boundary conditions

Fixed end at $x = 0$ $w(0,t) = 0$

Fixed end at $x = L$ $w(L,t) = 0$

Kinetic energy

$$T = \frac{1}{2} \int_0^L \mu \left(\frac{\partial w}{\partial t} \right)^2 dx + \frac{1}{2} m_1 \left[\frac{\partial w}{\partial t} (0,t) \right]^2 + \frac{1}{2} m_2 \left[\frac{\partial w}{\partial t} (L,t) \right]^2$$

Potential energy

$$V = \frac{1}{2} \int_0^L T \left(\frac{\partial u w}{\partial x} \right)^2 dx + \frac{1}{2} k_1 \left[\frac{\partial w}{\partial x} (0,t) \right]^2 + \frac{1}{2} k_2 \left[\frac{\partial w}{\partial x} (L,t) \right]^2$$

Natural boundary conditions

$$T(0) \frac{\partial w}{\partial x} (0,t) - k_1 w(0,t) = m_1 \frac{\partial^2 w}{\partial t^2} (0,t)$$

$$-T(L) \frac{\partial w}{\partial x} (L,t) - k_2 w(L,t) = m_2 \frac{\partial^2 w}{\partial t^2} (L,t)$$

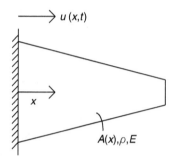

FIGURE 2.18 Elastic bar of Example 2.14.

Equation c becomes

$$\frac{1}{2}\int\limits_{t_1}^{t_2}\delta\left[\int\limits_{0}^{L}\rho A\left(\frac{\partial u}{\partial t}\right)^2 dx\right]dt \;=\; \frac{1}{2}\int\limits_{t_1}^{t_2}\left[\int\limits_{0}^{L}\rho A\delta\left(\frac{\partial u}{\partial t}\right)^2 dx\right]dt$$

$$=\frac{1}{2}\int\limits_{t_1}^{t_2}\left[\int\limits_{0}^{L}2\rho A\left(\frac{\partial u}{\partial t}\right)\delta\left(\frac{\partial u}{\partial t}\right)dx\right]dt \qquad (d)$$

$$=\frac{1}{2}\int\limits_{t_1}^{t_2}\left[\int\limits_{0}^{L}2\rho A\left(\frac{\partial u}{\partial t}\right)\frac{\partial}{\partial t}(\delta u)dx\right]dt$$

The order of integration in Equation d is interchanged. Integration by parts $\int f\,dg = fg - \int g\,df$ is applied to the integral over time with $f = \dfrac{\partial u}{\partial t}$, and $dg = \dfrac{\partial}{\partial t}(\delta u)dt$ gives

$$\frac{1}{2}\int\limits_{t_1}^{t_2}\delta\left[\int\limits_{0}^{L}\rho A\left(\frac{\partial u}{\partial t}\right)^2 dx\right]dt \;=\; \int\limits_{0}^{L}\rho A\left\{\int\limits_{t_1}^{t_2}\left(\frac{\partial u}{\partial t}\right)\left[\frac{\partial}{\partial t}(\delta u)\right]dt\right\}dx$$

$$(e)$$

$$=\int\limits_{0}^{L}\rho A\left\{\frac{\partial u}{\partial t}\delta u\Big|_{t=t_1}^{t=t_2} - \int\limits_{t_1}^{t_2}\left(\frac{\partial^2 u}{\partial t^2}\right)\delta u\,dt\right\}dx$$

The variation must vanish at t_1 and t_2. The order of integration may again be interchanged, leading to

$$\frac{1}{2}\int_{t_1}^{t_2}\delta\left[\int_0^L \rho A\left(\frac{\partial u}{\partial t}\right)^2 dx\right]dt = -\int_{t_1}^{t_2}\int_0^L \rho A\frac{\partial^2 u}{\partial t^2}\delta u\,dx\,dt \tag{f}$$

Now consider the potential energy term from Equation c

$$\frac{1}{2}\int_{t_1}^{t_2}\delta\left[\int_0^L EA\left(\frac{\partial u}{\partial x}\right)^2 dx\right]dt = \frac{1}{2}\int_{t_1}^{t_2}\int_0^L EA\,\delta\left(\frac{\partial u}{\partial x}\right)^2 dx\,dt$$

$$= \frac{1}{2}\int_{t_1}^{t_2}\int_0^L 2EA\left(\frac{\partial u}{\partial x}\right)\delta\left(\frac{\partial u}{\partial x}\right)dx\,dt \tag{g}$$

$$= \int_{t_1}^{t_2}\int_0^L EA\left(\frac{\partial u}{\partial x}\right)\frac{\partial}{\partial x}(\delta u)dx\,dt$$

Applying integration by parts to the inner integral of Equation g with $f = EA(\partial u/\partial x)$ and $dg = \frac{\partial}{\partial x}(\delta u)dx$ leads to

$$\frac{1}{2}\int_{t_1}^{t_2}\delta\left[\int_0^L EA\left(\frac{\partial u}{\partial x}\right)^2 dx\right]dt = \int_{t_1}^{t_2}\left\{\left[\left(EA\frac{\partial u}{\partial x}\right)(\delta u)\right]_{x=0}^{x=L}\right.$$

$$\left. -\int_0^L \frac{\partial}{\partial x}\left(EA\frac{\partial u}{\partial x}\right)\delta u\,dx\right\}dt$$

$$= \int_{t_1}^{t_2}\left\{E(L)A(L)\frac{\partial u}{\partial x}(L,t)\delta u(L,t)\right. \tag{h}$$

$$-E(0)A(0)\frac{\partial u}{\partial x}(0,t)\delta u(0,t)$$

$$\left. -\int_0^L \frac{\partial}{\partial x}\left(EA\frac{\partial u}{\partial x}\right)\delta u\,dx\right\}dt$$

Substitution of Equation f and Equation h into Equation c, and combining integrals, leads to

$$
-\int_{t_1}^{t_2}\int_0^L \rho A \frac{\partial^2 u}{\partial t^2}\,\delta u\,dx\,dt - \int_{t_1}^{t_2}\left\{ E(L)A(L)\frac{\partial u}{\partial x}(L)\delta u(L,t) \right.
$$

$$
\left. -E(0)A(0)\frac{\partial u}{\partial x}(0,t)\delta u(0,t) - \int_0^L \frac{\partial}{\partial x}\left(EA\frac{\partial u}{\partial x}\right)\delta u\,dx \right\}dt = 0
$$

(i)

$$
\int_{t_1}^{t_2}\left\{ -E(L)A(L)\frac{\partial u}{\partial x}(L,t)\delta u(L,t) + E(0)A(0)\frac{\partial u}{\partial x}(0,t)\delta u(0,t) \right.
$$

$$
\left. +\int_0^L \left[\frac{\partial}{\partial x}\left(EA\frac{\partial u}{\partial x}\right) - \rho A\frac{\partial^2 u}{\partial t^2}\right]\delta u\,dx \right\}dt = 0
$$

The DuBois–Reymond Lemma implies that the integral can vanish identically only if

$$
\frac{\partial}{\partial x}\left(EA\frac{\partial u}{\partial x}\right) - \rho A\frac{\partial^2 u}{\partial t^2} = 0
$$

(j)

as well as

$$
E(0)A(0)\frac{\partial u}{\partial x}(0,t)\delta u(0,t) = 0
$$

(k)

and

$$
E(L)A(L)\frac{\partial u}{\partial x}(L,t)\delta u(L,t) = 0
$$

(l)

Equation k and Equation l must both be satisfied in order for the left-hand side of Equation i to be identically zero. Equation k is satisfied if either $\delta u(0,t) = 0$ or $\frac{\partial u}{\partial x}(0,t) = 0$. The variation of the longitudinal displacement must satisfy any geometric constraints that the longitudinal displacement itself must satisfy. Since the left end of the bar is fixed $u(0,t) = 0$, Equation k is satisfied since $\delta u(0,t) = 0$.

Equation l is satisfied if either $\delta u(L,t) = 0$ or $\dfrac{\partial u}{\partial x}(L,t) = 0$. However, since the bar is free at $x = L$, $\delta u(L)$ cannot be prescribed. Thus, Equation l is satisfied only if $\dfrac{\partial u}{\partial x}(L,t) = 0$. (It is assumed that neither $E(L) = 0$ or $A(0) = 0$.)

Equation k and Equation l lead to boundary conditions, which must be satisfied by the longitudinal displacement. The mathematical problem for $u(x,t)$ is Equation j supplemented by

$$u(0,t) = 0 \tag{m}$$

$$\frac{\partial u}{\partial x}(L,t) = 0 \tag{n}$$

Example 2.14 shows that application of Hamilton's Principle for a continuous system leads to formulation of the partial differential equation governing the motion of the system, as well as specification of the boundary conditions that must be satisfied. For this second-order system, a pair of potential conditions is specified at each boundary. One or the other must be satisfied. The variation must satisfy all geometric constraints on the system response. A boundary condition specified by geometry is called a geometric boundary condition. Otherwise the boundary condition is called a natural boundary condition. It is so called, perhaps because it naturally occurs from the variational formulation, or perhaps because it can also be derived from application of basic laws of nature to the boundary of the region. The boundary condition $\dfrac{\partial u}{\partial x}(L,t) = 0$ is also a statement that there is no stress at the end of bar, $x = L$.

Example 2.15. Use Hamilton's Principle or Extended Hamilton's Principle to determine the boundary conditions that must be satisfied at the end $x = L$ of a uniform bar fixed at $x = 0$ for each of the following:

1. A discrete spring of stiffness k is attached to the bar at $x = L$, as in Figure 2.19a.
2. A particle of mass m is attached to the bar at $x = L$, as in Figure 2.19b.
3. A concentrated load P is applied to the free end of the bar, as in Figure 2.19c.
4. A spring of stiffness k in parallel with a viscous damper of damping coefficient c is attached to the bar at $x = L$, as illustrated in Figure 2.19d.

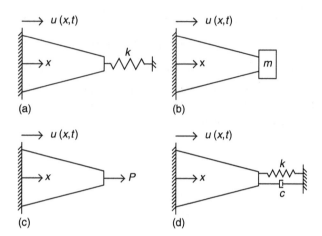

FIGURE 2.19 Systems of Example 2.15. (a) Potential energy functional is modified due to spring. (b) Kinetic energy functional is modified due to mass. (c) Applied load leads to non-conservative work term. (d) Viscous damper dissipates energy.

Solution: In each case, many of the steps used in the solution of Example 2.14 are used in application of Hamilton's Principle. Only the differences are specified in this example.

a. The kinetic energy of the system is the same as that of Equation a of Example 2.14, but the potential energy is modified to account for the discrete spring

$$V = \frac{1}{2} \int_0^L EA \left(\frac{\partial u}{\partial x} \right)^2 dx + \frac{1}{2} k[u(L,t)]^2 \qquad \text{(a)}$$

Application of Hamilton's Principle leads to

$$\frac{1}{2} \int_{t_1}^{t_2} \left[\delta \int_0^L \rho A \left(\frac{\partial u}{\partial t} \right)^2 dx - \delta \int_0^L EA \left(\frac{\partial u}{\partial x} \right)^2 dx - \delta k[u(L,t)]^2 \right] dt = 0 \quad \text{(b)}$$

The simplification of the first two terms in the integrand of the outer integral of Equation b is accomplished in Example 2.14. The variation of the potential energy term due to the discrete spring is

$$\frac{1}{2} \int_{t_1}^{t_2} k \delta [u(L,t)]^2 dt = \frac{1}{2} \int_{t_1}^{t_2} 2ku(L,t) \delta u(L,t) dt \qquad \text{(c)}$$

The algebra used in Example 2.14 is followed, modified by the inclusion of Equation c, until Equation i becomes

$$\int_{t_1}^{t_2} \left\{ -\left[E(L)A(L)\frac{\partial u}{\partial x}(L,t) + ku(L,t) \right] \delta u(L,t) \right.$$

$$+ E(0)A(0)\frac{\partial u}{\partial x}(0,t)\delta u(0,t) + \int_0^L \left[\frac{\partial}{\partial x}\left(EA\frac{\partial u}{\partial x} \right) \right.$$

$$\left. \left. - \rho A\frac{\partial^2 u}{\partial t^2} \right] \delta u\, dx \right\} dt = 0 \tag{d}$$

Application of the DuBois–Reymond Lemma to Equation d leads to the same differential equation and boundary condition at $x = 0$ as Example 2.14. The boundary condition at $x = L$, modified to include the effect of a discrete spring is

$$E(L)A(L)\frac{\partial u}{\partial x}(L,t) + ku(L,t) = 0 \tag{e}$$

b. The kinetic energy of the system with a discrete mass at the end of the bar leads to a kinetic energy of the form

$$T = \frac{1}{2}\int_0^L \rho A\left(\frac{\partial u}{\partial t} \right)^2 dx + \frac{1}{2}m\left(\frac{\partial u}{\partial t}(L,t) \right)^2 \tag{f}$$

The potential energy of the system is Equation a of Example 2.14, unaffected by the discrete mass. Application of Hamilton's Principle leads to

$$\frac{1}{2}\int_{t_1}^{t_2} \left[\delta \int_0^L \rho A\left(\frac{\partial u}{\partial t} \right)^2 dx + m\delta\left[\left(\frac{\partial u}{\partial t}(L,t) \right)^2 \right] \right.$$

$$\left. - \delta \int_0^L EA\left(\frac{\partial u}{\partial x} \right)^2 dx \right] dt = 0 \tag{g}$$

The variation of the kinetic energy term due to the discrete mass is

$$\frac{1}{2}\int_{t_1}^{t_2} m\delta\left[\left(\frac{\partial u}{\partial t}(L,t)\right)^2\right]dt = \frac{1}{2}\int_{t_1}^{t_2} 2m\left(\frac{\partial u}{\partial t}(L,t)\right)\delta\left(\frac{\partial u}{\partial t}(L,t)\right)dt$$

$$= \frac{1}{2}\int_{t_1}^{t_2} 2m\left(\frac{\partial u}{\partial t}(L,t)\right)\left[\frac{\partial}{\partial t}(\delta u(L,t))\right]dt$$

(h)

Applying integration by parts to Equation h with $f = \dfrac{\partial u}{\partial t}(L,t)$ and $dg = \left[\dfrac{\partial}{\partial t}(\delta u(L,t))\right]dt$ leads to

$$\frac{1}{2}\int_{t_1}^{t_2} m\delta\left[\left(\frac{\partial u}{\partial t}(L,t)\right)^2\right]dt$$

$$= m\frac{\partial u}{\partial t}(L,t)\delta u(L,t)\Big|_{t=t_1}^{t=t_2} - \int_{t_1}^{t_2} m\frac{\partial^2 u}{\partial t^2}(L,t)\delta u(L,t)dt$$

(i)

The variations are defined such that $\delta u(L,t_1) = \delta u(L,t_2) = 0$. Equation i becomes

$$\frac{1}{2}\int_{t_1}^{t_2} m\delta\left[\left(\frac{\partial u}{\partial t}(L,t)\right)^2\right]dt = -\int_{t_1}^{t_2} m\frac{\partial^2 u}{\partial t^2}(L,t)\delta u(L,t)dt$$

(j)

Use of Equation j in Equation g leads to

$$\int_{t_1}^{t_2}\left\{-\left[E(L)A(L)\frac{\partial u}{\partial x}(L,t) - m\frac{\partial^2 u}{\partial t^2}(L,t)\right]\delta u(L,t)\right.$$

$$\left. + E(0)A(0)\frac{\partial u}{\partial x}(0,t)\delta u(0,t)\int_0^L\left[\frac{\partial}{\partial x}\left(EA\frac{\partial u}{\partial x}\right) - \rho A\frac{\partial^2 u}{\partial t^2}\right]\delta u\, dx\right\}dt = 0$$

(k)

The appropriate boundary condition at $x = L$ is obtained from Equation k as

$$E(L)A(L)\frac{\partial u}{\partial x}(L,t) - m\frac{\partial^2 u}{\partial t^2}(L,t) = 0 \tag{l}$$

c. The concentrated load at the end of the bar is a non-conservative force. The virtual work of the force is

$$\delta W_{nc} = P\delta u(L,t) \tag{m}$$

Extended Hamilton's Principle, Equation 2.69, is applied with the kinetic and potential energies of Equation a and Equation b of Example 2.14. The result is

$$\frac{1}{2}\int_{t_1}^{t_2}\left[\delta\int_0^L \rho A\left(\frac{\partial u}{\partial t}\right)^2 dx - \delta\int_0^L EA\left(\frac{\partial u}{\partial x}\right)^2 dx + P\delta u(L,t)\right]dt = 0 \tag{n}$$

Using the procedures of Example 2.14, Equation n becomes

$$\int_{t_1}^{t_2}\left\{-\left[E(L)A(L)\frac{\partial u}{\partial x}(L,t) - P\right]\delta u(L,t)\right.$$
$$\left. + E(0)A(0)\frac{\partial u}{\partial x}(0,t)\delta u(0,t)\int_0^L\left[\frac{\partial}{\partial x}\left(EA\frac{\partial u}{\partial x}\right) - \rho A\frac{\partial^2 u}{\partial t^2}\right]\delta u\,dx\right\}dt = 0 \tag{o}$$

The appropriate boundary condition at $x = L$ is obtained from Equation o as

$$E(L)A(L)\frac{\partial u}{\partial x}(L,t) - P = 0 \tag{p}$$

d. The potential energy is that of Equation a of this example. The viscous damping force is non-conservative. The virtual work of the damping force when a variation is introduced is

$$\delta W_{nc} = -c\frac{\partial u}{\partial t}(L,t)\delta u(L,t) \tag{q}$$

Substitution of Equation a and Equation q into Extended Hamilton's Principle leads to

$$\frac{1}{2} \int_{t_1}^{t_2} \left[\delta \int_0^L \rho A \left(\frac{\partial u}{\partial t} \right)^2 dx - \delta \int_0^L EA \left(\frac{\partial u}{\partial x} \right)^2 dx - \delta k [u(L,t)]^2 \right.$$

$$\left. - c \frac{\partial u}{\partial t}(L,t)\delta u(L,t) \right] dt = 0 \qquad (r)$$

The algebraic methods of this example and Example 2.14 are used to rearrange Equation r into Equation d, but with a modification to account for the viscous damping term

$$\int_{t_1}^{t_2} \left\{ - \left[E(L)A(L) \frac{\partial u}{\partial x}(L,t) + ku(L,t) + c \frac{\partial u}{\partial t}(L,t) \right] \delta u(L,t) \right.$$

$$\left. + E(0)A(0) \frac{\partial u}{\partial x}(0,t)\delta u(0,t) \int_0^L \left[\frac{\partial}{\partial x} \left(EA \frac{\partial u}{\partial x} \right) - \rho A \frac{\partial^2 u}{\partial t^2} \right] \delta u\, dx \right\} dt = 0 \qquad (s)$$

The appropriate boundary condition at $x = L$ is obtained from Equation s as

$$E(L)A(L)\frac{\partial u}{\partial x}(L,t) + ku(L,t) + c \frac{\partial u}{\partial t}(L,t) = 0 \qquad (t)$$

2.11 EULER–BERNOULLI BEAMS

The Euler–Bernoulli beam is a model that includes the effect of bending (flexure), but ignores shear distortion and rotary inertia. The beam is made of an elastic material and stresses are within the elastic limit. Plane sections are assumed to remain plane and the curvature of the beam is assumed to be small.

Example 2.16. Use Hamilton's Principle to derive the governing problem for the transverse displacement $w(x,t)$ of an Euler–Bernoulli beam, including the required boundary conditions. Specify the boundary conditions for (a) a fixed-free beam, and (b) a fixed-pinned beam.

Solution: The potential energy of an Euler–Bernoulli beam is derived as Equation g of Example 1.5

$$V = \frac{1}{2} \int_0^L EI \left(\frac{\partial^2 w}{\partial x^2} \right)^2 dx \qquad (a)$$

The kinetic energy of the differential element of Figure 2.20 is

$$dT = \frac{1}{2} \left(\frac{\partial w}{\partial t} \right)^2 dm = \frac{1}{2} \rho A \left(\frac{\partial w}{\partial t} \right)^2 dx \qquad (b)$$

The total kinetic energy of the beam is

$$T = \frac{1}{2} \int_0^L \rho A \left(\frac{\partial w}{\partial t} \right)^2 dx \qquad (c)$$

Application of Hamilton's Principle leads to

$$\frac{1}{2} \int_{t_1}^{t_1} \left[\delta \int_0^L \rho A \left(\frac{\partial w}{\partial t} \right)^2 dx - \delta \int_0^L EI \left(\frac{\partial^2 w}{\partial x^2} \right) dx \right] dt = 0 \qquad (d)$$

The kinetic energy term is the same as for the motion of the longitudinal bar in Example 2.14. Using the same analysis it is shown that

$$\frac{1}{2} \int_{t_1}^{t_2} \delta \left[\int_0^L \rho A \left(\frac{\partial w}{\partial t} \right)^2 dx \right] dt = - \int_{t_1}^{t_2} \int_0^L \rho A \frac{\partial^2 w}{\partial t^2} \delta w \, dx \, dt \qquad (e)$$

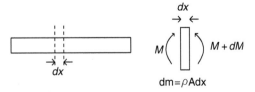

FIGURE 2.20 Differential element of Euler–Bernoulli beam.

Now consider

$$\frac{1}{2}\int_{t_1}^{t_2}\left(\delta\int_0^L EI\left(\frac{\partial^2 w}{\partial x^2}\right)^2 dx\right)dt = \frac{1}{2}\int_{t_1}^{t_2}\left\{\int_0^L EI\left[\delta\left(\frac{\partial^2 w}{\partial x^2}\right)^2\right]dx\right\}dt$$

$$= \int_{t_1}^{t_2}\int_0^L EI\left(\frac{\partial^2 w}{\partial x^2}\right)\delta\left(\frac{\partial^2 w}{\partial x^2}\right)dx\,dt \qquad (f)$$

$$= \int_{t_1}^{t_2}\int_0^L EI\left(\frac{\partial^2 w}{\partial x^2}\right)\left(\frac{\partial^2 \delta w}{\partial x^2}\right)dx\,dt$$

Application of integration by parts to the inner integral of the right-hand side of Equation f with $f = EI\dfrac{\partial^2 w}{\partial x^2}$ and $dg = \dfrac{\partial^2 \delta w}{\partial x^2}dx$ leads to

$$\frac{1}{2}\int_{t_1}^{t_2}\left(\delta\int_0^L EI\left(\frac{\partial^2 w}{\partial x^2}\right)^2 dx\right)dt = \int_{t_1}^{t_2}\left\{\left[\left(EI\frac{\partial^2 w}{\partial x^2}\right)\left(\delta\frac{\partial w}{\partial x}\right)\right]_{x=0}^{x=L}\right.$$

$$\left. -\int_0^L \frac{\partial}{\partial x}\left(EI\frac{\partial^2 w}{\partial x^2}\right)\left(\frac{\partial \delta w}{\partial x}\right)dx\right\}dt \qquad (g)$$

Application of integration by parts to the spatial integral on the right-hand side of Equation g, with $f = \dfrac{\partial}{\partial x}\left(EI\dfrac{\partial^2 w}{\partial x^2}\right)$ and $dg = \dfrac{\partial^2 \delta w}{\partial x}dx$, leads to

$$\frac{1}{2}\int_{t_1}^{t_2}\left(\delta\int_0^L EI\left(\frac{\partial^2 w}{\partial x^2}\right)^2 dx\right)dt = \int_{t_1}^{t_2}\left\{\left[\left(EI\frac{\partial^2 w}{\partial x^2}\right)\left(\delta\frac{\partial w}{\partial x}\right)\right]_{x=0}^{x=L}\right.$$

$$-\left[\left(\frac{\partial}{\partial x}\left(EI\frac{\partial^2 w}{\partial x^2}\right)\right)\delta w\right]_{x=0}^{x=L} \qquad (h)$$

$$\left. +\int_0^L \frac{\partial^2}{\partial x^2}\left(EI\frac{\partial^2 w}{\partial x^2}\right)\delta w\,dx\right\}dt$$

Substituting into Hamilton's Principle leads to

$$\int_{t_1}^{t_2} \left\{ \left[\left(E(L)I(L) \frac{\partial^2 w}{\partial x^2}(L,t) \right) \left(\delta \frac{\partial w}{\partial x}(L,t) \right) \right] \right.$$

$$- \left[\left(E(0)I(0) \frac{\partial^2 w}{\partial x^2}(0,t) \right) \left(\delta \frac{\partial w}{\partial x}(0,t) \right) \right] - \left[\left(\frac{\partial}{\partial x} \left(EI \frac{\partial^2 w}{\partial x^2} \right) \right) (L,t) \delta w(L,t) \right]$$

$$+ \left. \left[\left(\frac{\partial}{\partial x} \left(EI \frac{\partial^2 w}{\partial x^2} \right) \right) (0,t) \delta w(0,t) \right] - \int_0^L \left[\rho A \frac{\partial^2 w}{\partial t^2} + \frac{\partial^2}{\partial x^2} \left(EI \frac{\partial^2 w}{\partial x^2} \right) \right] \delta w \, dx \right\}$$

$$\times \, dt = 0$$

(i)

Application of the DuBois–Reymond Lemma to Equation i leads to

$$\rho A \frac{\partial^2 w}{\partial t^2} + \frac{\partial^2}{\partial x^2} \left(EI \frac{\partial^2 w}{\partial x^2} \right) = 0 \tag{j}$$

The appropriate boundary conditions are

$$\left(\frac{\partial}{\partial x} \left(EI \frac{\partial^2 w}{\partial x^2} \right) \right)(0,t) = 0 \quad \text{or} \quad \delta w(0,t) = 0 \tag{k}$$

$$\left(\frac{\partial}{\partial x} \left(EI \frac{\partial^2 w}{\partial x^2} \right) \right)(L,t) = 0 \quad \text{or} \quad \delta w(L,t) = 0 \tag{l}$$

$$EI \frac{\partial^2 w}{\partial x^2}(0,t) = 0 \quad \text{or} \quad \delta \left(\frac{\partial w}{\partial x} \right)(0,t) = 0 \tag{m}$$

$$EI \frac{\partial^2 w}{\partial x^2}(L,t) = 0 \quad \text{or} \quad \delta \left(\frac{\partial w}{\partial x} \right)(L,t) = 0 \tag{n}$$

Equation j is called the beam equation. Two boundary conditions are necessary at each end. The possible geometric boundary conditions are specification of w and $\partial w / \partial x$ at an end. The possible natural boundary conditions are specification of $EI \frac{\partial^2 w}{\partial x^2}$ and $\frac{\partial}{\partial x} \left(EI \frac{\partial^2 w}{\partial x^2} \right)$ at each end.

(a) The transverse displacement of a fixed-free beam satisfies two geometric boundary conditions at its fixed end

$$w(0,t) = 0 \tag{o}$$

$$\frac{\partial w}{\partial x}(0,t) = 0 \tag{p}$$

Equation o specifies that the displacement of a fixed end is zero, while Equation p specifies that the slope of the deflection curve is zero at the fixed end.

Since there are no geometric constraints at its free end, the transverse displacement must satisfy the natural boundary conditions of Equation 1 and Equation n

$$\frac{\partial}{\partial x}\left(EI\frac{\partial^2 w}{\partial x^2}\right)(L,t) = 0 \tag{q}$$

$$EI\frac{\partial^2 w}{\partial x^2}(L,t) = 0 \tag{r}$$

Equation q is a statement that there is no vertical shear stress at the free end, while Equation r is a statement that the bending stress due to bending is zero at the free end.

(b) The geometric boundary conditions at the fixed end of a fixed-pinned beam are the same as those for the fixed-free beam, Equation o and Equation p. The pinned end is restricted from transverse displacement, thus it must satisfy the geometric condition

$$w(L,t) = 0 \tag{s}$$

The deflection does not satisfy a second geometric condition at $x = L$. Thus, it must satisfy the natural boundary condition of Equation n

$$EI\frac{\partial^2 w}{\partial x^2}(L,t) = 0 \tag{t}$$

Example 2.17. The boundary value problem for the transverse displacement of the Euler–Bernoulli beam is the beam equation of Equation j of Example 2.16, subject to geometric and natural boundary conditions of Equation k through Equation n. Each of the following is a problem that is encountered later in this study. Use Hamilton's Principle or Extended Hamilton's Principle to determine how the Euler–Bernoulli beam model presented above is modified for each problem. For each of the following, assume the following variations are known

$$\frac{1}{2}\int\limits_{t_1}^{t_2}\delta\left[\int\limits_0^L \rho A\left(\frac{\partial w}{\partial t}\right)^2 dx\right]dt = -\int\limits_{t_1}^{t_2}\int\limits_0^L \rho A\frac{\partial^2 w}{\partial t^2}\delta w\,dx\,dt \qquad\text{(a)}$$

$$\frac{1}{2}\int\limits_{t_1}^{t_2}\left(\delta\int\limits_0^L EI\left(\frac{\partial^2 w}{\partial x^2}\right)^2 dx\right) = \int\limits_{t_1}^{t_2}\left\{\left[\left(EI\frac{\partial^2 w}{\partial x^2}\right)\left(\delta\frac{\partial w}{\partial x}\right)\right]_{x=0}^{x=L}\right.$$

$$-\left[\left(\frac{\partial}{\partial x}\left(EI\frac{\partial^2 w}{\partial x^2}\right)\right)\delta w\right]_{x=0}^{x=L} \qquad\text{(b)}$$

$$\left.+\int\limits_0^L \frac{\partial^2}{\partial x^2}\left(EI\frac{\partial^2 w}{\partial x^2}\right)\delta w\,dx\right\}dt$$

   a. A time-dependent force per unit length of $f(x,t)$ (Figure 2.21a).
   b. A uniform axial load P. Assume longitudinal displacement is small (Figure 2.21b).

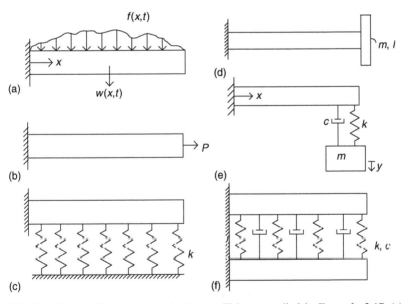

FIGURE 2.21 Modifications to Euler–Bernoulli beam studied in Example 2.17. (a) Force per unit length is non-conservative. (b) Axial load modifies potential energy functional. (c) Elastic foundation modifies potential energy functional. (d) Discrete inertia element attached to end of beam. (e) Discrete mass-spring-viscous damper system adds an extra degree-of-freedom. (f) Viscoelastic layer between beams modifies potential energy functional and leads to energy dissipation.

    c. The beam resting on an elastic foundation of stiffness per unit length
       k (Figure 2.21c).
    d. A body of mass m and moment of inertia I about the beam's neutral
       axis is attached to the end of the beam (Figure 2.21d).
    e. A particle of mass m suspended from the beam by a spring of stiffness
       k and viscous damper of damping coefficient c (Figure 2.21e).
    f. Two beams connected by a viscoelastic layer, as illustrated in
       Figure 2.21f. The layer is made of a material with a stiffness per
       unit length of k and a damping coefficient per unit length of c.

Solution: (a) The kinetic and potential energy functionals are the same as
Equation a and Equation c of Example 2.16. The external loading is a
nonconservative force, thus application of Extended Hamilton's Principle is
required. The total force acting on an element of the beam of thickness dx is
$f(x,t)d.x$ The virtual work of the force is

$$d(\delta W_{nc}) = f(x,t)dx\delta w \qquad (c)$$

The virtual work of the total distributed load is

$$\delta W_{nc} = \int d(\delta W_{nc}) = \int_0^L f(x,t)\delta w \, dx \qquad (d)$$

    Substitution into Extended Hamilton's Principle using Equation a and
Equation b leads to

$$\int_{t_1}^{t_2} [\delta T - \delta V + \delta W_{nc}] dt = 0$$

$$\int_{t_1}^{t_2} \left\{ -\int_0^L \rho A \frac{\partial^2 w}{\partial t^2} \delta w \, dx - \left[\left(EI \frac{\partial^2 w}{\partial x^2} \right) \left(\delta \frac{\partial w}{\partial x} \right) \right]_{x=0}^{x=L} \right.$$

$$+ \left[\left(\frac{\partial}{\partial x} \left(EI \frac{\partial^2 w}{\partial x^2} \right) \right) \delta w \right]_{x=0}^{x=L} - \int_0^L \frac{\partial^2}{\partial x^2} \left(EI \frac{\partial^2 w}{\partial x^2} \right) \delta w \, dx$$

$$\left. + \int_0^L f(x,t)\delta w \, dx \right\} dt = 0 \qquad (e)$$

$$\int_{t_1}^{t_2} \left\{ \left[\left(EI \frac{\partial^2 w}{\partial x^2} \right) \left(\delta \frac{\partial w}{\partial x} \right) \right]_{x=0}^{x=L} - \left[\left(\frac{\partial}{\partial x} \left(EI \frac{\partial^2 w}{\partial x^2} \right) \right) \delta w \right]_{x=0}^{x=L} \right.$$

$$\left. - \int_0^L \left[\rho A \frac{\partial^2 w}{\partial t^2} + \frac{\partial^2}{\partial x^2} \left(EI \frac{\partial^2 w}{\partial x^2} \right) - f(x,t) \right] \delta w \right\} dt = 0$$

Application of the DuBois–Reymond lemma to Equation e leads to the partial differential equation

$$\rho A \frac{\partial^2 w}{\partial t^2} + \frac{\partial^2}{\partial x^2}\left(EI\frac{\partial^2 w}{\partial x^2}\right) - f(x,t) = 0 \tag{f}$$

The boundary conditions are unaffected by the distributed load. The potential boundary conditions are those of Equation k through Equation n of Example 2.10.

(b) The slope of the elastic curve at any point is $\dfrac{\partial w}{\partial x}$. The component of the axial load in the transverse direction is $P\dfrac{\partial w}{\partial x}$. The virtual work done by the axial load within a differential segment of length dx is

$$d(\delta W) = -P\frac{\partial w}{\partial x}\delta\left(\frac{\partial w}{\partial x}\right) \tag{g}$$

The virtual work is negative, as the axial load is acting against the direction of the rotation. The total virtual work is

$$\delta W = -\int_0^L P\frac{\partial w}{\partial x}\delta\left(\frac{\partial w}{\partial x}\right)dx \tag{h}$$

The axial load does lead to stored strain energy in the beam. An alternate formulation is to calculate the change in length of a differential segment and express the strain energy in terms of this change in length.

When Hamilton's Principle is applied, the additional term resulting from the axial load is

$$\int_{t_1}^{t_2} \delta W\,dt = -\int_{t_1}^{t_2}\int_0^L P\frac{\partial w}{\partial x}\delta\left(\frac{\partial w}{\partial x}\right)dx\,dt \tag{i}$$

Integration by parts with $f = P(\partial w/\partial x)$ and $dg = (\partial w/\partial x)\delta w\,dx$ leads to

$$-\int_{t_1}^{t_2}\int_0^L P\frac{\partial w}{\partial x}\delta\left(\frac{\partial w}{\partial x}\right)dx\,dt = \int_{t_1}^{t_2}\left\{-P\frac{\partial w}{\partial x}\delta w\Big|_0^L + \int_0^L \frac{\partial}{\partial x}\left(P\frac{\partial w}{\partial x}\right)\delta w\,dx\right\}dt \tag{j}$$

Substitution into Hamilton's Principle leads to

$$-\int_{t_1}^{t_2}\int_0^L \rho A \frac{\partial^2 w}{\partial t^2} \delta w \, dx \, dt - \int_{t_1}^{t_2}\left\{\left[\left(EI\frac{\partial^2 w}{\partial x^2}\right)\left(\delta\frac{\partial w}{\partial x}\right)\right]_{x=0}^{x=L}\right.$$

$$-\left[\left(\frac{\partial}{\partial x}\left(EI\frac{\partial^2 w}{\partial x^2}\right)\right)\delta w\right]_{x=0}^{x=L} + \int_0^L \frac{\partial^2}{\partial x^2}\left(EI\frac{\partial^2 w}{\partial x^2}\right)\delta w \, dx \left.\right\} dt$$

$$+\int_{t_1}^{t_2}\left\{-P\frac{\partial w}{\partial x}\delta w\Big|_0^L + \int_0^L \frac{\partial}{\partial x}\left(P\frac{\partial w}{\partial x}\right)\delta w \, dx\right\} dt = 0$$

(k)

$$\int_{t_1}^{t_2}\left\{\left[\left(EI\frac{\partial^2 w}{\partial x^2}\right)\left(\delta\frac{\partial w}{\partial x}\right) - \left(\frac{\partial}{\partial x}\left(EI\frac{\partial^2 w}{\partial x^2}\right)\right)\delta w + P\frac{\partial w}{\partial x}\delta w\right]_{x=0}\right.$$

$$-\left[\left(EI\frac{\partial^2 w}{\partial x^2}\right)\left(\delta\frac{\partial w}{\partial x}\right) - \left(\frac{\partial}{\partial x}\left(EI\frac{\partial^2 w}{\partial x^2}\right)\right)\delta w + P\frac{\partial w}{\partial x}\delta w\right]_{x=L}$$

$$+\int_0^L \left[-\rho A\frac{\partial^2 w}{\partial t^2} - \frac{\partial^2}{\partial x^2}\left(EI\frac{\partial^2 w}{\partial x^2}\right) + \frac{\partial}{\partial x}\left(P\frac{\partial w}{\partial x}\right)\right]\delta w \, dx \left.\right\} dt = 0$$

Application of the DuBois–Reymond lemma to Equation k leads to the following differential equation

$$\rho A\frac{\partial^2 w}{\partial t^2} + \frac{\partial^2}{\partial x^2}\left(EI\frac{\partial^2 w}{\partial x^2}\right) - \frac{\partial}{\partial x}\left(P\frac{\partial w}{\partial x}\right) = 0 \tag{l}$$

The possible boundary conditions for the system are

$$EI\frac{\partial^2 w}{\partial x^2}(0,t) = 0 \quad \text{or} \quad \frac{\partial w}{\partial x}(0,t) = 0 \tag{m}$$

$$\frac{\partial}{\partial x}\left(EI\frac{\partial^2 w}{\partial x^2}\right)(0,t) + P\frac{\partial w}{\partial x}(0,t) = 0 \quad \text{or} \quad w(0,t) = 0 \tag{n}$$

$$EI\frac{\partial^2 w}{\partial x^2}(L,t) = 0 \quad \text{or} \quad \frac{\partial w}{\partial x}(L,t) = 0 \tag{o}$$

$$\frac{\partial}{\partial x}\left(EI\frac{\partial^2 w}{\partial x^2}\right)(L,t) + P\frac{\partial w}{\partial x}(L,t) = 0 \quad \text{or} \quad w(L,t) = 0 \tag{p}$$

The boundary conditions appropriate for a fixed-fixed beam are

$$w(0,t) = 0 \tag{q}$$

$$\frac{\partial w}{\partial x}(0,t) = 0 \tag{r}$$

$$w(L,t) = 0 \tag{s}$$

$$\frac{\partial w}{\partial x}(L,t) = 0 \tag{t}$$

The boundary conditions appropriate for a fixed-free beam are

$$w(0,t) = 0 \tag{u}$$

$$\frac{\partial w}{\partial x}(0,t) = 0 \tag{v}$$

$$\frac{\partial^2 w}{\partial x^2}(L,t) = 0 \tag{w}$$

$$\frac{\partial}{\partial x}\left(EI\frac{\partial^2 w}{\partial x^2}\right)(L,t) + P\frac{\partial w}{\partial x}(L,t) = 0 \tag{x}$$

(c) The potential energy of the elastic foundation is

$$V = \frac{1}{2}\int_0^L kw^2 dx \tag{y}$$

Then

$$\int_{t_1}^{t_2}\int_0^L \delta\left(\frac{1}{2}kw^2\right)dx\,dt = \int_{t_1}^{t_2}\int_0^L kw\delta w\,dx\,dt \tag{z}$$

Substitution into Equation c of Example 2.16 and using the Dubois–Reymond lemma leads to

$$EI\frac{\partial^4 w}{\partial x^4} + kw + \rho A\frac{\partial^2 w}{\partial t^2} = 0 \tag{aa}$$

(d) The kinetic energy of the discrete inertia element at the end of the beam is

$$T = \frac{1}{2}mv^2 + \frac{1}{2}I_p\omega^2 \tag{bb}$$

where the velocity of the body is

$$v = \frac{\partial w}{\partial t}(L,t) \tag{cc}$$

and its angular velocity due to the rotation of the beam is

$$\omega = \frac{\partial}{\partial t}\left(\frac{\partial w}{\partial x}\right)(L,t) \tag{dd}$$

Thus, the kinetic energy of the body is

$$T = \frac{1}{2}m\left(\frac{\partial w}{\partial t}(L,t)\right)^2 + \frac{1}{2}I_p\left(\frac{\partial^2 w}{\partial x\partial t}(L,t)\right)^2 \tag{ee}$$

The presence of the discrete body at the end of the beam leads to an additional term in the application of Hamilton's principle:

$$\int_{t_1}^{t_2}\delta\left[\frac{1}{2}m\left(\frac{\partial w}{\partial t}(L,t)\right)^2 + \frac{1}{2}I_p\left(\frac{\partial^2 w}{\partial x\partial t}(L,t)\right)^2\right]dt$$

$$= \int_{t_1}^{t_2}\left[m\frac{\partial w}{\partial t}(L,t)\delta\left(\frac{\partial w}{\partial t}(L,t)\right) + I_p\frac{\partial^2 w}{\partial x\partial t}(L,t)\delta\left(\frac{\partial^2 w}{\partial x\partial t}(L,t)\right)\right]dt \tag{ff}$$

Application of ntegration by parts, noting that $\delta w(L,t_1) = \delta w(L,t_2) = 0$, leads to

$$\int_{t_1}^{t_2} \left[m\frac{\partial w}{\partial t}(L,t)\delta\left(\frac{\partial w}{\partial t}(L,t)\right) \right] dt = m\frac{\partial w}{\partial t}(L,t)\delta w(L,t)\Big|_{t=t_1}^{t=t_2}$$

$$-\int_{t_1}^{t_2} m\frac{\partial^2 w}{\partial t^2}(L,t)\delta w(L,t)dt \qquad (gg)$$

$$= -\int_{t_1}^{t_2} m\frac{\partial^2 w}{\partial t^2}(L,t)\delta w(L,t)dt$$

and noting that $\delta\left(\frac{\partial w}{\partial t}(L,t_1)\right) = \delta\left(\frac{\partial w}{\partial t}(L,t_2)\right) = 0$

$$\int_{t_1}^{t_2} \left[I_p \frac{\partial^2 w}{\partial x\partial t}(L,t)\delta\left(\frac{\partial^2 w}{\partial x\partial t}(L,t)\right) \right] dt$$

$$= I_p \frac{\partial^2 w}{\partial x\partial t}\delta\left(\frac{\partial w}{\partial x}(L,t)\right)\Big|_{t=t_1}^{t=t_2} - \int_{t_1}^{t_2} I_p \frac{\partial^3 w}{\partial x\partial t^2}\delta\left(\frac{\partial w}{\partial x}(L,t)\right)dt$$

$$= -\int_{t_1}^{t_2} I_p \frac{\partial^3 w}{\partial x\partial t^2}\delta\left(\frac{\partial w}{\partial x}(L,t)\right)dt \qquad (hh)$$

Hamilton's Principle becomes

$$\int_{t_1}^{t_2} \left\{ -\left[\left(EI\frac{\partial^2 w}{\partial x^2}\right)\left(\delta\frac{\partial w}{\partial x}\right)\right]_{x=0}^{x=L} + \left[\left(\frac{\partial}{\partial x}\left(EI\frac{\partial^2 w}{\partial x^2}\right)\right)\delta w\right]_{x=0}^{x=L} \right.$$

$$-m\frac{\partial^2 w}{\partial t^2}(L,t)\delta w(L,t) - I_p\frac{\partial^3 w}{\partial x\partial t^2}(L,t)\delta\left(\frac{\partial w}{\partial x}(L,t)\right) \qquad (ii)$$

$$\left. -\int_0^L \left[\rho A\frac{\partial^2 w}{\partial t^2} + \frac{\partial^2}{\partial x^2}\left(EI\frac{\partial^2 w}{\partial x^2}\right)\right]\delta w \right\} dt = 0$$

The differential equation is that of the beam equation. The potential boundary conditions at $x = 0$ are given by Equation k and Equation m of Example 2.16. The boundary conditions at $x = L$ are of the form

$$\left[\frac{\partial}{\partial x}\left(EI\frac{\partial^2 w}{\partial x^2}(L,t)\right) - m\frac{\partial^2 w}{\partial t^2}(L,t)\right]\delta w(L,t) = 0 \qquad \text{(jj)}$$

$$\left[EI\frac{\partial^2 w}{\partial x^2}(L,t) + I_p\frac{\partial^3 w}{\partial x\partial t^2}(L,t)\right]\delta\left(\frac{\partial w}{\partial x}(L,t)\right) = 0 \qquad \text{(kk)}$$

(e) Let $y(t)$ be the displacement of the discrete mass-spring-viscous damper system. The total kinetic energy is the kinetic energy of the beam, plus the kinetic energy of the discrete mass

$$T = \frac{1}{2}\int_0^L \rho A\left(\frac{\partial w}{\partial t}\right)^2 dx + \frac{1}{2}m\left(\frac{dy}{dt}\right)^2 \qquad \text{(ll)}$$

The total potential energy of the system is

$$V = \frac{1}{2}\int_0^l EI\left(\frac{\partial^2 w}{\partial x^2}\right)^2 dx + \frac{1}{2}k[y - w(L,t)]^2 \qquad \text{(mm)}$$

The virtual work of the viscous damper is

$$\delta W_{nc} = -c\left(\dot{y} - \frac{\partial w}{\partial t}(L,t)\right)\delta y + c\left(\dot{y} - \frac{\partial w}{\partial t}(L,t)\right)\delta(w(L,t)) \qquad \text{(nn)}$$

Inclusion of the discrete mass in the kinetic energy leads to

$$\int_{t_1}^{t_2}\delta\left[\frac{1}{2}m\left(\frac{dy}{dt}\right)^2\right]dt = \int_{t_1}^{t_2}m\frac{dy}{dt}\delta\left(\frac{dy}{dt}\right)dt$$

$$= m\frac{dy}{dt}\delta y|_{t=t_1}^{t=t_2} - \int_{t_1}^{t_2}m\frac{d^2y}{dt^2}\delta y\, dt \qquad \text{(oo)}$$

$$= -\int_{t_1}^{t_2}m\frac{d^2y}{dt^2}\delta y\, dt$$

The addition to Equation b to include the potential energy of the discrete spring is

$$\int_{t_1}^{t_2} \delta \left[\frac{1}{2} k[y - w(L,t)]^2 \right] dt = \int_{t_1}^{t_2} \{ k[y - w(L,t)]\delta y - k[y - w(L,t)]\delta w(L,t) \} dt \quad \text{(pp)}$$

Application of Hamilton's Principle leads to

$$\int_{t_1}^{t_2} (\delta T - \delta V + \delta W_{nc}) dt = 0$$

$$-\int_{t_1}^{t_2}\int_0^L \rho A \frac{\partial^2 w}{\partial t^2} \delta w \, dx \, dt - \int_{t_1}^{t_2} m \frac{d^2 y}{dt^2} \delta y \, dt - \int_{t_1}^{t_2} \left\{ \left[\left(EI \frac{\partial^2 w}{\partial x^2} \right) \left(\delta \frac{\partial w}{\partial x} \right) \right]_{x=0}^{x=L} \right.$$

$$\left. - \left[\left(\frac{\partial}{\partial x} \left(EI \frac{\partial^2 w}{\partial x^2} \right) \right) \delta w \right]_{x=0}^{x=L} + \int_0^L \frac{\partial^2}{\partial x^2} \left(EI \frac{\partial^2 w}{\partial x^2} \right) \delta w \, dx \right\} dt$$

$$-\int_{t_1}^{t_2} \{ k[y - w(L,t)]\delta y - k[y - w(L,t)]\delta w(L,t) \} dt + \int_{t_1}^{t_2} \left[-c \left(\dot{y} - \frac{\partial w}{\partial t}(L,t) \right) \delta y \right.$$

$$\left. + c \left(\dot{y} - \frac{\partial w}{\partial t}(L,t) \right) \delta(w(L,t)) \right] dt = 0$$

$$\text{(qq)}$$

Equation qq is rearranged as

$$\int_{t_1}^{t_2} \left\{ \int_0^L \left[-\rho A \frac{\partial^2 w}{\partial t^2} - \frac{\partial^2}{\partial x^2} \left(EI \frac{\partial^2 w}{\partial x^2} \right) \right] \delta u \, dx - \left[m\ddot{y} + c \left(\dot{y} - \frac{\partial w}{\partial t}(L,t) \right) \right. \right.$$

$$\left. + k(y - w(L,t)) \right] \delta y + \left(-EI \frac{\partial^2 w}{\partial x^2}(L,t) \right) \delta \left(\frac{\partial w}{\partial x}(L,t) \right)$$

$$+ \left\{ \left(\frac{\partial}{\partial x} \left(EI \frac{\partial^2 w}{\partial x^2}(L,t) \right) + k[y - w(L,t)] + c \left[\dot{y} - \frac{\partial w}{\partial x}(L,t) \right] \right) \right\} \delta w(L,t)$$

$$\left. + \left(EI \frac{\partial^2 w}{\partial x^2}(0,t) \right) \delta \left(\frac{\partial w}{\partial x}(0,t) \right) - \frac{\partial}{\partial x} \left(EI \frac{\partial^2 w}{\partial x^2}(L,t) \right) \delta w(0,t) \right\} dt = 0$$

$$\text{(rr)}$$

The differential equations obtained from Equation rr

$$\rho A \frac{\partial^2 w}{\partial t^2} + \frac{\partial^2}{\partial x^2}\left(EI\frac{\partial^2 w}{\partial x^2}\right) = 0 \tag{ss}$$

$$m\ddot{y} + c\left(\dot{y} - \frac{\partial w}{\partial t}(L,t)\right) + k(y - w(L,t)) = 0 \tag{tt}$$

The boundary conditions for this system are of the form

$$\frac{\partial}{\partial x}\left(EI\frac{\partial^2 w}{\partial x^2}(0,t)\right) = 0 \quad \text{or} \quad w(0,t) = 0 \tag{uu}$$

$$EI\frac{\partial^2 w}{\partial x^2}(0,t) = 0 \quad \text{or} \quad \frac{\partial w}{\partial x}(0,t) = 0 \tag{vv}$$

$$\frac{\partial}{\partial x}\left(EI\frac{\partial^2 w}{\partial x^2}(L,t)\right) + k[y - w(L,t)] + c\left[\dot{y} - \frac{\partial w}{\partial x}(L,t)\right] \tag{ww}$$

$$\text{or} \quad w(L,t) = 0$$

$$EI\frac{\partial^2 w}{\partial x^2}(L,t) = 0$$

$$\text{or} \tag{xx}$$

$$\frac{\partial w}{\partial x}(L,t) = 0$$

(g) Let $w_1(x,t)$ and $w_2(x,t)$ represent the transverse displacements of each beam. The total kinetic energy of the system is

$$I = \frac{1}{2}\int_0^L \rho_1 A_1\left(\frac{\partial w_1}{\partial t}\right)^2 + \frac{1}{2}\int_0^L \rho_2 A_2\left(\frac{\partial w_2}{\partial t}\right)^2 \tag{yy}$$

such that

$$\int_{t_1}^{t_2} \delta T dt = \int_{t_1}^{t_2}\left[\int_0^L \rho_1 A_1\frac{\partial^2 w_1}{\partial t^2}\delta w_1 dx + \int_0^L \rho_2 A_2\frac{\partial^2 w_2}{\partial t^2}\delta w_2 dx\right] dt \tag{zz}$$

The total potential energy in the system is

$$
V = \frac{1}{2} \int_0^L E_1 I_1 \left(\frac{\partial^2 w_1}{\partial x^2} \right)^2 dx + \frac{1}{2} \int_0^L E_2 I_2 \left(\frac{\partial^2 w_2}{\partial x^2} \right)^2 dx + \frac{1}{2} \int_0^L k(w_2 - w_1)^2 dx
$$

(aaa)

such that

$$
\begin{aligned}
\int_{t_1}^{t_2} \delta V dt = \int_{t_1}^{t_2} & \left\{ \left[\left(E_1 I_1 \frac{\partial^2 w_1}{\partial x^2} \right) \left(\delta \frac{\partial w_1}{\partial x} \right) \right]_{x=0}^{x=L} - \left[\left(\frac{\partial}{\partial x} \left(E_1 I_1 \frac{\partial^2 w}{\partial x^2} \right) \right) \delta w \right]_{x=0}^{x=L} \right. \\
& + \int_0^L \frac{\partial^2}{\partial x^2} \left(E_1 I_1 \frac{\partial^2 w_1}{\partial x^2} \right) \delta w_1 dx + \left[\left(E_2 I_2 \frac{\partial^2 w_2}{\partial x^2} \right) \left(\delta \frac{\partial w_2}{\partial x} \right) \right]_{x=0}^{x=L} \\
& - \left[\left(\frac{\partial}{\partial x} \left(E_2 I_2 \frac{\partial^2 w_2}{\partial x^2} \right) \right) \delta w \right]_{x=0}^{x=L} + \int_0^L \frac{\partial^2}{\partial x^2} \left(E_2 I_2 \frac{\partial^2 w_2}{\partial x^2} \right) \delta w_2 dx \\
& \left. + \int_0^L k(w_2 - w_1) \delta(w_2 - w_1) dx \right\} dt
\end{aligned}
$$

(bbb)

The virtual work of the nonconservative forces in the viscoelastic layer is

$$
\delta W_{nc} = -c \left[\frac{\partial w_2}{\partial t} - \frac{\partial w_1}{\partial t} \right] \delta(w_2 - w_1)
\tag{ccc}
$$

Substitution of Equation ccc into Extended Hamilton's Principle, and setting the expressions multiplying variations δw_1 and δw_2 to zero, independently leads to

$$
\frac{\partial^2}{\partial x^2} \left(E_1 I_1 \frac{\partial^2 w_1}{\partial x^2} \right) + c \left(\frac{\partial w_2}{\partial x} - \frac{\partial w_1}{\partial x} \right) + k(w_2 - w_1) + \rho_1 A_1 \frac{\partial^2 w_1}{\partial t^2} = 0 \quad \text{(ddd)}
$$

$$
\frac{\partial^2}{\partial x^2} \left(E_2 I_2 \frac{\partial^2 w_2}{\partial x^2} \right) - c \left(\frac{\partial w_2}{\partial x} - \frac{\partial w_1}{\partial x} \right) - k(w_2 - w_1) + \rho_2 A_2 \frac{\partial^2 w_2}{\partial x^2} = 0 \quad \text{(eee)}
$$

The choices of boundary conditions on each beam are the same as Equation k through Equation n of Example 2.16.

2.12 TIMOSHENKO BEAMS

The Timoshenko model of a beam includes two effects neglected in the Euler–Bernoulli beam model. Rotary inertia refers to the kinetic energy of the beam due to rotation of the beam caused by bending. Shear deformation is the deformation of the beam caused by shear forces. The shear deformation is an angular distortion of the cross section due to shear. The shear deformation effect is illustrated in Figure 2.22.

Assume that the neutral axis displaces only in the transverse direction. Let $w(x,t)$ represent this transverse displacement. For small w, $\partial w/\partial x$ is the slope of the deflection curve. The slope is the sum of the angular rotation due to bending ψ, and the angle due to shear distortion λ, so that

$$\frac{\partial w}{\partial x} = \psi(x,t) + \lambda(x,t) \tag{2.144}$$

For small ψ, the bending moment is

$$M = EI\frac{\partial \psi}{\partial x} \tag{2.145}$$

Determination of the shear force is beyond this analysis, but it can be shown to be

$$S = \hat{k}GA\lambda \tag{2.146}$$

where $\hat{k}$ is a shape factor dependent upon the beam's cross section, G is the shear modulus of the material from which the beam is made, and A is its cross-sectional area.

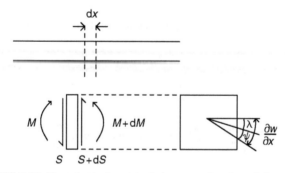

FIGURE 2.22 Timoshenko beam includes rotary inertia and shear distortion. ψ is due to bending; λ is due to shear distortion.

The kinetic energy of the Timoshenko beam is the kinetic energy due to the transverse displacement, plus the rotary inertia. The rotary inertia of a differential element of thickness dx is

$$dT_r = \frac{1}{2}dJ\left(\frac{\partial\psi}{\partial t}\right)^2 \tag{2.147}$$

The mass moment of inertia of the differential element is calculated as that of a thin disk of thickness dx, mass density ρ, and area moment of inertia $I(x)$,

$$dJ = \rho I\, dx \tag{2.148}$$

The total kinetic energy of the Timoshenko beam is

$$T = \frac{1}{2}\int_0^L \rho A\left(\frac{\partial w}{\partial t}\right)^2 dx + \frac{1}{2}\int_0^L \rho I\left(\frac{\partial\psi}{\partial t}\right)^2 dx \tag{2.149}$$

The bending moment and the shear force do work on the beam as deformation occurs. This work leads to strain energy stored in the beam. For a conservative system, the stored energy is equal to the work required to store the energy. The work done by the bending moment acting on the differential element in the free-body diagram of Figure 2.22 as it rotates through an angle $d\psi$ is

$$dW_b = -\frac{1}{2}M(x,t)d\psi$$

$$= -\frac{1}{2}M(x,t)\frac{\partial\psi}{\partial x}dx \tag{2.150}$$

$$= -\frac{1}{2}EI\left(\frac{\partial\psi}{\partial x}\right)^2 dx$$

The total work done by the internal bending moment is

$$W_b = -\frac{1}{2}\int_0^L EI\left(\frac{\partial\psi}{\partial x}\right)^2 dx \tag{2.151}$$

The work done by the shear force during its application is

$$dW_s = -\frac{1}{2}S(x,t)\lambda(x,t)dx$$

$$= -\frac{1}{2}\hat{\kappa}GA\lambda^2 dx$$

(2.152)

The total work done by the shear force is

$$W_s = -\frac{1}{2}\int_0^L \hat{\kappa}GA\lambda^2 dx$$

(2.153)

Assuming the work is reversible, the strain energy stored in the beam is

$$V = \frac{1}{2}\int_0^L EI\left(\frac{\partial\psi}{\partial x}\right)^2 dx + \frac{1}{2}\int_0^L \hat{\kappa}GA\lambda^2 dx$$

(2.154)

Noting from Equation 2.14 that $\lambda = \left(\dfrac{\partial w}{\partial x} - \psi\right)$, Equation 2.154 becomes

$$V = \frac{1}{2}\int_0^L EI\left(\frac{\partial\psi}{\partial x}\right)^2 dx + \frac{1}{2}\int_0^L \hat{\kappa}GA\left(\frac{\partial w}{\partial x} - \psi\right)^2 dx$$

(2.155)

The formulation of the kinetic and potential energy for a Timoshenko beam involves two dependent variables, $w(x,t)$ and $\psi(x,t)$. Variations of both dependent variables must be assumed when using Hamilton's Principle,

$$\int_{t_1}^{t_2}(\delta T - \delta V)dx = 0$$

$$\int_{t_1}^{t_2}\left\{\delta\int_0^L\left[\rho A\left(\frac{\partial w}{\partial t}\right)^2 + \rho I\left(\frac{\partial\psi}{\partial t}\right)^2\right]dx\right.$$

$$\left. -\delta\int_0^L\left[EI\left(\frac{\partial\psi}{\partial x}\right)^2 + \hat{\kappa}GA\left(\frac{\partial w}{\partial x} - \psi\right)^2\right]dx\right\}dt = 0$$

(2.156)

Techniques similar to those used in deriving equations for second-order systems and Euler–Bernoulli beams are used on Equation 2.156 leading to

$$
\int_{t_1}^{t_2} \left\{ \left. -\left(EI \frac{\partial \psi}{\partial x}\right) \delta\psi \right|_{x=0}^{x=L} - \left[\hat{\kappa}GA\left(\frac{\partial w}{\partial x} - \psi\right) \delta w \right]_{x=0}^{x=L} + \int_0^L \left[\frac{\partial}{\partial x}\left(\hat{\kappa}GA \frac{\partial w}{\partial x}\right) \right. \right.
$$

$$
\left. -\frac{\partial}{\partial x}(\kappa GA\psi) - \rho A \frac{\partial^2 w}{\partial t^2} \right] \delta w \, dx + \int_0^L \left[\frac{\partial}{\partial x}\left(EI \frac{\partial \psi}{\partial x}\right) + \hat{\kappa}GA \frac{\partial w}{\partial x} \right.
$$

$$
\left. \left. -\hat{\kappa}GA\psi - \rho I \frac{\partial^2 \psi}{\partial t^2} \right] \delta\psi \, dx \right\} dt = 0 \tag{2.157}
$$

Application of the DuBois–Reymond Lemma to Equation 2.157 leads to two coupled partial differential equations:

$$
\frac{\partial}{\partial x}\left(\hat{\kappa}GA \frac{\partial w}{\partial x}\right) - \frac{\partial}{\partial x}(\kappa GA\psi) = \rho A \frac{\partial^2 w}{\partial t^2} \tag{2.158}
$$

$$
\frac{\partial}{\partial x}\left(EI \frac{\partial \psi}{\partial x}\right) + \hat{\kappa}GA \frac{\partial w}{\partial x} - \hat{\kappa}GA\psi = \rho I \frac{\partial^2 \psi}{\partial t^2} \tag{2.159}
$$

The boundary conditions are of the form

$$
EI \frac{\partial \psi}{\partial x}(0,t) = 0 \quad \text{or} \quad \psi(0,t) = 0 \tag{2.160}
$$

$$
\hat{\kappa}GA\left(\frac{\partial w}{\partial x}(0,t) - \psi(0,t)\right) = 0 \quad \text{or} \quad w(0,t) = 0 \tag{2.161}
$$

$$
EI \frac{\partial \psi}{\partial x}(L,t) = 0 \quad \text{or} \quad \psi(L,t) = 0 \tag{2.162}
$$

$$
\hat{\kappa}GA\left(\frac{\partial w}{\partial x}(L,t) - \psi(L,t)\right) = 0 \quad \text{or} \quad w(L,t) = 0 \tag{2.163}
$$

The geometric boundary condition $w = 0$ is applied at an end where the transverse displacement is zero, while the geometric condition $\psi = 0$ is applied at an end where the rotation due to bending is zero (shear distortion may still occur). The natural boundary condition $EI(\partial\psi/\partial x) = 0$ is applied at an end where the bending moment is zero, while the boundary condition $\hat{\kappa}GA((\partial w/\partial x) - \psi) = 0$ is applied at an end where the shear force vanishes.

2.13 MEMBRANES

A membrane, illustrated in Figure 2.23, is a two-dimensional surface that undergoes transverse displacement. An internal tension, which is assumed to be constant, exists in the membrane as illustrated in Figure 2.24. Bending is neglected

Let $w(r,t)$ be the displacement of a particle on the surface of the membrane. Let ρ be the mass per unit area of the membrane. The kinetic energy of the membrane is

$$T = \frac{1}{2} \int_S \rho \left(\frac{\partial w}{\partial t} \right)^2 dS \tag{2.164}$$

The potential energy due to work of the tension is

$$V = \frac{1}{2} \int_S \hat{T} \left(\frac{\partial w}{\partial n} \right)^2 dS \tag{2.165}$$

where n is the direction normal to the boundary of a differential element of area dS on the surface of the membrane. The gradient of a scalar function is written as

$$\nabla w = \frac{\partial w}{\partial n} \mathbf{n} \tag{2.166}$$

Thus,

$$\nabla w \cdot \nabla w = \left(\frac{\partial w}{\partial n} \right)^2 \tag{2.167}$$

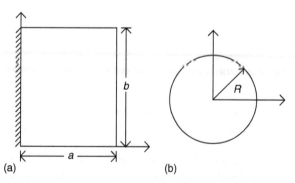

(a)   (b)

FIGURE 2.23 (a) Rectangular membrane. (b) Circular membrane.

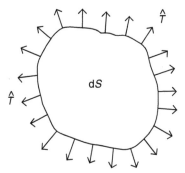

FIGURE 2.24 Uniform tension $\hat{T}$ is normal to surface at every point.

Application of Hamilton's Principle leads to

$$
\int_{t_1}^{t_2} \left[\delta \int_S \rho \left(\frac{\partial w}{\partial t} \right)^2 dS - \delta \int_S \hat{T} (\nabla w \nabla w) dS \right] dt = 0
$$

(2.168)

$$
\int_{t_1}^{t_2} \left\{ \int_S \rho \left(\frac{\partial w}{\partial t} \right) \delta \left(\frac{\partial w}{\partial t} \right) dS - \int_S \hat{T} (\nabla w) \delta (\nabla w) dS \right\} dt = 0
$$

It is shown, in a manner similar to that for the longitudinal bar problem, that by interchanging the order of integration, integrating by parts, noting that by definition variations must vanish at t_1 and t_2, and again interchanging the order of integration, that

$$
\int_{t_1}^{t_2} \left\{ \int_S \rho \left(\frac{\partial w}{\partial t} \right) \delta \left(\frac{\partial w}{\partial t} \right) dS \right\} dt = - \int_{t_1}^{t_2} \left(\int_S \rho \frac{\partial^2 w}{\partial t^2} \delta w \, dS \right) dt
$$

(2.169)

A useful vector identity is

$$
\nabla f \cdot \nabla g = \nabla \cdot (g \nabla f) - g \nabla^2 f
$$

(2.170)

Applying Equation 2.170 to Equation 2.167 with $f = w$ and $g = \delta w$ leads to

$$
\int_{t_1}^{t_2} \left[\int_S \hat{T} (\nabla w) \cdot \delta (\nabla w) dS \right] dt = \int_{t_1}^{t_2} \left\{ \int_S \hat{T} [\nabla \cdot (\delta w \nabla w) - \delta w \nabla^2 w] dS \right\} dt
$$

(2.171)

The Divergence Theorem is used to convert the surface integral into an integral over the boundary of the region C:

$$\int_S \nabla \cdot (\delta w \nabla w) dS = \int_C \frac{\partial w}{\partial n} \delta w \, dC \tag{2.172}$$

Substituting Equation 2.172 in Equation 2.171 and then substituting the resulting equation, as well as Equation 2.169 into Equation 2.168, leads to

$$\int_{t_1}^{t_2} \left(\int_S \left[-\rho \frac{\partial^2 w}{\partial t^2} + \hat{T} \nabla^2 w \right] \delta w \, dS - \int_C \frac{\partial w}{\partial n} \delta w \, dC \right) dt = 0 \tag{2.173}$$

Application off the DuBois–Reymond lemma to Equation 2.173 leads to

$$\hat{T} \nabla^2 w = \rho \frac{\partial^2 w}{\partial t^2} \tag{2.174}$$

The appropriate boundary conditions are

$$\frac{\partial w}{\partial n} = 0 \text{ on C or } \quad w = 0 \quad \text{on C} \tag{2.175}$$

One of the conditions of Equation 2.175 must be met everywhere on C. The boundary C may be composed of C_1, on which $w = 0$ is imposed, and C_2, on which $\partial w / \partial n = 0$ is imposed.

3 Linear Algebra

3.1 INTRODUCTION

Application of Lagrange's equations to the lagrangian of a mechanical system results in a differential equation, or a set of differential equations whose solution is the system response. Application of Lagrange's equations to a discrete system leads to a set of ordinary differential equations, while application of Lagrange's equations to a distributed parameter system leads to partial differential equations. Mathematical methods are employed to determine the solution of the equations, and hence the system response.

Linear algebra is the algebra of linear systems. It provides a foundation in which solutions to mathematical problems are developed. Success in obtaining a solution to a mathematical problem requires finding the specific solution from a possible set of solutions, the solution space. An understanding of the solution space and properties of elements of the solution space leads to the development of solution techniques. It is the spirit in which this review of linear algebra is presented.

An exact solution of a problem is a solution for the dependent variables that satisfies, without error, the mathematical problem for all possible values of the independent variables. An exact solution, while desirable, is not always possible. Approximate solutions are sought when an exact solution is not available. Approximate solutions are of two types. Variational methods are used to determine continuous functions of the independent variables which provide in some sense the "best approximation", chosen from a specified set, to the exact solution. Numerical solutions provide an approximation to the exact solution only at discrete values of independent variables. Linear algebra provides a framework in which these approximate solutions are developed, and a framework in which the error between the exact solution and an approximate solution is estimated.

Linear algebra provides a framework for developing solutions to linear problems. Modeling of engineering systems often leads to non-linear mathematical problems. Exact solutions of few non-linear problems exist. Often, assumptions are made such that the non-linear problem is approximated by a linear problem. Even if the assumptions that linearize the problem are not valid, some understanding of the solution can be obtained by studying the linearized problem. Indeed, knowledge of the behavior of the linearized system is necessary to understand the effect of non-linearities on the system behavior. Numerical methods developed to approximate solutions of non-linear problems are based upon numerical methods used to develop approximate solutions for linear problems. Approximate solutions of non-linear problems are often assumed to be perturbations of the linear solution. Thus, a knowledge of linear solutions is necessary to develop approximate solutions for non-linear problems.

This chapter provides a review of linear algebra and linear operator theory. Theorems are presented. Proofs are presented when the proofs are themselves instructive.

3.2 THREE-DIMENSIONAL SPACE

Every point in three-dimensional space has a unique set of x–y–z coordinates defined in a fixed Cartesian reference frame, as shown in Figure 3.1. The location of a particle at any instant of time is defined by the Cartesian coordinates of the point it occupies in space. A position vector, $\mathbf{r}$, defining the location of the particle, is a line segment drawn from the origin (0–0–0)

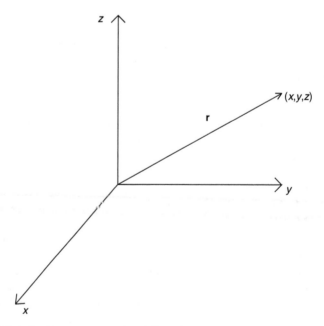

FIGURE 3.1 Position vector in a fixed Cartesian system.

of the Cartesian system to the particle. The position vector has a unique direction with respect to the Cartesian frame and a calculable length denoted by $|\mathbf{r}|$.

Let $\mathbf{r}_1$ and $\mathbf{r}_2$ represent two vectors defined in a Cartesian system. Vector addition in three-dimensional space is defined geometrically, as illustrated in Figure 3.2a. A vector parallel to $\mathbf{v}$ is drawn such that its tail coincides with the head of u. If $\mathbf{w} = \mathbf{r}_1 + \mathbf{r}_2$ then $\mathbf{w}$ is the vector drawn from the origin to the head of the vector parallel to $\mathbf{r}_2$. Figure 3.2b shows the process repeated, but with a vector parallel to $\mathbf{r}_1$ drawn with its tail coinciding with the head of $\mathbf{v}$. The resulting sum is the same as that obtained in Figure 3.2a. Thus, vector addition is commutative:

$$\mathbf{r}_1 + \mathbf{r}_2 = \mathbf{r}_2 + \mathbf{r}_1 \tag{3.1}$$

Figure 3.2 illustrates the triangle rule for vector addition. The vectors being added, and the resultants (sum), are depicted as sides of a triangle. Since the length of any side of a triangle must be less than the sum of the lengths of the other two sides,

$$|\mathbf{r}_1 + \mathbf{r}_2| \leq |\mathbf{r}_1| + |\mathbf{r}_2| \tag{3.2}$$

Equation 3.2 is called the triangle inequality.

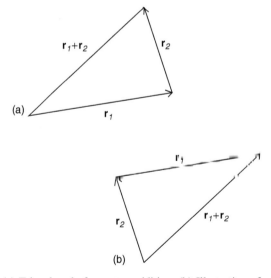

FIGURE 3.2 (a) Triangle rule for vector addition. (b) Illustration of commutivity of vector addition.

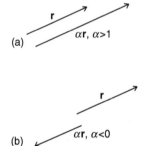

FIGURE 3.3 Illustration of multiplication of a vector by a scalar. (a) If a<0, the product vector is parallel to **r**, in the same direction. (b) If a>0, the product vector is parallel to **r** but in the opposite direction.

The concept of multiplication of a vector by a scalar is illustrated in Figure 3.3. Let α be any real value. The vector $\alpha\mathbf{r}$ is the vector parallel to **r**, whose length is α times the length of **r**. If α is positive, then the vector $\alpha\mathbf{r}$ is in the same direction as **r**. If α is negative, the vector $\alpha\mathbf{r}$ is in the direction opposite that of **r**. If α equals zero, then $\alpha\mathbf{r}$ is a vector whose length is zero and is called the zero vector, **0**.

The following properties follow from the definitions of vector addition and multiplication of a vector by a scalar:

  i. Associative law of addition $(\mathbf{r}_1 + \mathbf{r}_2) + \mathbf{r}_3 = \mathbf{r}_1 + (\mathbf{r}_2 + \mathbf{r}_3)$
  ii. Commutative law of addition $\mathbf{r}_1 + \mathbf{r}_2 = \mathbf{r}_2 + \mathbf{r}_1$
  iii. Addition of zero vector $\mathbf{0} + \mathbf{r}_1 = \mathbf{r}_1$
  iv. Multiplication by one $(1)\mathbf{r}_1 = \mathbf{r}_1$
  v. Negative vector $-\mathbf{r}_1 = (-1)\mathbf{r}_1$, $\mathbf{r}_1 + (-\mathbf{r}_1) = \mathbf{0}$
  vi. Distributive law of scalar multiplication $(\alpha + \beta)\mathbf{r}_1 = \alpha\mathbf{r}_1 + \beta\mathbf{r}_1$
  vii. Associative law of scalar multiplication $(\alpha\beta)\mathbf{r}_1 = \alpha(\beta\mathbf{r}_1)$
  viii. Distributive law of addition $\alpha(\mathbf{r}_1 + \mathbf{r}_2) = \alpha\mathbf{r}_1 + \alpha\mathbf{r}_2$

A unit vector is a vector whose length is one. Let **i**, **j**, and **k** be a set of unit vectors parallel to the x, y, and z coordinate axes respectively. Using the definitions of vector addition and multiplication of a vector by a scalar, the position vector for the particle can be written in terms of the unit vectors as

$$\mathbf{r} = x\mathbf{i} + y\mathbf{j} + z\mathbf{k} \tag{3.3}$$

Any vector in three-dimensional space may be written as a linear combination of the trio of unit vectors. The vectors **i**, **j**, and **k** form a basis for three-dimensional space.

The Pythagorean theorem is used to determine the length of the position vector as

$$|\mathbf{r}| = \sqrt{x^2 + y^2 + z^2} \tag{3.4}$$

A scalar function of two vectors, called the dot product, is defined by

$$\mathbf{r}_1 \cdot \mathbf{r}_2 = |\mathbf{r}_1||\mathbf{r}_2|\cos\theta \tag{3.5}$$

where θ is the angle made between $\mathbf{r}_1$ and $\mathbf{r}_2$. The dot product has the geometric interpretation that it is equal to the length of the projection of the vector $\mathbf{r}_1$ onto $\mathbf{r}_2$. This leads to the calculation of the dot product as

$$\mathbf{r}_1 \cdot \mathbf{r}_2 = x_1 x_2 + y_1 y_2 + z_1 z_2 \tag{3.6}$$

It is noted that when computed using Equation 3.6, the dot product has the following properties:

i. Commutative property $\mathbf{r}_1 \cdot \mathbf{r}_2 = \mathbf{r}_2 \cdot \mathbf{r}_1$
ii. Multiplication by scalar $(\alpha \mathbf{r}_1) \cdot \mathbf{r}_2 = \alpha(\mathbf{r}_1 \cdot \mathbf{r}_2)$
iii. Distributive property $(\mathbf{r}_1 + \mathbf{r}_2) \cdot \mathbf{r}_3 = \mathbf{r}_1 \cdot \mathbf{r}_3 + \mathbf{r}_2 \cdot \mathbf{r}_3$
iv. Non-negative property $\mathbf{r}_1 \cdot \mathbf{r}_1 \geq 0$ and $\mathbf{r}_1 \cdot \mathbf{r}_1 = 0$ if, and only if, $\mathbf{r}_1 = \mathbf{0}$

It is also noted that from Equation 3.5 and Equation 3.6 that

$$\mathbf{r} \cdot \mathbf{r} = |\mathbf{r}|^2 \tag{3.7}$$

Three-dimensional space, the definition of a vector, and the operations of vector addition and multiplication by a scalar, are generalized in Section 3.3 into the concept of vector spaces. The properties of vectors in a general vector space are defined the same as the properties satisfied by vectors in three-dimensional space. The scalar function of the dot product is generalized in Section 3.5 into the concept of inner products.

3.3 VECTOR SPACES

Vectors on three-dimensional space are defined, along with a definition of vector addition and multiplication by a scalar. These operations yield another vector in three-dimensional space. The definitions of these operations, along with the implicit definitions of addition and multiplication of real numbers, lead to eight properties that apply to all vectors in three-dimensional space.

Three-dimensional space is illustrated geometrically. However, it is only an example of the more general and abstract concept of a vector space.

Definition 3.1. A *vector space* is a collection of objects called *vectors*, together with defined operations of vector addition and scalar multiplication, that satisfy a set of 10 axioms. Let the collection of objects be collectively called V and let $\mathbf{u}$, $\mathbf{v}$, and $\mathbf{w}$ be arbitrary elements of V. Let α and β be arbitrary elements of an associated scalar field. Then, if V is a vector space,

    i. V is closed under addition, that is, $\mathbf{u}+\mathbf{v}$ is an element of V

    ii. V is closed under scalar multiplication, that is, $\alpha\mathbf{u}$ is an element of V

    iii. The associative law of addition, $(\mathbf{u}+\mathbf{v})+\mathbf{w}=\mathbf{u}+(\mathbf{v}+\mathbf{w})$, holds

    iv. The commutative law of addition, $\mathbf{u}+\mathbf{v}=\mathbf{v}+\mathbf{u}$, holds

    v. There exists a vector in V, called the zero vector $\mathbf{0}$, such that $\mathbf{u}+\mathbf{0}=\mathbf{u}$

    vi. There exists a vector in V, $-\mathbf{u}$, such that $-\mathbf{u}+\mathbf{u}=\mathbf{0}$

    vii. For the scalar 1, $(1)\mathbf{u}=\mathbf{u}$,

    viii. The distributive law of scalar multiplication, $(\alpha+\beta)\mathbf{u}=\alpha\mathbf{u}+\beta\mathbf{u}$, holds

    ix. The associative law of scalar multiplication, $(\alpha\beta)\mathbf{u}=\alpha(\beta\mathbf{u})$, holds

    x. The distributive law of addition, $\alpha(\mathbf{u}+\mathbf{v})=\alpha\mathbf{u}+\alpha\mathbf{v}$, holds

If the associated scalar field is the set of all real numbers, then the vector space is said to be a *real vector space*. If the associated scalar field is the set of complex numbers, then the vector space is called a *complex vector space*.

Example 3.1. Define R^n as the set of all ordered n-tuples of real numbers. Vectors $\mathbf{u}$ and $\mathbf{v}$ in R^n are represented as

$$\mathbf{u}=\begin{bmatrix} u_1 \\ u_2 \\ \vdots \\ u_n \end{bmatrix} \qquad \mathbf{v}=\begin{bmatrix} v_1 \\ v_2 \\ \vdots \\ v_n \end{bmatrix} \tag{a}$$

The operations of vector addition and scalar multiplication are defined such that

$$\mathbf{u} + \mathbf{v} = \begin{bmatrix} u_1 + v_1 \\ u_2 + v_2 \\ \vdots \\ u_n + v_n \end{bmatrix} \qquad \alpha\mathbf{u} = \begin{bmatrix} \alpha u_1 \\ \alpha u_2 \\ \vdots \\ \alpha u_n \end{bmatrix} \qquad \text{(b)}$$

Show that R^n is a vector space under these definitions of vector addition and multiplication by a scalar.

Solution: Certainly R^n is closed under vector addition and scalar multiplication. The zero vector is the vector whose components are all zero, and $-\mathbf{u}$ is defined as $(-1)\mathbf{u}$. It is easy to show that the other axioms defining a vector space hold for R^n under these definitions of vector addition and scalar multiplication. For example, consider the distributive law of scalar multiplication:

$$(\alpha + \beta)\mathbf{u} = \begin{bmatrix} (\alpha + \beta)u_1 \\ (\alpha + \beta)u_2 \\ \vdots \\ (\alpha + \beta)u_n \end{bmatrix} \qquad \text{(c)}$$

However, the distributive law holds for scalar multiplication of real numbers, that is $(\alpha + \beta)u_i = \alpha u_i + \beta u_i$. Thus,

$$(\alpha + \beta)\mathbf{u} = \begin{bmatrix} \alpha u_1 + \beta u_1 \\ \alpha u_2 + \beta u_2 \\ \vdots \\ \alpha u_n + \beta u_n \end{bmatrix} \qquad \text{(d)}$$

which using the definition of vector addition is written as

$$(\alpha + \beta)\mathbf{u} = \begin{bmatrix} \alpha u_1 \\ \alpha u_2 \\ \vdots \\ \alpha u_n \end{bmatrix} + \begin{bmatrix} \beta u_1 \\ \beta u_2 \\ \vdots \\ \beta u_n \end{bmatrix} = \alpha\mathbf{u} + \beta\mathbf{u} \qquad \text{(e)}$$

Since the 10 axioms of Definition 3.1 hold, R^n is a vector space.

The vector space R^n is a generalization of three-dimensional space R^3, discussed in Section 3.2. However, vectors in R^n lack the geometric representation of vectors in R^3. Equation a of Example 3.1 is an alternate to Equation 3.3 in writing a vector in R^3.

Example 3.2. Let $C^n[a, b]$ represent the set of functions of a real variable, call it x, that are n times continuously differentiable on the real number line within the interval, $a \leq x \leq b$. If $f(x)$ is an element of $C^n[a, b]$, then for any x, $a \leq x \leq b$, $f(x)$ is a real number. If $f(x)$ and $g(x)$ belong to $C^n[a, b]$ and α is a real scalar, then for any x, $a \leq x \leq b$, $f(x) + g(x)$ is the real number that is the scalar sum of $f(x) + g(x)$ and $\alpha f(x)$ is the real number defined by the scalar multiplication of α and $f(x)$. Show that $C^n[a, b]$ is a vector space under the defined operations of vector addition and scalar multiplication.

Solution: Clearly, if $f(x)$ and $g(x)$ are n times continuously differentiable then $f(x) + g(x)$ and $\alpha f(x)$ are also n times continuously differentiable. The zero function is defined as $f(x) = 0$ for all x, $a \leq x \leq b$ and $-f(x)$ is defined as $(-1)f(x)$. It is easy to show that the remaining axioms of Definition 3.1 are satisfied and $C^n[a,b]$ is a vector space under the given definitions of vector addition and scalar multiplication.

R^n and $C^n[a, b]$ are the two vector spaces used most extensively in this study. Other vector spaces used in applications in this study include:

1. $P^n[a, b]$ is the set of all polynomials of degree n, or less defined on the interval $a \leq x \leq b$. The definitions of vector addition and scalar multiplication for vectors in $P^n[a,b]$ are the same as used in the definition of $C^n[a,b]$.

2. $C^n(\Re)$ is the set of all functions n times continuously differentiable in the spatial volume defined by $\Re$. Vectors in $C^n(\Re)$ are functions of three spatial variables, perhaps of the form $f(x, y, z)$ or if cylindrical coordinates are used $f(r, \theta, z)$. Since $C^n(\Re)$ is really a generalization of $C^n[a,b]$, their definitions of vector addition and scalar multiplication are similar. Addition of two vectors in $C^n(\Re)$ is performed at each point in the spatial volume defined by $\Re$.

3. $S = R^k X C^n[a,b]$ is the set of elements of the form

$$\begin{bmatrix} f_1(x) \\ f_2(x) \\ \vdots \\ f_k(x) \end{bmatrix}$$

where $f_1(x)$, $f_2(x)$,..., $f_k(x)$ are each in $C^n[a, b]$. Definitions of vector addition and scalar multiplication are given by

$$\mathbf{f} + \mathbf{g} = \begin{bmatrix} f_1(x) \\ f_2(x) \\ \vdots \\ f_k(x) \end{bmatrix} + \begin{bmatrix} g_1(x) \\ g_2(x) \\ \vdots \\ g_k(x) \end{bmatrix} = \begin{bmatrix} f_1(x) + g_1(x) \\ f_2(x) + g_2(x) \\ \vdots \\ f_k(x) + g_k(x) \end{bmatrix}$$

$$\alpha\mathbf{f} = \begin{bmatrix} \alpha f_1(x) \\ \alpha f_2(x) \\ \vdots \\ \alpha f_k(x) \end{bmatrix}$$

Definition 3.2. Let S be a set of vectors contained in a vector space V. Then S is a *subspace* of V if S is a vector space in its own right.

In order to prove that S is a subspace of V, it must be shown that the 10 axioms of Definition 3.1 are satisfied by the vectors in S. However, since all vectors in S are also in V, and V is a vector space, then by hypothesis axioms (iii), (iv), (vii), (viii), (ix), and (x) are satisfied. Hence, it is only necessary to show that S is closed under vector addition, S is closed under scalar multiplication, the zero vector is in S, and for every $\mathbf{u}$ in S, $-\mathbf{u}$ is also in S.

Example 3.3. Determine whether each set of vectors is a subspace of R^3.

(a) Let S be the set of vectors in R^3, whose components sum to zero. That is, if $\mathbf{u}$ is in S, then $u_1 + u_2 + u_3 = 0$.
(b) Let S be the set of vectors in R^3, whose components sum to 1, $u_1 + u_2 + u_3 = 1$

Solution: (a) If $\mathbf{u}$ and $\mathbf{v}$ are in S, then

$$\mathbf{u} + \mathbf{v} = \begin{bmatrix} u_1 + v_1 \\ u_2 + v_2 \\ u_3 + v_3 \end{bmatrix} \tag{a}$$

Then, the sum of the components of $\mathbf{u} + \mathbf{v}$ is

$$(u_1 + v_1) + (u_2 + v_2) + (u_3 + v_3) = (u_1 + u_2 + u_3) + (v_1 + v_2 + v_3)$$

$$= 0 + 0 = 0 \tag{b}$$

Thus, S is closed under vector addition.

The sum of the components of $\alpha\mathbf{u}$ is $(\alpha u_1) + (\alpha u_2) + (\alpha u_3) = \alpha(u_1 + u_2 + u_3) = (\alpha)0 = 0$. Thus, S is closed under scalar multiplication.

The zero vector is in S as is $-\mathbf{u}$. Thus, since V is a vector space and axioms (i), (ii), (v), and (vi) of Definition 3.1 are satisfied, then S is a subspace of V.

(b) S is not a subspace of V. S is not closed under either vector addition or under scalar multiplication. If $\mathbf{u}$ and $\mathbf{v}$ are in S, then $u_1 + u_2 + u_3 = 1$ and $v_1 + v_2 + v_3 = 1$. Then, if $\mathbf{w} = \mathbf{u} + \mathbf{v}$, $w_1 + w_2 + w_3 = (u_1 + v_1) + (u_2 + v_2) + (u_3 + v_3) = 2$. Thus, by definition of S, $\mathbf{w}$ is not in S.

Example 3.4. The boundary-value problem for the non-dimensional temperature distribution in one-dimension, $\Theta(x)$, in a wall, with both sides fixed at the same temperature, and with a non-uniform internal heat generation, is of the form

$$\frac{d^2 \Theta}{dx^2} = u(x)$$

$$\Theta(0) = 0$$

$$\Theta(1) = 0$$

The solution must be twice differentiable and satisfy both boundary conditions. Let S be the set of functions in $C^2[0, 1]$ that satisfy the conditions $f(0) = 0$ and $f(1) = 0$. Is S, the set of possible solutions to the boundary value problem, a subspace of $C^2[0, 1]$.

Solution: If $f(x)$ and $g(x)$ are in S, then $f(0) = 0, f(1) = 0, g(0) = 0$, and $g(1) = 0$. Hence, $f(0) + g(0) = 0$ and $f(1) + g(1) = 0$. Thus, S is closed under addition. Additionally, S is closed under scalar multiplication as $\alpha f(0) = 0$. The function, $f(x) = 0$ satisfies $f(0) = 0$ and $f(1) = 0$ and hence, the zero vector is in S. Also, $-f(0) = 0$ and $-f(1) = 0$, thus $-f(x)$ is also in S. Hence, since axioms (i), (ii), (v), and (vi) of Definition 3.1 are satisfied, then S is a subspace of $C^2[0, 1]$.

3.4 LINEAR INDEPENDENCE

The mathematical solution to a problem involving vibrations of a n-degree-of-freedom system resides in a subspace of the vector space R^n, introduced in Example 3.1. One-degree-of-freedom systems discussed in Chapter 1 are special cases and do not require vector space analysis. The mathematical solution to a problem involving a continuous or distributed parameter system resides in a subspace of the vector space $C^n[a, b]$ of Example 3.2, where n is dependent on the type of continuous system.

In order to facilitate the solutions to these problems, it is imperative to understand properties of vector spaces and sets of vectors within the vector spaces. Approximate solutions are constructed from vectors in a subspace of a vector space, in which the exact solution resides. In order to be able to develop such solutions, it is imperative to understand properties of vector spaces and subspaces. Modal analysis, in which the response is assumed to be a linear combination of a set of vectors, is often used to determine the forced response of discrete and continuous systems. Approximate solutions, using variational methods, are obtained by assuming the solution as a linear combination of a set of predetermined vectors. The definitions and theorems in this section provide the foundation for these solutions.

Definition 3.3. A set of vectors is said to be *linearly independent* if, when a linear combination of the vectors is set equal to zero, all coefficients in the linear combination must be zero. If the vectors are not linearly independent, they are said to be *linearly dependent*.

Definition 3.3 states that if a set of vectors, $u_1, u_2, ..., u_k$, are linearly independent, then

$$\sum_{i=1}^{k} c_i \mathbf{u}_i = \mathbf{0} \tag{3.8}$$

implies that $c_1 = c_2 = ... = c_k = 0$. This implies that no vector in the set can be written as a linear combination of the other vectors.

Application of Equation 3.8 for a set of vectors in R^n leads to a set of n equations summarized by the matrix system

$$\mathbf{Uc} = \mathbf{0} \tag{3.9}$$

where $\mathbf{U}$ is a matrix with n rows and k columns, where its jth column is the vector $\mathbf{u}_j$ and $\mathbf{c}$ is a column vector of k rows, whose ith component is c_i. The vectors are linearly independent if, and only if, a non-trivial solution to the set of linear equations exists. This leads to the following theorem, presented without proof.

Theorem 3.1. *Let $\mathbf{u}_1, \mathbf{u}_2, ..., \mathbf{u}_k$ be a set of vectors in R^n. Let $\mathbf{U}$ be the $n \times k$ matrix whose ith column is $\mathbf{u}_j$. If $k > n$, a non-trivial solution of the system exists, and the set of vectors is linearly dependent. If $k = n$, the vectors are linearly dependent if, and only if, $\mathbf{U}$ is singular. If $k < n$, the vectors are linearly dependent only if every square submatrix of $\mathbf{U}$ is singular.*

Application of Equation 3.8 for a set of functions in $C[a, b], f_1(x), f_2(x), ..., f_k(x)$, leads to

$$c_1 f_1(x) + c_2 f_2(x) + \cdots + c_k f_k(x) = 0 \tag{3.10}$$

In order for a function to be identically zero on the interval $a \leq x \leq b$, it must be zero at every value of x. Differentiating Equation 3.10 $k-1$ times with respect to x, leads to

$$
\begin{aligned}
c_1 f_1'(x) + c_2 f_2'(x) + \cdots + c_k f_k'(x) &= 0 \\
c_1 f_1''(x) + c_2 f_2''(x) + \cdots + c_k f_k''(x) &= 0 \\
&\vdots \\
c_1 f_1^{(k-1)}(x) + c_2 f_2^{(k-1)}(x) + \cdots + c_k f_k^{(k-1)}(x) &= 0
\end{aligned}
\tag{3.11}
$$

Equation 3.10 and Equation 3.11 are summarized in matrix form as

$$\mathbf{W}\mathbf{c} = \mathbf{0} \tag{3.12}$$

where $\mathbf{W}$ is a $k \times k$ matrix called the Wronskian, defined by

$$
\mathbf{W} = \begin{bmatrix}
f_1(x) & f_2(x) & f_3(x) & \cdots & f_k(x) \\
f_1'(x) & f_2'(x) & f_3'(x) & \cdots & f_K'(x) \\
f_1''(x) & f_2''(x) & f_3''(x) & \cdots & f_K''(x) \\
\vdots & \vdots & \vdots & \ddots & \vdots \\
f_1^{(k-1)}(x) & f_2^{(k-1)}(x) & f_3^{(k-1)}(x) & \cdots & f_k^{(k-1)}(x)
\end{bmatrix}
\tag{3.13}
$$

and $\mathbf{c}$ is a column vector of k rows, whose ith component is c_i. Equation 3.12 has a non-trivial solution if, and only if, $\mathbf{W}$ is singular for every value of x, $a \leq x \leq b$. If there exists at least one value of x, such that $\mathbf{W}(x)$ is nonsingular, then the set of vectors is linearly independent. This leads to the following theorem, which is presented without formal proof.

Theorem 3.2. *A set of functions in $C[a, b]$, $f_1(x), f_2(x), \ldots, f_k(x)$ is linearly dependent if, and only if, the Wronskian for the set of functions, defined by Equation 3.13 is singular for all x, $a \leq x \leq b$.*

Example 3.5. Show that the set of functions

$$f_1(x) = 1 \qquad f_2(x) = x \qquad f_3(x) = x^2$$

is linearly independent in $C[0, 1]$.

Solution: The Wronskian of the set is determined as

$$W = \begin{bmatrix} 1 & x & x^2 \\ 0 & 1 & 2x \\ 0 & 0 & 2 \end{bmatrix}$$

The determinant of the Wronskian is

$$|W| = 2$$

Since the Wronskian is nonsingular for at least one x (actually it is non-singular for all x) the functions are linearly independent.

3.5 BASIS AND DIMENSION

A linear combination of a set of linearly independent vectors in a vector space V, because of closure, is an element of V. A change in any of the coefficients in the linear combination leads to a different element of V.

Definition 3.4. A set of vectors in a vector space V is said to *span* V if every vector in V can be written as a linear combination of the vectors in the set.

Theorem 3.4. *The span of a set of vectors in a vector space V is a subspace of V.*

Proof: Let $U = \{u_1, u_2,..., u_k\}$ be a set of vectors in V. A vector in $S = \text{span}\{U\}$ is written as

$$a = \sum_{i=1}^{k} a_i u_k$$

Since S is in V, axioms (iii), (iv), (vii), (viii), (ix), and (x) of Definition 3.1 are true, since V is a vector space. Consider the sum of two vectors in the span of S:

$$a + b = \sum_{i=1}^{k} a_i u_i + \sum_{i=1}^{k} b_i u_i = \sum_{i=1}^{k} (a_i + b_i) u_i$$

Hence, $a+b$ is in S, as it is written as a linear combination of the elements of U. Thus, U is closed under vector addition. Closure under scalar multiplication is similarly shown. The zero vector is obtained by setting each $a_i = 0$, $i = 1, 2,..., n$. The vector $-\mathbf{a}$ is obtained as

$$-\mathbf{a} = \sum_{i=1}^{k} (-a_i)\mathbf{u}_i$$

Hence, all 10 axioms of Definition 3.1 hold, and S is a subspace of V.

Example 3.6. $P^2[0, 1]$ is the space of all polynomials of degree two or less, defined on the interval, $0 \leq x \leq 1$. It is easy to show that $P^2[0, 1]$ is a vector space and is a subspace of $C^n[0, 1]$. Does the set

$$P = \{p_1(x), p_2(x), p_3(x)\} \qquad p_1(x) = 1, \quad p_2(x) = x, \quad p_3(x) = x^2 \qquad \text{(a)}$$

span $P^2[0, 1]$?

Solution: A linear combination of the set of vectors is

$$p(x) = a + bx + cx^2 \qquad \text{(b)}$$

Every polynomial of degree two or less can be obtained by varying the coefficients in the linear combination. Hence, the set spans $P^2[0, 1]$.

The span of a vector space is not unique. Indeed, the number of vectors in a span is also not unique. If the set P of Example 3.6 is augmented to include $p_4(x) = x^2 + 2x + 3$, then the augmented set also spans $P^2[0, 1]$. It is shown in Example 3.5 that the vectors in P are linearly independent. However, since $p_4(x) = 3p_1(x) + 2p_2(x) + p_3(x)$, the augmented set is not a set of linearly independent vectors.

Definition 3.5. A set of linearly independent vectors that spans a vector space V is called a *basis* for V.

The choice of basis vectors is not unique. However the number of vectors in the basis is unique. The unit vectors $\mathbf{i}$, $\mathbf{j}$, and $\mathbf{k}$, defined as being parallel to the coordinate axes in three dimensional space, form a basis for R^3. An alternate choice for a basis is a set of vectors parallel to coordinate axes obtained by rotating the x–y axes about the z axis, through a counterclockwise angle θ, as illustrated in Figure 3.3. These vectors are related to $\mathbf{i}$, $\mathbf{j}$, and $\mathbf{k}$ by $\mathbf{i} = \mathbf{i}' \sin \theta - \mathbf{j}' \cos \theta$, $\mathbf{j} = \mathbf{i}' \cos \theta - \mathbf{j}' \sin \theta$, and $\mathbf{k}' = \mathbf{k}$. The vectors $\mathbf{i}'$, $\mathbf{j}'$, and $\mathbf{k}'$ are also a basis for R^3.

By definition, the set P, defined in Example 3.6, spans $P^2[0, 1]$. Indeed, any set of three linearly independent polynomials of degree two or less constitute a basis for $P^2[0, 1]$. Not all vectors in $P^2[0, 1]$ can be written as a linear combination of only two linearly independent polynomials. A set of four polynomials of degree two or less is not linearly independent. Thus, the number of elements in the basis for $P^2[0, 1]$ is exactly three.

Definition 3.6. The number of vectors in the basis of a vector space V is called the *dimension* of V.

The vector space R^n has dimension n. The space $P^n[a, b]$ has dimension $n+1$. These are examples of *finite-dimensional vector spaces*.

It is not possible to express every element in $C^2[0, 1]$ by a linear combination of a finite number of elements of $C^2[0, 1]$. A basis for $C^2[0, 1]$ contains an infinite, but countable, number of elements. This is an example of an *infinite dimensional vector space*. Any set of vectors, from which any element of an infinite dimensional vector space can be written, is said to be *complete* in that space. The subspace of Example 3.4 is also an infinite dimensional vector space. It is important to determine a complete set of vectors for such a space.

The expression of a vector in an infinite-dimensional vector space, in terms of members of a complete set (or basis), is an infinite series. Thus, questions of convergence arise. Such questions include whether the series converges at every point within the interval $[a, b]$, and if so, then to what does it converge. While these questions are interesting, they are not within the scope of this study. Theorems regarding convergence of series will be presented without proof, and subsequently used. Occasionally questions of convergence will not be considered. This does not mean that they are not important, only that their consideration does not improve the understanding of the material being presented. Such a case is in the proof of Theorem 3.3. If the number of vectors in U is not finite, then questions about series convergence arise when considering the closure of S under vector addition and scalar multiplication. However, it is assumed that the series for **a** and **b** converge, and hence their sum converges.

Example 3.7. The determination of the natural frequencies and mode shapes for a non-uniform beam requires the solution of a variable-coefficient differential equation of the form

$$\frac{d^2}{dx^2}\left(EI(x)\frac{d^2w}{dx^2}\right) - \rho A(x)\omega^2 w = 0 \tag{a}$$

where x, the independent variable, is a coordinate along the neutral axis of the beam, measured from its left support, $w(x)$, the dependent variable, is the transverse deflection of the beam; E is the elastic modulus of the material from which the beam is made; ρ is the mass density of the material; $I(x)$ is the area moment of inertia of the non-uniform cross section; $A(x)$ is the area of the cross section; and ω is the natural frequency that is to be determined. After non-dimensionalization, Equation a becomes

$$\frac{d^2}{dx^2}\left(\alpha(x)\frac{d^2w}{dx^2}\right) - \beta(x)\omega^2 w = 0 \tag{b}$$

The non-dimensional boundary conditions for a fixed-fixed beam are

$$w(0) = 0 \qquad \frac{dw}{dx}(0) = 0$$

$$w(1) = 0 \qquad \frac{dw}{dx}(1) = 0 \tag{c}$$

Since the beam is non-uniform, determination of an exact solution for the mode shapes and natural frequencies is difficult. Thus, an approximate solution is sought. The exact solution resides in a subspace of $C^4[0, 1]$, call it S, such that if $f(x)$ is in S, then

$$f(0) = 0 \qquad \frac{df}{dx}(0) = 0$$

$$f(1) = 0 \qquad \frac{df}{dx}(1) = 0 \tag{d}$$

An approximate solution is sought from $P^6[0, 1]$, the space of polynomials of degree six or less. In order to satisfy the boundary conditions, the approximate solution must also lie in S. Thus, it is desired to seek an approximate solution from the subspace of S and $P^6[0, 1]$ defined as $Q = S \cap P^6[0,1]$.

Determine the dimension of Q and determine a basis for Q.

Solution: Let $f(x)$ be an element of Q. Since $f(x)$ is in $P^6[0, 1]$ then it is of the form

$$f(x) = a_6 x^6 + a_5 x^5 + a_4 x^4 + a_3 x^3 + a_2 x^2 + a_1 x + a_0 \tag{e}$$

Since $f(x)$ is also in S, it must satisfy the boundary conditions. To this end

$$f(0) = 0 = a_0 \tag{f}$$

$$\frac{df}{dx}(0) = 0 = a_1 \tag{g}$$

$$f(1) = 0 = a_6 + a_5 + a_4 + a_3 + a_2 + a_1 + a_0 \tag{h}$$

$$\frac{df}{dx}(1) = 0 = 6a_6 + 5a_5 + 4a_4 + 3a_3 + 2a_2 + a_1 \qquad \text{(i)}$$

Equation f and Equation g obviously imply that $a_0 = 0$ and $a_1 = 0$. Equation h and Equation i are manipulated to give

$$a_2 = a_4 + 2a_5 + 3a_6 \qquad \text{(j)}$$

$$a_3 = -2a_4 - 3a_5 - 4a_6 \qquad \text{(k)}$$

Thus, an arbitrary element of Q is of the form

$$f(x) = a_6 x^6 + a_5 x^5 + a_4 x^4 - (2a_4 + 3a_5 + 4a_6)x^3$$
$$+ (a_4 + 2a_5 + 3a_6)x^2 \qquad \text{(l)}$$

$$f(x) = a_6(x^6 - 4x^3 + 3x^2) + a_5(x^5 - 3x^3 + 2x^2) + a_4(x^4 - 2x^3 + x^2) \quad \text{(m)}$$

Since $f(x)$ is a linear combination of three linearly independent functions, the dimension of Q is three and a basis for Q is

$$f_1(x) = x^6 - 4x^3 + 3x^2 \qquad \text{(n)}$$

$$f_2(x) = x^5 - 3x^3 + 2x^2 \qquad \text{(o)}$$

$$f_3(x) = x^4 - 2x^3 + x^2 \qquad \text{(p)}$$

3.6 INNER PRODUCTS

A function of a vector is an operation that can be performed on all vectors in a vector space. The operation may involve more than one vector from the vector space. If the result of the operation is a scalar, the function is called a *scalar function*. An example of a scalar function is the dot product defined for two vectors in R^3. Properties satisfied by the dot product are summarized in Section 3.2. The concept of dot product can be generalized into the following definition.

Definition 3.7. Let **u** and **v** be arbitrary vectors in a vector space V. An inner product of **u** and **v**, written (**u**,**v**), is a scalar function which satisfies the following properties:

i. $(\mathbf{u}, \mathbf{v}) = (\overline{\mathbf{v}, \mathbf{u}})$ (Commutative property)
ii. $(\mathbf{u} + \mathbf{v}, \mathbf{w}) = (\mathbf{u}, \mathbf{w}) + (\mathbf{v}, \mathbf{w})$ (Distributive property)
iii. $(\alpha\mathbf{u}, \mathbf{v}) = \alpha(\mathbf{u}, \mathbf{v})$ (Associative property of scalar multiplication)
iv. $(\mathbf{u}, \bar{\mathbf{u}}) \geq 0$ and $(\mathbf{u}, \bar{\mathbf{u}}) = 0$ if, and only if, $\mathbf{u} = \mathbf{0}$ (Non-negativity property),

where α is an arbitrary scalar and an overbar denotes a complex conjugate.

Example 3.8. Which of the following constitutes a valid definition of an inner product for vectors in R^3?

$$(\mathbf{u}, \mathbf{v}) = u_1 v_1 + u_2 v_2 + u_3 v_3 \tag{a}$$

$$(\mathbf{u}, \mathbf{v}) = u_1 v_1 + 2u_2 v_2 + u_3 v_3 \tag{b}$$

$$(\mathbf{u}, \mathbf{v}) = u_1 v_1 - u_2 v_2 + u_3 v_3 \tag{c}$$

$$(\mathbf{u}, \mathbf{v}) = u_1 v_1 + u_2 v_3 + u_3 v_2 \tag{d}$$

$$(\mathbf{u}, \mathbf{v}) = u_1 v_1 + u_2 v_2 + u_3^2 v_3^2 \tag{e}$$

Solution: In order to determine whether each constitutes a valid inner product on R^3, it must be determined if all of the properties of Definition 3.7 are satisfied.

a. This is the same as the definition of the dot product of two vectors, from which the definition of inner product is a generalization, thus Equation a defines a valid inner product for R^3.
b. Equation b defines a valid inner product for R^3. Its satisfaction of the properties of a valid inner product is shown below:

i. $(\mathbf{u},\mathbf{v}) = u_1 v_1 + 2u_2 v_2 + u_3 v_3 = v_1 u_1 + 2v_2 u_2 + v_3 u_3 = (\mathbf{v},\mathbf{u})$
ii. $(\mathbf{u} + \mathbf{v},\mathbf{w}) = (u_1 + v_1)w_1 + 2(u_2 + v_2)w_2 + (u_3 + v_3)w_3$
$\qquad = u_1 w_1 + v_1 w_1 + 2u_2 w_2 + 2v_2 w_2 + u_3 w_3 + v_3 w_3$
$\qquad = (u_1 w_1 + 2u_2 w_2 + u_3 w_3) + (v_1 w_1 + 2v_2 w_2 + v_3 w_3)$
$\qquad = (\mathbf{u},\mathbf{w}) + (\mathbf{v},\mathbf{w})$
iii. $(\alpha\mathbf{u},\mathbf{v}) = (\alpha u_1)v_1 + 2(\alpha u_2)v_2 + (\alpha u_3 v_3)$
$\qquad = \alpha(u_1 v_1 + 2u_2 v_2 + u_3 v_3)$
$\qquad = \alpha(\mathbf{u},\mathbf{v})$

iv. $(\mathbf{u},\mathbf{u}) = u_1^2 + 2u_2^2 + u_3^2$. Since this is the sum of non-negative terms it is non-negative for any $\mathbf{u}$. In addition, it is clear that $(\mathbf{u}, \mathbf{u}) \geq 0$ if, and only if, $\mathbf{u} = 0$.

c. Equation c is not a definition of a valid inner product for R^3. Consider, for example, the vector

$$\mathbf{u} = \begin{bmatrix} 1 \\ 1 \\ 0 \end{bmatrix}$$

Using the proposed definition of the inner product

$$(\mathbf{u},\mathbf{u}) = (1)(1) - (1)(1) + (0)(0) = 0$$

Thus, there exists a $\mathbf{u} \neq 0$ such that $(\mathbf{u},\mathbf{u}) = 0$ and property (iv) is violated.

d. Equation d does not represent a valid inner product for R^3. Consider the vector

$$\mathbf{u} = \begin{bmatrix} 0 \\ 1 \\ 0 \end{bmatrix}$$

Using the proposed definition of inner product

$$(\mathbf{u},\mathbf{u}) = (0)(0) + (1)(0) + (0)(1) = 0$$

Thus, there exists a $\mathbf{u} \neq 0$ such that $(\mathbf{u}, \mathbf{u}) = 0$ and property (iv) is violated.

e. Equation e does not define a valid definition of inner product for R^3. Consider property (ii)

$$(\mathbf{u} + \mathbf{v},\mathbf{w}) = (u_1 + v_1)w_1 + (u_2 + v_2)w_2 + (u_3 + v_3)^2 w_3^2$$

It is easy to find a set of vectors for which this is not equal to $(\mathbf{u}, \mathbf{w}) + (\mathbf{v}, \mathbf{w})$.

The definition of an inner product is not unique. There are many valid definitions of an inner product for any vector space. The inner product for R^n, defined by

$$(\mathbf{u},\mathbf{v}) = \sum_{i=1}^{n} u_i v_i$$

is called the *standard inner product for R^n*. The inner product on $C[a, b]$, defined by

$$(\mathbf{f},\mathbf{g}) = \int_a^b f(x)\, g(x)\mathrm{d}x$$

is called the *standard inner product for $C[a, b]$*.

Theorem 3.5. *(Cauchy–Schwartz Inequality). Let $(\mathbf{u},\mathbf{v})$ represent a valid inner product defined for a real vector space V. Then,*

$$(\mathbf{u},\mathbf{v})^2 \le (\mathbf{u},\mathbf{u})(\mathbf{v},\mathbf{v}) \tag{3.14}$$

Proof: The proof of the Cauchy–Schwartz inequality involves the use of properties of inner products and the algebra of inner products, and is thus instructive in its own right. From property (iv) of Definition 3.7, for any real α and β it is noted that

$$(\alpha\mathbf{u} - \beta\mathbf{v}, \alpha\mathbf{u} - \beta\mathbf{v}) \ge 0 \tag{3.15}$$

Use of property (iii) of Definition 3.7 leads to

$$(\alpha\mathbf{u},\alpha\mathbf{u}) + (\alpha\mathbf{u} - \beta\mathbf{v}) + (-\beta\mathbf{v},\alpha\mathbf{u}) + (-\beta\mathbf{v} - \beta\mathbf{v}) \ge 0 \tag{3.16}$$

Since V is a real vector space, property (i) of Definition 3.7 becomes the commutative property. Using properties (i) and (iii) leads to

$$\alpha^2(\mathbf{u},\mathbf{u}) - 2\alpha\beta(\mathbf{u},\mathbf{v}) + \beta^2(\mathbf{v},\mathbf{v}) \ge 0 \tag{3.17}$$

Specifically, define $\alpha = (\mathbf{v},\mathbf{v})^{1/2}$ and $\beta = (\mathbf{u},\mathbf{u})^{1/2}$, whose use in Equation 3.17 leads to

$$(\mathbf{v},\mathbf{v})(\mathbf{u},\mathbf{u}) - 2(\mathbf{u},\mathbf{u})^{1/2}(\mathbf{v},\mathbf{v})^{1/2}(\mathbf{u},\mathbf{v}) + (\mathbf{u},\mathbf{u})(\mathbf{v},\mathbf{v}) \ge 0$$
$$(\mathbf{u},\mathbf{u})^{1/2}(\mathbf{v},\mathbf{v})^{1/2} \ge (\mathbf{u},\mathbf{v}) \tag{3.18}$$

which upon squaring becomes Equation 3.14.

Definition 3.8. Two vectors $\mathbf{u}$ and $\mathbf{v}$ in a vector space V are said to be orthogonal with respect to a defined inner product if $(\mathbf{u}, \mathbf{v}) = 0$.

The use of the term orthogonality is a generalization of the concept from R^3, in which two vectors are geometrically perpendicular, or orthogonal, when their dot product is zero. However, note that, in general, the determination of orthogonality of two vectors is dependent upon the inner product. Orthogonality of two vectors with respect to one inner product defined on a vector space V does not guarantee orthogonality with respect to any other valid inner product.

Example 3.9. Consider two vectors in R^3 defined as

$$\mathbf{u} = \begin{bmatrix} 1 \\ 3 \\ -2 \end{bmatrix} \quad \mathbf{v} = \begin{bmatrix} -2 \\ 1 \\ 2 \end{bmatrix}$$

Determine whether these vectors are orthogonal with respect to the inner products defined in (a) and (b) of Example 3.8.

Solution: (a) Checking orthogonality with respect to the inner product of (a) leads to

$$\begin{aligned} (\mathbf{u},\mathbf{v}) &= u_1 v_1 + u_2 v_2 + u_3 v_3 \\ &= (1)(-2) + (3)(1) + (-2)(2) = -1 \end{aligned}$$

Thus, the vectors are not orthogonal with respect to this inner product.

(b) Checking orthogonality with respect to the inner product of (b) leads to

$$\begin{aligned} (\mathbf{u}, \mathbf{v}) &= u_1 v_1 + 2u_2 v_2 + u_3 v_3 \\ &= (1)(-2) + 2(3)(1) + (-2)(2) - 0 \end{aligned}$$

Thus, the vectors are orthogonal with respect to this inner product.

3.7 NORMS

It is often important to have a measure of the "length" of a vector. This is especially important when estimating the error in approximating one vector by another vector. A measure of the error might be the "length" of the error vector, which is defined as the difference between the exact vector an approximate vector.

The calculation to determine the geometric length of a vector in R^3 is an operation performed on the vector, which results in a scalar. Thus, the length is a scalar function of the vectors that satisfies certain properties. This concept can be extended to develop a definition for the length of a vector in a general vector space.

Definition 3.9. A *norm* of a vector **u** in a vector space V, written $\|u\|$ is a function of the vector, whose result is a real scalar and that satisfies the following properties:

 i. $\|u\| \geq 0$ and $\|u\| = 0$ if, and only if, $\mathbf{u} = \mathbf{0}$
 ii. $\|\alpha u\| = |\alpha| \, \|u\|$ for any scalar α in the associated scalar field of V
 iii. $\|u + v\| \leq \|u\| + \|v\|$, the triangle inequality

The norm is a generalization of the concept of length of a vector. The first property requires the length of the vector to be non-negative and that only the zero vector may have a length of zero. The second property is a scaling property. If the vector is multiplied by a scalar, then the length of the resulting product is proportional to the length of the original vector with the constant of proportionality being the absolute value of the scalar. The third property is a generalization of the triangle rule for vector addition in three-dimensional space, in which a geometric representation of the addition of two vectors is obtained by placing the head of one vector at the tail of the other, with the resultant vector being drawn from the tail of the first vector to the head of the second; the two vectors and their resultant forming a triangle. The triangle inequality is then a statement that the lengths of any side of a triangle must be less than the sum of the lengths of the other two sides.

Occasionally a function satisfies all requirements to be called a norm, except that vectors other than the zero vector will have a norm of zero. This case is covered in the following definition.

Definition 3.10. A *semi-norm* of a vector **u** in a vector space V, $s(\mathbf{u})$ is a function of the vector whose result is a real scalar and that satisfies the following properties:

 i. $s(\mathbf{u}) \geq 0$ and $s(\mathbf{u}) = 0$
 ii. $s(\alpha \mathbf{u}) = |\alpha| s(\mathbf{u})$ for any scalar α in the associated scalar field of V
 iii. $s(\mathbf{u} + \mathbf{v}) \leq s(\mathbf{u}) + s(\mathbf{v})$, the triangle inequality

Example 3.10. Show that the following is a valid definition of a norm on $C[a, b]$

$$\|f\|_\infty = \max_{a \le x \le b} |f(x)| \tag{a}$$

Solution: The proof that Equation a is a valid definition of a norm requires that property (i) through property (iii) is shown to hold for all $f(x)$ in $C[a, b]$. Clearly, from its definition of Equation (a) $\|f\|_\infty \ge 0$ for all $f(x)$ and $\|0\|_\infty = 0$. Since $f(x)$ is a continuous function, if $f(x) \ne 0$ for some x_0, $a \le x_0 \le b$, then there is a finite region around x_0 where $f(x) \ne 0$. Thus, it is easy to argue that any $f(x) \ne 0$ must have a maximum greater than zero.

Considering property (ii),

$$\begin{aligned}
\|\alpha f\|_\infty &= \max_{a \le x \le b} |\alpha f(x)| \\
&= \max_{a \le x \le b} |\alpha| |f(x)| \\
&= |\alpha| \max_{a \le x \le b} |f(x)| \\
&= |\alpha| \|f\|_\infty
\end{aligned} \tag{b}$$

Finally, considering the triangle inequality,

$$\begin{aligned}
\|f + g\|_\infty &= \max_{a \le x \le b} |f(x) + g(x)| \le \max_{a \le x \le b} (|f(x)| + |g(x)|) \\
&= \max_{a \le x \le b} |f(x)| + \max_{a \le x \le b} |g(x)| \\
&= \|f\|_\infty + \|g\|_\infty
\end{aligned} \tag{c}$$

Hence since the three properties of Definition 3.9 are satisfied, $\|f\|_\infty$, defined by Equation (a), constitutes a valid norm on $C[0, 1]$.

The definition of a norm for a vector space is not unique. One possible definition of a norm for $C[a, b]$ is presented in Equation a of Example 3.10. Another scalar function defined for vectors in $C[a, b]$, which satisfies the properties of Definition 3.9 is

$$\|f\|_1 = \int_a^b |f(x)| dx$$

When comparing the closeness of two vectors, the use of the $\|f\|_\infty$ determines whether the error between the two functions is small at every value of x, whereas the use of the $\|f\|_1$ norm determines whether the functions are close in an average sense.

Example 3.11. Show that the function

$$s(\mathbf{u}) = [u_1^2 - 2u_1 u_2 + u_2^2]^{1/2} \tag{a}$$

defined for a vector $\mathbf{u}$ in R^2 is a semi-norm for R^2 but is not a valid norm.

Solution: Note that

$$s(\mathbf{u}) = [(u_1 - u_2)^2]^{1/2} = |u_1 - u_2|$$

and that if

$$\mathbf{u} = \begin{bmatrix} 1 \\ 1 \end{bmatrix}$$

Then, $s(\mathbf{u}) = (1 - 1) = 0$. Thus, since $s(\mathbf{u}) = 0$ for $\mathbf{u} \neq \mathbf{0}$, then if $s(\mathbf{u})$ is a semi-norm, it is not a norm. Clearly, $s(\mathbf{u}) \geq 0$ and $s(\mathbf{0}) = 0$. Then consider

$$s(\alpha \mathbf{u}) = [(\alpha u_1)^2 - 2(\alpha u_1)(\alpha u_2) + (\alpha u_2)^2]^{1/2}$$
$$= |\alpha| s(\mathbf{u})$$

and

$$s(\mathbf{u} + \mathbf{v}) = [(u_1 + v_1)^2 - 2(u_1 + v_1)(u_2 + v_2) + (u_2 + v_2)^2]^{1/2}$$

$$= [u_1^2 + 2u_1 v_1 + v_1^2 - 2u_1 u_2 - 2u_1 v_2 - 2v_1 u_2 - 2v_1 v_2 + u_2^2 + 2u_2 v_2 + v_2^2]^{1/2}$$

$$= [u_1^2 - 2u_1 u_2 + u_2^2 + v_1^2 - 2v_1 v_2 + v_2^2]^{1/2}$$

$$\leq [u_1^2 - 2u_1 u_2 + u_2^2]^{1/2} + [v_1^2 - 2v_1 v_2 + v_2^2]^{1/2}$$

$$= s(\mathbf{u}) + s(\mathbf{v})$$

Hence, $s(\mathbf{u})$ satisfies all properties of a semi-norm.

Theorem 3.6. Let $(\mathbf{u}, \mathbf{v})$ be a valid inner product defined on a vector space V. Then, $\|\mathbf{u}\| = (\mathbf{u}, \bar{\mathbf{u}})^{1/2}$ is a valid norm on V. Such a norm is called an inner product generated norm.

Proof: Since $(\mathbf{u}, \mathbf{v})$ represents a valid inner product on V, it satisfies the properties of Definition 3.7. To prove that $\|\mathbf{u}\|$ represents a valid norm, it must be shown that the properties of Definition 3.9 are satisfied:

i. $\|\mathbf{u}\| = (\mathbf{u}, \bar{\mathbf{u}})^{1/2} \geq 0$ and $\|\mathbf{u}\| = 0$ if, and only if, $\mathbf{u} = \mathbf{0}$ is true due to Property (iv) of Definition 3.7

ii. $\|\alpha\mathbf{u}\| = (\alpha\mathbf{u}, \alpha\mathbf{u})^{1/2} = [\alpha\bar{\alpha}(\mathbf{u}, \bar{\mathbf{u}})]^{1/2} = |\alpha|(\mathbf{u}, \bar{\mathbf{u}})^{1/2} = |\alpha|\|\mathbf{u}\|$

iii. Use of the Cauchy–Schwartz inequality, Equation 3.14, leads to

$$\|\mathbf{u} + \mathbf{v}\| \le [(\mathbf{u}, \mathbf{u}) + 2(\mathbf{u}, \bar{\mathbf{u}})^{1/2}(\mathbf{v}, \bar{\mathbf{v}})^{1/2} + (\mathbf{v}, \bar{\mathbf{v}})]^{1/2}$$
$$= [[(\mathbf{u}, \bar{\mathbf{u}})^{1/2} + (\mathbf{v}, \bar{\mathbf{v}})^{1/2}]^2]^{1/2}$$
$$= \|\mathbf{u}\| + \|\mathbf{v}\|$$

The $\|f\|_\infty$ norm, and the $\|f\|_1$ norm, have already been defined for $C[a,b]$. Theorem 3.4 shows that another valid norm for $C[a,b]$ is the inner-product-generated norm

$$\|f\| = \left[\int_a^b [f(x)]^2 dx \right]^{1/2} \tag{3.19}$$

This norm will be used without subscript and is referred to as the standard inner-product-generated norm for $C[a, b]$.

Similarly, the standard inner-product-generated norm for R^n is

$$\|\mathbf{u}\| = \left[\sum_{i=1}^n u_i^2 \right]^{1/2} \tag{3.20}$$

3.8 GRAM–SCHMIDT ORTHONORMALIZATION METHOD

The dot product defined in three-dimensional space has a geometric interpretation as the length of the projection of one vector onto another. If the two vectors are mutually orthogonal, the dot product of the vectors is zero. It is convenient in R^3 to use a set of basis vectors that are mutually orthogonal to represent any vector in R^3. To this end, a basis of vectors that are mutually orthogonal is defined as

$$\mathbf{i} = \begin{bmatrix} 1 \\ 0 \\ 0 \end{bmatrix} \quad \mathbf{j} = \begin{bmatrix} 0 \\ 1 \\ 0 \end{bmatrix} \quad \mathbf{k} - \begin{bmatrix} 0 \\ 0 \\ 1 \end{bmatrix}$$

Any vector in R^3 can be written as

$$\mathbf{u} = u_1\mathbf{i} + u_2\mathbf{j} + u_3\mathbf{k}$$

An orthogonal basis is convenient for the representation of the vector space. It is often easy to obtain a linearly independent basis for a vector space.

For example, it is clear that the vectors $p_1(x) = 1$, $p_2(x) = x$, and $p_3(x) = x^2$ form a basis for $P^2[a, b]$. However, this basis is not an orthogonal basis for $P^2[a, b]$ with respect to the standard inner product. Thus, it is necessary to develop a scheme to determine an orthogonal basis from a set of linearly independent vectors.

Definition 3.11. A set of vectors is said to be *normalized* with respect to a norm if the norm of every vector in the set is one.

Definition 3.12. A set of vectors is said to be *orthonormal* with respect to an inner product if the vectors in the set are mutually orthogonal with respect to the inner product, and the set is normalized with respect to the inner-product-generated norm.

If a set of vectors $\mathbf{u}_1$, $\mathbf{u}_2$, $\cdots$, $\mathbf{u}_n$ is orthonormal with respect to an inner product, then

$$(\mathbf{u}_i, \mathbf{u}_j) = \delta_{ij} = \begin{cases} 0 & i \neq j \\ 1 & i = j \end{cases} \tag{3.21}$$

for all $i, j = 1, 2, \ldots, n$.

Theorem 3.7. (Gram–Schmidt Orthonormalization) *Let $\mathbf{u}_1$, $\mathbf{u}_2$, $\mathbf{u}_3$,... be a finite, or countably infinite, set of linearly independent vectors in a vector space V, with a defined inner product $(\mathbf{u}, \mathbf{v})$. Let S be the subspace of V spanned by the vectors. There exists an orthonormal set of vectors $\mathbf{v}_1$, $\mathbf{v}_2$, $\mathbf{v}_3$,..., product whose span is also S. The members of the orthonormal basis are calculated sequentially according to*

$$\mathbf{w}_1 = \mathbf{u}_1 \qquad\qquad \mathbf{v}_1 = \frac{\mathbf{w}_1}{||\mathbf{w}_1||}$$

$$\mathbf{w}_2 = \mathbf{u}_2 - (\mathbf{u}_2, \mathbf{v}_1)\mathbf{v}_1 \qquad \mathbf{v}_2 = \frac{\mathbf{w}_2}{||\mathbf{w}_2||}$$

$$\vdots \qquad\qquad\qquad \vdots \tag{3.22}$$

$$\mathbf{w}_n = \mathbf{u}_n - \sum_{i=1}^{n-1}(\mathbf{u}_n, \mathbf{v}_i)\mathbf{v}_i \quad \mathbf{v}_n = \frac{\mathbf{w}_n}{||\mathbf{w}_n||}$$

$$\vdots \qquad\qquad\qquad \vdots$$

Proof: Since an orthogonal set of vectors is linearly independent, then, if Equation 3.22 generates an orthonormal set, $\mathbf{v}_1, \mathbf{v}_2, \ldots, \mathbf{v}_n$... are linearly independent. Equation 3.22 also shows that each of the vectors in the proposed orthonormal set is a linear combination of the vectors in the basis for S, thus they are also in S. If S is a space of dimension n, then the proposed set of vectors is composed of n linearly independent vectors, and thus span S. If S is an infinite dimensional space, then it can be shown that the proposed set is complete in S, but that is beyond the scope of this text.

First note that

$$||\mathbf{v}_i|| = \left|\left| \frac{\mathbf{w}_i}{||\mathbf{w}_i||} \right|\right| = \frac{1}{||\mathbf{w}_i||} ||\mathbf{w}_i|| = 1$$

Hence, the set is normalized. It only remains to show that $\mathbf{w}_1$, $\mathbf{w}_2$,... form an orthogonal set of vectors. This is done by induction. First it is shown that $\mathbf{w}_2$ is orthogonal to $\mathbf{v}_1$,

$$(\mathbf{w}_2,\mathbf{v}_1) = (\mathbf{u}_2 - (\mathbf{u}_2,\mathbf{v}_1),\mathbf{v}_1) = (\mathbf{u}_2,\mathbf{v}_1) - (\mathbf{u}_2,\mathbf{v}_1)(\mathbf{v}_1,\mathbf{v}_1)$$

However, $||\mathbf{v}_1|| = 1 = (\mathbf{v}_1, \mathbf{v}_1)^{1/2}$. Thus

$$(\mathbf{w}_2,\mathbf{v}_1) = (\mathbf{u}_2,\mathbf{v}_1) - (\mathbf{u}_2,\mathbf{v}_1) = 0$$

Now, assume

$$(\mathbf{v}_i,\mathbf{v}_j) = \delta_{ij} \quad \text{for } i,j = 1,2,...,k-1$$

and consider for $i<k$,

$$(\mathbf{w}_k,\mathbf{v}_i) = \left(\mathbf{u}_k - \sum_{j=1}^{k-1}(\mathbf{u}_k,\mathbf{v}_j)\mathbf{v}_j,\mathbf{v}_i \right) = (\mathbf{u}_k,\mathbf{v}_i) - \sum_{j=1}^{k-1}(\mathbf{u}_k,\mathbf{v}_j)(\mathbf{v}_j,\mathbf{v}_i)$$

Since $(\mathbf{v}_i, \mathbf{v}_i) = 0$ for $i \neq j$ and for $i, j \leq k$, the only non zero sum in the sum occurs when $j = i$. Thus,

$$(\mathbf{w}_k,\mathbf{v}_j) = (\mathbf{u}_k,\mathbf{v}_i) - (\mathbf{u}_k,\mathbf{v}_i)(\mathbf{v}_i,\mathbf{v}_i) = (\mathbf{u}_k,\mathbf{v}_i) - (\mathbf{u}_k,\mathbf{v}_i) = 0$$

The theorem is thus proved by induction.

Example 3.12. In Example 3.7 a basis is derived for S, the intersection of $P^6[0, 1]$ with the subspace of $C^4[0, 1]$ defined as those functions that satisfy the boundary conditions for the differential equation governing the vibrations of a fixed-fixed beam. A basis for S was found to be

$$f_1(x) = x^6 - 4x^3 + 3x^2$$

$$f_2(x) = x^5 - 3x^3 + 2x^2$$

$$f_3(x) = x^4 - 2x^3 + x^2$$

Use the Gram–Schmidt procedure to determine a basis for S that is orthonormal with respect to the standard inner product for $C^4[0, 1]$.

Solution: Normalizing f_1,

$$\|f_1\| = \left[\int_0^1 (x^6 - 4x^3 + 3x^2)^2 dx \right]^{1/2} = 0.171$$

Then

$$v_1(x) = \frac{f_1(x)}{\|f_1(x)\|}$$

$$= 5.84x^6 - 23.4x^3 + 17.5x^2$$

Calculating w_2,

$$w_2(x) = f_2(x) - (f_2(x), v_1(x))v_1(x)$$

where

$$(f_2(x), v_1(x)) = \int_0^1 (x^5 - 3x^3 + 2x^2)(5.84x^6 - 23.4x^3 + 17.5x^2)dx = 0.09968$$

Thus,

$$w_2(x) = x^5 - 3x^3 + 2x^2 - 0.009968(5.84x^6 - 23.4x^3 + 17.5x^2)$$

$$= -0.5823x^6 + x^5 - 0.6708x^3 + 0.253x^2$$

Normalizing

$$\|w_2(x)\| = \left[\int_0^1 (-0.5823x^6 + x^5 - 0.6708x^3 + 0.253x^2)^2 dx\right]^{1/2} = 0.00459$$

$$v_2(x) = \frac{1}{0.00459}(-0.5923x^6 + x^5 - 0.6708x^3 + 0.253x^2)$$

$$v_2(x) = -127.74x^6 + 219.4x^5 - 147.2x^3 + 55.5x^2$$

Calculating $w_3(x)$,

$$w_3(x) = f_3(x) - (f_3(x), v_1(x))v_1(x) - (f_3(x), v_2(x))v_2(x)$$

where

$$(f_3(x), v_1(x)) = \int_0^1 (x^4 - 2x^3 + x^2)(5.84x^6 - 23.4x^3 + 17.5x^2)dx = 0.03962$$

$$(f_3(x), v_2(x)) = \int_0^1 (x^4 - 2x^3 + x^2)(-127.74x^6 + 219.4x^5 - 147.2x^3 + 55.5x^2)dx$$

$$= 0.00419$$

Thus,

$$w_3(x) = x^4 - 2x^3 + x^2 - 0.03962(5.84x^6 - 23.4x^3 + 17.5x^2)$$

$$-0.00419(-127.74x^6 + 219.4x^5 - 147.2x^3 + 55.5x^2)$$

$$w_3(x) = 0.3035x^6 - 0.9186x^5 + x^4 - 0.4580x^3 + 0.0732x^2$$

Finally,

$$\|w_3(x)\| = \left[\int_0^1 (0.3034x^6 - 0.9182x^5 + x^4 - 0.4580x^3 + 0.0732x^2)^2 dx\right]^{1/2}$$

$$= 0.0003444$$

$$v_3(x) = 881.2x^6 - 2667.3x^5 + 2903.6x^4 - 1329.9x^3 + 212.4x^2$$

The representation of a vector by an orthonormal set in an infinite dimensional vector space is complicated by questions of convergence, which are beyond the scope of this text. However some basic results and definitions are presented.

3.9 ORTHOGONAL EXPANSIONS

The Gram–Schmidt Theorem, Theorem 3.7, shows that an orthonormal basis for any vector space can be obtained with respect to a valid inner product $(\mathbf{u}, \mathbf{v})$. Let S be a finite-dimensional vector space of dimension n, and let $\mathbf{v}_1, \mathbf{v}_2,...,\mathbf{v}_n$ be a set of orthonormal vectors that span S with respect to a valid inner product on S. Then, any $\mathbf{u}$ in S can be written as a linear combination of the vectors in the orthonormal basis.

$$\mathbf{u} = \sum_{i=1}^{n} \alpha_i \mathbf{v}_i \tag{3.23}$$

Taking the inner product of both sides of Equation 3.23, with $\mathbf{v}_j$ for an arbitrary $j = 1, 2,..., n$ leads to

$$(u,v_j) = \left(\sum_{i=1}^{n} \alpha_i v_i, v_j \right) \tag{3.24}$$

which, using properties (ii) and (iii) of Definition 3.7, becomes

$$(u,v_j) = \sum_{i=1}^{n} \alpha_i (v_i, v_j) \tag{3.25}$$

Since the basis is an orthonormal basis, the only non-zero term in the summation on the right-hand side of Equation 3.25 corresponds to $i = j$. Simplification, thus leads to

$$\alpha_j = (u,v_j) \tag{3.26}$$

Use of Equation 3.26 in Equation 3.25, leads to

$$u = \sum_{i=1}^{n} (u,v_i) v_i \tag{3.27}$$

Equation 3.27 provides an expansion for **u** in terms of the vectors in an orthonormal basis. The coefficient α_i, given by Equation 3.26, is the component of **u** for the vector v_i.

Example 3.13. The function

$$f(x) = 2x^6 - x^5 + 2x^4 - 9x^3 + 6x^2$$

is a member of Q, the vector space defined in Example 3.7 and Example 3.12. Expand $f(x)$ in terms of the orthonormal basis for Q, determined in Example 3.12.

Solution: Application of Equation 3.27 with $n = 3$ leads to

$$f(x) = (f(x),v_1(x))v_1(x) + (f(x),v_2(x))v_2(x) + (f(x),v_3(x))v_3(x) \qquad \text{(a)}$$

where

$$(f(x),v_1(x)) = \int_0^1 (2x^6 - x^5 + 2x^4 - 9x^3 + 6x^2)(5.84x^6 - 23.4x^3$$

$$+ 17.5x^2)dx$$

$$= 0.322 \qquad \text{(b)}$$

$$(f(x),v_2(x)) = \int_0^1 (2x^6 - x^5 + 2x^4 - 9x^3 + 6x^2)(-127.74x^6 + 219.4x^5$$

$$- 147.2x^3 + 55.5x^2)dx$$

$$= 0.00382 \qquad \text{(c)}$$

$$(f(x),v_3(x)) = \int_0^1 (2x^6 - x^5 + 2x^4 - 9x^3 + 6x^2)(881.2x^6 - 2667.3x^5$$

$$+ 2903.6x^4 - 1329.9x^3 + 212.4x^2)dx$$

$$= 0.0006888 \qquad \text{(d)}$$

Thus,

$$2x^6 - x^5 + 2x^4 - 9x^3 + 6x^2 = 0.322(5.84x^6 - 23.4x^3 + 17.5x^2)$$

$$+0.00382(-127.74x^6 + 219.4x^5 - 147.2x^3 + 55.5x^2) \qquad \text{(e)}$$

$$+0.0006888(881.2x^6 - 2667.3x^5 + 2903.6x^4 - 1329.9x^3 + 212.4x^2)$$

The development of the expansion of a vector in an infinite dimensional vectors space is complicated by questions of completeness and convergence. A detailed discussion of these topics is beyond the scope of this text. However, several definitions and concepts are presented and theorems are presented without proof.

Definition 3.13. Let V be an infinite dimensional vector space with a defined norm. A sequence of vectors $u_1, u_2, u_3, \ldots$ is said to *converge* to a vector u if

$$\lim_{n \to \infty} \|u_n - u\| = 0 \qquad (3.28)$$

Definition 3.14. Let V be an infinite dimensional vector space with a defined norm and let $u_1, u_2, u_3, \ldots$ be a sequence of vectors in V. The sequence is said to be a *Cauchy sequence* if, for each $\varepsilon > 0$, there exists an N, such that if m, $n \geq N$ then $\|u_m - u_n\| < \varepsilon$.

A convergent sequence is also a Cauchy sequence, but the converse is not always true. Consider for example, the sequence of real numbers defined by $x_n = 1/n$. This sequence of real numbers is a Cauchy sequence, but it does not converge to a real number.

Definition 3.15. Let V be an infinite dimensional vector space with a defined norm. The vector space is said to be *complete* if every Cauchy sequence in the space converges to an element of the space.

A vector space can be complete with respect to one norm, but not with respect to another. For example, the space $C[a, b]$, is complete with respect to the norm defined by $\|f\|_2 = \left[\int_a^b [f(x)]^2 dx \right]^{1/2}$, but $C[a, b]$ is not complete whenusing the norm $\|f\|_\infty = \max_{a \leq x \leq b} |f(x)|$.

Definition 3.16. An infinite dimensional space with an inner product defined on it, and is complete with respect to the inner-product-generated norm, is called a *Hilbert space*.

Definition 3.17. An orthonormal set of vectors, S, is said to be *complete* in a Hilbert space if there is no other set of vectors of which S is a subset.

The Gram–Schmidt theorem implies that there is an orthonormal basis for every Hilbert space, which is then complete in the space. However, not every orthonormal set is complete. The following theorem speaks to completeness of the orthonormal set.

Theorem 3.8. (Parseval's identity) *Let u_1, u_2, u_3, ... be an orthonormal set in a Hilbert space. Then, if*

$$\|u\| = \left[\sum_{i=1}^{\infty} (u,u_i)^2 \right]^{1/2} \tag{3.29}$$

for every u in the Hilbert space, then the set is complete.

Finally, we arrive at the following theorem, which shows how to establish an expansion of a vector in terms of a complete orthonormal set in a Hilbert space.

Theorem 3.9. *Let u_1, u_2, u_3, ... be a complete orthonormal set in a Hilbert space V, and let u be an arbitrary element of V. Then,*

$$u = \sum_{i=1}^{\infty} (u,u_i)u_i \tag{3.30}$$

Example 3.14. It is shown in Chapter 5 that the set of functions defined by

$$f_i(x) = \sqrt{2}\,\sin(i\pi x) \tag{a}$$

is a complete orthonormal set on the Hilbert space V, defined as the set of all functions in $C[0,1]$ such that if $y(x)$ is in V, then $y(0) = 0$ and $y(1) = 0$, with the standard inner product for $C[0,1]$. Expand $f(x) = x^2 - 1$, which is a member of V, in terms of this orthonormal set.

Solution: The appropriate expansion is

$$x^2 - 1 = \sum_{i=1}^{''} \alpha_i \sqrt{2}\,\sin(i\pi x) \tag{b}$$

where

$$\alpha_i = \int_0^1 (x^2 - 1)\sqrt{2}\,\sin(i\pi x)\,dx = \frac{2\sqrt{2}}{i^3\pi^3}\left[(-1)^i - 1 - \frac{1}{2}i^2\pi^2 \right] \tag{c}$$

Thus,

$$x^2 - 1 = \frac{4}{\pi^3} \sum_{i=1}^{\infty} \frac{1}{i^3} \left[(-1)^i - 1 - \frac{1}{2} i^2 \pi^2 \right] \sin(i\pi x) \tag{d}$$

3.10 LINEAR OPERATORS

Mathematical modeling of physical systems leads to the formulation of a mathematical problem, whose solution leads to required information regarding the physical system. It is convenient to examine these equations using a consistent formulation. Let **u** represent a vector of dependent variables, which is an element of a vector space D, and let **f** represent a vector which is an element of a vector space R. The relationship between **u** and **f**, obtained from mathematical modelling, is written as

$$\mathbf{Lu} = \mathbf{f} \tag{3.31}$$

where **L** is an *operator*. A formal definition of an operator follows.

Definition 3.18. An operator **L** is a function, which an element of a vector space, D, called the domain of **L**, is mapped into an element of a vector space R, called the range of **L**.

Equation 3.31 is the general form of an operator equation. Given a vector **f** and the definition of **L**, it is desired to find the vector or vectors for which Equation 3.31 is satisfied. The vectors **u**, which satisfy Equation 3.31, are said to be solutions of the equation. Before considering how the solutions to Equation 3.31 are obtained, two basic questions are considered. (1) Does a solution exist? (2) If so, how many solutions exist? The first question can be phrased as "For a specific **f** in R, is there at least one **u** in D, which solves Equation 3.31". An alternate form of the second question is, "If a solution exists, is it unique?" If, for each **f** in R, a unique solution **u**, a vector in D, exists, then there is a one-to-one correspondence between the elements of D and R. In this case, an inverse operator, $\mathbf{L}^{-1}$, exists, whose domain is R and whose range is D, such that

$$\mathbf{u} = \mathbf{L}^{-1}\mathbf{f} \tag{3.32}$$

Definition 3.19. If **L** is an operator defined, such that there is a one-to-one correspondence between the elements of its domain D, and elements of its range R, then the inverse of **L**, denoted as $\mathbf{L}^{-1}$, exists, whose domain is R, and whose range is D, and is defined such that if $\mathbf{Lu} = \mathbf{f}$, then $\mathbf{u} = \mathbf{L}^{-1}\mathbf{f}$.

One method of finding the solution of Equation 3.31 is to determine the inverse of **L** and apply Equation 3.32. Note that substitution for **f** from Equation 3.31 into Equation 3.32 leads to

$$L^{-1}(L u) = u \qquad (3.33)$$

If $R = D$, then u is an element of the domain of L^{-1}, then Equation 3.31 and Equation 3.32 lead to

$$L(L^{-1}u) = u \qquad (3.34)$$

Definition 3.20. An operator L is said to be a *linear operator* if, for each u and v in D, and for all scalars α and β,

$$L(\alpha u + \beta v) = \alpha L u + \beta L v \qquad (3.35)$$

An equation of the form of Equation 3.31, in which L is linear, is said to be a linear equation. If L is not a linear operator, then Equation 3.31 is a non-linear equation. The focus of this study is on linear equations.

A $n \times m$ matrix is a linear operator whose, domain is R^m and whose range is a subspace of R^n. Consider a set of linear equations to solve for a set of m variables $x_1, x_2, x_3, ..., x_m$, as illustrated below:

$$
\begin{aligned}
a_{1,1}x_1 + a_{1,2}x_2 + a_{1,3}x_3 + \cdots + a_{1,m}x_m &= y_1 \\
a_{2,1}x_2 + a_{2,2}x_2 + a_{2,3}x_3 + \cdots + a_{2,m}x_m &= y_2 \\
\vdots \qquad \vdots \qquad \vdots \qquad\qquad \vdots \qquad\ \ \vdots & \\
a_{n,1}x_1 + a_{n,2}x_2 + a_{n,3}x_3 + \cdots + a_{n,m}x_m &= y_n
\end{aligned}
\qquad (3.36)
$$

The matrix formulation of this set of equations is

$$
\begin{bmatrix}
a_{1,1} & a_{1,2} & a_{1,3} & \cdots & a_{1,m} \\
a_{2,1} & a_{2,2} & a_{2,3} & \cdots & a_{2,m} \\
\vdots & \vdots & \vdots & \ddots & \vdots \\
a_{n,1} & a_{n,2} & a_{n,3} & \cdots & a_{n,m}
\end{bmatrix}
\begin{bmatrix}
x_1 \\ x_2 \\ x_3 \\ \vdots \\ x_m
\end{bmatrix}
=
\begin{bmatrix}
y_1 \\ y_2 \\ \vdots \\ y_n
\end{bmatrix}
\qquad (3.37)
$$

Equation 3.37 is summarized by

$$A x = y \qquad (3.38)$$

where A is the matrix operator that represents the coefficients of the equations of Equation 3.36, arranged in n rows and m columns. The element in the ith row and jth column is identified as $a_{i,j}$. The solution x is a vector in R^m, while the input vector y is a vector in R^n.

When $n > m$, the number of linear equations represented by Equation 3.37 is greater than the dimension of the solution vector. If more than m equations are independent, then a solution does not exist for all y. That is, the range of

A is not all of R^n, When $n < m$, the number of equations is less than the dimension of the solution vector. In this case, a solution exists, but is not unique; the range is all of R^n, but for each **y** in R^n there is more than one **x** in R^m that solves Equation 3.38.

When $n = m$, and the number of equations is equal to the dimension of the solution vector, then the domain of **A** is R^n and the range of **A** is a subspace of R^n. It can be shown that the range is all of R^n when the matrix **A** is nonsingular. That is, its determinant is not equal to zero. When the matrix is non-singular, then $\mathbf{A}^{-1}$ exists and has the property.

$$\mathbf{A}^{-1}(\mathbf{A}x) = x \tag{3.39}$$

An associative property can be used on Equation 3.39, giving $(\mathbf{A}^{-1}\mathbf{A})\mathbf{x} = \mathbf{x}$. The operator in parentheses defines the $n \times n$ identity matrix (a matrix with ones along the diagonal and zeros for all off-diagonal elements).

When the determinant of a square matrix is zero, the matrix is said to be singular. In this case, the range of **A** is only a subset of R^n. That is, a solution does not exist for all **y** in R^n. When a solution does exist for a system with a singular matrix, the solution is not unique.

There are many examples of linear operators defined for infinite dimensional vector spaces. Some are illustrated in Table 3.1. The proof of existence and uniqueness of solutions of equations of the form of Equation 3.31, when the domain of **L** is an infinite dimensional vector space, is beyond the scope of this study. However, the problems considered are formulated from the mathematical modeling of a physical system. The physics of the system often dictate that a solution must exist and it must be unique. For example, the temperature distribution in a solid must be continuous and single-valued, requiring a unique solution. If the mathematical problem correctly models the physics, then unique solutions should exist.

Example 3.15. The non-dimensional differential equation for the steady-state temperature distribution in the thin rod shown below, subject to an internal heat generation, is

$$\frac{d^2\Theta}{dx^2} - Bi\Theta = \alpha + \beta \sin(\pi x) \tag{a}$$

where α and β are non-dimensional constants, and $Bi = hp/kA$ is the Biot number. The left end of the rod is maintained at a constant temperature, while the right end is insulated. The temperature distribution satisfies the boundary conditions:

$$\Theta(0) = 0 \tag{b}$$

TABLE 3.1

Examples of Operators on Infinite Dimensional Vector Spaces

$D(L)$	Lu	Inner Product Definition (u,v)	Vibrations Application
$u(x)\in C^2[0,1]$ such that $u(0)=0$ and $du/dx(1)=0$	$-\frac{d}{dx}\left(A(x)u(x)\right)$	$\int\limits_0^1 u(x)v(x)A(x)dx$	Vibrations of bar of area $A(x)$, fixed at $x=0$ and free at $x=1$
$u(x)\in C^4[0,1]$ such that $u(0)=0,\ \frac{du}{dx}(0)=0$ and $A(x)\frac{d^2u}{dx^2}(1)=0,\ \frac{d}{dx}\left(A(x)\frac{d^2u}{dx^2}\right)\Big|_{x=1}=0$	$\frac{d^2}{dx^2}\left(A(x)\frac{d^2u}{dx^2}\right)$	$\int\limits_0^1 u(x)v(x)A(x)dx$	Vibrations of beam of area $A(x)$ fixed at $x=0$ and free at $x=1$
$\mathbf{u}=\begin{bmatrix} f(x) \\ a \end{bmatrix}$ where a is a real number and $f(x)\in C^4[0,1]$ such that $f(0)=0,\ \frac{df}{dx}(0)=0$ and $A(x)\frac{d^2f}{dx^2}(1)=0\ \frac{d}{dx}\left(A(x)\frac{d^2f}{dx^2}\right)\Big|_{x=1}=ka$	$\begin{bmatrix} \dfrac{d^2}{dx^2}\left(A(x)\dfrac{d^2f}{dx^2}\right) \\ ca \end{bmatrix}$	$\int\limits_0^1 \mathbf{v}^{\mathbf{T}}\mathbf{u}A(x)dx$	Vibrations of beam fixed at $x=0$ and with mass-spring system attached at $x=1$

(continued)

TABLE 3.1 (Continued)

D(L)	Lu	Inner Product Definition (u,v)	Vibrations Application
$\mathbf{u} = \begin{bmatrix} f(x) \\ g(x) \end{bmatrix}$ where $f(x)$ and $g(x)$ are both in a subspace of $C^4[0,1]$ such that $f(0)=0$, $\frac{df}{dx}(0)=0$ $A(x)\frac{d^2 f}{dx^2}(1)=0$ $\frac{d}{dx}\left(A(x)\frac{d^2 f}{dx^2}\right)\big|_{x=1}=0$	$\begin{bmatrix} \dfrac{d^2}{dx^2}\left(A(x)\dfrac{d^2}{dx^2}\right) & -c \\[2ex] -c & \dfrac{d^2}{dx^2}\left(A(x)\dfrac{d^2}{dx^2}\right) \end{bmatrix}$	$\int_0^1 \mathbf{v}^T \mathbf{u} A(x)\,dx$	Vibrations of two identical beams in parallel connected by and elastic layer
$\mathbf{u} = \begin{bmatrix} f(x) \\ g(x) \end{bmatrix}$ where $f(x)$ and $g(x)$ are both in a subspace of $C^4[0,1]$ such that $f(0)=0$, $\frac{df}{dx}(0)=0$ and $f(1)=0$, $\frac{df}{dx}(1)=0$	$\begin{bmatrix} \dfrac{d^2}{dx^2} & -\dfrac{d}{dx} \\[2ex] \dfrac{d}{dx} & 1-\eta\dfrac{d^2}{dx^2} \end{bmatrix}$	$\int_0^1 \mathbf{v}^T \mathbf{u} A(x)\,dx$	Vibrations of a Timoshenko beam fixed at both ends
$u=f(\mathbf{r})$ where $\mathbf{r}$ is a position vector defined from an origin to a point on a surface S such that $f=0$ on the boundary of S	$-\nabla^2 u$	$\int_S f(\mathbf{r})g(\mathbf{r})\,dS$	Vibrations of a membrane clamped along its edges

$$\frac{d\Theta}{dx}(1) = 0 \tag{c}$$

a. Write this problem in the operator form of Equation 3.31 and define D and R.
b. Show, by obtaining the solution of the differential equation, the existence and uniqueness of the solution.

Solution: (a) Equation a can be written in the form $L\Theta = f$, where $L\Theta = d^2\Theta/dx^2 - Bi\Theta$, and $f(x) = \alpha + \beta \sin(\pi x)$. The domain D is the subspace of $C^2[0, 1]$ of all functions $g(x)$, such that $g(0) = 0$ and $dg/dx(1) = 0$. The range R is the set of all elements of $PC[0, 1]$, the space of all piecewise continuous functions defined on $[0, 1]$.

c. The homogeneous solution of Equation a is of the form

$$\Theta_h(x) = C_1 \cosh(\sqrt{Bi}x) + C_2 \sinh(\sqrt{Bi}x) \tag{d}$$

where C_1 and C_2 are arbitrary constants of integration. The particular solution of Equation a is

$$\Theta_p(x) = -\frac{\alpha}{Bi} - \frac{\beta}{\pi^2 + Bi} \sin(\pi x) \tag{e}$$

The general solution of Equation a is

$$\Theta(x) = C_1 \cosh(\sqrt{Bi}x) + C_2 \sinh(\sqrt{Bi}x) - \frac{\alpha}{Bi}$$

$$-\frac{\beta}{\pi^2 + Bi} \sin(\pi x) \tag{f}$$

Application of the boundary condition of Equation b through Equation d leads to

$$0 = C_1 - \frac{\alpha}{Bi} \Rightarrow C_1 = \frac{\alpha}{Bi} \tag{g}$$

Application of the boundary condition of Equation c leads to

$$0 = \frac{\alpha}{\sqrt{Bi}} \sinh(\sqrt{Bi}) + C_2\sqrt{Bi}\cosh(\sqrt{Bi}) + \frac{\beta\pi}{\pi^2 + Bi}$$

$$C_2 = -\frac{\beta\pi}{\sqrt{Bi}(\pi^2 + Bi)\cosh(\sqrt{Bi})} - \frac{\alpha}{Bi}\tanh(\sqrt{Bi}) \qquad \text{(h)}$$

Use of Equation g and Equation h in Equation d leads to

$$\Theta(x) = \frac{\alpha}{Bi}\cosh(\sqrt{Bi}x)$$

$$+ \left[\frac{\beta\pi}{\sqrt{Bi}(\pi^2 + Bi)\cosh(\sqrt{Bi})} - \frac{\alpha}{Bi}\tanh(\sqrt{Bi})\right]\sinh(\sqrt{Bi}x)$$

$$- \frac{\alpha}{Bi} - \frac{\beta}{\pi^2 + \beta}\sin(\pi x) \qquad \text{(i)}$$

Equation i satisfies the differential equation, Equation a, as well as the boundary conditions of Equation b and Equation c. Thus, a solution exists. If a second solution, $\hat{\Theta}(x)$, exists such that $\hat{\Theta}(x) \neq \Theta(x)$ for some values of x, then the temperature is multi-valued at those values, a physical impossibility. Thus, if Equation a through Equation c are a true mathematical model of the physical system, the solution must be unique.

3.11 ADJOINT OPERATORS

Linear operators have a related operator, called its adjoint, which has important properties.

Definition 3.21. Let V be a vector space with an inner product $(\mathbf{u}, \mathbf{v})$ defined for all $\mathbf{u}$ and $\mathbf{v}$ in V. Let $\mathbf{L}$ be a linear operator whose domain is D, a subspace of V, and whose range is R, also a subspace of V. The adjoint of $\mathbf{L}$ with respect to the defined inner product, written $\mathbf{L}^*$, is an operator whose domain is R, and whose range is D, such that

$$(\mathbf{Lu,v}) = (\mathbf{u,L^*v}) \qquad (3.40)$$

Definition 3.22. If $D = R$ and $\mathbf{L}^* = \mathbf{L}$, then $\mathbf{L}$ is said to be self-adjoint, that is

$$(\mathbf{Lu,v}) = (\mathbf{u,Lv}) \qquad (3.41)$$

for all $\mathbf{u}$ and $\mathbf{v}$ in D.

Example 3.16. Let $\mathbf{A}$ be a $n \times n$ matrix of the form

$$\mathbf{A} = \begin{bmatrix} a_{1,1} & a_{1,2} & a_{1,3} & \cdots & a_{1,n} \\ a_{2,1} & a_{2,2} & a_{2,3} & \cdots & a_{2,n} \\ a_{3,1} & a_{3,2} & a_{3,3} & \cdots & a_{3,n} \\ \vdots & \vdots & \vdots & \ddots & \vdots \\ a_{n,1} & a_{n,2} & a_{n,3} & \cdots & a_{n,n} \end{bmatrix} \qquad \text{(a)}$$

a. Determine $\mathbf{A}^*$ with respect to the standard inner product on R^n.
b. Under what conditions is $\mathbf{A}$ self-adjoint with respect to the standard inner product on R^n?

Solution: (a) Let $\mathbf{u}$ and $\mathbf{v}$ be arbitrary vectors in R^n. Then, by definition of the standard inner product for R^n

$$(\mathbf{A}u,\mathbf{v}) = \sum_{i=1}^{n} (\mathbf{A}u)_i \mathbf{v}_i = \sum_{i=1}^{n} \sum_{j=1}^{n} a_{i,j} u_j v_i \qquad \text{(b)}$$

Since $\mathbf{A}^*$ is an operator whose domain and range are R^n, it has a matrix representation of the form

$$\mathbf{A}^* = \begin{bmatrix} a_{1,1}^* & a_{1,2}^* & a_{1,3}^* & \cdots & a_{1,n}^* \\ a_{2,1}^* & a_{2,2}^* & a_{2,3}^* & \cdots & a_{2,n}^* \\ a_{3,1}^* & a_{3,2}^* & a_{3,3}^* & \cdots & a_{3,n}^* \\ \vdots & \vdots & \vdots & \ddots & \vdots \\ a_{n,1}^* & a_{n,2}^* & a_{n,3}^* & \cdots & a_{n,n}^* \end{bmatrix} \qquad \text{(c)}$$

Then,

$$(\mathbf{u},\mathbf{A}^*\mathbf{v}) = \sum_{i=1}^{n} u_i (\mathbf{A}^*\mathbf{v})_i = \sum_{i=1}^{n} \sum_{j=1}^{n} u_i a_{i,j}^* v_j \qquad \text{(d)}$$

Interchanging the names of the indices in Equation e leads to

$$(\mathbf{u},\mathbf{A}^*\mathbf{v}) = \sum_{j=1}^{n} \sum_{i=1}^{n} a_{j,i}^* u_j v_i \tag{e}$$

In order for the expression in Equation b to be equal to the expression in Equation e for all possible $\mathbf{u}$ and $\mathbf{v}$, it is required that

$$a_{i,j}^* = a_{j,i} \quad \begin{cases} i = 1,2,\ldots,n \\ j = 1,2,\ldots,n \end{cases} \tag{f}$$

Thus,

$$\mathbf{A}^* = \begin{bmatrix} a_{1,1} & a_{2,1} & a_{3,1} & \cdots & a_{n,1} \\ a_{1,2} & a_{2,2} & a_{3,3} & \cdots & a_{n,2} \\ a_{11,3} & a_{2,3} & a_{3,3} & \cdots & a_{n,3} \\ \vdots & \vdots & \vdots & \ddots & \vdots \\ a_{1,n} & a_{2,n} & a_{3,n} & \cdots & a_{n,n} \end{bmatrix} \tag{g}$$

Thus, the adjoint of $\mathbf{A}$ with respect to the standard inner product on R^n, is the matrix obtained by interchanging the rows and columns of $\mathbf{A}$. Such a matrix is called the transpose matrix, $\mathbf{A}^T$. Thus, $\mathbf{A}^* = \mathbf{A}^T$.

(b) From Equation g, it is clear that $\mathbf{A}$ is self-adjoint with respect to the standard inner product if

$$a_{i,j} = a_{j,i} \quad \begin{cases} i = 1,2,\ldots,n \\ j = 1,2,\ldots,n \end{cases} \tag{h}$$

Such a matrix, whose columns can be interchanged with its rows without changing the matrix, is called a symmetric matrix. Thus a $n \times n$ matrix $\mathbf{A}$ is self-adjoint with respect to R^n if, and only if, $\mathbf{A}$ is a symmetric matrix.

Example 3.17. The non-dimensional differential equation governing the deflection, $w(x)$, of a beam due to a distributed load-per-unit length, $f(x)$ is

$$\frac{d^2}{dx^2}\left(\alpha(x)\frac{d^2 w}{dx^2}\right) = f(x) \tag{a}$$

where $\alpha(x)$ is a function describing the variation of material and geometric parameters across the span of the beam. The boundary conditions are dependent upon the end constraints. For a beam fixed at $x = 0$ and free at $x = L$, the appropriate boundary conditions are

$$w(0) = 0 \tag{b1}$$

$$\frac{dw}{dx}(0) = 0 \tag{b2}$$

$$\frac{d^2w}{dx^2}(1) = 0 \tag{b3}$$

$$\frac{d}{dx}\left(\alpha(x)\frac{d^2w}{dx^2}\right)(1) = 0 \tag{b4}$$

Equation a may be written in the form of Equation 3.31 with $Lw = d^2(\alpha(x)(d^2w/dx^2))/dx^2$. Define V as $C^4[0, 1]$. The domain of L is the subspace of V, such that if $g(x)$ is in D, then $g(x)$ satisfies the boundary conditions of Equation b. Consider the range of V to be the same as its domain. The standard inner product on V is

$$(f(x),g(x)) = \int_0^1 f(x)g(x)dx \tag{c}$$

Show that L is self-adjoint on D with respect to the inner product of Equation e.

Solution: If L is self-adjoint on D, then $(\mathbf{Lf}, \mathbf{g}) = (\mathbf{g}, \mathbf{Lf})$ for all $f(x)$ and $g(x)$ in D. That is,

$$\int_0^1 \frac{d^2}{dx^2}\left(\alpha(x)\frac{d^2f}{dx^2}\right)g(x)dx = \int_0^1 f(x)\frac{d^2}{dx^2}\left(\alpha(x)\frac{d^2g}{dx^2}\right)dx \tag{d}$$

Recall the integration by parts formula $\int u\,dv = uv - \int v\,du$. Using integration by parts on the integral on the left hand side of Equation d, with $u = g(x)$, and $dv = (d^2(\alpha(d^2f/dx^2))/dx^2)dx$, gives

$$\int_0^1 \frac{d^2}{dx^2}\left(\alpha(x)\frac{d^2f}{dx^2}\right)g(x)dx = g(x)\frac{d}{dx}\left(\alpha\frac{d^2f}{dx^2}\right)\Big|_{x=0}^{x=1} - \int_0^1 \frac{d}{dx}\left(\alpha\frac{d^2f}{dx^2}\right)\frac{dg}{dx}dx \tag{e}$$

Using integration by parts on the remaining integral of Equation e, with $u = dg/dx$ and $dv = d(\alpha(d^2f/dx^2))/dx$, leads to

$$\int_0^1 \frac{d^2}{dx^2}\left(\alpha(x)\frac{d^2f}{dx^2}\right)g(x)dx$$

$$= g(x)\frac{d}{dx}\left(\alpha\frac{d^2f}{dx^2}\right)\Big|_{x=0}^{x=1} - \left[\frac{dg}{dx}\alpha(x)\frac{d^2f}{dx^2}\right]_{x=0}^{x=1} + \int_0^1 \alpha(x)\frac{d^2f}{dx^2}\frac{d^2g}{dx^2}dx$$

$$= g(1)\frac{d}{dx}\left(\alpha\frac{d^2f}{dx^2}\right)(1) - g(0)\frac{d}{dx}\left(\alpha\frac{d^2f}{dx^2}\right)(0) - \frac{dg}{dx}(1)\alpha(1)\frac{d^2f}{dx^2}(1)$$

$$- \frac{dg}{dx}(0)\alpha(0)\frac{d^2f}{dx^2}(0) + \int_0^1 \alpha(x)\frac{d^2f}{dx^2}\frac{d^2g}{dx^2}dx \qquad\qquad (f)$$

Since $f(x)$ and $g(x)$ are both in D $d(\alpha(d^2f/dx^2))/dx(1) = 0$, $g(0) = 0$, $(d^2f/dx^2)(1) = 0$ and $(dg/dx)(0) = 0$. Thus, Equation f reduces to

$$\int_0^1 \frac{d^2}{dx^2}\left(\alpha(x)\frac{d^2f}{dx^2}\right)g(x)dx = \int_0^1 \alpha(x)\frac{d^2f}{dx^2}\frac{d^2g}{dx^2}dx \qquad\qquad (g)$$

Using similar steps, it can be shown that for all $f(x)$ and $g(x)$ in D,

$$\int_0^1 f(x)\frac{d^2}{dx^2}\left(\alpha(x)\frac{d^2g}{dx^2}\right)dx = \int_0^1 \alpha(x)\frac{d^2f}{dx^2}\frac{d^2g}{dx^2}dx \qquad\qquad (h)$$

Thus, Equation a is proved, and L is self-adjoint on D with respect to the standard inner product.

Example 3.18. The non-dimensional partial differential equation for the steady-state temperature distribution, $\Theta(\mathbf{r})$ in a bounded three-dimensional region, V, is

$$\nabla^2\Theta = f(\mathbf{r}) \qquad\qquad (a)$$

where $\mathbf{r}$ is the position vector from the origin of the coordinate system to a point in the body. The surface of the region is described by $S(\mathbf{r}) = 0$. The surface of the body is open, and heat transfer occurs though convection leading to a boundary condition of the form

$$\nabla\Theta\bullet\mathbf{n} = -Bi\Theta \quad \text{on } S \tag{b}$$

Let V be the space of all functions defined in V. The Laplacian is a linear operator. Define D as the subspace of V, consisting of all functions $g(\mathbf{r})$ that satisfy the boundary condition of Equation b. The standard inner product on V is

$$(f(\mathbf{r}),g(\mathbf{r})) = \int_V f(\mathbf{r})g(\mathbf{r}) \, dV \tag{c}$$

Show that $L = \nabla^2$ is a self-adjoint operator on D with respect to the standard inner product for V.

Solution: If $\mathbf{L}$ is self-adjoint, then for any vectors $f(\mathbf{r})$ and $g(\mathbf{r})$ both in D,

$$\int_V (\nabla^2 f)g \, dV = \int_V f(\nabla^2 g) \, dV \tag{d}$$

Recall the vector identity

$$\nabla\bullet(g\nabla f) = \nabla f\bullet\nabla g + g\nabla^2 f \tag{e}$$

Using the identity of Equation e in the integral on the left-hand side of Equation d leads to

$$\int_V (\nabla^2 f)g \, dV = \int_V [-\nabla f\bullet\nabla g + \nabla\bullet(g\nabla f)] \, dV \tag{f}$$

The divergence theorem implies

$$\int_V \nabla\bullet(g\nabla f) \, dV = \int_S (g\nabla f)\bullet n \, dS \tag{g}$$

which, when used in Equation g, leads to

$$\int_V (\nabla^2 f)g \, dV = -\int_V \nabla f\bullet\nabla g \, dV + \int_S g\nabla f\bullet n \, dS \tag{h}$$

Since both f and g are in D, $\nabla F\cdot\mathbf{n} = -Bif$ on S and $\nabla g\cdot\mathbf{n} = -Big$ on S.

Thus, Equation h can be written as

$$\int_V (\nabla^2 f) g \, dV = - \int_V \nabla f \bullet \nabla g \, dV - Bi \int_S g f \, dS \tag{i}$$

In a similar fashion, it can be shown that

$$\int_V f(\nabla^2 g) \, dV = - \int_V \nabla f \bullet \nabla g \, dV - Bi \int_S g f \, dS \tag{j}$$

The quality of the right-hand sides of Equation i and Equation j proves that L is self-adjoint.

Example 3.19. A Fredholm integral equation is of the form

$$\int_0^x f(x) k(x,y) \, dy = g(x) \tag{a}$$

Equation a can be formulated as $Lf = g$, where $Lf = \int_0^x f(x) k(x,y) \, dy = g(x)$. The domain and range of L is $C[0, a]$. Assuming the adjoint operator is of the form $L^* f = \int_0^x f(x) k^*(x,y) \, dy$, determine the form of $k^*(x, y)$ using the standard inner product $(f,g) = \int_0^a f(x) g(x) \, dx$.

Solution: If L^* is the adjoint of L with respect to the standard inner product, then

$$\int_0^a \left(\int_0^x f(x) k(x,y) \, dy \right) g(x) \, dx = \int_0^a f(x) \left(\int_0^x k^*(x,y) g(y) \, dy \right) dx \tag{b}$$

Working with the left-hand side of Equation b, by interchanging the order of integration, leads to

$$\int_0^a \left(\int_0^x f(y) k(x,y) \, dy \right) g(x) \, dx = \int_0^a \int_0^y f(x) k(x,y) g(y) \, dx \, dy \tag{c}$$

Renaming the variables of integration, replacing x by λ and y by τ in Equation c, leads to

$$(Lf,g) = \int_0^a \int_0^\tau f(\lambda)g(\tau)k(\lambda,\tau)d\lambda d\tau \tag{d}$$

Performing similar steps on the right-hand side of Equation a leads to

$$(f,L^*g) = \int_0^a \int_0^x f(y)g(x)k^*(x,y)dydx = \int_0^a \int_0^\tau f(\lambda)g(\tau)k^*(\tau,\lambda)d\lambda d\tau \tag{e}$$

It is clear that Equation d and Equation e are equivalent for all f and g if, and only if, $k^*(x, y) = k(y, x)$ for all x and y, such that $0 \leq x \leq a$ and $0 \leq y \leq a$.

3.12 POSITIVE DEFINITE OPERATORS

Positive definiteness is a property of many linear operators that, if proven for the operator, has many ramifications regarding approximate and exact solutions of an equation involving the operator.

Definition 3.23. Let **L** be a linear operator, whose domain D, is a subspace of a vector space V, on which an inner product $(\mathbf{u}, \mathbf{v})$ is defined. **L** is positive definite with respect to the inner product if $(\mathbf{Lu}, \mathbf{u}) \geq 0$ for all $\mathbf{u}$ in D and $(\mathbf{Lu}, \mathbf{u}) = 0$ if, and only if, $\mathbf{u} = 0$.

Some operators studied will not satisfy the "only if" clause in order for Definition 3.23 to apply. For such cases, Definition 3.23 is modified as

Definition 3.24. Let **L** be a linear operator whose domain D is a subspace of a vector space V, on which an inner product $(\mathbf{u}, \mathbf{v})$ is defined. **L** is positive semi-definite with respect to the inner product if $(\mathbf{Lu}, \mathbf{u}) \geq 0$ for all $\mathbf{u}$ in D and $(\mathbf{Lu}, \mathbf{u}) \geq 0$, when $\mathbf{u} = 0$, but there are vectors $\mathbf{v} \neq 0$, such that $(\mathbf{Lu}, \mathbf{u}) = 0$.

Example 3.20 The stiffness matrix for a two-degree-of freedom system is

$$\mathbf{K} = \begin{bmatrix} 4 & -2 \\ -2 & 3 \end{bmatrix} \tag{a}$$

Determine if **K** is positive definite with respect to the standard inner product for R^2.

Solution: Let $\mathbf{u}$ be an arbitrary vector in R^2. Then,

$$\mathbf{Ku} = \begin{bmatrix} 4 & -2 \\ -2 & 3 \end{bmatrix} \begin{bmatrix} u_1 \\ u_2 \end{bmatrix} = \begin{bmatrix} 4u_1 - 2u_2 \\ -2u_1 + 3u_2 \end{bmatrix} \tag{b}$$

and

$$(\mathbf{Ku,u}) = (4u_1 - 2u_2)u_1 + (-2u_1 + 3u_2)u_2 = 4u_1^2 - 4u_1u_2 + 3u_2^2$$

$$= 2u_1^2 + 2(u_1^2 - 2u_1u_2 + u_2^2) + u_2^2 = 2u_1^2 + 2(u_1 - u_2)^2 + u_2^2 \tag{c}$$

Thus, from Equation c, since $(\mathbf{Ku}, \mathbf{u})$ is the sum of three non-negative terms, $(\mathbf{Ku}, \mathbf{u}) \geq 0$ for all $\mathbf{u}$. Clearly the right-hand side of Equation c is zero if $\mathbf{u} = 0$ and if either component of $\mathbf{u}$ is not zero then $(\mathbf{Ku}, \mathbf{u})$ is greater than zero. Thus, $\mathbf{K}$ is positive definite according to Definition 3.23.

Example 3.21. Consider the operator defined in Example 3.17, which is used to solve for the deflection of a beam. Show that the operator is positive definite on the domain specified in Example 3.17 with respect to the standard inner product for C^4 [0, 1]. Note that $\alpha(x) > 0$ for all x, $0 \leq x \leq 1$.

Solution: Algebraic manipulation of (Lf, g) for arbitrary f and g, both in D, led to Equation g of Example 3.17, which is repeated below:

$$(Lf,g) = \int_0^1 \frac{d^2}{dx^2}\left(\alpha(x)\frac{d^2f}{dx^2}\right)g(x)dx = \int_0^1 \alpha(x)\frac{d^2f}{dx^2}\frac{d^2g}{dx^2}dx \tag{a}$$

Using Equation a with $g = f$ leads to

$$(Lf,f) = \int_0^1 \alpha(x)\left(\frac{d^2f}{dx^2}\right)^2 dx \tag{h}$$

The integrand of the integral on the right-hand side of Equation b is non-negative for all x within the range of integration. Thus, $(Lf, f) \geq 0$ for all f. If $f(x) = 0$, $d^2f/dx^2 = 0$ and the definite integral in Equation b are zero. The integral is zero for any f, such that $d^2f/dx^2 = 0$ for all x, $0 \leq x \leq 1$. The only continuous function that has a second derivative identically equal to zero is a linear function of the form

$$f(x) = c_1 + c_2 x \tag{c}$$

However, f must be in D, and thus satisfy the boundary conditions of the fixed-free beam given in Equation b of Example 3.17. Applying $f(0) = 0$ leads to $c_1 = 0$. Then, application of $df/dx(0) = 0$ requires $c_2 = 0$. Thus, the only linear function in D is $f(x) = 0$. Thus, the only $f(x)$ in D, such that $(Lf, f) = 0$ is $f(x) = 0$.

From the above, and Definition 3.23, it is clear that L is positive definite.

Example 3.22. Reconsider the heat transfer problem of Example 3.18. Define $L\Theta = -\nabla^2\Theta$. Determine if L is positive definite with respect to the standard inner product for $C^2(V)$ on the domain D, defined such that if $f(\mathbf{r})$ is in D, then (a) $f(\mathbf{r}) = 0$ everywhere on S, the surface of V (this condition corresponds to the surface at a prescribed temperature), (b) $\nabla f \cdot \mathbf{n} = 0$ everywhere on S (this corresponds to a completely insulated surface), and (c) $\nabla f \cdot \mathbf{n} = -Bi(f)$ everywhere on S (this corresponds to a condition of heat transfer by convection from the ambient to the surface).

Solution: Using the same algebra as in Example 3.18, an equation similar to that of Equation h of Example 3.18 can be obtained as

$$(Lf, g) = \int_V (-\nabla^2 f)g\,dV = \int_V \nabla f \bullet \nabla g\,dV - \int_S g\nabla f \bullet \mathbf{n}\,dS \tag{a}$$

Applying Equation a with $g = f$ leads to

$$(Lf, f) = \int_V \nabla f \bullet \nabla f\,dV - \int_S f\nabla f \bullet \mathbf{n}\,dS = \int_V |\nabla f|^2\,dV - \int_S f\nabla f \bullet \mathbf{n}\,dS \tag{b}$$

The integrand of the volume integral of Equation b is non-negative everywhere in V, thus the integral is clearly non-negative. It is also equal to zero when $f = 0$. However, it could also be zero when ∇f is zero everywhere in V. This may occur if $f(\mathbf{r}) = C$, a non-zero constant.

a. If $f(\mathbf{r}) = 0$ everywhere on S, then the integrand of the surface integral of Equation b is zero everywhere on S and the integral is zero. Since f is prescribed to be zero on S, if $f(\mathbf{r}) = C$, then C must be zero. Thus, the only f in D, such that $(Lf, f) = 0$, is $f = 0$. Hence, L is positive definite on D.

b. If $\nabla f \cdot \mathbf{n} = 0$ everywhere on S, then the surface integral of Equation a is identically zero. However, the surface condition is satisfied by $f(\mathbf{r}) = C$ for any value of C. Thus, there exists $f(\mathbf{r}) \neq 0$, such that $(Lf, f) = 0$. Hence, L is positive semi-definite on D.

c. If $\nabla f \cdot \mathbf{n} = -Bi(f)$ everywhere on S, then Equation b can be rewritten as

$$(Lf, f) = \int_V |\nabla f|^2 dV + Bi \int_S f^2 dS \qquad (c)$$

Clearly, the right-hand side of Equation c is non-negative. From the boundary condition, if $\nabla f = 0$ everywhere in V, which implies that $f(\mathbf{r}) = C$, then $f = 0$ everywhere on S, which implies $C = 0$, and thus the right-hand side of Equation c is zero only when $f = 0$ and L is positive definite on D.

3.13 ENERGY INNER PRODUCTS

Theorem 3.9. *If L is a positive definite and self-adjoint linear operator with respect to a defined inner product on a domain D, then an inner product, called an energy inner product, can be defined by*

$$(\mathbf{u}, \mathbf{v})_L = (L\mathbf{u}, \mathbf{v}) \qquad (3.42)$$

Proof: It is required to show that property (i) through (iv) of definition 3.7 hold for all **u**, **v**, and **w** in D when Equation 3.42 is used as an inner product definition. To this end,
Property (i)

$$(\mathbf{u}, \mathbf{v})_L = (L\mathbf{u}, \mathbf{v}) \qquad \text{Definition of proposed inner product}$$
$$= (\mathbf{u}, L\mathbf{v}) \qquad \text{By hypothesis } L \text{ is self-adjoint}$$
$$= (\mathbf{v}, L\mathbf{u}) \qquad \text{By hypothesis } (\mathbf{u}, \mathbf{v}) \text{ is a valid inner product}$$
$$\qquad\qquad\qquad \text{thus Property (i) of Definition 3.7 applies}$$
$$= (\mathbf{v}, \mathbf{u})_L \qquad \text{Definition of proposed inner product}$$

Property (ii)

$$(\mathbf{u} + \mathbf{v}, \mathbf{w})_L = (\mathbf{L}(\mathbf{u} + \mathbf{v}), \mathbf{w}) \qquad \text{Definition of proposed inner product}$$
$$= ((\mathbf{Lu} + \mathbf{Lv}), \mathbf{w}) \qquad \qquad \text{Linearity of } \mathbf{L}$$
$$= (\mathbf{Lu}, \mathbf{w}) + (\mathbf{Lv}, \mathbf{w}) \qquad \text{Property (ii) of Definition 3.7}$$
$$= (\mathbf{u}, \mathbf{w})_L + (\mathbf{v}, \mathbf{w})_L \qquad \text{Definition of proposed inner product}$$

Property (iii)

$$(\alpha\mathbf{u}, \mathbf{v})_L = (\mathbf{L}(\alpha\mathbf{u}), \mathbf{v}) \qquad \text{Definition of proposed inner product}$$
$$= (\alpha\mathbf{Lu}, \mathbf{v}) \qquad \qquad \text{Linearity of } \mathbf{L}$$
$$= (\alpha(\mathbf{Lu}), \mathbf{v}) \qquad \text{Property (iii) of Definition 3.7}$$
$$= \alpha(\mathbf{u}, \mathbf{v}) \qquad \text{Definition of proposed inner product}$$

Property (iv)

$$(\mathbf{u}, \mathbf{u})_L = (\mathbf{Lu}, \mathbf{u}) \qquad \qquad \text{Definition of proposed inner product}$$

$$\begin{cases} \geq 0 \quad \text{for all } \mathbf{u} \text{ in D} \\ = 0 \text{ if and only if } \mathbf{u} = 0 \end{cases} \qquad \mathbf{L} \text{ is positive definite}$$

Thus, since Property (i) through (iv) of Definition 3.xx are satisfied by the proposed definition of Equation 3.42, the energy inner product is a valid inner product on D.

Corollary. *If* $\mathbf{L}$ *is a positive definite and self-adjoint linear operator with respect to an inner product* (u, v), *defined for its domain* D, *then an energy norm defined by*

$$\|\mathbf{u}\|_r = [(\mathbf{u}, \mathbf{u})]^{1/2} \tag{3.43}$$

is a valid norm on D.

Proof: Since $\mathbf{L}$ is positive definite and self-adjoint with respect to (u, v), then by Theorem 3.9, the energy inner product is a valid inner product on D. Then, by Theorem, any inner product can generate a norm of the form of Equation 3.43.

The term "energy" inner product is applied to inner products of the form of Equation 3.43, as positive definite and self-adjoint operators that arise from the mathematical modeling of physical systems can often be derived using energy methods. In such cases $(\mathbf{Lu}, \mathbf{u})$ is related to some form of energy.

Example 3.23. The stiffness matrix of Example 3.20 is the stiffness matrix obtained in the modeling of the two degree-of-freedom system shown below. Since the matrix is symmetric, it is self-adjoint with respect to the standard inner product on R^2. It is shown in Example 3.20 that the matrix is positive definite. (a) Determine the form of the energy inner product and confirm the commutivity property of the inner energy inner product. (b) If u_1 and u_2 represent the displacements of the two blocks, measured from equilibrium, show how the energy norm relates to the potential energy of the system.

Solution: Using the stiffness matrix of Example 3.20

$$(\mathbf{u},\mathbf{v})_K = (\mathbf{Ku},\mathbf{v}) = (4u_1 - 2u_2)v_1 + (-2u_1 + 3u_2)v_2$$
$$= 4u_1 v_1 - 2u_2 v_1 - 2u_1 v_2 + 3u_2 v_2 \tag{a}$$

The commutativity of the energy inner product is confirmed by

$$(\mathbf{v},\mathbf{u})_K = (\mathbf{u},\mathbf{Kv}) = u_1(4v_1 - 2v_2) + u_2(-2v_1 + 3v_2)$$
$$= 4u_1 v_1 - 2u_1 v_2 + -2u_2 v_1 + 3u_2 v_2 = (\mathbf{u},\mathbf{v})_K \tag{b}$$

(b) The potential energy of the system when the blocks have displacements u_1 and u_2, respectively, is

$$V = \frac{1}{2}2u_1^2 + \frac{1}{2}2(u_2 - u_1)^2 + \frac{1}{2}u_2^2 = \frac{1}{2}\left[2u_1^2 + 2(u_2 - u_1)^2 + u_2^2\right] \tag{c}$$

Comparing Equation c with Equation b of Example 3.20 leads to

$$V = \frac{1}{2}(\mathbf{Ku},\mathbf{u}) - \frac{1}{2}(\mathbf{u},\mathbf{u})_K = \frac{1}{2}\|\mathbf{u}\|_K^2 \tag{d}$$

4 Operators Used in Vibration Problems

Methods of derivation of ordinary and partial differential equations governing vibrations of discrete and continuous systems are developed in Chapter 2. Chapter 3 provides a framework for the discussion of the response of the system using linear algebra and linear operators and characterizes properties of linear operators. Specific vibrations problems are considered in this chapter. The differential equations governing these problems are derived using the methods of Chapter 2. The differential equations are analyzed in the context of Chapter 3. For each problem, an appropriate vector space for analysis of the problem and its response is determined, the problem is written in a form using linear operators with the domain and range of each operator specified, and the conditions under which the operators are self-adjoint and/or positive definite are determined.

This analysis paves the way for determination of the free and forced response as considered in Chapter 5 through Chapter 8.

4.1 SUMMARY OF BASIC THEORY

A method for deriving the differential equations of motion for mechanical systems is presented in Chapter 2. The resulting differential equations for a linear system without gyroscopic effects can be written in the form

$$\mathbf{K}\mathbf{w} + \mathbf{C}\frac{\partial \mathbf{w}}{\partial t} + \mathbf{M}\frac{\partial^2 \mathbf{w}}{\partial t^2} = \mathbf{F} \tag{4.1}$$

where $\mathbf{w}$ is a vector of dependent variables, $\mathbf{K}$ is a stiffness operator, $\mathbf{C}$ is a damping operator, $\mathbf{M}$ is a mass operator, and $\mathbf{F}$ is a force vector. For a discrete system, $\mathbf{w}$ is a vector of generalized coordinates, $\mathbf{M}$, $\mathbf{C}$, and $\mathbf{K}$ are symmetric matrices, and $\mathbf{F}$ is a discrete vector. For a continuous system, $\mathbf{w}$ is either a single continuous function of spatial coordinates and time or a vector of continuous functions or a combination of continuous and discrete functions.

A foundation on which to evaluate linear operators was developed in Chapter 3. A domain and range of each operator is defined in terms of subspaces of a comprehensive vector space. For discrete systems, the comprehensive vector space is R^n, the space of vectors which are n-tuples of real numbers. For continuous systems, the comprehensive vector space is $C^n[a,b]$, the space of functions which are n times differentiable on the interval $[a,b]$. For combinations of discrete and continuous systems, the space $R^n \times C^m[a,b]$, whose elements are n-tuples of continuous function that are an m-times differentiable on $[a,b]$. Other vector spaces are defined where appropriate.

Inner products are defined for each vector space. The standard inner product for R^n is defined as

$$(\mathbf{f}, \mathbf{g}) = \mathbf{g}^T \mathbf{f} = \sum_{i=1}^{n} f_i g_i \tag{4.2}$$

The standard inner product for $C^n[a,b]$ is

$$(f, g) = \int_a^b f(x)g(x)\mathrm{d}x \tag{4.3}$$

Inner-product-generated norms are defined.

Properties of linear operators are considered with respect to a defined inner product. An operator $\mathbf{L}$ is self-adjoint with respect to a defined inner product if

$$(\mathbf{L}\mathbf{f}, \mathbf{g}) = (\mathbf{f}, \mathbf{L}\mathbf{g}) \tag{4.4}$$

for all $\mathbf{f}$ and $\mathbf{g}$ in its domain. An operator is positive definite if

$$(\mathbf{L}\mathbf{f}, \mathbf{f}) \geq 0 \tag{4.5}$$

for all $\mathbf{f}$ in the domain of the operator and the inner product is zero if and only if $\mathbf{f} = \mathbf{0}$. If Equation 4.5 is satisfied but there exists a vector other than the zero vector for which the inner product is equal to zero, the operator is positive semidefinite. When an operator is self-adjoint and positive definite with respect to an inner product, then an energy inner product is defined as

$$(\mathbf{f}, \mathbf{g})_\mathbf{L} = (\mathbf{L}\mathbf{f}, \mathbf{g}) \tag{4.6}$$

The discussion in this chapter focuses on how the linear operators encountered in the course of the derivation of the differential equations meet the above definitions of self-adjointness and positive definiteness and the definition of energy inner products.

4.2 DIFFERENTIAL EQUATIONS FOR DISCRETE SYSTEMS

Consider a linear n-degree-of-freedom system with chosen generalized coordinates x_1, x_2,..., x_n. It is shown in Chapter 2 that for a nongyroscopic system, the quadratic forms of the kinetic energy, potential energy, and Rayleigh's dissipation function are used in Lagrange's equations to formulate the differential equations in the matrix form

$$\mathbf{M}\ddot{x} + \mathbf{C}\dot{x} + \mathbf{K}x = \mathbf{F} \tag{4.7}$$

where $\mathbf{M}$ is the $n \times n$ mass matrix, $\mathbf{C}$ is the $n \times n$ damping matrix, $\mathbf{K}$ is the $n \times n$ stiffness matrix, and $\mathbf{F}$ is the $n \times 1$ force vector. Since the lagrangian is developed using quadratic forms for kinetic energy and potential energy, the mass and stiffness matrices are symmetric. Use of Lagrange's equations for a damped system with the dissipation function used to express energy dissipation due to viscous damping leads to a symmetric damping matrix.

It is shown in Chapter 3 that a $n \times n$ matrix is a linear operator whose domain and range are each R^n. Each of these matrices, $\mathbf{M}$, $\mathbf{K}$, and $\mathbf{C}$, as well as others formulated from products of these matrices, are considered in this context.

4.3 STIFFNESS MATRIX

The stiffness matrix is derived from the quadratic form of potential energy

$$V = \frac{1}{2} \sum_{i=1}^{n} \sum_{j=1}^{n} k_{i,j} x_i x_j \tag{4.8}$$

The stiffness matrix $\mathbf{K}$ is the $n \times n$ matrix in which the element in the ith row and jth column is $k_{i,j}$. Equation 4.8 can be written in terms of the standard inner product on R^n as

$$V = \frac{1}{2}(\mathbf{K}\mathbf{x}, \mathbf{x}) \tag{4.9}$$

The difference in potential energies at two system configurations is equal to the work done by all conservative forces. Thus the potential energy of the system is the same for a given $\mathbf{x}$ regardless of how the system attained this configuration. Consider an alternate means of the system attaining a given configuration through successive application of systems of static loads. Let $f_{1,j}, f_{2,j}, \ldots, f_{n,j}$ be a set of forces applied statically such that $f_{k,j}$ is applied to the particle hose displacement is x; and that the resulting system configuration is $x_k = x_k \delta_{k,j}$.

The process described is illustrated in the example of three machines placed along the span of a massless beam in Figure 4.1. The loads $f_{1,1}, f_{2,1}, f_{3,1}, \ldots, f_{n,1}$ are applied first, resulting in $x_1 = x_1$, $x_2 = 0$, $x_3 = 0, \ldots$, $x_n = 0$. The work done during application of this set of loads is $W_{0 \to 1} = \frac{1}{2} f_{1,1} x_1$. Next, with the first set of loads still applied, statically apply the set of loads $f_{1,2}, f_{2,2}, f_{3,2}, \ldots, f_{n,2}$ such that after application of this set $x_1 = x_1$, $x_2 = x_2$, $x_3 = 0$, $x_4 = 0, \ldots$, $x_n = 0$. Noting that during application of this set of forces, only forces applied to the particle whose displacement is x_2 do any work and that the force $f_{2,1}$ is present when application begins and $f_{2,2}$ is being applied leads to $W_{1 \to 2} = f_{2,1} x_2 + \frac{1}{2} f_{2,2} x_2$. This process is continued. On the kth step, the forces are $f_{1,k}, f_{2,k}, f_{3,k}, \ldots, f_{n,k}$ such that after their application, $x_1 = x_2, x_2 = x_2, \ldots, x_{k-1} = x_{k-1}, x_k = x_k, x_{k+1} = 0, x_{k+2} = 0, \ldots, x_n = 0$ and that $W_{k-1 \to k} = f_{k,1} x_k + f_{k,2} x_k + \cdots + f_{k,k-1} x_k + \frac{1}{2} f_{k,k} x_k$. After n steps, the desired configuration is attained. The total work done through successive application of these forces is

$$W_{0 \to n} = W_{0 \to 1} + W_{1 \to 2} + W_{2 \to 3} + \cdots + W_{n-2 \to n-1} + W_{n-1 \to n}$$

$$= \frac{1}{2} f_{1,1} x_1 + f_{2,1} x_2 + \frac{1}{2} f_{2,2} x_2 + \cdots + \sum_{\ell=1}^{k-1} f_{k,\ell} x_k + \frac{1}{2} f_{k,k} x_k + \ldots$$

$$+ \sum_{\ell=1}^{n-1} f_{n,\ell} x_n + \frac{1}{2} f_{n,n} x_n \tag{4.10}$$

The Principle of Work and Energy is used to show that the potential energy (stored energy) in the system after application of these forces is the total work done by the forces during their application

$$V = W_{0 \to n} \tag{4.11}$$

The system is linear in that multiplication of each force in a set of forces by a constant value, say α, leads to the resulting displacement being multiplied by α. Thus if a set of forces, $k_{1,j}, k_{2,j}, \ldots, k_{n,j}$, is defined such that their application leads to $x_k = \delta_{k,j}$, then the forces required to obtain the displacement $x_k = x_k \delta_{k,j}$ are $f_{i,k} = k_{i,k} x_k$, $i = 1, 2, \ldots, n$.

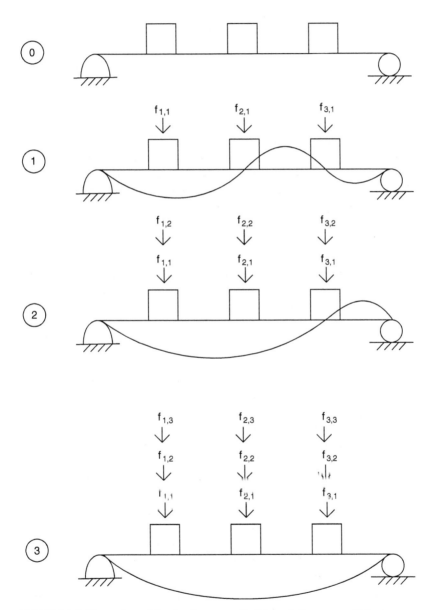

FIGURE 4.1 Illustration of the development of stiffness influence coefficients.

The order in which the set of forces is applied is irrelevant in the final configuration of the system. The work done by application of the first two sets of forces, written in terms of the ks is

$$W_{0\rightarrow 2} = \frac{1}{2}k_{1,1}x_1^2 + k_{2,1}x_1x_2 + \frac{1}{2}k_{2,2}x_2^2 \qquad (4.12)$$

If the order of application of the sets of forces is reversed, the work done is

$$W_{0\rightarrow 2} = \frac{1}{2}k_{1,1}x_1^2 + k_{1,2}x_2x_1 + \frac{1}{2}k_{2,2}x_2^2 \qquad (4.13)$$

Since the order of application leads to the same system configuration and the potential energy is independent of how the configuration is achieved, the right-hand sides of Equation 4.12 and Equation 4.13 must be the same for all x_1 and x_2. Thus, $k_{1,2} = k_{2,1}$. This argument can be generalized by reversing the order of application of any two sets of forces, say steps i and j, to show that

$$k_{i,j} = k_{j,i} \qquad (4.14)$$

The potential energy for this arbitrary system configuration can be written using Equation 4.8 and Equation 4.9 and the definition of the ks as

$$V = \frac{1}{2}\sum_{i=1}^{n}\sum_{j=1}^{n}k_{i,j}x_ix_j \qquad (4.15)$$

Equation 4.15 is identical to Equation 4.9 and thus the ks as defined above are the elements of the stiffness matrix. In addition, Equation 4.14 shows that the stiffness matrix is symmetric.

The elements of the stiffness matrix can be obtained using the potential energy formulation, as in Equation 4.9, or through the method illustrated above. In this method, the ks are called stiffness influence coefficients. The columns of the stiffness matrix are obtained sequentially. The first column is the set of forces applied to the particles whose displacements are the chosen generalized coordinates whose static application leads to a configuration with $x_1 = 1$, $x_2 = 0$, $x_3 = 0,\ldots,x_n = 0$. The second column is obtained in the same way, except that $x_1 = 0$ and $x_2 = 1$. This process is completed until all columns are obtained. If a generalized coordinate is an angular coordinate, a moment is applied rather than a force. This procedure is illustrated in the following example.

Example 4.1. Use stiffness influence coefficients to derive the stiffness matrix for the three-degree-of-freedom system shown in Figure 4.2.

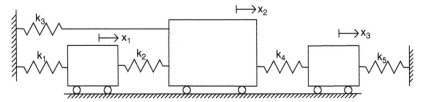

FIGURE 4.2 Three-degree-of-freedom system for Example 4.1.

Solution: The first column is obtained by determining the set of forces, as illustrated in Figure 4.3a, leading to $x_1 = 1$, $x_2 = 0$, $x_3 = 0$. Summing forces to zero leads to $k_{1,1} = k_1 + k_2$, $k_{2,1} = -k_2$, $k_{3,1} = 0$. Similar analyses using the free-body diagrams of Figure 4.3 are used to determine the remaining columns, leading to

$$\mathbf{K} = \begin{bmatrix} k_1 + k_2 & -k_2 & 0 \\ -k_2 & k_2 + k_3 + k_4 & -k_4 \\ 0 & -k_4 & k_4 + k_5 \end{bmatrix} \quad \text{(a)}$$

It is shown in Example 3.16 that a symmetric matrix is self-adjoint with respect to the standard inner product on R^n. The symmetry of the stiffness matrix is guaranteed by Equation 4.14 and thus $\mathbf{K}$ is self-adjoint with respect to the standard inner product on R^n.

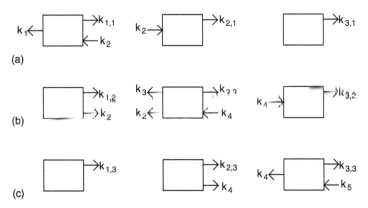

FIGURE 4.3 Steps in determination of stiffness matrix for system of Figure 4.2 and Example 4.1 using stiffness influence coefficients.

FIGURE 4.4 An example of a system in which the stiffness matrix is positive semidefinite.

The stiffness matrix is positive definite with respect to the standard inner product for R^n if the quadratic form of potential energy, given by Equation 4.9, is positive for all possible system configurations. If the only sources of potential energy are discrete springs, as in Example 4.1, the total potential energy for any system configuration is non-negative—the potential energy in a spring is proportional to the square of the change in length of the spring. It is possible, however, for a system in which springs are the only source of potential energy to have a system configuration in which the potential energy is zero. Such is the case for the two-degree-of-freedom system of Figure 4.4. The total potential energy is $V = \frac{1}{2}k(x_2 - x_1)^2$. The potential energy is zero for any system configuration in which $x_1 = x_2$. Such a system is called an unconstrained system. A system in which the potential energy is positive for any system configuration is constrained. The stiffness matrix for a constrained system is positive definite while the stiffness matrix for an unconstrained system is positive semidefinite.

A system in which gravity is a source of potential energy may have a negative potential energy if there exists a system configuration in which the potential energy due to gravity is less than the potential energy due to gravity when the system is in equilibrium is not positive definite. The inverted pendulum of Figure 4.5 is an example in such a system. The mass center of the pendulum is below its equilibrium position for all possible system configurations.

Since the stiffness matrix is symmetric for all linear systems, a potential energy inner product may be defined for those systems which are also positive definite with respect to the standard inner product for R^n. In such cases, the potential energy inner product is

$$(\mathbf{x}, \mathbf{y})_{\mathbf{K}} = (\mathbf{K}\mathbf{x}, \mathbf{y}) \tag{4.16}$$

Equation 4.9, Equation 4.10, and Equation 4.16 show that the potential energy for a system configuration defined by $\mathbf{x}$ is

$$V = \frac{1}{2}(\mathbf{x}, \mathbf{x})_{\mathbf{K}} \tag{4.17}$$

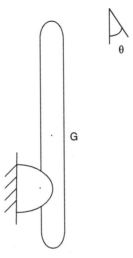

FIGURE 4.5 Example of a system with an unstable equilibrium position and has a stiffness matrix that is not positive definite or positive semidefinite.

4.4 MASS MATRIX

The mass matrix is derived from the quadratic form of kinetic energy

$$T = \frac{1}{2} \sum_{i=1}^{n} \sum_{j=1}^{n} m_{i,j} \dot{x}_i \dot{x}_j \qquad (4.18)$$

The mass matrix, $\mathbf{M}$, is the $n \times n$ matrix in which the element in the ith row and jth column is $m_{i,j}$. Equation 4.18 can be written in terms of the standard inner product on R^n as

$$T = \frac{1}{2}(\mathbf{M}\dot{\mathbf{x}}, \dot{\mathbf{x}}) \qquad (4.19)$$

Similar to stiffness influence coefficients developed for determination of the stiffness matrix, inertia influence coefficients may be developed as an alternate for determination of the mass matrix. When an impulse is applied to a particle in a discrete system, an instantaneous change in velocity occurs. The principle of impulse and momentum is used to obtain the induced velocities of particles in the system. The kinetic energy for a given system configuration is a function of the time derivatives of the generalized coordinates, related of course to particle velocities. A system of impulses may be applied to specify the induced velocity of every particle in the system.

The jth column of inertia influence coefficients $m_{i,j}$ $i = 1,2,\ldots,n$ is the set of impulses, applied to the particles whose displacements are the chosen generalized coordinates, such that the induced velocities are $\dot{x}_i = \delta_{i,j}$. The columns of inertia influence coefficients are determined sequentially.

Noting that the work done to a system by application of an impulse of magnitude I to a particle which has a velocity v^- before application of the impulse and a velocity of v^+ immediately after application of the impulse is $W = I(v^+ - v^-)$, it can be shown that the total energy imparted to the system through sequential application of systems of impulses is the same as Equation 4.18. Thus the inertia influence coefficients are the elements of the mass matrix.

By reversing the order of application of systems of impulses, it can be shown in a manner analogous to that used for the stiffness matrix that in order for the system to have the same kinetic energy, $m_{i,j} = m_{j,i}$ for all $i,j = 1,2,\ldots,n$. Since the mass matrix is guaranteed to be symmetric, Example 3.16 implies that it is self-adjoint with respect to the standard inner product on R^n.

Equation 4.18 expresses the kinetic energy of an n-degree-of-freedom system at an arbitrary instant in terms of the chosen generalized coordinates. Equation 4.18 is a rearrangement of the sum of the kinetic energies of all particles and rigid bodies in the system. By definition, the kinetic energy of a particle and rigid body must be positive. Thus Equation 4.19 results from the summation of terms, each of which are always positive. Hence, Equation 4.19 is positive for all choices of $\dot{x}$. Thus from Equation 4.19, the mass matrix is positive definite with respect to the standard inner product on R^n. Unlike the stiffness matrix, there is no system in which the kinetic energy can be zero for any system configuration.

Since $\mathbf{M}$ is positive definite and self-adjoint with respect to the standard inner product on R^n, a kinetic energy inner product is defined according to

$$(\mathbf{x}, \mathbf{y})_{\mathbf{M}} = (\mathbf{Mx}, \mathbf{y}) \qquad (4.20)$$

Equation 4.19 and Equation 4.20 show that the kinetic energy at an arbitrary instant is

$$T = \frac{1}{2}(\dot{\mathbf{x}}, \dot{\mathbf{x}})_{\mathbf{M}} \qquad (4.21)$$

4.5 FLEXIBILITY MATRIX

The stiffness matrix $\mathbf{K}$ is a symmetric matrix and thus self-adjoint with respect to the standard inner product on R^n. However, in certain cases, the stiffness matrix is not positive definite. When the system is unconstrained, as is the system of Figure 4.4, the stiffness matrix is positive semidefinite. Stiffness influence coefficients are used to show that the stiffness matrix for this system is $\mathbf{K} = \begin{bmatrix} k & -k \\ -k & k \end{bmatrix}$. The second row of the stiffness matrix is the negative of the first row. Hence the rows are not independent and thus the stiffness matrix is singular. Indeed this representation of the stiffness matrix can be used to define an unconstrained system as a system whose stiffness matrix is singular. Conversely, only unconstrained systems have singular stiffness matrices.

An inverse of a matrix exists when the matrix is nonsingular. Thus the inverse of the stiffness matrix exists when the system is constrained. The flexibility matrix **A**, when it exists, is defined as the inverse of the stiffness matrix

$$\mathbf{A} = \mathbf{K}^{-1} \tag{4.22}$$

Equation 4.7 can be premultiplied by the flexibility matrix leading to

$$\mathbf{A}\mathbf{M}\ddot{\mathbf{x}} + \mathbf{A}\mathbf{C}\dot{\mathbf{x}} + \mathbf{x} = \mathbf{A}\mathbf{F} \tag{4.23}$$

Equation 4.23 is the flexibility matrix formulation for the differential equations governing the motion of a linear n-degree-of-freedom system. Equation 4.22 shows that knowledge of the flexibility matrix is sufficient to model the system.

The inverse of a symmetric matrix is symmetric. Thus, **A** is self-adjoint with respect to the standard inner product on R^n. The inverse of a positive-definite matrix is also positive definite. Thus, when **K** is positive definite, so is **A**. This infers that when **K** may be used to define a potential energy inner product, **A** may be used to define a flexibility inner product as

$$(\mathbf{x}, \mathbf{y})_A = (\mathbf{A}\mathbf{x}, \mathbf{y}) \tag{4.24}$$

The flexibility matrix can be determined using flexibility influence coefficients. A column of flexibility influence coefficients, $a_{i,j}$ $i = 1,2,...,n$, are defined as the static displacements for each of the generalized coordinates that are obtained when a unit load is applied to the particle whose displacement is x_j.

Example 4.2. Use flexibility influence coefficients to determine the flexibility matrix for the system of Figure 4.6 using the indicated generalized coordinates.

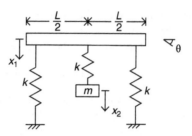

FIGURE 4.6 System of Example 4.2.

Solution: The first column is obtained by applying a unit force to the left end of the bar. The resulting displacement of the left end of the bar is $a_{1,1}$; the clockwise angular rotation of the bar is $a_{1,2}$; the displacement of the particle is $a_{1,3}$. Setting the summation of forces acting on the bar to zero leads to

$$ka_{1,1} + k(a_{1,1} + La_{2,1}) - k\left[a_{3,1} - \left(a_{1,1} + \frac{L}{2}a_{2,1}\right)\right] - 1 = 0$$

(a)

$$3ka_{1,1} + 3k\frac{L}{2}a_{2,1} - ka_{3,1} = 1$$

Setting the sum of the counterclockwise moments about the mass center of the bar to zero leads to

$$-(ka_{1,1})\left(\frac{L}{2}\right) + (ka_{1,1} + kLa_{2,1})\left(\frac{L}{2}\right) + \frac{L}{2} = 0 \quad k\frac{L^2}{2}a_{2,1} = -\frac{L}{2}$$

(b)

Setting the sum of the forces acting on the particle to zero leads to

$$k\left[a_{3,1} - \left(a_{1,1} + \frac{L}{2}a_{2,1}\right)\right] = 0$$

(c)

Equation a, Equation b, and Equation c are solved simultaneously, leading to $a_{1,1} = 1/2k$, $a_{1,2} = -1/kL$, $a_{1,3} = 1/2k$. The second and third columns of the flexibility matrix are obtained by applying the conservation laws to the free-body diagrams of Figure 4.7b and Figure 4.7c, respectively.

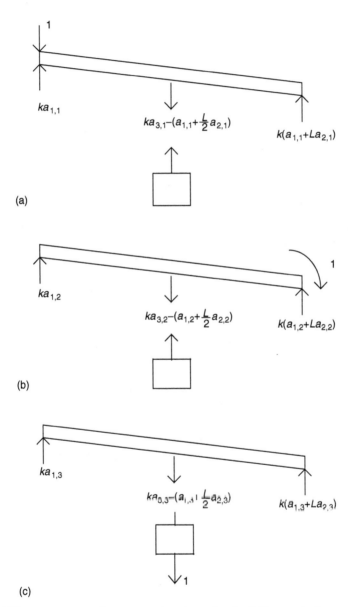

FIGURE 4.7 Free-body diagrams used to determine flexibility matrix for system of Example 4.2 (a) first column (b) second column (c) third column.

The flexibility matrix is obtained as

$$
\mathbf{A} =
\begin{bmatrix}
\dfrac{1}{k} & -\dfrac{1}{kL} & \dfrac{1}{2k} \\[2ex]
-\dfrac{1}{kL} & \dfrac{2}{kL^2} & 0 \\[2ex]
\dfrac{1}{2k} & 0 & \dfrac{3}{2k}
\end{bmatrix}
\tag{d}
$$

The flexibility matrix is useful in developing multi-degree-of-freedom models for continuous systems.

Example 4.3. Three machines are placed along the span of a pinned-pinned beam as shown in Figure 4.8. Using the displacements of the machines as generalized coordinates, determine the flexibility matrix for the system.

Solution: The deflection of a particle a distance z along the neutral axis, measured from the left support, due to a concentrated unit load statically applied at a distance b from the left support is

$$
y(z) = \frac{L}{6EI}\left(1 - \frac{b}{L}\right)\left[\frac{b}{L}\left(2 - \frac{b}{L}\right)\frac{z}{L} - \left(\frac{z}{L}\right)^2\right]
\tag{a}
$$

Equation a is applicable only for $z \le b$. Since the beam is subject to a concentrated load at $z = b$, the shear force in the beam is discontinuous at $z = b$. While the moment, slope, and deflection are continuous, they have a change in mathematical description at $z = b$. For this reason, the third column of the flexibility matrix is calculated first and the guaranteed symmetry of the matrix is used to determine the flexibility coefficients which would evaluate $y(z)$ for $z > b$.

The third column of the flexibility matrix is obtained by applying a unit concentrated load at $b = 3L/4$, as illustrated in Figure 4.9 for which

$$
y(z) = \frac{L^3}{24EI}\left[\frac{15}{16}\frac{z}{L} - \left(\frac{z}{L}\right)^2\right]
\tag{b}
$$

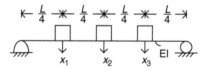

FIGURE 4.8 A three-degree-of-freedom system is used to model vibrations of three machines along the span of a beam.

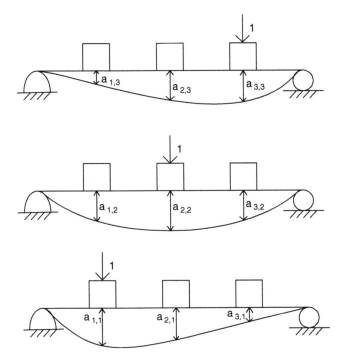

FIGURE 4.9 Determination of flexibility matrix for system of Example 4.3.

The third column of flexibility influence coefficients are obtained using Equation b as

$$a_{1,3} = y\left(\frac{L}{4}\right) = \frac{7L^3}{768EI} \tag{c}$$

$$a_{2,3} = y\left(\frac{L}{2}\right) = \frac{11L^3}{768EI} \tag{d}$$

$$u_{3,3} = y\left(\frac{3L}{4}\right) = \frac{3L^3}{256EI} \tag{e}$$

The second column of flexibility influence coefficients are calculated by applying a unit concentrated load at $b = L/2$, for which

$$y(z) = \frac{L^3}{12EI}\left[\frac{3}{4}\frac{z}{L} - \left(\frac{z}{L}\right)^2\right] \tag{f}$$

Due to symmetry of the beam, it is noted that $a_{1,2} = a_{3,2}$, and from symmetry of the flexibility matrix, $a_{3,2} = a_{2,3} = 11L^3/768EI$. Thus, $a_{1,2} = 11L^3/768EI$. Then

$$a_{2,2} = y\left(\frac{L}{2}\right) = \frac{L^3}{48EI} \tag{g}$$

The first column is determined through symmetry of the matrix, $a_{2,1} = a_{1,2} = 11L^3/768EI$ and $a_{3,1} = a_{1,3} = 7L^3/768EI$, and due to symmetry of the beam, $a_{1,1} = a_{3,3} = 3L^3/256EI$.

The flexibility matrix for this three-degree-of-freedom model is

$$\mathbf{A} = \frac{L^3}{EI}\begin{bmatrix} \dfrac{3}{256} & \dfrac{11}{768} & \dfrac{7}{768} \\[2mm] \dfrac{11}{768} & \dfrac{1}{48} & \dfrac{11}{768} \\[2mm] \dfrac{7}{768} & \dfrac{11}{768} & \dfrac{3}{256} \end{bmatrix} \tag{h}$$

4.6 $\mathbf{M}^{-1}\mathbf{K}$ AND AM

Alternate formulations of the equations governing the vibrations of a linear n-degree-of-freedom system are obtained by pre-multiplying Equation 4.7 by the inverse of either the mass matrix or the stiffness matrix. First pre-multiplication by $\mathbf{M}^{-1}$ leads to

$$\ddot{\mathbf{x}} + \mathbf{M}^{-1}\mathbf{C}\dot{\mathbf{x}} + \mathbf{M}^{-1}\mathbf{K}\mathbf{x} = \mathbf{M}^{-1}\mathbf{F} \tag{4.25}$$

Premultiplication of Equation 4.7 by $\mathbf{A}$ the flexibility matrix which is the inverse of the stiffness matrix leads to

$$\mathbf{A}M\ddot{\mathbf{x}} + \mathbf{A}C\dot{\mathbf{x}} + \mathbf{x} = \mathbf{A}\mathbf{F} \tag{4.26}$$

It has been shown that the mass matrix, the stiffness matrix, and the flexibility matrix are all symmetric matrices. However, the product of two symmetric matrices is not necessarily a symmetric matrix. Thus, $\mathbf{AM}$ and $\mathbf{M}^{-1}\mathbf{K}$ are not necessarily self-adjoint with respect to the standard inner product. However, three energy inner products have been identified, the kinetic energy inner product of Equation 4.21, the potential energy inner product of Equation 4.17, and the flexibility inner product of Equation 4.24.

Consider for any **x** and **y** in R^n the following kinetic energy inner product:

$$(\mathbf{M}^{-1}\mathbf{Kx}, \mathbf{y})_\mathbf{M} = (\mathbf{MM}^{-1}\mathbf{Kx}, \mathbf{y}) \qquad (4.27)$$

The right-hand side of Equation 4.27 is a result of the definition of the kinetic energy inner product. Noting that $\mathbf{MM}^{-1}\mathbf{K} = \mathbf{K}$ and using the self-adjointness of **K**, Equation 4.27 leads to

$$(\mathbf{M}^{-1}\mathbf{Kx}, \mathbf{y})_\mathbf{M} = (\mathbf{Kx}, \mathbf{y}) = (\mathbf{x}, \mathbf{Ky}) = (\mathbf{x}, \mathbf{MM}^{-1}\mathbf{Ky}) \qquad (4.28)$$

Using the self-adjointness of **M**, Equation 4.28 becomes

$$(\mathbf{M}^{-1}\mathbf{Kx}, \mathbf{y})_\mathbf{M} = (\mathbf{Mx}, \mathbf{M}^{-1}\mathbf{Ky}) = (\mathbf{x}, \mathbf{M}^{-1}\mathbf{Ky})_\mathbf{M} \qquad (4.29)$$

Equation 4.29 shows that $\mathbf{M}^{-1}\mathbf{K}$ is self-adjoint with respect to the kinetic energy inner product.

A similar process is used to show that $\mathbf{M}^{-1}\mathbf{K}$ is also self-adjoint with respect to the potential energy inner product:

$$
\begin{aligned}
(\mathbf{M}^{-1}\mathbf{Kx}, \mathbf{y})_\mathbf{K} &= (\mathbf{KM}^{-1}\mathbf{Kx}, \mathbf{y}) \quad &&\text{definition of potetnial energy inner product} \\
&= (\mathbf{M}^{-1}\mathbf{Kx}, \mathbf{Ky}) \quad &&\text{\textbf{K} is self adjoint with respect to standard} \\
& &&\text{inner product} \\
&= (\mathbf{Kx}, \mathbf{M}^{-1}\mathbf{Ky}) \quad &&\mathbf{M}^{-1} \text{ is self adjoint with respect to standard} \\
& &&\text{inner product} \\
&= (\mathbf{x}, \mathbf{M}^{-1}\mathbf{Ky})_\mathbf{K} \quad &&\text{definition of potenial energy inner product}
\end{aligned}
$$
$$(4.30)$$

Assuming **K** is positive definite, it is easy to show that $\mathbf{M}^{-1}\mathbf{K}$ is positive definite with respect to the kinetic energy inner product. Rewriting the first line of Equation 4.28 with $\mathbf{y} = \mathbf{x}$ leads to

$$(\mathbf{M}^{-1}\mathbf{Kx}, \mathbf{x})_\mathbf{M} = (\mathbf{Kx}, \mathbf{x}) \qquad (4.31)$$

Equation 4.31 shows that if **K** is positive definite with respect to the standard inner product, then $\mathbf{M}^{-1}\mathbf{K}$ is positive definite with respect to the kinetic energy inner product.

Now consider the self-adjointness of **AM** with respect to the stiffness inner product.

$(\mathbf{AMx}, \mathbf{y})_K$ $= (\mathbf{KAMx}, \mathbf{y})$ definition of potential energy scalar product

              $= (\mathbf{Mx}, \mathbf{y})$ defintion of flexibility matrix

              $= (\mathbf{x}, \mathbf{My})$ $\mathbf{M}$ is self-adjoint with respect to standard inner product

              $= (\mathbf{x}, \mathbf{KAMy})$ $\mathbf{KA} = \mathbf{I}$

              $= (\mathbf{KAx}, \mathbf{AMy})$ $\mathbf{K}$ is self adjoint with respect to standard inner product

              $= (\mathbf{x}, \mathbf{AMy})_K$ definition of potetnial energy inner product

$$(4.32)$$

Equation 4.32 shows that $\mathbf{AM}$ is self-adjoint with respect to the potential energy inner product. Furthermore, it can be shown that $\mathbf{AM}$ is self-adjoint with respect to the kinetic energy inner product and positive definite with respect to the potential energy inner product.

4.7 FORMULATION OF PARTIAL DIFFERENTIAL EQUATIONS FOR CONTINUOUS SYSTEMS

Let $\mathbf{w}(x,y,z,t)$ represent a vector of dependent variables used to describe the motion of a system of deformable bodies. The general form of the partial differential equation governing the motion of a linear undamped continuous system is

$$\mathbf{K}\mathbf{w}(x, y, z, t) + \mathbf{M}\frac{\partial^2 \mathbf{w}}{\partial t^2} = \mathbf{F}(x, y, z, t) \qquad (4.33)$$

where $\mathbf{K}$, called the stiffness operator, is a matrix of partial differential operators involving only spatial derivatives of $\mathbf{w}$; $\mathbf{M}$, called the inertia operator, is a matrix of spatial functions; and $\mathbf{F}$ is a vector of functions which may depend upon both spatial coordinates and time. The forms of $\mathbf{K}$, $\mathbf{M}$, and $\mathbf{F}$ are determined through derivation of Equation 4.33 from extended Hamilton's Principle. Note that since $\mathbf{M}$ is a matrix of spatial functions, Equation 4.33 may be written as

$$\mathbf{K}\mathbf{w}(x, y, z, t) + \frac{\partial^2}{\partial t^2}(\mathbf{M}\mathbf{w}) = \mathbf{F}(x, y, z, t) \qquad (4.34)$$

Equation 4.34 is supplemented by boundary conditions which are derived from application of extended Hamilton's principle. In addition, initial conditions of the form

$$\mathbf{w}(x, y, z, 0) = \mathbf{w}_0(x, y, z) \tag{4.35}$$

$$\frac{\partial \mathbf{w}}{\partial t}(x, y, z, 0) = \dot{\mathbf{w}}_0(x, y, z)$$

are necessary to specify the solution.

For simplicity in explanation, consider a problem in which the system has only one deformable body and the problem is one-dimensional. In which case, $\mathbf{w}$ is a single function of spatial variables and time and is written as $w(x,t)$, the inertia operator $\mathbf{M}$ is a function of x, and the stiffness operator $\mathbf{K}$ is a partial differential operator. Although reference is made to specific inner products, the results can be extended to any inner product defined for system of deformable bodies.

The solution for $w(x,t)$ is sought on an interval in x, say $0 \leq x \leq L$. The domain of the inertia operator is $C[0,L]$, the space of continuous functions on the interval $[0,L]$. The domain of the stiffness operator is more restrictive as the operation $\mathbf{K}w$ requires differentiation of $w(x,t)$ and requires $w(x,t)$ to satisfy boundary conditions at $x = 0$ and $x = L$. The domain of $\mathbf{K}$ is defined as the subspace of $C^n[0,L]$ of functions $f(x)$ that satisfy the boundary conditions at $x = 0$ and $x = L$. The value of n is equal to the order of the partial differential operator.

For example, in solving for the longitudinal vibrations of a bar fixed at $x = 0$ and free at $x = L$, the domain of $\mathbf{K}$ is defined as, S, the subspace of $C^2[0,L]$, such that if $f(x)$ is in S then $f(0) = 0$ and $df/dx(L) = 0$.

It is convenient to nondimensionalize the partial differential equations governing the vibrations of continuous systems. A partial differential equation is nondimensionalized through the introduction of nondimensional dependent and independent variables. Nondimensional variables are introduced such as

$$w^* = \frac{w}{L} \tag{4.36}$$

$$x^* = \frac{x}{L} \tag{4.37}$$

$$t^* = \frac{t}{T} \tag{4.38}$$

where the variable with an * is a nondimensional variable, L is a characteristic length in the system such as the length of a beam, and T is a parameter with the dimension of time that is chosen for convenience. Nondimensional functions of the spatial variable are also introduced, such as

$$\alpha(x^*) = \frac{\rho(x)A(x)}{\rho_0 A_0} \tag{4.39}$$

where ρ_0 and A_0 are reference values of the density and the area, perhaps the values at $x = 0$.

Equation 4.36 through Equation 4.39 are substituted into the governing equation. The chain rule is used to convert derivative with respect to dimensional variables into derivative with respect to nondimensional variables:

$$\frac{\partial}{\partial x} = \frac{\partial}{\partial x^*}\frac{dx^*}{dx} = \frac{1}{L}\frac{\partial}{\partial x^*} \tag{4.40}$$

The partial differential equation is rewritten such that all variables, independent and dependent, and all parameters are nondimensional. Then, by agreement, in order to reduce the tedium of writing $*$ on every variable, all of the $*$s are dropped, and it is understood that from that point on, all variables are dimensionless.

The interval over which the nondimensional variable x is defined is $[0,1]$. The standard inner product on $C^2[0,1]$ is

$$(f,g) = \int_0^1 f(x)g(x)dx \tag{4.41}$$

It is generally a trivial exercise to show that the inertia operator is self-adjoint and positive definite with respect to this inner product. Thus, a kinetic energy inner product is defined. The focus of the remainder of this chapter is the determination as to whether the stiffness operator is self-adjoint and positive definite with respect to this inner product defined for all functions in S, the domain of $\mathbf{K}$. If so, then a potential energy inner product is defined.

It is necessary to work with the operator $\mathbf{M}^{-1}\mathbf{K}$ in determination of the free response of a system. If both $\mathbf{M}$ and $\mathbf{K}$ are positive definite and self-adjoint with respect to the standard inner product, it can easily be shown that $\mathbf{M}^{-1}\mathbf{K}$ is self-adjoint with respect to both the kinetic energy inner product and the potential energy inner product and that $\mathbf{M}^{-1}\mathbf{K}$ is positive definite with respect to the kinetic energy inner product. The same form of proof is used as in Section 4.5 for discrete systems.

4.8 SECOND-ORDER PROBLEMS

Vibrations of the elastic bar of Figure 4.10, whose governing partial differential equation is derived in Example 2.14, is an example of a system whose vibrations are governed by a one-dimensional partial differential equation which is second-order in both space and time. Let $u(x,t)$ be the longitudinal displacement of the a particle that is a distance x from the left end of the bar. The equation for the free response of $u(x,t)$ is derived in Example 2.14, as

$$\frac{\partial}{\partial x}\left(EA\frac{\partial u}{\partial x}\right) = \rho A\frac{\partial^2 u}{\partial t^2} \tag{4.42}$$

Equation 4.42 is nondimensionalized through introduction of the following nondimensional variables of the form of Equation 4.36 through Equation 4.38. The spatially dependent functions are nondimensionalized according to

$$\alpha(x^*) = \frac{E(x)A(x)}{E_0 A_0} \tag{4.43a}$$

$$\beta(x^*) = \frac{\rho(x)A(x)}{\rho_0 A_0} \tag{4.43b}$$

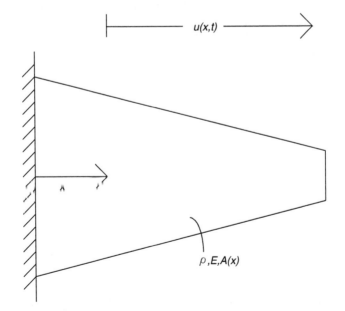

FIGURE 4.10 Variable area bar with $u(x,t)$ as longitudinal displacement.

where E_0, A_0, and ρ_0 are reference values, perhaps at $x = 0$, of the elastic modulus, cross section area, mass density, respectively. Substitution of Equation 4.36 through Equation 4.38 and Equation 4.43a,b into Equation 4.42 leads to

$$\frac{\partial}{\partial x^*}\left[\alpha(x^*)\frac{\partial u^*}{\partial x^*}\right] = \frac{\rho_0 L^2}{E_0 T^2}\beta(x^*)\frac{\partial^2 u^*}{\partial t^{*2}} \tag{4.44}$$

The characteristic time is chosen for convenience as

$$T = L\sqrt{\frac{\rho_0}{E_0}} \tag{4.45}$$

It is customary to drop the * from the nondimensional variables once the nondimensional formulation is complete. Using Equation 4.45 in Equation 4.44 and dropping the *s leads to

$$\frac{\partial}{\partial x}\left[\alpha(x)\frac{\partial u}{\partial x}\right] = \beta(x)\frac{\partial^2 u}{\partial t^2} \tag{4.46}$$

Second-order systems include longitudinal vibrations of elastic bars, transverse vibrations of strings, and torsional vibrations of shafts. The non-dimensional formulation of each of these problems is given in Table 4.1.

Equation 4.46 is of the form of Equation 4.33 with the stiffness and inertia operators defined as

$$\mathbf{K}u = -\frac{\partial}{\partial x}\left[\alpha(x)\frac{\partial u}{\partial x}\right] \tag{4.47}$$

$$\mathbf{M}u = \beta(x)u \tag{4.48}$$

At any instant of time, the solution for the response is in a subspace, S, of $C^2[0,1]$ defined by homogeneous boundary conditions satisfied by $u(x,t)$. The self-adjointness of $\mathbf{M}$ is tested by considering

$$(\mathbf{M}f, g) = \int_0^1 [\beta(x)f(x)]g(x)\,dx = \int_0^1 f(x)[\beta(x)g(x)]\,dx = (f, \mathbf{M}g) \tag{4.49}$$

Equation 4.49 shows that $\mathbf{M}$ is self-adjoint with respect to the standard inner product on $C^2[0,1]$.

TABLE 4.1
Nondimensionalizations Used for Continuous Systems

Problem	Dimensional Equation	Nondimensional Variables	Nondimensional Equation
Longitudinal motion of bar	$\frac{\partial}{\partial x}\left(EA\frac{\partial u}{\partial x}\right) = \rho A\frac{\partial^2 u}{\partial t^2}$	$x^* = x/L$ $u^* = u/L$ $t^* = \sqrt{E_0/\rho_0 L^2}\, t$ $\alpha(x) = E(x)A(x)/E_0 A_0$ $\beta(x) = \rho(x)A(x)/\rho_0 A_0$	$\frac{\partial}{\partial x}\left(\alpha\frac{\partial u}{\partial x}\right) = \beta\frac{\partial^2 u}{\partial t^2}$
Transverse displacement of string	$\frac{\partial}{\partial x}\left(T\frac{\partial y}{\partial x}\right) = \mu\frac{\partial^2 y}{\partial t^2}$	$x^* = x/L$ $y^* = y/L$ $t^* = \sqrt{T_0/\mu_0 L^2}\, t$ $\alpha(x) = T(x)/T_0$ $\beta(x) = \mu(x)/\mu_0$	$\frac{\partial}{\partial x}\left(\alpha\frac{\partial y}{\partial x}\right) = \beta\frac{\partial^2 y}{\partial t^2}$
Torsional oscillations of shaft	$\frac{\partial}{\partial x}\left(JG\frac{\partial\theta}{\partial x}\right) = \rho J\frac{\partial^2\theta}{\partial t^2}$	$x^* = x/L$ $\theta^* = \theta$ $t^* = \sqrt{G_0/\rho_0 L^2}\, t$ $\alpha(x) = J(x)G(x)/J_0 G_0$ $\beta(x) = \rho(x)J(x)/\rho_0 J_0$	$\frac{\partial}{\partial x}\left(\alpha\frac{\partial u}{\partial x}\right) = \beta\frac{\partial^2 u}{\partial t^2}$
Transverse displacement of Euler–Bernoulli beam	$\frac{\partial^2}{\partial x^2}\left(EI\frac{\partial^2 w}{\partial x^2}\right) + \rho A\frac{\partial^2 w}{\partial t^2}$	$x^* = x/L$	$\frac{\partial^2}{\partial x^2}\left(\alpha\frac{\partial^2 w}{\partial x^2}\right) + \beta\frac{\partial^2 w}{\partial t^2} = 0$

(continued)

TABLE 4.1 *(Continued)*

Problem	Dimensional Equation	Nondimensional Variables	Nondimensional Equation
		$w^* = w/L$ $t^* = \sqrt{E_0 J_0/\rho_0 A_0 L^4}\, t$ $\alpha(x) = E(x)I(x)/E_0 I_0$ $\beta(x) = \rho(x)A(x)/\rho_0 A_0$	
Timoshenko beam	$\kappa GA\left(\dfrac{\partial^2 w}{\partial x^2} - \dfrac{\partial \psi}{\partial x}\right) = \rho A\dfrac{\partial^2 w}{\partial t^2}$ $EI\dfrac{\partial^2 \psi}{\partial x^2} + \kappa GA\left(\dfrac{\partial w}{\partial x} - \psi\right)$ $= \rho I\dfrac{\partial^2 \psi}{\partial t^2}$	$w^* = w/L$ $\psi^* = \psi$ $x^* = x/L$ $t^* = t\sqrt{\kappa G/\rho L^2}\, t$ $\eta_1 = EI/\kappa GL^2 A$ $\eta_2 = I/AL^2$	$\dfrac{\partial^2 w}{\partial x^2} - \dfrac{\partial \psi}{\partial x} = \dfrac{\partial^2 w}{\partial t^2}$ $\eta_1\dfrac{\partial^2 \psi}{\partial x^2} + \dfrac{\partial w}{\partial x} - \psi$ $= \eta_2\dfrac{\partial^2 \psi}{\partial t^2}$

The positive-definiteness of **M** is tested by considering

$$(\mathbf{M}f, f) = \int_0^1 [\beta(x)f(x)]f(x)dx = \int_0^1 \beta(x)[f(x)]^2 dx \qquad (4.50)$$

Since $\beta(x) > 0$ for all x and $f(x)$ is a continuous function, Equation 4.50 shows that $(\mathbf{M}f, f) \geq 0$ for all $f(x)$ in $C^2[0,1]$, $(\mathbf{M}(0), 0) = 0$, and there is no $f(x)$ other than $f(x) = 0$, for which $(\mathbf{M}f, f) = 0$. Hence **M** is positive definite with respect to the standard inner product on $C^2[0,1]$.

Since **M** is positive definite and self-adjoint with respect to the standard inner product on $C^2[0,1]$, a kinetic energy inner product can be defined as

$$(f, g)_{\mathbf{M}} = (\mathbf{M}f, g) = \int_0^1 \beta(x)f(x)g(x)dx \qquad (4.51)$$

Now consider the possible self-adjointness of **K** by considering

$$(\mathbf{K}f, g) = -\int_0^1 \frac{\partial}{\partial x}\left[\alpha(x)\frac{\partial f}{\partial x}\right]g(x)dx \qquad (4.52)$$

Integration by parts ($\int u dv = uv - \int v du$) is applied to the right-hand side of Equation 4.52, with $u = g(x)$ and $dv = \partial/\partial x[\alpha(x)\partial f/\partial x]dx$ leading to

$$(\mathbf{K}f, g) = -\alpha(1)g(1)\frac{\partial f}{\partial x}(1) + \alpha(0)g(0)\frac{\partial f}{\partial x}(0) + \int_0^1 \alpha(x)\frac{\partial f}{\partial x}\frac{\partial g}{\partial x}dx \qquad (4.53)$$

Application of Integration by parts with $u = \alpha(x)dg/dx$ and $dv = \frac{\partial f}{\partial x}dx$ to the integral in Equation 4.53 leads to

$$(\mathbf{K}f, g) = -\alpha(1)g(1)\frac{\partial f}{\partial x}(1) + \alpha(0)g(0)\frac{\partial f}{\partial x}(0) + \alpha(1)f(1)\frac{\partial g}{\partial x}(1)$$

$$- \alpha(0)f(0)\frac{\partial g}{\partial x}(0) - \int_0^1 f(x)\frac{\partial}{\partial x}\left[\alpha(x)\frac{\partial g}{\partial x}\right]dx \qquad (4.54)$$

It is clear from Equation 4.54 that **K** is self-adjoint if

$$\alpha(1)g(1)\frac{\partial f}{\partial x}(1) - \alpha(0)g(0)\frac{\partial f}{\partial x}(0) - \alpha(1)f(1)\frac{\partial g}{\partial x}(1) + \alpha(0)f(0)\frac{\partial g}{\partial x}(0) = 0$$

(4.55)

The subspace, S, of $C^2[0,1]$ from which the solution for a bar fixed at $x = 0$ and free at $x = 1$ is defined as containing all $f(x)$ in $C^2[0,1]$ with $f(0) = 0$ and $\frac{\partial f}{\partial x}(1) = 0$. It is clear that Equation 4.55 is satisfied for all $f(x)$ and $g(x)$ is S, and thus **K** is self-adjoint with respect to the standard inner product on S. Clearly, all combinations of fixed and free ends lead to self-adjointness of **K** with respect to the standard inner product on an appropriately defined S.

Consider the case where the right end of the bar is attached to a spring as illustrated in Figure 4.11. The appropriate boundary condition obtained in Example 2.15 is

$$E(L)A(L)\frac{\partial u}{\partial x}(L, t) + ku(L, t) = 0$$

(4.56)

Use of the nondimensional variables of Equation 4.36 through Equation 4.38 and functions of Equation 4.40 through Equation 4.41 in Equation 4.56 leads to

$$\frac{\partial u}{\partial x}(1, t) + \eta_1 u(1, t) = 0$$

(4.57)

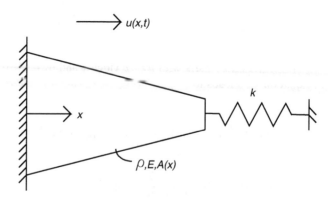

FIGURE 4.11 Variable area bar with discrete linear spring.

where all variables in Equation 4.57 are nondimensional and

$$\eta_1 = \frac{kL}{E_0 L_0 \alpha(1)} \tag{4.58}$$

For a bar fixed at $x = 0$ and attached to a linear spring at $x = 1$, the appropriate definition of S is all function in $C^2[0,1]$ such that $f(0) = 0$ and $f(x)$ satisfies Equation 4.57. In this case, the left-hand side of Equation 4.55 reduces to

$$\alpha(1)g(1)[-\eta_1 f(1)] - \alpha(1)f(1)[-\eta_1 g(1)] = 0 \tag{4.59}$$

Thus Equation 4.55 is true and **K** is self-adjoint for this condition as well.

The positive definiteness of **K** is considered by letting $g = f$ in Equation 4.53,

$$(\mathbf{K}f,f) = -\alpha(1)f(1)\frac{\partial f}{\partial x}(1) + \alpha(0)f(0)\frac{\partial f}{\partial x}(0) + \int_0^1 \alpha(x)\left[\frac{\partial f}{\partial x}\right]^2 dx \tag{4.60}$$

First, consider the integral term of Equation 4.60. The integrand is non-negative for all x, hence the integral is non-negative. It is clear that if $f(x,t) = 0$, then the integral is zero. The integral is also zero for any $f(x,t)$ such that $\partial f/\partial x = 0$ for all x, $0 \le x \le 1$, which is true only if $f(x,t)$ is independent of x, $f(x,t) = g(t)$. However, if one end, say $x = 0$, is fixed, then $f(0,t) = 0$, which implies that if $f(x,t) = g(t)$, then $g(t) = 0$ and, of course, $f(x,t) = 0$. The same can be shown if one end is constrained by a spring. Thus for these cases, there is no $f(x,t)$ in the appropriately defined S such that the integral is zero for any $f(x,t)$ in S other than $f(x,t) = 0$.

If both ends of the bar are free, then $f(x,t) = g(t)$ satisfies $\partial f/\partial x(0) = 0$ and $\partial f/\partial x(1) = 0$. Thus there is an $f(x,t) \ne 0$ in S such that the integral is zero.

Now consider the boundary terms. Clearly, if an end is fixed or free, then the boundary term in Equation 4.60 which corresponds to that end vanishes. If the end $x = 0$ is constrained with a spring, the appropriate boundary condition from Table 4.2 is of the form $\partial f/\partial x(0) - \eta_0 f(0) = 0$. The appropriate boundary condition when the end at $x = 1$ is constrained by a spring is Equation 4.57. Using these in Equation 4.60 leads to

$$(\mathbf{K}f,f) = \alpha(1)\eta_1[f(1)]^2 + \alpha(0)\eta_0[f(0)]^2 + \int_0^1 \alpha(x)\left[\frac{\partial f}{\partial x}\right]^2 dx \tag{4.61}$$

Thus, all boundary terms in the evaluation of $(\mathbf{K}f, f)$ are non-negative.

TABLE 4.2

Nondimensional Boundary Conditions for Continuous Systems

End Condition	First Boundary Condition at $x = 0$	Second Boundary Condition at $x = 1$	Nondimensional Parameters
	A. Longitudinal Motion of Bar		
Fixed	$u = 0$	$u = 0$	
Free	$\dfrac{\partial u}{\partial x} = 0$	$\dfrac{\partial u}{\partial x} = 0$	
Attached spring	$\dfrac{\partial u}{\partial x} = \kappa u$	$\dfrac{\partial u}{\partial x} = -\kappa u$	$\kappa = \dfrac{kL}{EA}$
Attached viscous damper	$\dfrac{\partial u}{\partial x} = \zeta \dfrac{\partial u}{\partial t}$	$\dfrac{\partial u}{\partial x} = -\zeta \dfrac{\partial u}{\partial t}$	$\zeta = \dfrac{c}{A\sqrt{\rho E}}$
Attached mass	$\dfrac{\partial u}{\partial x} = -\mu \dfrac{\partial^2 u}{\partial t^2}$	$\dfrac{\partial u}{\partial x} = \mu \dfrac{\partial^2 u}{\partial t^2}$	$\mu = \dfrac{m}{\rho A L}$
	B. Angular Oscillations of Shaft		
Fixed	$\theta = 0$	$\theta = 0$	
Free	$\dfrac{\partial \theta}{\partial x} = 0$	$\dfrac{\partial \theta}{\partial x} = 0$	
Attached torsional spring	$\dfrac{\partial \theta}{\partial x} = \kappa \theta$	$\dfrac{\partial \theta}{\partial x} = -\kappa \theta$	$\kappa = \dfrac{k_t L}{JG}$
Attached torsional viscous damper	$\dfrac{\partial \theta}{\partial x} = \zeta \dfrac{\partial \theta}{\partial t}$	$\dfrac{\partial \theta}{\partial x} = -\zeta \dfrac{\partial \theta}{\partial t}$	$\zeta = \dfrac{c_t}{J\sqrt{\rho G}}$
Attached disk	$\dfrac{\partial \theta}{\partial x} = -\eta \dfrac{\partial^2 \theta}{\partial t^2}$	$\dfrac{\partial \theta}{\partial x} = \eta \dfrac{\partial^2 \theta}{\partial t^2}$	$\eta = \dfrac{I}{\rho J L}$
	C. Euler–Bernoulli Beam		
Fixed	$w = 0$	$\partial w / \partial x = 0$	
Pinned	$w = 0$	$\partial^2 w / \partial x^2 = 0$	
Free	$\partial^2 w / \partial x^2 = 0$	$\dfrac{\partial}{\partial x}\left(\alpha \dfrac{\partial^2 w}{\partial x^2}\right) = 0$	$\alpha = E(x)I(x)/E_0 I_0$
Linear spring	$\partial^2 w / \partial x^2 = 0$	$\dfrac{\partial}{\partial x}\left(\alpha \dfrac{\partial^2 w}{\partial x^2}\right) = -\kappa w \quad x = 0$	$\alpha = E(x)I(x)/E_0 I_0$
		$\dfrac{\partial}{\partial x}\left(\alpha \dfrac{\partial^2 w}{\partial x^2}\right) = \kappa w \quad x = 1$	$\kappa = kL^3/E_0 I_0$
Viscous damper	$\partial^2 w / \partial x^2 = 0$	$\dfrac{\partial}{\partial x}\left(\alpha \dfrac{\partial^2 w}{\partial x^2}\right) = -\zeta \dfrac{\partial w}{\partial t} \quad x = 0$	$\alpha = E(x)I(x)/E_0 I_0$
		$\dfrac{\partial}{\partial x}\left(\alpha \dfrac{\partial^2 w}{\partial x^2}\right) = \zeta \dfrac{\partial w}{\partial t} \quad x = 1$	$\zeta = cL/\sqrt{\rho_0 A_0 E_0 I_0}$
Attached mass	$\partial^2 w / \partial x^2 = 0$	$\dfrac{\partial}{\partial x}\left(\alpha \dfrac{\partial^2 w}{\partial x^2}\right) = -\mu \dfrac{\partial^2 w}{\partial t^2} \quad x = 0$	$\alpha = E(x)I(x)/E_0 I_0$
		$\dfrac{\partial}{\partial x}\left(\alpha \dfrac{\partial^2 w}{\partial x^2}\right) = \mu \dfrac{\partial^2 w}{\partial t^2} \quad x = 1$	$\mu = m/\rho_0 A_0 L$

(continued)

TABLE 4.2 (Continued)

End Condition	First Boundary Condition at $x = 0$	Second Boundary Condition at $x = 1$	Nondimensional Parameters
Attached mass with moment of inertia	$\frac{\partial^2 w}{\partial x^2} = -\beta \frac{\partial^3 w}{\partial x \partial t^2}$ $x = 0$	$\frac{\partial}{\partial x}\left(\alpha \frac{\partial^2 w}{\partial x^2}\right) = -\mu \frac{\partial^2 w}{\partial t^2}$ $x = 0$	$\alpha = E(x)I(x)/E_0 I_0$
	$\frac{\partial^2 w}{\partial x^2} = \beta \frac{\partial^3 w}{\partial x \partial t^2}$ $x = 1$	$\frac{\partial}{\partial x}\left(\alpha \frac{\partial^2 w}{\partial x^2}\right) = \mu \frac{\partial^2 w}{\partial t^2}$ $x = 1$	$\mu = m/\rho_0 A_0 L$
			$\mu = J/\rho_0 A_0 L3$

From the above discussion, it is clear that **K** is positive definite with respect to the standard inner product for $C^2[0,1]$ unless the bar is free at both ends, in which case **K** is positive semidefinite with respect to the standard inner product. This case is similar to the unconstrained case for discrete systems.

Thus, since **K** is self-adjoint and positive definite with respect to the standard inner product, a potential energy inner product is defined as

$$(f, g)_K = (Kf, g) = \int_0^1 \left\{ -\frac{\partial}{\partial x}\left[\alpha(x)\frac{\partial f}{\partial x}\right]\right\} g(x)\,dx \qquad (4.62)$$

Now consider the operator

$$M^{-1}K(f) = -\frac{1}{\beta(x)}\frac{\partial}{\partial x}\left[\alpha(x)\frac{\partial f}{\partial x}\right] \qquad (4.63)$$

Unless $\beta(x) = 1$, $M^{-1}K$ is not self-adjoint with respect to the standard inner product for $C^2[0,1]$. The same proofs as used for discrete systems in Equation 4.26 and Equation 4.27 are used to show that $M^{-1}K$ is self-adjoint with respect to both the kinetic energy inner product and the potential energy inner product. In addition, as long as **K** is positive definite with respect to the standard inner product, $M^{-1}K$ is positive definite with respect to the kinetic energy inner product.

4.9 EULER–BERNOULLI BEAM

The differential equation for the transverse displacement, $w(x,t)$, of an Euler–Bernoulli beam, illustrated in Figure 4.12, subject to a transverse load per unit length $F(x,t)$ is derived in Example 2.17 as

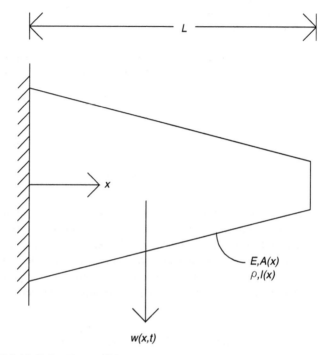

FIGURE 4.12 Euler–Bernoulli beam.

$$\frac{\partial^2}{\partial x^2}\left[E(x)I(x)\frac{\partial^2 w}{\partial x^2}\right] + \rho(x)A(x)\frac{\partial^2 w}{\partial t^2} = F(x,t) \qquad (4.64)$$

where E is the elastic modulus of the beam, ρ is its mass density, $A(x)$ is its cross section area, and $I(x)$ is its cross section moment of inertia.

Equation 4.64 is nondimensionalized using the nondimensional variables of Equation 4.36 through Equation 4.38 and

$$\alpha(x) = \frac{E(x)I(x)}{E_0 I_0} \qquad (4.65)$$

$$\beta(x) = \frac{\rho(x)A(x)}{\rho_0 A_0} \qquad (4.66)$$

The nondimensional formulation of Equation 4.64 with $f(x,t) = 0$ is

$$-\frac{\partial^2}{\partial x^2}\left[\alpha(x)\frac{\partial^2 w}{\partial x^2}\right] = \beta(x)\frac{\partial^2 w}{\partial t^2} \qquad (4.67)$$

where the *s have been dropped from the nondimensional variables and, for convenience, T is chosen as

$$T = L^2 \sqrt{\frac{\rho_0 A_0}{E_0 I_0}} \tag{4.68}$$

Equation 4.68 is of the form of Equation 4.33 with

$$\mathbf{K}w = \frac{\partial^2}{\partial x^2}\left[\alpha(x)\frac{\partial^2 w}{\partial x^2}\right] \tag{4.69}$$

$$\mathbf{M}w = \beta(x)w \tag{4.70}$$

Boundary conditions for end specifications are presented in Table 4.2.

The nondimensional spatial variable ranges from 0 to 1, $0 \le x \le 1$. The solution of Equation 4.67 subject to appropriate initial conditions must be fourth-order differentiable. Thus the domain of $\mathbf{M}$ is $C^4[0,1]$. The domain of $\mathbf{K}$ is the subspace of $C^4[0,1]$, specified by the boundary conditions outlined in Table 4.2 for specific end conditions.

The confirmation that $\mathbf{M}$ is self-adjoint and positive definite with respect to the standard inner product for $C^4[0,1]$ is exactly the same as that for a second-order continuous system as illustrated in Section 4.7. Thus, since $\mathbf{M}$ is positive definite and self-adjoint with respect to the standard inner product, a kinetic energy inner product may be defined as in Equation 4.51.

Consider

$$(\mathbf{K}f, g) = \int_0^1 \frac{\partial^2}{\partial x^2}\left[\alpha(x)\frac{\partial^2 f}{\partial x^2}\right]g(x)dx \tag{4.71}$$

Application of integration by parts to the right-hand side of Equation 4.71 with $u = g(x)$ and $dv = \partial^2/\partial x^2[\alpha(x)(\partial^2 f/\partial x^2)]dx$ leads to

$$(\mathbf{K}f, g) = g(1)\frac{\partial}{\partial x}\left[\alpha(x)\frac{\partial^2 f}{\partial x^2}\right]_{x=1} - g(0)\frac{\partial}{\partial x}\left[\alpha(x)\frac{\partial^2 f}{\partial x^2}\right]_{x=0}$$

$$- \int_0^1 \frac{\partial}{\partial x}\left[\alpha(x)\frac{\partial^2 f}{\partial x^2}\right]\frac{\partial g}{\partial x}dx \tag{4.72}$$

Application of integration by parts on the integral in Equation 4.72 with

$u = \partial g/\partial x$ and $dv = \dfrac{\partial}{\partial x}\left[\alpha(x)\dfrac{\partial^2 f}{\partial x^2}\right]dx$ leads to

$$(\mathbf{K}f, g) = g(1)\frac{\partial}{\partial x}\left[\alpha(x)\frac{\partial^2 f}{\partial x^2}\right]_{x=1} - g(0)\frac{\partial}{\partial x}\left[\alpha(x)\frac{\partial^2 f}{\partial x^2}\right]_{x=0}$$

$$-\alpha(1)\frac{\partial g}{\partial x}(1)\frac{\partial^2 f}{\partial x^2}(1) + \alpha(0)\frac{\partial g}{\partial x}(0)\frac{\partial^2 f}{\partial x^2}(0) + \int_0^1 \alpha(x)\frac{\partial^2 f}{\partial x^2}\frac{\partial^2 g}{\partial x^2}dx$$

$$(4.73)$$

It is noted that an expression for $(f, \mathbf{K}g)$ may be determined by switching $f(x)$ and $g(x)$ in the expression for $(\mathbf{K}f, g)$. Thus switching f with g in Equation 4.73 leads to

$$(f, \mathbf{K}g) = f(1)\frac{\partial}{\partial x}\left[\alpha(x)\frac{\partial^2 g}{\partial x^2}\right]_{x=1} - f(0)\frac{\partial}{\partial x}\left[\alpha(x)\frac{\partial^2 g}{\partial x^2}\right]_{x=0}$$

$$-\alpha(1)\frac{\partial f}{\partial x}(1)\frac{\partial^2 g}{\partial x^2}(1) + \alpha(0)\frac{\partial f}{\partial x}(0)\frac{\partial^2 g}{\partial x^2}(0) + \int_0^1 \alpha(x)\frac{\partial^2 f}{\partial x^2}\frac{\partial^2 g}{\partial x^2}dx$$

$$(4.74)$$

The operator $\mathbf{K}$ is self-adjoint with respect to the standard inner product for $C^4[0,1]$ if $(\mathbf{K}f, g) = (f, \mathbf{K}g)$ for all f and g in an appropriately defined S. The right-hand sides of Equation 4.73 and Equation 4.74 are equal for all $f(x,t)$ and $g(x,t)$ in S if

$$g(1)\frac{\partial}{\partial x}\left[\alpha(x)\frac{\partial^2 f}{\partial x^2}\right]_{x=1} - g(0)\frac{\partial}{\partial x}\left[\alpha(x)\frac{\partial^2 f}{\partial x^2}\right]_{x=0} - \alpha(1)\frac{\partial g}{\partial x}(1)\frac{\partial^2 f}{\partial x^2}(1)$$

$$+ \alpha(0)\frac{\partial g}{\partial x}(0)\frac{\partial^2 f}{\partial x^2}(0) = f(1)\frac{\partial}{\partial x}\left[\alpha(x)\frac{\partial^2 g}{\partial x^2}\right]_{x=1} - f(0)\frac{\partial}{\partial x}\left[\alpha(x)\frac{\partial^2 g}{\partial x^2}\right]_{x=0}$$

$$- \alpha(1)\frac{\partial f}{\partial x}(1)\frac{\partial^2 g}{\partial x^2}(1) + \alpha(0)\frac{\partial f}{\partial x}(0)\frac{\partial^2 g}{\partial x^2}(0)$$

$$(4.75)$$

Suppose, for example, that the beam is fixed at $x = 0$, $w(0,t) = 0$, $\partial w/\partial x(0,t) = 0$, and attached to a linear spring at $x = 1$, $\partial^2 w/\partial t^2(1, t) = 0$, $\partial/\partial x[\alpha(x)$ $\partial^2 w/\partial x^2]_{x=1} = \eta_1 w(1, t)$. The appropriately defined S contains all $f(x,t)$ in $C^4[0,1]$ such that $f(0,t) = 0, \partial f/\partial x(0,t) = 0, \partial^2 f/\partial x^2(1,t) = 0$, and $\partial/\partial x$ $[\alpha(x)\partial^2 f/\partial x^2]_{x=1} = \eta_1 f(1,t)$. Evaluation of both sides of Equation 4.75 leads to

$\eta_1 g(1) f(1)$, and hence the operator is self-adjoint. It can be shown that **K** is self-adjoint with respect to the standard inner product on $C^4[0,1]$ for all end conditions of Table 4.2.

Setting $g = f$ in Equation 4.73 leads to

$$(\mathbf{K}f,f) = f(1)\frac{\partial}{\partial x}\left[\alpha(x)\frac{\partial^2 f}{\partial x^2}\right]_{x=1} - f(0)\frac{\partial}{\partial x}\left[\alpha(x)\frac{\partial^2 f}{\partial x^2}\right]_{x=0}$$

$$-\alpha(1)\frac{\partial f}{\partial x}(1)\frac{\partial^2 f}{\partial x^2}(1) + \alpha(0)\frac{\partial f}{\partial x}(0)\frac{\partial^2 f}{\partial x^2}(0) + \int_0^1 \alpha(x)\left[\frac{\partial^2 f}{\partial x^2}\right]^2 dx$$

$$\tag{4.76}$$

The integral on the right-hand side of Equation 4.76 is non-negative and is equal to zero when $f(x,t) = 0$. The integral is also zero when f is of the form $f(x,t) = a(t) + xb(t)$. The only end conditions for which such a function is in S are for a beam free at one end but not fixed at its other end. For example, the subspace defined for a pinned-free beam contains functions of the form $f(x,t) = xb(t)$.

The boundary terms in Equation 4.76 are either identically zero or can be shown to be positive for all end conditions. Thus, **K** is positive definite and self-adjoint with respect to the standard inner product for $C^4[0,1]$ unless it is free at one end and not fixed at its other end. In the latter case, **K** is positive semidefinite with respect to the standard inner product for $C^4[0,1]$.

It has been shown that **K** is self-adjoint with respect to the standard inner product for $C^4[0,1]$. Thus for those subspaces for which **K** is also positive definite, a potential energy inner product is defined as

$$(f,g)_{\mathbf{K}} = (\mathbf{K}f,g) = \int_0^1 \frac{\partial^2}{\partial x^2}\left[\alpha(x)\frac{\partial^2 f}{\partial x^2}\right]g\,dx \tag{4.77}$$

The operator $\mathbf{M}^{-1}\mathbf{K}$ is defined as

$$\mathbf{M}^{-1}\mathbf{K}w = \frac{1}{\beta(x)}\frac{\partial^2}{\partial x^2}\left[\alpha(x)\frac{\partial^2 w}{\partial x^2}\right] \tag{4.78}$$

It has been shown that kinetic energy and potential energy inner product can be defined for the conservative Euler–Bernoulli beam. Using the same analysis as for discrete systems and wave equation systems, it can be shown that $\mathbf{M}^{-1}\mathbf{K}$ is self-adjoint with respect to both the kinetic energy inner product and the potential energy inner product. In addition, $\mathbf{M}^{-1}\mathbf{K}$ is positive definite with respect to the kinetic energy inner product.

Example 4.4. The partial differential equation governing the free transverse vibrations of an Euler–Bernoulli beam with a constant axial load P is derived in Example 2.17 as

$$\frac{\partial^2}{\partial x^2}\left[EI(x)\frac{\partial^2 w}{\partial x^2}\right] - P\frac{\partial^2 w}{\partial x^2} + \rho A(x)\frac{\partial^2 w}{\partial t^2} = 0 \tag{a}$$

(a) Choose the value of T such that when Equation 4.36 through Equation 4.38 and Equation 4.41 and Equation 4.42a,b are substituted into Equation a, the resulting nondimensional equation is

$$\varepsilon\frac{\partial^2}{\partial x^2}\left[\alpha(x)\frac{\partial^2 w}{\partial x^2}\right] - \frac{\partial^2 w}{\partial x^2} + \beta(x)\frac{\partial^2 w}{\partial t^2} = 0 \tag{b}$$

Identify ε.

(b) Discuss the self-adjointness and positive definiteness of the operator **K** when written in a nondimensional form.

(a) Substitution of Equation 4.36 through Equation 4.38 and Equation 4.41 and Equation 4.42a,b into Equation a leads to

$$\frac{E_0 I_0}{L^3}\frac{\partial^2}{\partial x^2}\left[\alpha(x)\frac{\partial^2 w}{\partial x^2}\right] - \frac{P}{L}\frac{\partial^2 w}{\partial x^2} + \frac{\rho_0 A_0 L}{T^2}\beta(x)\frac{\partial^2 w}{\partial t^2} = 0 \tag{c}$$

where *s have been dropped form nondimensional variables. Multiplying Equation c by L/P leads to

$$\frac{E_0 I_0}{PL^2}\frac{\partial^2}{\partial x^2}\left[\alpha(x)\frac{\partial^2 w}{\partial x^2}\right] - \frac{\partial^2 w}{\partial x^2} + \frac{\rho_0 A_0 L^2}{PT^2}\beta(x)\frac{\partial^2 w}{\partial t^2} = 0 \tag{d}$$

Choosing

$$\varepsilon = \frac{E_0 I_0}{PL^2} \tag{e}$$

$$T = L\sqrt{\frac{\rho_0 A_0}{P}} \tag{f}$$

and substituting into Equation d leads to Equation b.

(b) The operator **K** is defined as

$$\mathbf{K}w = \varepsilon\frac{\partial^2}{\partial x^2}\left[\alpha(x)\frac{\partial^2 w}{\partial x^2}\right] - \frac{\partial^2 w}{\partial x^2} \tag{g}$$

The self-adjointness of **K** with respect to the standard inner product for $C^4[0,1]$ is determined from examining

$$(Kf,g) = \int_0^1 \left\{ \varepsilon \frac{\partial^2}{\partial x^2} \left[\alpha(x) \frac{\partial^2 f}{\partial x^2} \right] - \frac{\partial^2 f}{\partial x^2} \right\} g \, dx$$

$$= \varepsilon \int_0^1 \frac{\partial^2 f}{\partial x^2} \left[\alpha(x) \frac{\partial^2 f}{\partial x^2} \right] g \, dx - \int_0^1 \frac{\partial^2 f}{\partial x^2} g \, dx \tag{h}$$

Integration by parts, applied twice, to the first integral on the right-hand side of Equation h leads to the right-hand side of Equation 4.73. Integration by parts on the second integral on the right-hand side of Equation h with $u = g$ and $dv = \partial^2 f/\partial x^2 dx$, leads to

$$\int_0^1 \frac{\partial^2 f}{\partial x^2} g \, dx = g(1) \frac{\partial f}{\partial x}(1) - g(0) \frac{\partial f}{\partial x}(0) - \int_0^1 \frac{\partial f}{\partial x} \frac{\partial g}{\partial x} \, dx \tag{i}$$

Using Equation i in Equation h leads to

$$(Kf,g) = \varepsilon \left\{ g(1) \frac{\partial}{\partial x} \left[\alpha(x) \frac{\partial^2 f}{\partial x^2} \right]_{x=1} - g(0) \frac{\partial}{\partial x} \left[\alpha(x) \frac{\partial^2 f}{\partial x^2} \right]_{x=0} \right.$$
$$\left. -\alpha(1) \frac{\partial g}{\partial x}(1) \frac{\partial^2 f}{\partial x^2}(1) + \alpha(0) \frac{\partial g}{\partial x}(0) \frac{\partial^2 f}{\partial x^2}(0) + \int_0^1 \alpha(x) \frac{\partial^2 f}{\partial x^2} \frac{\partial^2 g}{\partial x^2} \, dx \right\}$$
$$- \left[g(1) \frac{\partial f}{\partial x}(1) - g(0) \frac{\partial f}{\partial x}(0) - \int_0^1 \frac{\partial f}{\partial x} \frac{\partial g}{\partial x} \, dx \right] \tag{j}$$

Switching f and g in Equation j leads to

$$(f,Kg) = \varepsilon \left\{ f(1) \frac{d}{\partial x} \left[\alpha(x) \frac{d^2 g}{\partial x^2} \right]_{x=1} - f(0) \frac{d}{\partial x} \left[\alpha(x) \frac{d^2 g}{\partial x^2} \right]_{x=0} \right.$$
$$\left. -\alpha(1) \frac{\partial f}{\partial x}(1) \frac{\partial^2 g}{\partial x^2}(1) + \alpha(0) \frac{\partial f}{\partial x}(0) \frac{\partial^2 g}{\partial x^2}(0) + \int_0^1 \alpha(x) \frac{\partial^2 f}{\partial x^2} \frac{\partial^2 g}{\partial x^2} \, dx \right\}$$
$$- \left[f(1) \frac{\partial g}{\partial x}(1) - f(0) \frac{\partial g}{\partial x}(0) - \int_0^1 \frac{\partial f}{\partial x} \frac{\partial g}{\partial x} \, dx \right] \tag{k}$$

The integrals in Equation j and Equation k are identical. Thus, **K** is self-adjoint if

$$
\varepsilon\left\{ g(1)\frac{\partial}{\partial x}\left[\alpha(x)\frac{\partial^2 f}{\partial x^2}\right]_{x=1} - g(0)\frac{\partial}{\partial x}\left[\alpha(x)\frac{\partial^2 f}{\partial x^2}\right]_{x=0} \right.
$$

$$
\left. -\alpha(1)\frac{\partial g}{\partial x}(1)\frac{\partial^2 f}{\partial x^2}(1) + \alpha(0)\frac{\partial g}{\partial x}(0)\frac{\partial^2 f}{\partial x^2}(0) \right\} - \left[g(1)\frac{\partial f}{\partial x}(1) - g(0)\frac{\partial f}{\partial x}(0) \right]
$$

$$
= \varepsilon\left\{ f(1)\frac{\partial}{\partial x}\left[\alpha(x)\frac{\partial^2 g}{\partial x^2}\right]_{x=1} - f(0)\frac{\partial}{\partial x}\left[\alpha(x)\frac{\partial^2 g}{\partial x^2}\right]_{x=0} - \alpha(1)\frac{\partial f}{\partial x}(1)\frac{\partial^2 g}{\partial x^2}(1) \right.
$$

$$
\left. +\alpha(0)\frac{\partial f}{\partial x}(0)\frac{\partial^2 g}{\partial x^2}(0) \right\} - \left[f(1)\frac{\partial g}{\partial x}(1) - f(0)\frac{\partial g}{\partial x}(0) \right]
\tag{l}
$$

The boundary conditions for a fixed-fixed beam are all geometric and Equation l is satisfied.

The natural boundary conditions at $x = 1$ for a fixed-free beam derived in Example 2.17 and nondimensionalized using Equation 4.36 through Equation 4.38 and Equation 4.65 and Equation 4.66 are

$$
\alpha(1)\frac{\partial^2 w}{\partial x^2}(1,t) = 0
\tag{m}
$$

$$
\varepsilon\frac{\partial}{\partial x}\left(\alpha\frac{\partial^2 w}{\partial x^2} \right)(1,t) = \frac{\partial w}{\partial x}(1,t)
\tag{n}
$$

Use of the boundary conditions in Equation l leads to

$$
g(1)\frac{\partial f}{\partial x}(1) - g(1)\frac{\partial f}{\partial x}(1) = f(1)\frac{\partial g}{\partial x}(1) - f(1)\frac{\partial g}{\partial x}(1)
$$

and thus the stiffness operator is self-adjoint.

The positive definiteness of the stiffness operator is examined by letting $f = g$ in Equation j:

$$(\mathbf{K}f,f) = \varepsilon\left\{f(1)\frac{\partial}{\partial x}\left[\alpha(x)\frac{\partial^2 f}{\partial x^2}\right]_{x=1} - f(0)\frac{\partial}{\partial x}\left[\alpha(x)\frac{\partial^2 f}{\partial x^2}\right]_{x=0}\right.$$

$$\left. -\alpha(1)\frac{\partial f}{\partial x}(1)\frac{\partial^2 f}{\partial x^2}(1) + \alpha(0)\frac{\partial f}{\partial x}(0)\frac{\partial^2 f}{\partial x^2}(0) + \int_0^1 \alpha(x)\frac{\partial^2 f}{\partial x^2}\frac{\partial^2 f}{\partial x^2}dx\right\} \quad \text{(o)}$$

$$-\left[f(1)\frac{\partial f}{\partial x}(1) - f(0)\frac{\partial f}{\partial x}(0) - \int_0^1 \frac{\partial f}{\partial x}\frac{\partial f}{\partial x}dx\right]$$

All of the boundary terms are zero for a fixed–fixed beam, in which case Equation o reduces to

$$(\mathbf{K}f,f) = \varepsilon\int_0^1 \alpha(x)\frac{\partial^2 f}{\partial x^2}\frac{\partial^2 f}{\partial x^2}dx + \left[\int_0^1 \frac{\partial f}{\partial x}\frac{\partial f}{\partial x}dx\right] \quad \text{(p)}$$

Clearly, $(\mathbf{K}f, f)\geq 0$ and $(\mathbf{K}0,0) = 0$. The inner product could be zero if the domain of the stiffness operator contains an $f(x)$ such that $\partial f/\partial x = 0 \Rightarrow f(x,t) = c(t)$ for all x in $[0,1]$. However, since $f(0,t) = 0$, $c(t) = 0$. Thus, such a function is not contained in the domain of $\mathbf{K}$ and hence $\mathbf{K}$ is positive definite.

Application of the boundary conditions for a fixed-free beam to Equation o leads to

$$(\mathbf{K}f,f) = \varepsilon\left\{f(1)\frac{\partial}{\partial x}\left[\alpha(x)\frac{\partial^2 f}{\partial x^2}\right]_{x=1} + \int_0^1 \alpha(x)\frac{\partial^2 f}{\partial x^2}\frac{\partial^2 f}{\partial x^2}dx\right\}$$

$$-\left[f(1)\frac{\partial f}{\partial x}(1) - \int_0^1 \frac{\partial f}{\partial x}\frac{\partial f}{\partial x}dx\right]$$

$$= f(1)\frac{\partial f}{\partial x}(1) + \int_0^1 \alpha(x)\frac{\partial^2 f}{\partial x^2}\frac{\partial^2 f}{\partial x^2}dx - f(1)\frac{\partial f}{\partial x}(1) + \int_0^1 \frac{\partial f}{\partial x}\frac{\partial f}{\partial x}dx \quad \text{(q)}$$

Equation q shows that the stiffness operator is positive definite with respect to the standard inner product.

4.10 TIMOSHENKO BEAMS

The Timoshenko beam is an extension of the Euler–Bernoulli beam in which the effects of shear deformation and rotary inertia are included in the governing equations. The equations governing the motion of a uniform Timoshenko beam are derived in Section 2.12 as

$$\kappa GA \left(\frac{\partial^2 w}{\partial x^2} - \frac{\partial \psi}{\partial x} \right) = \rho A \frac{\partial^2 w}{\partial t^2} \tag{4.79}$$

$$EI \frac{\partial^2 \psi}{\partial x^2} + \kappa GA \left(\frac{\partial w}{\partial x} - \psi \right) = \rho I \frac{\partial^2 \psi}{\partial t^2} \tag{4.80}$$

Nondimensional variables are introduced according to

$$w^* = \frac{w}{L} \tag{4.81}$$

$$\psi^* = \psi \tag{4.82}$$

$$x^* = \frac{x}{L} \tag{4.83}$$

$$t^* = \frac{t}{T} \tag{4.84}$$

Substitution of Equation 4.81 through Equation 4.84 into Equation 4.79 and Equation 4.80 and dropping the *s from nondimensional variables leads to

$$\frac{\partial^2 w}{\partial x^2} - \frac{\partial \psi}{\partial x} = \frac{\partial^2 w}{\partial t^2} \tag{4.85}$$

$$\eta_1 \frac{\partial^2 \psi}{\partial x^2} + \frac{\partial w}{\partial x} - \psi = \eta_2 \frac{\partial^2 \psi}{\partial t^2} \tag{4.86}$$

where T has been chosen for convenience as

$$T = L \sqrt{\frac{\rho}{\kappa G}} \tag{4.87}$$

and

$$\eta_1 = \frac{EI}{\kappa G L^2 A} \tag{4.88}$$

$$\eta_2 = \frac{I}{AL^2} \qquad (4.89)$$

It is noted that η_2 is equal to the square of the ratio of the radius of gyration of the beam's cross section to the length of the beam.

Let Q be the vector space of $R^2 \times C^2[0,1]$. An element in Q is of the form $\mathbf{v} = \begin{bmatrix} v_1(x) \\ v_2(x) \end{bmatrix}$. Equation 4.85 and Equation 4.86 are written in the form of Equation 4.33, with

$$\mathbf{Kv} = - \begin{bmatrix} \dfrac{\partial^2}{\partial x^2} & -\dfrac{\partial}{\partial x} \\ \dfrac{\partial}{\partial x} & \eta_1 \dfrac{\partial^2}{\partial x^2} - 1 \end{bmatrix} \begin{bmatrix} w(x) \\ \psi(x) \end{bmatrix} \qquad (4.90)$$

$$\mathbf{Mv} = \begin{bmatrix} 1 & 0 \\ 0 & \eta_2 \end{bmatrix} \begin{bmatrix} w(x) \\ \psi(x) \end{bmatrix} \qquad (4.91)$$

The standard inner product on Q is defined as

$$(\mathbf{f},\mathbf{g}) = \int_0^1 \mathbf{g}^T \mathbf{f} dx = \int_0^1 [f_1(x)g_1(x) + f_2(x)g_2(x)]dx \qquad (4.92)$$

The self-adjointness of $\mathbf{M}$ is determined by examining

$$(\mathbf{Mf},\mathbf{g}) = \int_0^1 \mathbf{g}^T \mathbf{Mf} dx = \int_0^1 \{[f_1(x)]g_1(x) + [\eta_2 f_2(x)]g_2(x)\}dx$$

$$= \int_0^1 \{f_1(x)g_1(x) + f_2(x)\eta_2 g_2(x)\}dx = \int_0^1 (\mathbf{Mg})^T \mathbf{f} dx = (\mathbf{f},\mathbf{Mg}) \quad (4.93)$$

Equation 4.93 shows that $\mathbf{M}$ is self-adjoint with respect to the standard inner product on Q. The positive definiteness of $\mathbf{M}$ is confirmed by letting $\mathbf{g} = \mathbf{f}$ in Equation 4.93. Since $\mathbf{M}$ is positive definite and self-adjoint with respect to the standard inner product on Q, a kinetic energy inner product may be defined as

$$(\mathbf{f},\mathbf{g})_\mathbf{M} = (\mathbf{Mf},\mathbf{g}) \qquad (4.94)$$

Let S be a subspace of Q defined appropriately for the end conditions. The boundary conditions for each end condition are summarized in Table 4.3. The self-adjointness of $\mathbf{K}$ with respect to the standard inner product for S is determined by examining

$$(\mathbf{K}f,\mathbf{g}) = \int_0^1 \mathbf{g}^T \mathbf{K}f\,dx$$

$$= -\int_0^1 \left\{ g_1(x)\left[\frac{\partial^2 f_1}{\partial x^2} - \frac{\partial f_2}{\partial x}\right] + g_2(x)\left[\frac{\partial f_1}{\partial x} + \eta_1\frac{\partial^2 f_2}{\partial x^2} - f_2\right] \right\} dx \quad (4.95)$$

Integration by parts applied to several integrals leads to

$$(\mathbf{K}f,\mathbf{g}) = -g_1\frac{\partial f_1}{\partial x}\bigg|_0^1 + g_1 f_2\bigg|_0^1 + \int_0^1\left(\frac{\partial g_1}{\partial x}\frac{\partial f_1}{\partial x} - \frac{\partial g_1}{\partial x}f_2\right)dx$$

$$-g_2\eta_1\frac{\partial f_2}{\partial x}\bigg|_0^1 - g_2 f_1\bigg|_0^1$$

$$+ \int_0^1\left(\eta_1\frac{\partial g_2}{\partial x}\frac{\partial f_2}{\partial x} + \frac{\partial g_2}{\partial x}f_1 + g_2 f_2\right)dx \quad (4.96)$$

TABLE 4.3
Nondimensional Boundary Conditions for a Uniform Timoshenko Beam

End Condtion	Boundary Condition 1	Boundary Condition 2
Fixed	$w = 0$	$\psi = 0$
Pinned	$w = 0$	$\dfrac{\partial \psi}{\partial x} = 0$
Free	$\dfrac{\partial \psi}{\partial x} = 0$	$\dfrac{\partial w}{\partial x} - \psi = 0$

Now consider

$$(\mathbf{f},\mathbf{K}g) = \int_0^1 (\mathbf{K}g)^T \mathbf{f} dx = -\int_0^1 \left\{ f_1(x) \left[\frac{\partial^2 g_1}{\partial x^2} - \frac{\partial g_2}{\partial x} \right] \right. $$

$$\left. + f_2(x) \left[\frac{\partial g_1}{\partial x} + \eta_1 \frac{\partial^2 g_2}{\partial x^2} - g_2 \right] \right\} dx \tag{4.97}$$

Integration by parts on several integral leads to

$$(\mathbf{f},\mathbf{K}g) = -f_1 \frac{\partial g_1}{\partial x} \bigg|_0^1 + \int_0^1 \left(\frac{\partial g_1}{\partial x} \frac{\partial f_1}{\partial x} \right) dx + \eta_1 f_2 \frac{\partial g_2}{\partial x} \bigg|_0^1$$

$$+ \int_0^1 \left(\eta_1 \frac{\partial g_2}{\partial x} \frac{\partial f_2}{\partial x} \right) dx - \int_0^1 \left\{ f_1(x) \left[-\frac{\partial g_2}{\partial x} \right] + f_2(x) \left[\frac{\partial g_1}{\partial x} - g_2 \right] \right\} dx \tag{4.98}$$

Subtracting Equation 4.98 from Equation 4.96 leads to

$$(\mathbf{K}f,\mathbf{g}) - (\mathbf{f},\mathbf{K}g) = -g_1 \frac{\partial f_1}{\partial x} \bigg|_0^1 + g_1 f_2 \bigg|_0^1 - g_2 \eta_1 \frac{\partial f_2}{\partial x} \bigg|_0^1 - g_2 f_1 \bigg|_0^1$$

$$+ f_1 \frac{\partial g_1}{\partial x} \bigg|_0^1 - \eta_1 f_2 \frac{\partial g_2}{\partial x} \bigg|_0^1 \tag{4.99}$$

Clearly, the right-hand side Equation 4.99 is zero for a fixed–fixed beam and the stiffness operator is self-adjoint. The boundary conditions for a pinned end are such that if f is in the domain of the stiffness operator, then $f_1 = 0$ and $\partial f_2/\partial x = 0$ at that end. Requiring both f and g to satisfy these conditions also clearly leads to the boundary terms at that end being equal to zero. The nondimensional boundary conditions at a free end require that if f is in the domain of the stiffness operator, then $\partial f_1/\partial x - f_2 = 0$ and $\partial f_2/\partial x = 0$. Substitution of these conditions in Equation 4.99 again leads to the boundary terms at a free end being equal to zero. Thus, the stiffness operator is self-adjoint with respect to the inner product of Equation 4.92 for all combinations of beams that have ends that are fixed, simply supported, or free.

The positive definiteness of $\mathbf{K}$ is considered by letting $g = f$ in Equation 4.98.

$$(\mathbf{f}, \mathbf{K}f) = -f_1 \left.\frac{\partial f_1}{\partial x}\right|_0^1 + \int_0^1 \left(\frac{\partial f_1}{\partial x}\right)^2 dx - \eta_1 f_2 \left.\frac{\partial f_2}{\partial x}\right|_0^1 + \int_0^1 \eta_1 \left(\frac{\partial f_2}{\partial x}\right)^2 dx$$

$$- \int_0^1 \left\{ f_1(x)\left[-\frac{\partial f_2}{\partial x}\right] + f_2(x)\left[\frac{\partial f_1}{\partial x} - f_2\right] \right\} dx \qquad (4.100)$$

Using integration by parts on $\int_0^1 f_1(x)\partial f_2/\partial x\, dx$ leads to

$$(\mathbf{f}, \mathbf{K}f) = -f_1 \left.\frac{\partial f_1}{\partial x}\right|_0^1 - \eta_1 f_2 \left.\frac{\partial f_2}{\partial x}\right|_0^1 + f_1 f_2 \Big|_0^1 + \int_0^1 \left[\left(\frac{\partial f_1}{\partial x}\right)^2 - 2 f_2 \frac{\partial f_1}{\partial x} + f_2^2 \right.$$

$$\left. + \eta_1 \left(\frac{\partial f_2}{\partial x}\right)^2 \right] dx = -f_1 \left.\frac{\partial f_1}{\partial x}\right|_0^1 - \eta_1 f_2 \left.\frac{\partial f_2}{\partial x}\right|_0^1 + f_1 f_2 \Big|_0^1$$

$$+ \int_0^1 \left[\left(\frac{\partial f_1}{\partial x} - f_2\right)^2 + \eta_1 \left(\frac{\partial f_2}{\partial x}\right)^2 \right] dx \qquad (4.101)$$

The integral in Equation 4.101 is clearly non-negative. It is not hard to show that the boundary terms will vanish or are positive for all possible boundary conditions. The integral is zero and all boundary terms zero for the vector $f = \begin{bmatrix} 1 & 0 \end{bmatrix}^T$ when the stiffness operator for a free-free beam is considered. Thus, $\mathbf{K}$ is positive definite for all end conditions except the free–free condition in which case it is positive semi definite.

4.11 SYSTEMS WITH MULTIPLE DEFORMABLE BODIES

Carbon nanotubes are cylinders of tightly bonded carbon atoms. The radius of a nanontube is that of a carbon atom, 0.34 nm. Multi-scale modeling has been used to predict behavior of nanotubes under certain assumptions. Continuous models may be used to study vibrations of long nanotubes. A multi-wall nanotube consists of layers of carbon atoms. A van der Waals bond exists between the atoms. When the tube is at rest, the equilibrium configuration is such that the atoms in layers are separated by the bond length of a carbon atom. If the nanotube is undergoing a transverse motion, an outer layer may move relative to an inner layer as the equilibrium distance is disturbed. The resulting changes in van der Waals forces are determined using a Lennard–Jones potential function and can be approximated as those from a linear elastic layer.

As a model problem, consider a series of n uniform Euler–Bernoulli beams connected through elastic layers, modeled by layers of continuous springs. Each beam in the system, illustrated in Figure 4.13, has the same length and the beams are supported identically. Let $w_i(x,t)$ represent the transverse displacement of the ith beam in the series which has mass density ρ_i, elastic modulus E_i, cross section area A_i, and area moment of inertia I_i. Let k_i be the stiffness per length between the ith and $(i+1)$st2 layer. For generality, the structure itself may be on an elastic foundation with k_0 and k_n the stiffnesses per unit length of layers connecting the first and nth beams to a foundation.

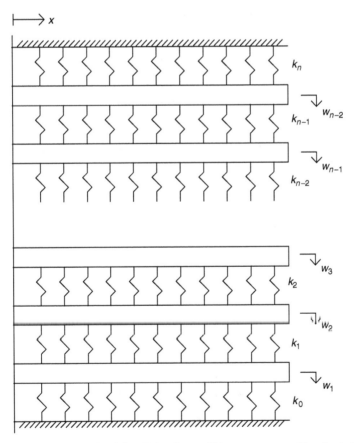

FIGURE 4.13 Series of fixed-free Euler–Bernoulli beams connected by elastic layers modeled as linear springs.

The differential equations governing the free responses of the beams are

$$E_1 I_1 \frac{\partial^4 w_1}{\partial x^4} + k_0 w_1 + k_1(w_1 - w_2) + \rho_1 A_1 \frac{\partial^2 w_1}{\partial t^2} = 0 \tag{4.102}$$

$$E_i I_i \frac{\partial^4 w_i}{\partial x^4} + k_i(w_i - w_{i+1}) - k_{i-1}(w_{i-1} - w_i) + \rho_i A_i \frac{\partial^2 w_i}{\partial t^2} = 0,$$

$$i = 2,3,\ldots,n-1 \tag{4.103}$$

$$E_n I_n \frac{\partial^4 w_n}{\partial x_4} + k_n w_n - k_{n-1}(w_{n-1} - w_n) + \rho_n A_n \frac{\partial^2 w_n}{\partial t^2} = 0 \tag{4.104}$$

Nondimensional variables are introduced as

$$w_i^* = \frac{w_i}{L} \tag{4.105}$$

$$x^* = \frac{x}{L} \tag{4.106}$$

$$t^* = \frac{t}{T} \tag{4.107}$$

Substitution of Equation 4.105 through Equation 4.107 into Equation 4.102 through Equation 4.104 leads to

$$\frac{\partial^4 w_1}{\partial x^4} + \kappa_0 w_1 + \kappa_1(w_1 - w_2) + \frac{\partial^2 w_1}{\partial t^2} = 0 \tag{4.108}$$

$$\phi_i \frac{\partial^4 w_i}{\partial x^4} + \kappa_i(w_i - w_{i+1}) - \kappa_{i-1}(w_{i-1} - w_i) + \mu_i \frac{\partial^2 w_i}{\partial t^2} = 0$$

$$i = 2,3,\ldots,n-1 \tag{4.109}$$

$$\phi_n \frac{\partial^4 w_n}{\partial x^4} + \kappa_n w_n - \kappa_{n-1}(w_{n-1} - w_n) + \mu_n \frac{\partial^2 w_n}{\partial t^2} = 0 \tag{4.110}$$

where

$$T = L^2 \sqrt{\frac{\rho_1 A_1}{E_1 I_1}} \tag{4.111}$$

$$\kappa_i = \frac{k_i L^3}{E_1 I_1} \quad i = 0,1,2,\ldots,n \tag{4.112}$$

$$\phi_i = \frac{E_i I_i}{E_1 I_1} \quad i = 1,2,\ldots,n \tag{4.113}$$

$$\mu_i = \frac{\rho_i A_i}{\rho_1 A_1} \quad i = 1,2,\ldots,n \tag{4.114}$$

Equation 4.108 through Equation 4.110 are written in the form of Equation 4.33, as

$$
\begin{bmatrix}
\dfrac{\partial^4}{\partial x^4} + \kappa_0 + \kappa_1 & -\kappa_1 & 0 & 0 & \cdots & 0 \\[2mm]
-\kappa_1 & \phi_2 \dfrac{\partial^4}{\partial x^4} + \kappa_1 + \kappa_2 & -\kappa_2 & 0 & \cdots & 0 \\[2mm]
0 & -\kappa_2 & \phi_3 \dfrac{\partial^4}{\partial x^4} + \kappa_2 + \kappa_3 & -\kappa_3 & \cdots & 0 \\[2mm]
\vdots & \vdots & \vdots & \vdots & \ddots & \vdots \\[2mm]
0 & 0 & 0 & 0 & \cdots & -\kappa_{n-1} \\[2mm]
0 & 0 & 0 & 0 & \cdots & \phi_n \dfrac{\partial^4}{\partial x^4} + \kappa_{n-1} + \kappa_n
\end{bmatrix}
$$

$$
\times
\begin{bmatrix}
w_1 \\ w_2 \\ w_3 \\ \vdots \\ w_{n-1} \\ w_n
\end{bmatrix}
+
\begin{bmatrix}
1 & 0 & 0 & 0 & \cdots & 0 \\
0 & \mu_2 & 0 & 0 & \cdots & 0 \\
0 & 0 & \mu_3 & 0 & \cdots & 0 \\
0 & 0 & 0 & \mu_4 & \cdots & 0 \\
\vdots & \vdots & \vdots & \vdots & \ddots & 0 \\
0 & 0 & 0 & 0 & \ddots & \mu_n
\end{bmatrix}
\begin{bmatrix}
\dfrac{\partial^2 w_1}{\partial t^2} \\[2mm]
\dfrac{\partial^2 w_2}{\partial t^2} \\[2mm]
\dfrac{\partial^2 w_3}{\partial t^2} \\[2mm]
\dfrac{\partial^2 w_4}{\partial t^2} \\[2mm]
\vdots \\[2mm]
\dfrac{\partial^2 w_n}{\partial t^2}
\end{bmatrix}
=
\begin{bmatrix}
0 \\ 0 \\ 0 \\ 0 \\ \vdots \\ 0
\end{bmatrix}
$$

$$(4.115)$$

Let Q be the vector space $R^n \times C^4[0,1]$ and S be the domain of the stiffness operator, a subspace of Q appropriately defined for the specific end conditions. The standard inner product for Q and S is $(\mathbf{f},\mathbf{g}) = \int_0^1 \mathbf{g}^T \mathbf{f} dx$. It can be shown, in a manner analogous to that for the Timoshenko beam of Section 4.10, that $\mathbf{M}$ is self-adjoint and positive definite with respect to the standard inner product on Q. The self-adjointness of $\mathbf{K}$ is determined by considering

$$(\mathbf{K}f, \mathbf{g}) = \int_0^1 \mathbf{g}^T \mathbf{K} f \, dx = \int_0^1 \left[g_1(x) \left[\frac{\partial^4 f_1}{\partial x^4} + \kappa_0 f_1 + \kappa_1 (f_1 - f_2) \right] \right.$$

$$+ \sum_{i=2}^{n-1} \left\{ g_i(x) \left[\phi_i \frac{\partial^4 f_i}{\partial x^4} + \kappa_i (f_i - f_{i+1}) - \kappa_{i-1}(f_{i-1} - f_i) \right] \right\} \qquad (4.116)$$

$$+ g_n(x) \left[\phi_n \frac{\partial^4 f_n}{\partial x^4} + \kappa_n f_n - \kappa_{n-1}(f_{n-1} - f_n) \right] \right] dx$$

Note that

$$(\mathbf{f}, \mathbf{K}g) = \int_0^1 (\mathbf{K}g)^T \mathbf{f} \, dx = \int_0^1 \left[f_1(x) \left[\frac{\partial^4 g_1}{\partial x^4} + \kappa_0 g_1 + \kappa_1 (g_1 - g_2) \right] \right.$$

$$+ \sum_{i=2}^{n-1} \left\{ f_i(x) \left[\phi_i \frac{\partial^4 g_i}{\partial x^4} + \kappa_i (g_i - g_{i+1}) \right. \right.$$

$$\qquad (4.117)$$

$$\left. \left. - \kappa_{i-1}(g_{i-1} - g_i) \right] \right\} + f_n(x) \left[\phi_n \frac{\partial^4 g_n}{\partial x^4} + \kappa_n g_n \right.$$

$$\left. \left. - \kappa_{n-1}(g_{n-1} - g_n) \right] \right] dx$$

Subtraction of Equation 4.117 from Equation 4.116 leads to

$$(\mathbf{K}f, \mathbf{g}) - (\mathbf{f}, \mathbf{K}g) = \int_0^1 \sum_{i=1}^n \left[g_i(x) \frac{\partial^4 f_i}{\partial x^4} - f_i(x) \frac{\partial^4 g_i}{\partial x^4} \right] dx$$

$$\qquad (4.118)$$

$$= \sum_{i=1}^n \int_0^1 \left[g_i(x) \frac{\partial^4 f_i}{\partial x^4} - f_i(x) \frac{\partial^4 g_i}{\partial x^4} \right] dx$$

Each integral in the summation of Equation 4.118, using integration by parts twice, can be shown to be equal to zero for all end conditions in Table 4.2. Thus, $\mathbf{K}$ is self-adjoint with respect to the standard inner product on S.

The stiffness operator is written as

$$
\mathbf{K} = \begin{bmatrix}
\dfrac{\partial^4}{\partial x^4} & 0 & 0 & 0 & \cdots & 0 \\
0 & \phi_2 \dfrac{\partial^4}{\partial x^4} & 0 & 0 & \cdots & 0 \\
0 & 0 & \phi_3 \dfrac{\partial^4}{\partial x^4} & 0 & \cdots & 0 \\
0 & 0 & 0 & \phi_4 \dfrac{\partial^4}{\partial x^4} & \cdots & 0 \\
\vdots & \vdots & \vdots & \vdots & \ddots & \vdots \\
0 & 0 & 0 & 0 & \cdots & \phi_n \dfrac{\partial^4}{\partial x^4}
\end{bmatrix}
$$

$$
+ \begin{bmatrix}
\kappa_0 + \kappa_1 & -\kappa_1 & 0 & 0 & \cdots & 0 \\
-\kappa_1 & \kappa_1 + \kappa_2 & -\kappa_2 & 0 & \cdots & 0 \\
0 & -\kappa_2 & \kappa_2 + \kappa_3 & -\kappa_3 & \cdots & 0 \\
0 & 0 & -\kappa_3 & \kappa_3 + \kappa_4 & \cdots & 0 \\
\vdots & \vdots & \vdots & \vdots & \ddots & \vdots \\
0 & 0 & 0 & 0 & \cdots & \kappa_{n-1} + \kappa_n
\end{bmatrix} \tag{4.119}
$$

The stiffness operator is positive definite as each operator matrix is positive definite. The first matrix operator can easily be shown to be positive definite with respect to the standard inner product on an appropriately defined S unless the end conditions are such that an Euler–Bernoulli beam with such end conditions is positive semidefinite. The second matrix is a symmetric matrix of discrete stiffnesses. This matrix is positive definite with respect to the standard inner product on R^n unless the matrix corresponds to an unconstrained system, which occurs only when both $\kappa_0 = 0$ and, $\kappa_n = 0$, in which case the system is positive semidefinite.

The above discussion shows that the stiffness operator is positive definite with respect to the standard inner product on S unless the beams have one end free without their opposite ends fixed and both $\kappa_0 = 0$ and $\kappa_n = 0$. If these conditions occur, the system is positive semidefinite.

Since $\mathbf{K}$ is self-adjoint and positive definite with respect to the standard inner product on an appropriately defined S, a kinetic energy inner product may be defined as

$$(\mathbf{f},\mathbf{g})_{\mathbf{K}} = (\mathbf{K}f,\mathbf{g}) \tag{4.120}$$

The operator $\mathbf{M}^{-1}\mathbf{K}$ is determined as

$$\begin{bmatrix} \dfrac{\partial^4}{\partial x^4}+\kappa_0+\kappa_1 & -\dfrac{\kappa_1}{\mu_2} & 0 & 0 & \cdots & 0 \\[2ex] -\kappa_1 & \dfrac{1}{\mu_2}\left(\phi_2\dfrac{\partial^4}{\partial x^4}+\kappa_1+\kappa_2\right) & -\dfrac{\kappa_2}{\mu_3} & 0 & \cdots & 0 \\[2ex] 0 & -\dfrac{\kappa_2}{\mu_2} & \dfrac{1}{\mu_3}\left(\phi_3\dfrac{\partial^4}{\partial x^4}+\kappa_2+\kappa_3\right) & -\dfrac{\kappa_3}{\mu_4} & \cdots & 0 \\[2ex] \vdots & \vdots & \vdots & \vdots & \ddots & \vdots \\[2ex] 0 & 0 & 0 & 0 & \cdots & -\dfrac{\kappa_{n-1}}{\mu_n} \\[2ex] 0 & 0 & 0 & 0 & \cdots & \dfrac{1}{\mu_n}\left(\phi_n\dfrac{\partial^4}{\partial x^4}+\kappa_{n-1}+\kappa_n\right) \end{bmatrix} \tag{4.121}$$

The operator $\mathbf{M}^{-1}\mathbf{K}$ is self-adjoint with respect to the kinetic energy inner product and the potential energy inner product and positive definite with respect to the kinetic energy inner product.

4.12 CONTINUOUS SYSTEMS WITH ATTACHED INERTIA ELEMENTS

The shaft of Figure 4.14 has a rigid thin disk of mass moment of inertia I attached to its free end. The problem governing the angular displacement of the shaft is formulated using extended Hamilton's principle. The kinetic energy of the system, including the kinetic energy of the disk, is

$$T = \frac{1}{2}\int_0^L \rho J\left(\frac{\partial\theta}{\partial t}\right)^2 dx + \frac{1}{2}I\left(\frac{\partial\theta}{\partial t}(L,t)\right)^2 \tag{4.122}$$

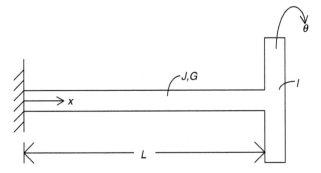

FIGURE 4.14 Shaft with attached disk.

The potential energy at an arbitrary instant is

$$V = \frac{1}{2} \int_0^L JG \left(\frac{\partial \theta}{\partial x} \right)^2 dx \qquad (4.123)$$

Application of extended Hamilton's principle eventually leads to

$$\int_{t_1}^{t_2} \left\{ \int_0^L \left[\rho J \frac{\partial^2 \theta}{\partial t} - \frac{\partial}{\partial x} \left(JG \frac{\partial \theta}{\partial x} \right) \right] \delta\theta dx + J(L)G(L) \frac{\partial \theta}{\partial x}(L) \delta\theta(L) \right.$$

$$\qquad (4.124)$$

$$\left. + I \frac{\partial^2 \theta}{\partial t^2}(L,t)\delta\theta(L) - J(0)G(0) \frac{\partial \theta}{\partial x}(0)\delta\theta(0) \right\} dt = 0$$

Since Equation 4.124 must be true for all possible varied paths, the governing partial differential equation is

$$\frac{\partial}{\partial x} \left(JG \frac{\partial \theta}{\partial x} \right) = \rho J \frac{\partial^2 \theta}{\partial t^2} \qquad (4.125)$$

The boundary condition at $x = 0$ is either $\partial\theta/\partial x(0,t) = 0$ or $\theta(0,t) = 0$. The boundary condition at $x = L$ is either $J(L)G(L)\frac{\partial\theta}{\partial x}(L,t) + I\frac{\partial^2\theta}{\partial t^2}(L,t) = 0$ or $\theta(L,t) = 0$.

Following the approaches of previous sections, the inertia operator could be taken as $\mathbf{M}\theta = \rho J\theta$. Indeed, this operator is positive definite and self-adjoint with respect to the standard inner product on $C^2[0,L]$. However, if the stiffness operator is then taken as $\mathbf{K}\theta = -\partial/\partial x(JG\partial\theta/\partial x)$, then examination of its self-adjointness leads to $(\mathbf{K}f,g)-(g,\mathbf{K}f)=I\left[f(\partial^2 g/\partial t^2)-g(\partial^2 f/\partial t^2)\right]$, which cannot be shown to be equal to zero for all f and g in an appropriately defined subspace of $C^2[0,L]$. Indeed, the definition of the domain of $\mathbf{K}$ is even in question.

The problem arises because the kinetic energy inner product $(\mathbf{M}\dot{\theta},\dot{\theta})$, where $\dot{\theta}=\partial\theta/\partial t$, defined from the standard inner product is not yield twice the total kinetic energy of the system as in all previous problems. The kinetic energy of the discrete mass is not included. Mathematically, the problem arises because of the presence of time derivatives in one of the boundary conditions.

An alternate formulation of the kinetic energy is

$$T = \frac{1}{2}\int_0^L \left[\rho J\left(\frac{\partial\theta}{\partial t}\right)^2 + I\left(\frac{\partial\theta}{\partial t}\right)^2 \delta(x-L)\right]dx \qquad (4.126)$$

where $\delta(x-L)$ is the Dirac delta function or the unit impulse function. A property of the Dirac delta function is

$$\int_0^L f(x)\delta(x-a)dx = f(a) \quad a \leq L \qquad (4.127)$$

Thus,

$$\int_0^L I\left(\frac{\partial\theta}{\partial t}\right)^2 \delta(x-L)dx = I\left(\frac{\partial\theta}{\partial t}(L,t)\right)^2 \qquad (4.128)$$

When this formulation is used, application of extended Hamilton's principle leads to

$$\int_{t_1}^{t_2}\left\{\int_0^L \left[(\rho J + I\delta(x-L))\frac{\partial^2\theta}{\partial t^2} - \frac{\partial}{\partial x}\left(JG\frac{\partial\theta}{\partial x}\right)\right]\delta\theta dx + J(L)G(L)\frac{\partial\theta}{\partial x}(L)\delta\theta(L)\right.$$

$$\left. -J(0)G(0)\frac{\partial\theta}{\partial x}(0)\delta\theta(0) \right\}dt = 0 \qquad (4.129)$$

The differential equation is now written as

$$\frac{\partial}{\partial x}\left(JG\frac{\partial \theta}{\partial x}\right) = [\rho J + I\delta(x-L)]\frac{\partial^2 \theta}{\partial t^2} \qquad (4.130)$$

When Equation 4.130 is used the boundary conditions for a bar fixed at $x = 0$ and free at $x = L$ are $\theta(0,t) = 0$ and $\partial\theta/\partial x(L,t) = 0$.

The following nondimensional variables are defined

$$\theta^* = \theta \qquad (4.131)$$

$$x^* = \frac{x}{L} \qquad (4.132)$$

$$t^* = \frac{t}{T} \qquad (4.133)$$

$$\alpha(x) = \frac{J(x)G(x)}{J_0 G_0} \qquad (4.134)$$

$$\beta(x) = \frac{\rho(x)J(x)}{\rho_0 J_0} \qquad (4.135)$$

Substitution of Equation 4.131 through Equation 4.135 into Equation 4.130 leads to the nondimensional differential equation

$$\frac{\partial}{\partial x}\left[\alpha(x)\frac{\partial \theta}{\partial x}\right] = [\beta(x) + \mu\delta(x-1)]\frac{\partial^2 \theta}{\partial t^2} \qquad (4.136)$$

in which the *s have been dropped from the nondimensional variables and

$$T = L\sqrt{\frac{\rho_0}{G_0}} \qquad (4.137)$$

$$\mu = \frac{I}{\rho_0 J_0 L} \qquad (4.138)$$

The inertia operator is defined as

$$\mathbf{M}\theta = \beta(x) + \mu\delta(x-1) \qquad (4.139)$$

The inertia operator is self-adjoint and positive definite with respect to the standard inner product on $C^2[0,1]$. The kinetic energy inner product is defined as

$$(f,g)_M = (Mf,g) = \int_0^1 [\beta(x) + \mu\delta(x-1)] f(x)g(x)dx$$

$$= \int_0^1 \beta(x)f(x)g(x)dx + \mu f(1)g(1) \qquad (4.140)$$

The stiffness operator is self-adjoint and, for a constrained system, positive definite with respect to the standard inner product for S, the subspace of $C^2[0,1]$, such that if f is in S then $f(0) = 0$ and $\frac{\partial f}{\partial x}(0) = 0$. Then a potential energy inner product is defined as

$$(f,g)_K = \int_0^1 \frac{-\partial}{\partial x}\left[\alpha(x)\frac{\partial f}{\partial x}\right]g(x)dx \qquad (4.141)$$

The operator $\mathbf{M}^{-1}\mathbf{K}$ is defined as

$$\mathbf{M}^{-1}\mathbf{K}\theta = \frac{-1}{\beta(x) + \mu\delta(x-1)}\frac{\partial}{\partial x}\left[\alpha(x)\frac{\partial\theta}{\partial x}\right] \qquad (4.142)$$

is self-adjoint with respect to the kinetic energy inner product and the potential energy inner product and positive definite with respect to the kinetic energy inner product.

Example 4.5. A thin disk of mass m and radius r is attached to the end of the bar of Figure 4.15. Formulate the kinetic energy inner product for (a) torsional oscillations of the bar; (b) longitudinal oscillations of the bar; and (c) transverse oscillations.

Solution: (a) The kinetic energy inner product for torsional oscillations of the bar is formulated in Equation 4.140. The polar moment of inertia of the disk is $I_D = 1/2mr^2$, which is used to define the parameter μ of Equation 4.138.

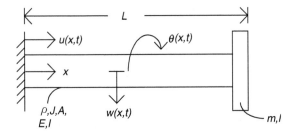

FIGURE 4.15 Bar of Example 4.5 has three independent modes of vibration.

(b) The kinetic energy when the bar and attached mass are undergoing longitudinal motion is

$$T = \frac{1}{2} \int_0^L \left[\rho A \left(\frac{\partial u}{\partial t} \right)^2 + m \left(\frac{\partial u}{\partial t}(L,t) \right) \delta(x-L) \right] dx \qquad (a)$$

The longitudinal motion of the bar is considered in Example 2.14. When Equation a is used for the kinetic energy, the use of Hamilton's Principle leads to

$$\frac{\partial}{\partial x} \left(EA \frac{\partial u}{\partial x} \right) = [\rho A + m\delta(x-L)] \frac{\partial^2 u}{\partial t^2} \qquad (b)$$

When nondimensionalized using the variables of Equation 4.36 through Equation 4.38 Equation b becomes

$$\frac{\partial}{\partial x} \left(\alpha \frac{\partial u}{\partial x} \right) = [\beta(x) + \mu \delta(x-1)] \frac{\partial^2 u}{\partial t^2} \qquad (c)$$

where

$$\mu = \frac{m}{\rho_0 A_0 L} \qquad (d)$$

The appropriate definition of the kinetic energy inner product for the longitudinal motion of the bar is

$$(f,g) = \int_0^L f(x)g(x)dx + \mu f(1)g(1) \qquad (e)$$

(c) The kinetic energy of a beam with an end mass is

$$T = \frac{1}{2} \int_0^L \left[\rho A \left(\frac{\partial w}{\partial t} \right)^2 + m \left(\frac{\partial w}{\partial t}(L,t) \right)^2 \delta(x-L) \right] dx \qquad \text{(f)}$$

Application of Hamilton's Principle to the lagrangian for the Euler–Bernoulli beam with an end inertia element leads to

$$\frac{\partial^2}{\partial x^2} \left(EI \frac{\partial^2 w}{\partial x^2} \right) + \rho A \frac{\partial^2 w}{\partial t^2} + m \frac{\partial^2 w}{\partial t^2}(L,t)\delta(x-L) = 0 \qquad \text{(g)}$$

Use of the nondimensional variables of Equation 4.36 through Equation 4.38 in Equation g leads to

$$\frac{\partial^2}{\partial x^2} \left(\alpha \frac{\partial^2 w}{\partial x^2} \right) + \beta \frac{\partial^2 w}{\partial t^2} + \mu \frac{\partial^2 w}{\partial t^2}(1,t)\delta(x-1) = 0 \qquad \text{(h)}$$

The appropriate inner product is that of Equation e.

4.13 COMBINED CONTINUOUS AND DISCRETE SYSTEMS

Consider the system of Figure 4.16 in which a particle of mass m is attached through a linear spring to the end of a beam. The kinetic energy of the system is

$$T = \frac{1}{2} \int_0^L \rho A \left(\frac{\partial w}{\partial t} \right)^2 + \frac{1}{2} m \left(\frac{dy}{dt} \right)^2 \qquad \text{(4.143)}$$

where $w(x,t)$ is the transverse deflection of the beam and $y(t)$ is the displacement of the particle, measured from the system's equilibrium position. The potential energy of the system is

$$V = \frac{1}{2} \int_0^L EI \left(\frac{\partial^2 w}{\partial x^2} \right)^2 dx + \frac{1}{2} k[y - w(L,t)]^2 \qquad \text{(4.144)}$$

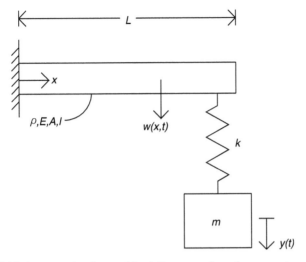

FIGURE 4.16 An example of a combined discrete and continuous system.

An alternate expression for the potential energy is

$$V = \frac{1}{2} \int_0^L \left\{ EI \left(\frac{\partial^2 w}{\partial x^2} \right)^2 + k[y - w(x,t)]^2 \delta(x-L) \right\} dx \qquad (4.145)$$

where $\delta(x-L)$ is the Dirac delta function. Using the latter formulation for potential energy application of extended Hamilton's principle leads to

$$\int_{t_1}^{t_2} \left\{ \int_0^1 \left[\frac{\partial^2}{\partial x^2} \left(EI \frac{\partial^2 w}{\partial x^2} \right) + k[w(L,t) - y]\delta(x-L) + \rho A \frac{\partial^2 w}{\partial t^2} \right] \delta w \, dx \right.$$
$$\left. + \left\{ m \frac{d^2 y}{dt^2} + k[y - w(L,t)\delta(x-L)] \right\} \delta y \right\} dt = 0 \qquad (4.146)$$

The resulting differential equations are

$$\frac{\partial^2}{\partial x^2} \left(EI \frac{\partial^2 w}{\partial x^2} \right) - k[y - w]\delta(x-L) + \rho A \frac{\partial^2 w}{\partial t^2} = 0 \qquad (4.147)$$

$$k[y - w(L,t)] + m \frac{d^2 y}{dt^2} = 0 \qquad (4.148)$$

The following nondimensional variables are introduced:

$$w^* = \frac{w}{L} \tag{4.149}$$

$$y^* = \frac{y}{L} \tag{4.150}$$

$$x^* = \frac{x}{L} \tag{4.151}$$

$$t^* = \frac{t}{T} \tag{4.152}$$

$$\alpha(x) = \frac{EI}{E_0 I_0} \tag{4.153}$$

$$\beta(x) = \frac{\rho A}{\rho_0 A_0} \tag{4.154}$$

Substitution of Equation 4.149 through Equation 4.154 into Equation 4.147 and Equation 4.148 leads to

$$\frac{\partial^2}{\partial x^2}\left[\alpha(x)\frac{\partial^2 w}{\partial x^2}\right] + \kappa(y-w)\delta(x-1) + \beta(x)\frac{\partial^2 w}{\partial t^2} = 0 \tag{4.155}$$

$$\kappa[y - w(1,t)] + \mu\frac{d^2 y}{dt^2} = 0 \tag{4.156}$$

where the *s have been dropped from the nondimensional variables and

$$T = L^2\sqrt{\frac{\rho A}{EI}} \tag{4.157}$$

$$\kappa = \frac{kL^3}{EI} \tag{4.158}$$

$$\mu = \frac{m}{\rho AL} \tag{4.159}$$

Let Q be the vector space such that a vector $\mathbf{v}$ in Q is of the form

$$\mathbf{v} = \begin{bmatrix} v_1(x) \\ v_2 \end{bmatrix} \tag{4.160}$$

where $v_1(x)$ is an element of $C^4[0,1]$ and v_2 is a real number. The solution vector $\mathbf{q} = \begin{bmatrix} w(x,t) \\ y \end{bmatrix}$ is an element of Q. The standard inner product defined for Q is

$$(\mathbf{f},\mathbf{g}) = \int\limits_0^1 f_1(x)g_1(x)\mathrm{d}x + f_2 g_2 \tag{4.161}$$

The inertia operator for this system is

$$\mathbf{M} = \begin{bmatrix} \beta(x) & 0 \\ 0 & \mu \end{bmatrix} \tag{4.162}$$

Testing for self-adjointness of $\mathbf{M}$ with respect to the standard inner product for Q leads to

$$(\mathbf{M}f,\mathbf{g}) = \int\limits_0^1 \beta(x)f_1(x)g_1(x)\mathrm{d}x + \mu f_2 g_2 \tag{4.163}$$

Equation 4.163 is used to show that $\mathbf{M}$ is positive definite and self-adjoint with respect to the standard inner product on Q. Thus a kinetic energy inner product is defined such that $(\mathbf{f},\mathbf{g})_M = (\mathbf{Mf},\mathbf{g})$.

The stiffness operator cannot be easily represented by a matrix. It is an operator $\mathbf{K}$ whose domain and range are both S, an appropriately defined subspace of Q, such that if

$$\mathbf{K}f = \begin{bmatrix} \dfrac{\partial^2}{\partial x^2}\left[\alpha(x)\dfrac{\partial^2 f_1}{\partial x^2}\right] - \kappa(f_2 - f_1)\delta(x-1) \\ \kappa[f_2 - f_1(1)] \end{bmatrix} \tag{4.164}$$

The self-adjointness of **K** is tested by examining

$$
(\mathbf{Kf},\mathbf{g}) = \int_0^1 \left\{ \frac{\partial^2}{\partial x^2}\left[\alpha(x)\frac{\partial^2 f_1}{\partial x^2}\right] - \kappa(f_2 - f_1)\delta(x-1) \right\} g_1(x)\,dx
$$

$$
+ \kappa[f_2 - f_1(L)]g_2
$$

$$
= \int_0^1 \frac{\partial^2}{\partial x^2}\left[\alpha(x)\frac{\partial^2 f_1}{\partial x^2}\right]g_1(x)\,dx + \kappa[f_1(L)g_1(L) + f_2 g_2 - f_2 g_1(L)
$$

$$
- f_1(L)g_2]
$$

(4.165)

In addition

$$
(\mathbf{f},\mathbf{Kg}) = \int_0^1 \left\{ \frac{\partial^2}{\partial x^2}\left[\alpha(x)\frac{\partial^2 g_1}{\partial x^2}\right] - \kappa(g_2 - g_1)\delta(x-1) \right\} f_1(x)\,dx + \kappa[g_2
$$

$$
- g_1(L)]f_2
$$

$$
= \int_0^1 \frac{\partial^2}{\partial x^2}\left[\alpha(x)\frac{\partial^2 g_1}{\partial x^2}\right]f_1(x)\,dx + \kappa[g_1(L)f_1(L) + g_2 f_2 - g_2 f_1(L)
$$

$$
- g_1(L)f_2]
$$

(4.166)

Thus,

$$
(\mathbf{Kf},\mathbf{g}) - (\mathbf{f},\mathbf{Kg}) = \int_0^1 \frac{\partial^2}{\partial x^2}\left[\alpha(x)\frac{\partial^2 f_1}{\partial x^2}\right]g_1(x)\,dx
$$

$$
- \int_0^1 \frac{\partial^2}{\partial x^2}\left[\alpha(x)\frac{\partial^2 g_1}{\partial x^2}\right]f_1(x)\,dx
$$

(4.167)

which can be shown to be zero using the analysis for the self-adjointness of the stiffness operator for an Euler–Bernoulli beam. In addition, the conditions under which **K** is positive definite with respect to the standard inner product on S are those for which the stiffness operator for the Euler–Bernoulli beam is positive definite.

Since **K** is self-adjoint and positive definite a potential energy inner product is defined such that $(\mathbf{f},\mathbf{g})_\mathbf{K} = (\mathbf{Kf},\mathbf{g})$.

The operator $\mathbf{M}^{-1}\mathbf{K}$ is defined such that

$$
(\mathbf{M}^{-1}\mathbf{K})\mathbf{f} = \begin{bmatrix} \dfrac{1}{\beta(x)} & 0 \\[2mm] 0 & \dfrac{1}{\mu} \end{bmatrix} \begin{bmatrix} \dfrac{\partial^2}{\partial x^2}\left[\alpha(x)\dfrac{\partial^2 f_1}{\partial x^2}\right] - \kappa(f_2 - f_1)\delta(x-1) \\[4mm] \kappa[f_2 - f_1(1)] \end{bmatrix}
$$

$$
= \begin{bmatrix} \dfrac{1}{\beta(x)}\left\{\dfrac{\partial^2}{\partial x^2}\left[\alpha(x)\dfrac{\partial^2 f_1}{\partial x^2}\right] - \kappa(f_2 - f_1)\delta(x-1)\right\} \\[4mm] \dfrac{1}{\mu}\{\kappa[f_2 - f_1(1)]\} \end{bmatrix}
$$

$$\tag{4.168}$$

$\mathbf{M}^{-1}\mathbf{K}$ is self-adjoint with respect to the kinetic energy inner product as well as the potential energy inner product and is positive definite with respect to the kinetic energy inner product.

4.14 MEMBRANES

The partial differential equation for the transverse vibrations of a membrane, $w(\mathbf{r},t)$, is derived in Section 2.14 as

$$
F\nabla^2 w(\mathbf{r},t) = \rho\frac{\partial^2 w}{\partial t^2} \tag{4.169}
$$

where F is the uniform tension in the membrane, ρ is its mass density, and ∇^2 is the Laplacian operator. The coordinate system used in analysis of the vibrations is dependent upon the shape of the membrane's boundary. In general, the displacement is a function of the components of a two dimensional position vector measured from a specified origin. The region is enclosed by a closed curve of the form $B(\mathbf{r}) = 0$. For a rectangular membrane, $\mathbf{r} = x\mathbf{i} + y\mathbf{j}$, and the origin is usually measured at one corner of the membrane. For a circular membrane, $\mathbf{r} = r\cos\theta\mathbf{i} + r\sin\theta\mathbf{j}$, and the origin is usually measured at the center of the membrane.

Equation 4.169 is nondimensionalized through introduction of

$$
w^* = \frac{w}{L} \tag{4.170}
$$

$$\mathbf{r}^* = \frac{\mathbf{r}}{L} \tag{4.171}$$

$$t^* = \frac{t}{T} \tag{4.172}$$

where L is a characteristic length, the radius of a circular membrane or the length of a side of a rectangular membrane, and T is a characteristic time. Substitution of Equation 4.170 through Equation 4.172 into Equation 4.169 leads to

$$\nabla^2 w = \frac{\partial^2 w}{\partial t^2} \tag{4.173}$$

where the *s have been dropped from nondimensional variables and

$$T = L\sqrt{\frac{\rho}{F}} \tag{4.174}$$

Equation 4.174 is written in the form of Equation 4.33, with $\mathbf{M} = 1$ and $\mathbf{K}w = -\nabla^2 w$. Variational formulation of the problem shows that the possible boundary conditions are that $w = 0$ on the boundary or $\partial w/\partial n = 0$ on the boundary. The boundary of the two-dimensional region, described by the closed curve $B(\mathbf{r}) = 0$, may be expressed as the superposition of two curves, $B_1(\mathbf{r})$, defined as the part of the boundary over which $w = 0$ and $B_2(\mathbf{r})$, defined as the part of the boundary over which $\partial w/\partial n = 0$.

A solution of Laplace's equation exists in the vector space $C^2(R_B)$, functions which are twice differentiable in a two-dimensional region enclosed by the closed curve $B(\mathbf{r}) = 0$. The standard inner product on this region is

$$(f(\mathbf{r}), g(\mathbf{r})) = \int_{R_B} f(\mathbf{r})g(\mathbf{r})dR \tag{4.175}$$

If the region is rectangular, defined in the x–y plane, defined by $0 \le x \le 1, 0 \le y \le \eta$ (the characteristic length for nondimensionalization is the length of a side parallel to the x axis), then Equation 4.175 becomes

$$(f(x,y), g(x,y)) = \int_0^1 \int_0^\eta f(x,y)g(x,y)dydx \tag{4.176}$$

If the region is bounded by circle $r = 1$, then Equation 4.176 is written using polar coordinates as

$$(f(r,\theta),g(r,\theta)) = \int_0^{2\pi} \int_0^1 f(r,\theta)g(r,\theta)r\,dr\,d\theta \qquad (4.177)$$

Let S be the subspace of $C^2(R_B)$ defined appropriately for the boundary conditions.

It is shown in Example 3.16 that $-\nabla^2$ is positive definite and self-adjoint with respect to the standard inner product on S. Hence a potential energy inner product is defined as

$$(f,g)_\mathbf{K} = (\mathbf{K}f, g) = \int_{R_B} (-\nabla^2 f)g\,dR \qquad (4.178)$$

Since $\mathbf{M} = 1$, the kinetic energy inner product is the same as the standard inner product for $C^2(R_B)$.

5 Free Vibrations of Conservative Systems

5.1 NORMAL MODE SOLUTION

The differential equations for the free response of a linear conservative system can be written in the form of

$$\mathbf{Kw} + \mathbf{M\ddot{w}} = 0 \tag{5.1}$$

where $\mathbf{K}$ is the stiffness operator and $\mathbf{M}$ is the inertia operator. For a continuous system, $\mathbf{\ddot{w}} = \partial^2 \mathbf{w}/\partial t^2$. The free response of a conservative single-degree-of-freedom system is shown in Chapter 1 to be periodic with a constant natural frequency. In this spirit, a solution of Equation 5.1, called the normal-mode solution, is sought in the form

$$\mathbf{w} = \mathbf{v}e^{i\omega t} \tag{5.2}$$

Recalling that $e^{i\omega t} = \cos(\omega t) + i\sin(\omega t)$, it is clear that ω, as expressed in Equation 5.2, is a natural frequency. The vector $\mathbf{v}$ is called the mode shape vector.

Substitution of Equation 5.2 into Equation 5.1 leads to

$$\mathbf{Kv}e^{i\omega t} - \omega^2 \mathbf{Mv}e^{i\omega\omega} = 0 \tag{5.3}$$

or

$$\mathbf{Kv} = \omega^2 \mathbf{Mv} \tag{5.4}$$

Pre-multiplying Equation 5.4 by $\mathbf{M}^{-1}$, as it is guaranteed to exist, leads to

$$\mathbf{M}^{-1}\mathbf{Kv} = \omega^2 \mathbf{v} \tag{5.5}$$

An obvious solution to Equation 5.5 is $\mathbf{v} = 0$, called the trivial solution. Determination of the free response requires determination of nontrivial solutions of Equation 5.5. Nontrivial solutions exist only for certain values of the parameter ω, which are the system's natural frequencies. The corresponding nontrivial solutions are the mode shape vectors.

287

Equation 5.5 is a specific example of a general class of problems called eigenvalue problems. The eigenvalue of a linear operator **L** are the values of the parameter λ such that there exist nontrivial solutions, **y**, for the equation

$$\mathbf{L}\mathbf{y} = \lambda\mathbf{y} \qquad (5.6)$$

The corresponding nontrivial solutions are called the eigenvectors.

Comparison of Equation 5.5 and Equation 5.6 shows that the natural frequencies are the square roots of the eigenvalues of $\mathbf{M}^{-1}\mathbf{K}$ and the mode shape vectors are the corresponding eigenvectors.

If **y** is an eigenvector of **L** corresponding to an eigenvalue λ, thus satisfying Equation 5.6, consider the vector $c\mathbf{y}$ for any $c \neq 0$. Taking **L** operating on $c\mathbf{y}$ leads to

$$
\begin{aligned}
\mathbf{L}(c\mathbf{y}) &= c\mathbf{L}\mathbf{y} && \text{by linearity of } \mathbf{L} \\
&= c(\lambda\mathbf{y}) && \text{Equation 5.6} \\
&= \lambda(c\mathbf{y}) && \text{associative property of scalar multiplication}
\end{aligned} \qquad (5.7)
$$

Equation 5.7 shows if **y** is an eigenvector of **L** corresponding to an eigenvalue λ, then $c\mathbf{y}$ is also an eigenvector of **L** corresponding to λ. Thus an eigenvector is unique to at most a multiplicative constant.

The stiffness and inertia operators for a discrete system are matrices. The eigenvalue problem to determine the natural frequencies of a discrete system is called an algebraic eigenvalue problem. The stiffness operator for a continuous system is a differential operator or perhaps a matrix of differential operators. The inertia operator for a continuous system is a scalar function or a matrix of scalar functions. The eigenvalue problem for a continuous system is called a differential eigenvalue problem. There are many similarities in the properties of eigenvalues and eigenvectors between conservative discrete and continuous systems. For that reason, the general properties of eigenvalues and eigenvectors freedom are considered before specifically being applied to discrete or continuous systems. However, to illustrate these properties, the following two simple examples are used throughout the general discussion.

Example 5.1. The differential equations governing the motion of the three-degrees-of-system of Figure 5.1 are

$$
\begin{bmatrix} m & 0 & 0 \\ 0 & 2m & 0 \\ 0 & 0 & 2m \end{bmatrix}
\begin{bmatrix} \ddot{x}_1 \\ \ddot{x}_2 \\ \ddot{x}_3 \end{bmatrix}
+
\begin{bmatrix} 3k & -k & 0 \\ -k & 2k & -k \\ 0 & -k & 3k \end{bmatrix}
\begin{bmatrix} x_1 \\ x_2 \\ x_3 \end{bmatrix}
=
\begin{bmatrix} 0 \\ 0 \\ 0 \end{bmatrix} \qquad (a)
$$

Determine the system's natural frequencies and mode shapes.

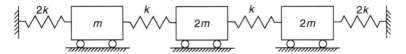

FIGURE 5.1 Three-degree-of-freedom system for Example 5.1.

Solution: The natural frequencies are the square roots of the eigenvalues of $\mathbf{M}^{-1}\mathbf{K}$, where

$$\mathbf{M}^{-1}\mathbf{K} = \begin{bmatrix} \dfrac{1}{m} & 0 & 0 \\[2mm] 0 & \dfrac{1}{2m} & 0 \\[2mm] 0 & 0 & \dfrac{1}{2m} \end{bmatrix} \begin{bmatrix} 3k & -k & 0 \\ -k & 2k & -k \\ 0 & -k & 3k \end{bmatrix}$$

$$= \begin{bmatrix} 3\dfrac{k}{m} & -\dfrac{k}{m} & 0 \\[2mm] -\dfrac{k}{2m} & \dfrac{k}{m} & -\dfrac{k}{2m} \\[2mm] 0 & -\dfrac{k}{2m} & \dfrac{3k}{2m} \end{bmatrix} \qquad (b)$$

The eigenvalues of $\mathbf{M}^{-1}\mathbf{K}$ are calculated by determining the values of λ such that $|\mathbf{M}^{-1}\mathbf{K} - \lambda\mathbf{I}| = 0$, where $\mathbf{I}$ is the 3×3 identity matrix. To this end,

$$\begin{vmatrix} 3\dfrac{k}{m} - \lambda & -\dfrac{k}{m} & 0 \\[2mm] -\dfrac{k}{2m} & \dfrac{k}{m} - \lambda & -\dfrac{k}{2m} \\[2mm] 0 & -\dfrac{k}{2m} & \dfrac{3k}{2m} - \lambda \end{vmatrix} = 0 \qquad (c)$$

Expansion of the determinant in Equation c leads to

$$-\lambda^3 + \frac{11}{2}\phi\lambda^2 - \frac{33}{4}\phi\lambda + 3 = 0 \qquad (d)$$

where $\phi = (k/m)$. The solutions of Equation d are

$$\lambda_1 = 0.5373\phi, \qquad \lambda_2 = 1.724\phi, \qquad \lambda_3 = 3.238\phi \qquad \text{(e)}$$

The natural frequencies are obtained from Equation e as

$$\omega_1 = 0.733\sqrt{\frac{k}{m}} \qquad \omega_2 = 1.313\sqrt{\frac{k}{m}} \qquad \omega_3 = 1.800\sqrt{\frac{k}{m}} \qquad \text{(f)}$$

The system has three linearly independent mode shape vectors, one corresponding to each natural frequency. A mode shape vector is determined by finding a nontrivial solution to $\mathbf{M}^{-1}\mathbf{K}\mathbf{v}_i - \omega_i^2\mathbf{v}_i = 0$. A mode shape vector for the first mode is thus obtained from

$$\begin{bmatrix} 3\phi - 0.5373\phi & -\phi & 0 \\[2mm] -\dfrac{\phi}{2} & \phi - 0.5373\phi & -\dfrac{\phi}{2} \\[2mm] 0 & -\dfrac{\phi}{2} & \dfrac{3\phi}{2} - 0.5373\phi \end{bmatrix} \begin{bmatrix} v_1 \\ v_2 \\ v_3 \end{bmatrix} = \begin{bmatrix} 0 \\ 0 \\ 0 \end{bmatrix} \qquad \text{(g)}$$

The first and third equations from the system of Equation g lead to

$$v_1 = \frac{\phi}{3\phi - 0.5373\phi} v_2 = 0.4061 v_2 \qquad \text{(h)}$$

$$v_3 = \frac{\frac{\phi}{2}}{\frac{3\phi}{2} - 0.5373\phi} v_2 = 0.5194 v_2 \qquad \text{(i)}$$

Arbitrarily setting $v_2 = C_1$, a constant, leads to a mode shape vector corresponding to the system's lowest natural frequency, as

$$\mathbf{v}_1 = C_1 \begin{bmatrix} 0.4061 \\ 1 \\ 0.5194 \end{bmatrix} \qquad \text{(j)}$$

Similar analyses lead to mode shape vectors corresponding to the second and third modes as

$$\mathbf{v_2} = C_2 \begin{bmatrix} 0.7837 \\ 1 \\ -2.232 \end{bmatrix} \qquad \mathbf{v_3} = C_3 \begin{bmatrix} -4.207 \\ 1 \\ -0.2877 \end{bmatrix} \tag{k}$$

Example 5.2. The nondimensional partial differential equation for the free response of a uniform bar fixed at $x = 0$ and free at $x = 1$ is

$$\frac{\partial^2 u}{\partial x^2} = \frac{\partial^2 u}{\partial t^2} \tag{a}$$

and that time is nondimensionalized by $t^* = (t/T)$ where $t = L\sqrt{\rho/E}$. Determine the dimensional natural frequencies and their corresponding mode shapes.

Solution: Use of the normal mode solution in Equation a leads to

$$\frac{d^2 v}{dx^2} = -\omega^2 v \tag{b}$$

The boundary condition at the fixed end is analyzed by

$$u(x,0) = 0 = v(0)e^{i\omega t} \tag{c}$$

Since Equation c must hold for all t, $v(0) = 0$. Similarly it is shown that $dv/dx(1) = 0$

The square of the natural frequency is assumed to be positive. Thus, Equation b is rewritten as

$$\frac{d^2 v}{dx^2} + \omega^2 v = 0 \tag{d}$$

The general solution of Equation d is

$$v(x) = A_1 \cos(\omega x) + A_2 \sin(\omega x) \tag{e}$$

Application of the boundary condition at $x = 0$ to Equation e leads to $A_1 = 0$. Application of the boundary condition at $x = 1$ then leads to

$$\frac{dv}{dx}(1) = 0 = \omega A_2 \cos(\omega) \tag{f}$$

The choice of $\omega = 0$ leads to the trivial solution. Any value of ω of the form

$$\omega_k = (2k-1)\frac{\pi}{2} \tag{g}$$

for an integer k satisfies $\cos(\omega_k) = 0$. Thus the natural frequencies are given by Equation g. These natural frequencies are nondimensional. The dimensional natural frequencies are

$$\omega_k = \frac{(2k-1)\frac{\pi}{2}}{T} = \frac{(2k-1)\pi}{2L}\sqrt{\frac{E}{\rho}} \tag{h}$$

There are an infinite, but countable, number of natural frequencies. A mode shape for each natural frequency is of the form

$$v_k(x) = A_k \sin\left[(2k-1)\frac{\pi}{2}x\right] \tag{i}$$

Any scalar multiple of the right-hand side of Equation i is also a mode shape vector for the bar.

5.2 PROPERTIES OF EIGENVALUES AND EIGENVECTORS

5.2.1 Eigenvalues of Self-Adjoint Operators

Let $\mathbf{L}$ be a self-adjoint operator with respect to an inner product defined for all elements in the domain of $\mathbf{L}$. Let $\mathbf{L}^*$ be the adjoint of $\mathbf{L}$ be the operator defined according such that $(\mathbf{L}\mathbf{u}, \mathbf{v}) = (\mathbf{u}, \mathbf{L}^*\mathbf{v})$ for all $\mathbf{u}$ in the domain of $\mathbf{L}$ and $\mathbf{v}$ in the domain of $\mathbf{L}^*$.

Let λ be an eigenvalue of $\mathbf{L}$ with a corresponding eigenvector $\mathbf{u}$ and let μ be an eigenvalue of $\mathbf{L}^*$ with corresponding eigenvector $\mathbf{v}$,

$$\mathbf{L}\mathbf{u} = \lambda\mathbf{u} \tag{5.8}$$

$$\mathbf{L}^*\mathbf{v} = \mu\mathbf{v} \tag{5.9}$$

Taking the inner product of both sides of Equation 5.8 with $\mathbf{v}$ leads to

$$(\mathbf{L}\mathbf{u},\mathbf{v}) = (\lambda\mathbf{u},\mathbf{v}) \tag{5.10}$$

Using properties of inner products as well as the definition of the adjoint in Equation 5.10 leads to

$$(\mathbf{u}, \mathbf{L}^* \mathbf{v}) = \lambda(\mathbf{u}, \mathbf{v}) \qquad (5.11)$$

Using Equation 5.9 in Equation 5.11 followed by properties of inner products leads to

$$(\mathbf{u}, \mu\mathbf{v}) = \lambda(\mathbf{u}, \mathbf{v})$$

$$\bar{\mu}(\mathbf{u}, \mathbf{v}) = \lambda(\mathbf{u}, \mathbf{v}) \qquad (5.12)$$

$$(\bar{\mu} - \lambda)(\mathbf{u}, \mathbf{v}) = 0$$

Equation 5.12 implies that either $\bar{\mu} = \lambda$ or $(\mathbf{u}, \mathbf{v}) = 0$. That is, either an eigenvalue of $\mathbf{L}$ is the complex conjugate of an eigenvalue of its adjoint operator or their eigenvectors are mutually orthogonal. If $\mathbf{L}$ is self-adjoint with respect to this inner product, then Equation 5.12 proves that if $\mathbf{u}$ and $\mathbf{v}$ are eigenvectors corresponding to the same eigenvalue λ, then $\bar{\lambda} = \lambda$, which implies that λ is real and that eigenvectors corresponding to distinct eigenvalues are mutually orthogonal. Thus, the following theorem is proved.

Theorem 5.1. *Let $\mathbf{L}$ be a linear operator whose domain is the vector space S, such that $\mathbf{L}$ is self-adjoint operator with respect to an inner product $(\mathbf{u}, \mathbf{v})$ defined for all $\mathbf{u}$ and $\mathbf{v}$ in S. Then (i) The eigenvalues of $\mathbf{L}$ are real and (ii) Eigenvectors corresponding to distinct' eigenvectors of $\mathbf{L}$ are mutually orthogonal with respect to $(\mathbf{u}, \mathbf{v})$.*

The more general case of non-self-adjoint operators is considered in Chapter 7.

A self-adjoint operator generates a set of orthogonal eigenvectors, each of which are unique to, at best, a multiplicative constant. If λ is an eigenvalue of multiplicity n, then there are up to n linearly independent eigenvectors corresponding to λ. It is possible to mathematically define a problem in which there are less than n linearly independent eigenvectors. However, for the vibrations problems considered, there will be n linearly independent eigenvectors corresponding to λ. The Gram–Schmidt process may be used to render a set on n orthogonal eigenvectors corresponding to λ Theorem 5.1 shows that these eigenvectors are orthogonal to eigenvectors corresponding to all other eigenvalues.

Eigenvectors corresponding to an eigenvalue are at best unique to a multiplicative constant. It is often convenient to refer to a specific eigenvector. This can be accomplished by normalizing the eigenvector, or requiring the inner product of an eigenvector with itself to be one. Then, for a self-adjoint operator, a set of orthonormal eigenvectors can be generated. That is, if λ_i and λ_j are eigenvalues of a self-adjoint operator with corresponding normalized eigenvectors $\mathbf{u}_i$ and $\mathbf{u}_j$, respectively, then

$$(\mathbf{u}_i, \mathbf{u}_j) = \delta_{i,j} \qquad (5.13)$$

Let $\mathbf{u}_i$ be a normalized eigenvector of $\mathbf{L}$ corresponding to an eigenvalue λ_i such that

$$\mathbf{L}\mathbf{u}_i = \lambda_i \mathbf{u}_i \qquad (5.14)$$

Taking the inner product of Equation 5.14 with $\mathbf{u}_i$ leads to

$$(\mathbf{L}\mathbf{u}_i, \mathbf{u}_i) = (\lambda_i \mathbf{u}_i, \mathbf{u}_i) = \lambda_i(\mathbf{u}_i, \mathbf{u}_i) = \lambda_i \qquad (5.15)$$

Example 5.3. Apply Theorem 5.1 to the system of (a) Example 5.1 and (b) Example 5.2 and develop a set of orthonormal eigenvectors for each problem.

Solution: (a) The operator $\mathbf{M}^{-1}\mathbf{K}$ is self-adjoint with respect to both the kinetic energy inner product $(\mathbf{M}\mathbf{u},\mathbf{v})$ and the potential energy inner product $(\mathbf{K}\mathbf{u}, \mathbf{v})$. Thus, as illustrated in Example 5.1, all of its eigenvalues are real. Theorem 5.1 implies that the eigenvectors are mutually orthogonal with respect to the kinetic energy inner product. Their orthogonality is demonstrated below.

$$(\mathbf{v}_2, \mathbf{v}_1)_M = C_1^2 \begin{bmatrix} 0.4061 & 1 & 0.5194 \end{bmatrix} \begin{bmatrix} m & 0 & 0 \\ 0 & 2m & 0 \\ 0 & 0 & 2m \end{bmatrix} \begin{bmatrix} 0.7837 \\ 1 \\ -2.232 \end{bmatrix}$$

$$= C_1^2 \begin{bmatrix} 0.4061 & 1 & 0.5194 \end{bmatrix} \begin{bmatrix} 0.7837m \\ 2m \\ -4.464m \end{bmatrix} \qquad (a)$$

$$= C_1^2[(0.4061)(0.7837m) + (1)(2m) + (0.5194)(-4.464m)]$$

$$\approx 0$$

The evaluation of Equation a is not identically zero due to numerical round off. Similarly, it is shown that $(\mathbf{v}_1, \mathbf{v}_3)_M \approx 0$ and $(\mathbf{v}_2, \mathbf{v}_3)_M \approx 0$.

The mode shape vectors are normalized by requiring that $(\mathbf{v}_i, \mathbf{v}_i)_M = 1$. To this end,

$$(\mathbf{v}_1, \mathbf{v}_1)_M = 1$$

$$C_1^2 \begin{bmatrix} 0.4061 & 1 & 0.5194 \end{bmatrix} \begin{bmatrix} m & 0 & 0 \\ 0 & 2m & 0 \\ 0 & 0 & 2m \end{bmatrix} \begin{bmatrix} 0.4061 \\ 1 \\ 0.5194 \end{bmatrix} = 1$$

(b)

$$C_1^2 [(0.4061)(0.4061m) + (1)(2m) + (0.5194)(1.388m)] = 1$$

$$C_1 = \frac{0.6081}{\sqrt{m}}$$

Similar calculations lead to $C_2 = 0.2819/\sqrt{m}$ and $C_3 = 0.2244/\sqrt{m}$. Thus, the normalized eigenvectors are

$$\mathbf{v}_1 = \frac{1}{\sqrt{m}} \begin{bmatrix} 0.2469 \\ 0.6081 \\ 0.3158 \end{bmatrix} \quad \mathbf{v}_2 = \frac{1}{\sqrt{m}} \begin{bmatrix} 0.2210 \\ 0.2820 \\ -0.6293 \end{bmatrix} \quad \mathbf{v}_3 = \frac{1}{\sqrt{m}} \begin{bmatrix} -0.9435 \\ 0.2252 \\ -0.0647 \end{bmatrix}$$

(c)

The normalized mode shape vectors are graphically illustrated in (Figure 5.2)(a). A node is a point in a system which for a specific mode has a displacement of zero. (Figure 5.2) shows that there are no nodes corresponding to the first mode, the second mode has a node in the spring connecting the two blocks of mass 2m and the third mode has two nodes, one in each of the springs which connect two blocks.

(b) The eigenvalues of the operator **L** are of the form

$$\lambda_k = \left[(2k - 1) \frac{\pi}{2} \right]^2$$

(d)

And the corresponding mode shapes are

$$v_k(x) = A_k \sin \left[(2k - 1) \frac{\pi}{2} x \right]$$

(e)

The eigenvalues are all real. Since **L** is self-adjoint with respect to the standard inner product on $C^2[0,1]$, eigenvector orthogonality is shown by

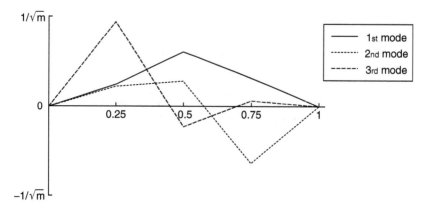

FIGURE 5.2 Graphical illustration of normalized mode shape vectors for system of Example 5.1.

$$(\nu_m, \nu_n) = \int_0^1 \nu_m(x)\nu_n(x)dx$$

$$= \int_0^1 A_m \sin\left[(2m-1)\frac{\pi}{2}x\right] A_n \sin\left[(2n-1)\frac{\pi}{2}x\right] dx$$

$$= \frac{1}{2}A_m A_n \int_0^1 \{\cos[(m-n)\pi x] - \cos[(m+n-1)\pi x]\}dx$$

$$= \frac{1}{2\pi}A_m A_n \left\{\frac{1}{m-n}\sin[(m-n)\pi x]\Big|_{x=1}^{x=1} - \frac{1}{m-n+1}\sin[(m+n-1)\pi x]\Big|_{x=0}^{x=1}\right\} = 0$$

$$(f)$$

The eigenvectors are normalized by

$$(\nu_k, \nu_k) = 1$$

$$\int_0^1 A_k^2 \sin^2\left[(2k-1)\frac{\pi}{2}x\right] dx = 1$$

$$(g)$$

$$A_k^2 \int_0^1 \frac{1}{2}\{1 - \cos[(2k-1)\pi x]dx\} = 1$$

$$A_k = \sqrt{2}$$

Thus, the normalized eigenvectors are

$$v_k(x) = \sqrt{2} \sin\left[(2k-1)\frac{\pi}{2}x\right] \qquad \text{(h)}$$

5.2.2 POSITIVE DEFINITE OPERATORS

Let **L** be a positive definite linear operator with respect to any valid inner product. Let λ be an eigenvalue of **L** with a corresponding eigenvector **u**.

$$\mathbf{Lu} = \lambda\mathbf{u} \qquad (5.16)$$

Taking the inner product of both sides of Equation 5.16 with **u** leads to

$$(\mathbf{Lu},\mathbf{u}) = (\lambda\mathbf{u},\mathbf{u})$$
$$(\mathbf{Lu},\mathbf{u}) = \lambda(\mathbf{u},\mathbf{u}) \qquad (5.17)$$

By definition, $\mathbf{u} \neq 0$, and since **L** is positive definite, Equation 5.14 may be rearranged leading to

$$\lambda = \frac{(\mathbf{Lu},\mathbf{u})}{(\mathbf{u},\mathbf{u})} \qquad (5.18)$$

Since $(\mathbf{u},\mathbf{v})$ is a valid inner product for vectors is the domain of **L**, the denominator of Equation 5.18 is positive, and since **L** is positive definite with respect to this inner product, the numerator must also be positive. Thus the eigenvalue λ must be positive. This proves the following theorem.

Theorem 5.2. *All eigenvalues of a positive definite operator are positive.*
Suppose **L** is positive semi-definite with respect to an inner product. That is there exists a vector **u**, not equal to the zero vector, such that $(\mathbf{Lu},\mathbf{u}) = 0$. This will be true if there exists a $\mathbf{u} \neq 0$ such that $\mathbf{Lu} = 0$. In this case, $\lambda = 0$ is an eigenvalue of **L**. The proof that $\lambda = 0$ is always an eigenvalue of a positive semi-definite operator uses the expansion theorem which is presented later. However, for convenience, the following theorem is now stated.

Theorem 5.3. $\lambda = 0$ *is an eigenvalue of a positive semi-definite operator.*

5.2.3 Expansion Theorem

The eigenvectors of a self-adjoint operator have been shown to be orthogonal and are therefore linearly independent. If $\mathbf{L}$ is an operator on S, a vector space of dimension n and n linearly independent eigenvectors are determined, then these eigenvectors span S. Every element of S can be written as a linear combination of these eigenvectors.

Let $\mathbf{L}$ be a self-adjoint operator defined on S a vector space of dimension n. Let $\mathbf{v}_1$, $\mathbf{v}_2$, $\ldots \mathbf{v}_n$ be the orthonormal eigenvectors of $\mathbf{L}$. Then for any $\mathbf{u}$ in S there exists a linear combination

$$\mathbf{u} = \sum_{i=1}^{n} \alpha_i \mathbf{v}_i \tag{5.19}$$

Taking the inner product of Equation 5.19 with $\mathbf{v}_j$ for any $j = 1, 2,.., n$ leads to

$$(\mathbf{u},\mathbf{v}_j) = \left(\sum_{i=1}^{n} \alpha_i \mathbf{v}_i, \mathbf{v}_j \right) \tag{5.20}$$

Application of properties of inner products to the right-hand side of Equation 5.20 leads to

$$(\mathbf{u},\mathbf{v}_j) = \sum_{i=1}^{n} \alpha_i (\mathbf{v}_i,\mathbf{v}_j) \tag{5.21}$$

The set of eigenvectors are orthonormal with respect to this inner product, thus

$$(\mathbf{u},\mathbf{v}_i) = \sum_{i=1}^{n} \alpha_i \delta_{i,j} \tag{5.22}$$

Equation 5.22 reduces to

$$\alpha_j = (\mathbf{u},\mathbf{v}_j) \tag{5.23}$$

Substitution of Equation 5.23 into Equation 5.19 completes the proof of the following theorem.

Theorem 5.4. *(Expansion Theorem) If* $\mathbf{L}$ *is a linear operator whose domain is* S, *an n dimensional vector space and* $\mathbf{L}$ *is self-adjoint with respect to an inner product* $(\mathbf{u},\mathbf{v})$. *Let* $\mathbf{v}_1$, $\mathbf{v}_2$, $\ldots \mathbf{v}_n$ *be the orthonormal set of eigenvectors for* $\mathbf{L}$, *then any vector in S has the representation*

$$\mathbf{u} = \sum_{i=1}^{n} (\mathbf{u}, \mathbf{v}_i) \mathbf{v}_i \tag{5.24}$$

If **L** is a linear operator whose domain S is an infinite dimensional vector space, then in order to prove an analogous version of the expansion theorem, questions of convergence and completeness must be addressed. For example, in the derivation of Theorem 5.4, between Equation 5.20 and Equation 5.21, the operations of summation and inner product evaluation are interchanged. If the summation is over an infinite number of terms, the switch in operations is valid only if the series converges. A more fundamental question is whether a set of orthonormal eigenvectors is complete on S. That is, can every vector in S be written as a linear combination of the orthonormal eigenvectors? Clearly the set of eigenvectors spans at least a subspace of S. However, it is questioned whether any additional vector or vectors need to be added to the set of eigenvectors in order for the set to be complete on S. These issues can be addressed and the following theorem stated.

Theorem 5.5. *(Expansion Theorem for Infinite Dimensional Vector Spaces) Let **L** be a linear operator whose domain is an infinite dimensional vector space S and is self-adjoint with respect to an inner product $(\mathbf{u}, \mathbf{v})$. Then the set of orthonormal eigenvectors $\mathbf{v}_1, \mathbf{v}_2, \ldots, \mathbf{v}_{k-1}, \mathbf{v}_k, \mathbf{v}_{k+1}, \ldots$ is complete in S and any vector $\mathbf{u}$ in S has the representation*

$$\mathbf{u} = \sum_{i=1}^{\infty} (\mathbf{u}, \mathbf{v}_i) \mathbf{v}_i \tag{5.25}$$

Knowledge of the expansion theorem allows finalization of the proof of Theorem 5.3 for self-adjoint operators. Let **L** a positive semi-definite and self-adjoint operator with respect to an inner product $(\mathbf{u}, \mathbf{v})$. Let $\mathbf{v}_1, \mathbf{v}_2, \ldots, \mathbf{v}_1, \mathbf{v}_2, \ldots, \mathbf{v}_{k-1}, \mathbf{v}_k, \mathbf{v}_{k+1} \ldots$, be the set of orthonormal eigenvectors corresponding to eigenvalues $\lambda_1, \lambda_2, \ldots, \lambda_{k-1}, \lambda_k, \lambda_{k+1} \ldots$. Let $\mathbf{u} \neq 0$ be a vector such that $(\mathbf{L}\mathbf{u}, \mathbf{u}) = 0$. By the Expansion Theorem there exists a set of coefficients $\alpha_1, \alpha_2, \ldots, \alpha_{k-1}, \alpha_k, \alpha_{k+1} \ldots$, such that

$$\mathbf{u} = \sum_i \alpha_i \mathbf{v}_i \tag{5.26}$$

Using the expansion of Equation 5.25, evaluation of $(\mathbf{L}\mathbf{u}, \mathbf{u})$ leads to

$$(\mathbf{L}\mathbf{u}, \mathbf{u}) = \left(\mathbf{L}(\sum_i \alpha_i \mathbf{v}_i) \sum_j \alpha_j \mathbf{v}_j \right) \tag{5.27}$$

Using the linearity of $\mathbf{L}$ and properties of inner products, Equation 5.27 becomes

$$(\mathbf{Lu},\mathbf{u}) = \sum_i \sum_j \alpha_i \alpha_j (\mathbf{Lv}_i, \mathbf{v}_j) \tag{5.28}$$

Since $\mathbf{v}_i$ is an eigenvector of $\mathbf{L}$ corresponding to the eigenvalue λ_i, $\mathbf{Lv}_i = \lambda_i \mathbf{v}_i$. Thus Equation 5.28 is rewritten as

$$(\mathbf{Lu},\mathbf{u}) = \sum_i \sum_j \alpha_i \alpha_j \lambda_i (\mathbf{v}_i, \mathbf{v}_j) \tag{5.29}$$

Noting that the eigenvectors are orthonormal leads to

$$(\mathbf{Lu},\mathbf{u}) = \sum_i \sum_j \alpha_i \alpha_j \lambda_i \delta_{i,j} = \sum_i \lambda_i \alpha_i^2 \tag{5.30}$$

Since $\mathbf{u} \neq 0$, there must be at least one i, such that $\alpha_i \neq 0$. Thus, in order for the right-hand side of Equation 5.30 to be zero, the eigenvalue corresponding to that value of i must be zero. Thus, Theorem 5.3 is proved when $\mathbf{L}$ is a self-adjoint operator.

Example 5.4. (a) Expand the vector $\mathbf{u} = \begin{bmatrix} 1 \\ -1 \\ 0 \end{bmatrix}$ using the normalized

eigenvectors for the system of Example 5.1 and Example 5.3a. (b) Expand the function $f(x) = x - (1/2)x^2$ using the normalized eigenvectors of the system of Example 5.2 and Example 5.3b.

Solution: (a) The normalized eigenvectors are given in Equation c of Example 5.3. Application of the Expansion Theorem leads to

$$\mathbf{u} = (\mathbf{u},\mathbf{v}_1)_M \mathbf{v}_1 + (\mathbf{u},\mathbf{v}_2)_M \mathbf{v}_2 + (\mathbf{u},\mathbf{v}_3)_M \mathbf{v}_3 \tag{a}$$

where

$$(\mathbf{u},\mathbf{v}_1)_M = \frac{1}{\sqrt{m}} \begin{bmatrix} 0.2469 & 0.6081 & 0.3158 \end{bmatrix} \begin{bmatrix} m & 0 & 0 \\ 0 & 2m & 0 \\ 0 & 0 & 2m \end{bmatrix} \begin{bmatrix} 1 \\ -1 \\ 0 \end{bmatrix}$$

$$= -0.9693\sqrt{m} \tag{b}$$

$$(\mathbf{u},\mathbf{v}_2)_M = \frac{1}{\sqrt{m}} \begin{bmatrix} 0.2209 & 0.2819 & -0.6292 \end{bmatrix} \begin{bmatrix} m & 0 & 0 \\ 0 & 2m & 0 \\ 0 & 0 & 2m \end{bmatrix} \begin{bmatrix} 1 \\ -1 \\ 0 \end{bmatrix}$$

$$= -0.3429\sqrt{m} \tag{c}$$

$$(\mathbf{u},\mathbf{v}_3)_M = \frac{1}{\sqrt{m}} \begin{bmatrix} -0.9439 & 0.2244 & -0.00646 \end{bmatrix} \begin{bmatrix} m & 0 & 0 \\ 0 & 2m & 0 \\ 0 & 0 & 2m \end{bmatrix} \begin{bmatrix} 1 \\ -1 \\ 0 \end{bmatrix}$$

$$= -1.3927\sqrt{m} \tag{d}$$

Thus,

$$\begin{bmatrix} 1 \\ -1 \\ 0 \end{bmatrix} = -0.9693 \begin{bmatrix} 0.2469 \\ 0.6081 \\ 0.3158 \end{bmatrix} - 0.3429 \begin{bmatrix} 0.2209 \\ 0.2819 \\ -0.6292 \end{bmatrix}$$

$$- 1.3927 \begin{bmatrix} -0.9439 \\ 0.2244 \\ -0.0646 \end{bmatrix} \tag{e}$$

(b) The normalized eigenvectors for Example 5.2 are given by Equation h of Example 5.3. It is noted that $f(x) = x - (1/2)x^2$ is in the domain of L as $f(0) = 0$ and $(df/dx)(1) = 0$. Application of the Expansion Theorem leads to

$$x - \frac{1}{2}x^2 = \sum_{k=1}^{\infty} \left\{ \int_0^1 \left(x - \frac{1}{2}x^2 \right) \sqrt{2} \sin\left[(2k-1)\frac{\pi}{2}x \right] dx \right\} \sqrt{2} \sin\left[(2k-1)\frac{\pi}{2}x \right] \tag{f}$$

Evaluation of the integral leads to

$$x - \frac{1}{2}x^2 = \frac{16}{\pi^3} \sum_{k=1}^{\infty} \frac{\sin\left[(2k-1)\frac{\pi}{2}x \right]}{8k^3 - 12k^2 + 6k - 1} \tag{g}$$

5.2.4 SUMMARY

It is shown in Chapter 4 that for conservative systems whose differential equations are derived from Hamilton's Principle that

- The inertia operator $\mathbf{M}$ is self-adjoint with respect to either the standard inner product for the vector space or an inner product which is modified to account for the kinetic energy of a discrete mass.
- The inertia operator is positive definite with respect to this inner product.
- Since the inertia operator is positive definite and self-adjoint, a kinetic energy inner product is defined.
- The stiffness operator $\mathbf{K}$ is self-adjoint with respect to an inner product defined on the domain of $\mathbf{K}$.
- The stiffness matrix is singular for unrestrained systems.
- The stiffness matrix is positive definite unless unrestrained in which case it is positive semi-definite.
- Since the stiffness operator is self-adjoint and positive definite (unless unrestrained) a potential energy inner product is defined.
- The operator $\mathbf{M}^{-1}\mathbf{K}$ is self-adjoint with respect to the kinetic energy inner product and the potential energy inner product.
- The operator $\mathbf{M}^{-1}\mathbf{K}$ is positive definite with respect to a potential energy inner product unless the system is unrestrained.

The use of the normal mode solution shows that the natural frequencies of the system are the square roots of the eigenvalues of $\mathbf{M}^{-1}\mathbf{K}$ and the mode shape vectors are their corresponding eigenvectors. Study of this section reveals the following:

- All eigenvalues of a self-adjoint operator are real.
- All eigenvalues of a positive definite operator are positive.
- The smallest eigenvalue of a positive semidefinite operator is zero.
- Eigenvectors corresponding to distinct eigenvalues of a self-adjoint operator are mutually orthogonal with respect to the inner product for which self-adjointness is ascertained.
- Eigenvectors may be normalized by requiring $(\mathbf{u}_i, \mathbf{u}_i) = 1$.
- If an eigenvector is normalized, then $(\mathbf{L}\mathbf{u}_i, \mathbf{u}_i) = \lambda_i$.
- Any vector in the domain of $\mathbf{L}$ has an eigenvector expansion of the form of Equation 5.24 (for a finite dimensional vector space) or Equation 5.25 (for an infinite dimensional vector space).

The results of Chapter 4 and Sections 5.1 and 5.2 are used to deduce the following:

- Since $M^{-1}K$ is self-adjoint and positive definite with respect to the kinetic energy inner product, all of its eigenvalues are real and positive, which implies all natural frequencies are real.
- If the system is unrestrained, its lowest natural frequency is zero.
- Mode shapes corresponding to distinct natural frequencies are orthogonal with respect to the kinetic energy inner product.
- Mode shapes corresponding to distinct natural frequencies are orthogonal with respect to the potential energy inner product.
- Mode shapes may be normalized with respect to the kinetic energy inner product, $(v_i, v_i)_M = 1$.
- When mode shapes are normalized with respect to the kinetic energy inner product, then $(v_i, v_i)_K = \omega_i^2$.
- Any vector in the domain of K has an eigenvector expansion of the form

$$v = \sum_i (v, v_i)_M v_i \tag{5.31}$$

5.3 RAYLEIGH'S QUOTIENT

Equation 5.18 expresses the eigenvalue as the ratio of two inner products

$$\lambda = \frac{(Lu, u)}{(u, u)} \tag{5.32}$$

If L is a self-adjoint positive definite operator, the inner product in the denominator is an energy inner product.

Let w denote any vector in the domain of L. A functional of w is defined as

$$R(w) - \frac{(Lw, w)}{(w, w)} \tag{5.33}$$

The functional $R(w)$ is called Rayleigh's Quotient. It is clear from Equation 5.32 that if v is an eigenvector of L corresponding to an eigenvalue λ, then $R(v) = \lambda$.

Suppose L is a self-adjoint operator and let $v_1, v_2,..., v_{k-1}, v_k, v_{k+1},...$ be the orthonormal set of eigenvectors corresponding to the eigenvalues. λ_1, $\lambda_2,...,\lambda_{k-1}, \lambda_k, \lambda_{k+1},....$ By the Expansion Theorem there exists a set of coefficients $\alpha_1, \alpha_2,...,\alpha_{k-1}, \alpha_k, \alpha_{k+1},....$ such that

$$\mathbf{w} = \sum_i \alpha_i \mathbf{v}_i \tag{5.34}$$

Substitution of Equation 5.34 into Equation 5.33 leads to

$$R(\mathbf{w}) = \frac{\left(\mathbf{L}\left(\sum_i \alpha_i \mathbf{v}_i\right), \sum_j \alpha_j \mathbf{v}_j\right)}{\left(\sum_i \alpha_i \mathbf{v}_i, \sum_j \alpha_j \mathbf{v}_j\right)} \tag{5.35}$$

Application of linearity of **L**, properties of inner products, definition of the eigenvalue–eigenvector relation, and orthonormality of the eigenvectors leads to the following sequence of simplifications to Equation 5.35:

$$R(\mathbf{w}) = \frac{\sum_i \sum_j \alpha_i \alpha_j (\mathbf{L}\mathbf{v}_i, \mathbf{v}_j)}{\sum_j \sum_i \alpha_i \alpha_j (\mathbf{v}_i, \mathbf{v}_j)}$$

$$= \frac{\sum_i \sum_j \alpha_i \alpha_j (\lambda_i \mathbf{v}_i, \mathbf{v}_j)}{\sum_i \sum_j \alpha_i \alpha_j \delta_{i,j}} \tag{5.36}$$

$$= \frac{\sum_i \lambda_i \alpha_i^2}{\sum_i \alpha_i^2}$$

Equation 5.36 illustrates that $R(\mathbf{w})$ is a function of the coefficients used in the expansion of $\mathbf{w}$ in terms of the orthonormal eigenvectors.

Rayleigh's Quotient is rendered stationary when

$$\delta R = 0$$

$$\sum_i \frac{\partial R}{\partial \alpha_i} \delta \alpha_i = 0 \tag{5.37}$$

The vectors for which $R(\mathbf{w})$ and the corresponding stationary values are obtained by setting

$$\frac{\partial R}{\partial \alpha_1} = \frac{\partial R}{\partial \alpha_2} = \ldots = \frac{\partial R}{\partial \alpha_{k-1}} = \frac{\partial R}{\partial \alpha_k} = \frac{\partial R}{\partial \alpha_{k+1}} = \ldots = 0 \qquad (5.38)$$

To this end

$$\frac{\partial R}{\partial \alpha_k} = \frac{\left(\sum_i \alpha_i^2\right)\left[\frac{\partial}{\partial \alpha_k}\left(\sum_j \lambda_j \alpha_j^2\right)\right] - \left(\sum_i \lambda_i \alpha_i^2\right)\left[\frac{\partial}{\partial \alpha_k}\left(\sum_j \alpha_j^2\right)\right]}{\left(\sum_j \alpha_j^2\right)^2} \qquad (5.39)$$

Setting $(\partial R/\partial \alpha_k) = 0$ leads to

$$\left(\sum_i \alpha_i^2\right)\left(\sum_j \lambda_j \alpha_j \delta_{j,k}\right) - \left(\sum_i \lambda_i \alpha_i^2\right)\left(\sum_j \alpha_j \delta_{j,k}\right) = 0$$

$$\alpha_k\left[\lambda_k\left(\sum_i \alpha_i^2\right) - \left(\sum_i \lambda_i \alpha_i^2\right)\right] = 0 \qquad (5.40)$$

Equation 5.40 is satisfied if either $\alpha_k = 0$ or $\lambda_k\left(\sum_i \alpha_i^2\right) - \left(\sum_i \lambda_i \alpha_i^2\right) = 0$.
However, the stationary value of R occurs when Equation 5.40 is satisfied for all values of k. Consider the choice of $\alpha_k = \delta_{j,m}$ for some value of m. Then Equation 5.40 is satisfied for all $k \neq m$ as $\alpha_k = 0$. For $k = m$, Equation 5.40 is also satisfied because $\lambda_k\left(\sum_i \alpha_i^2\right) - \left(\sum_i \lambda_i \alpha_i^2\right) = \lambda_m \sum_i \delta_{i,m} - \sum_i \lambda_i \delta_{i,m} = \lambda_m - \lambda_m = 0$. Thus, there is one solution corresponding to each m of this form.

Each solution considered above corresponds to $\mathbf{w}$ equal to an eigenvector. Thus, $R(\mathbf{w})$ is rendered stationary if $\mathbf{w}$ is an eigenvector of $\mathbf{L}$.

Consider a solution in which two different coefficients, say α_m and α_n, are both non zero and all other coefficients are zero. Satisfaction of Equation 5.40 for $k = m$ requires

$$0 = \lambda_m(\alpha_m^2 + \alpha_n^2) - (\lambda_m \alpha_m^2 + \lambda_n \alpha_n^2) = (\lambda_m - \lambda_n)\alpha_n^2 \qquad (5.41)$$

Unless $\lambda_m = \lambda_n$, $\alpha_n = 0$. Thus the existence of such a solution is not possible. If $\lambda_m = \lambda_n$, then $\mathbf{w} = \alpha_m \mathbf{v}_m + \alpha_n \mathbf{v}_n$, a liner combination of eigenvectors corresponding to the same eigenvalue, which by definition is also an eigenvector of $\mathbf{L}$ corresponding to the eigenvalue λ_m.

The following theorem has thus been proven.

Theorem 5.6. *Let* **L** *be a self-adjoint operator with respect to an inner product* **(u,v).** *Rayleigh's Quotient, Equation 5.33, is stationary if and only if* **w** *is an eigenvector of* **L.** *Furthermore, if the eigenvalues of* **L** *are chosen such that* $\lambda_1 \leq \lambda_2 \leq \ldots \leq \lambda_{k-1} \leq \lambda_k \leq \lambda_{k+1} \leq \ldots$, *then min* $(R(\mathbf{W})) = \lambda_1$.

5.4 SOLVABILITY CONDITIONS

Consider a linear operator **L** for which $\lambda = 0$ is an eigenvalue. Then $L\mathbf{u} = \mathbf{0}$ has a nontrivial solution, $\mathbf{u}_0$, the eigenvector corresponding to the zero eigenvalue. Consider the nonhomogeneous problem

$$\mathbf{Lu} = \mathbf{f} \tag{5.42}$$

for a specified **f**. If **u** is a solution of Equation 5.42, then

$$\mathbf{L}(\mathbf{u} + c\mathbf{u}_0) = \mathbf{Lu} + \mathbf{L}(c\mathbf{u}_0) = \mathbf{Lu} + c\mathbf{Lu}_0 = \mathbf{f} \tag{5.43}$$

Equation 5.43 shows that if a solution to Equation 5.42 exists, then it is not unique; $\mathbf{u} + c\mathbf{u}_0$ is also a solution of Equation 5.43 for any nonzero scalar c.

The question of existence is equivalent to determine if **f** is in the range of **L**. Assume **L** is self-adjoint with respect to an inner product, **(u, v).** Assuming a solution **u** of Equation 5.42, taking the inner product of Equation (1) with $\mathbf{u}_0$, the homogeneous solution, gives

$$(\mathbf{Lu}, \mathbf{u}_0) = (\mathbf{f}, \mathbf{u}_0) \tag{5.44}$$

But since **L** is self-adjoint and $\mathbf{Lu} = \mathbf{0}$,

$$(\mathbf{Lu}, \mathbf{u}_0) = (\mathbf{u}, \mathbf{Lu}_0) = (\mathbf{u}, \mathbf{0}) = 0 \tag{5.45}$$

Thus Equation 5.44 requires

$$(\mathbf{f}, \mathbf{u}_0) = 0 \tag{5.46}$$

Equation 5.46 defines a solvability condition which states that if **L** is a self-adjoint operator with an eigenvalue of zero, then a solution of the nonhomogeneous problem $\mathbf{Lu} = \mathbf{f}$ has a solution if and only if **f** is orthogonal to the eigenvector corresponding to the zero eigenvalue.

Now suppose **L** is not self-adjoint, and $\mathbf{L}^*$ is the adjoint of **L**. Let $\mathbf{v}_0$ be the eigenvector of $\mathbf{L}^*$ corresponding to its eigenvalue, $\mu = 0$,

$$\mathbf{L}^* \mathbf{v}_0 = \mathbf{0} \tag{5.47}$$

Taking the inner product of Equation 5.47 with $\mathbf{v}_0$, using the definition of the adjoint operator, and Equation 5.42 leads to

$$(\mathbf{Lu},\mathbf{v_0}) = (\mathbf{f},\mathbf{v_0})$$
$$(\mathbf{u},\mathbf{L}^*\mathbf{v_0}) = (\mathbf{f},\mathbf{v_0}) \tag{5.48}$$
$$(\mathbf{u},0) = (\mathbf{f},\mathbf{v_0})$$

The solvability condition obtained from Equation 5.48 is

$$(\mathbf{f},\mathbf{v_0}) = 0 \tag{5.49}$$

The following theorem has been proved:

Theorem 5.7. *(Fredholm Alternative) If* $\mathbf{Lu} = \mathbf{0}$ *has a nontrivial solution,* $\mathbf{u_0}$ *(*$\lambda = 0$ *is an eigenvalue of* $\mathbf{L}$*) then* $\mathbf{Lu} = \mathbf{f}$ *has a solution if and only if* $(\mathbf{f},\mathbf{v_0}) = 0$ *where* $\mathbf{v_0}$ *is the nontrivial solution of* $\mathbf{L}^*\mathbf{v} = \mathbf{0}$*. When a solution of* $\mathbf{Lu} = \mathbf{f}$ *exists, it is not unique as* $\mathbf{u} + c\mathbf{u_0}$ *is also a solution for any scalar c.*

Example 5.5. Let $\mathbf{L}$ be a linear operator whose domain is a vector space S and is self-adjoint with respect to an inner product $(\mathbf{u},\mathbf{v})$. Let $\mathbf{u}$ be an eigenvector of $\mathbf{L}$ corresponding to an eigenvalue λ. Consider the eigenvalue problem

$$\hat{\mathbf{L}}\mathbf{v} = \kappa\mathbf{v} \tag{a}$$

where

$$\hat{\mathbf{L}} = \mathbf{L} + \varepsilon\mathbf{L}_1 \tag{b}$$

is a perturbed operator in which ε is a small nondimensional parameter.

Expansions for the eigenvalue and eigenvector of the perturbed operator are sought as

$$\kappa = \lambda + \varepsilon\lambda_1 + \varepsilon^2\lambda_2 \tag{c}$$

and

$$\mathbf{v} = \mathbf{u} + \varepsilon\mathbf{u}_1 + \varepsilon^2\mathbf{u}_2 \tag{d}$$

Use the Fredholm Alternative to determine the first-order perturbations for κ and $\mathbf{v}$.

Solution: Substitution of Equation b through Equation d into Equation a leads to

$$(\mathbf{L} + \varepsilon \mathbf{L}_1)(\mathbf{u} + \varepsilon \mathbf{u}_1) = (\lambda + \varepsilon \lambda_1)(\mathbf{u} + \varepsilon \mathbf{u}_1) \tag{e}$$

Using the linearity of L, Equation e can be written as

$$\mathbf{L}\mathbf{u} + \varepsilon \mathbf{L}\mathbf{u}_1 + \varepsilon \mathbf{L}_1 \mathbf{u} + \varepsilon^2 \mathbf{L}_1 \mathbf{u}_1 = \lambda \mathbf{u} + \varepsilon \lambda \mathbf{u}_1 + \varepsilon \lambda_1 \mathbf{u} + \varepsilon^2 \lambda_1 \mathbf{u}_1 \tag{f}$$

Noting $\mathbf{L}\mathbf{u} = \lambda \mathbf{u}$ Equation f becomes

$$\varepsilon(\mathbf{L}\mathbf{u}_1 + \mathbf{L}_1 \mathbf{u}) + O(\varepsilon^2) = \varepsilon[\lambda \mathbf{u}_1 + \lambda_1 \mathbf{u}] + O(\varepsilon^2) \tag{g}$$

Since ε is an independent dimensionless parameter, powers of ε are linearly independent, leading to

$$\mathbf{L}\mathbf{u}_1 + \mathbf{L}_1 \mathbf{u} = \lambda \mathbf{u}_1 + \lambda_1 \mathbf{u}$$
$$\mathbf{L}\mathbf{u}_1 - \lambda \mathbf{u}_1 = \lambda_1 \mathbf{u} - \mathbf{L}_1 \mathbf{u} \tag{h}$$

Equation h is a nonhomogeneous equation to solve for $\mathbf{u}_1$. But $\mathbf{L}\mathbf{u}_1 - \lambda \mathbf{u}_1 = 0$ has a nontrivial solution, $\mathbf{u}$. Thus 0 is an eigenvalue of the operator $\mathbf{L}\mathbf{u}_1 - \lambda \mathbf{u}_1$ and a solution of Equation h exists only if an appropriate solvability condition is satisfied. Application of the Fredholm Alternative leads to the appropriate solvability condition as

$$(\lambda_1 \mathbf{u} - \mathbf{L}_1 \mathbf{u}, \mathbf{u}) = 0 \tag{i}$$

Equation i is rearranged to solve for λ_1 as

$$\lambda_1 = \frac{(\mathbf{L}_1 \mathbf{u}, \mathbf{u})}{(\mathbf{u}, \mathbf{u})} \tag{j}$$

If the eigenvector is normalized, then $(\mathbf{u}, \mathbf{u}) = 1$ and Equation j becomes

$$\lambda_1 = (\mathbf{L}_1 \mathbf{u}, \mathbf{u}) \tag{k}$$

The eigenvector perturbation is obtained through use of the expansion theorem. Let $\mathbf{v}_k$ represent the set of normalized eigenvectors of L corresponding to a set of eigenvalues μ_k. Assume that the eigenvalue and eigenvector perturbation are sought for μ_m and $\mathbf{v}_m$. The expansion theorem is used to expand $\mathbf{u}_1$, as

$$\mathbf{u}_1 = \sum_{k \neq m} \alpha_k \mathbf{v}_k \tag{1}$$

The summation in Equation 1 excludes m because the perturbation must be orthogonal to $\mathbf{u} = \mathbf{v}_m$. If not, the contribution in Equation 1 from $\alpha_m \mathbf{v}_m$ is already included as part of $\mathbf{u}$.

Substitution of Equation 1 into Equation h leads to

$$\sum_{k \neq m} \alpha_k \mathbf{L} \mathbf{v}_k - \mu_m \sum_{k \neq m} \alpha_k \mathbf{v}_k = \lambda_1 \mathbf{v}_m - \mathbf{L}_1 \mathbf{v}_m$$
$$\sum_{k \neq m} \alpha_k \mu_k \mathbf{v}_k - \mu_m \sum_{k \neq m} \alpha_k \mathbf{v}_k = \lambda_1 \mathbf{v}_m - \mathbf{L}_1 \mathbf{v}_m \tag{m}$$

Taking the inner product of Equation m with $\mathbf{v}_i$ for some $i \neq m$ and using orthonormality of the eigenvectors leads to

$$\alpha_i = \frac{(\mathbf{L}_1 \mathbf{v}_m, \mathbf{v}_i)}{\mu_m - \mu_i} \tag{n}$$

Substitution in Equation 1 leads to the eigenvector perturbation as

$$\mathbf{u}_1 = \sum_{k \neq m} \frac{(\mathbf{L}_1 \mathbf{v}_m, \mathbf{v}_k)}{\mu_m - \mu_k} \mathbf{v}_k \tag{o}$$

5.5 FREE RESPONSE USING THE NORMAL MODE SOLUTION

5.5.1 GENERAL FREE RESPONSE

The differential equations governing the free response $\mathbf{w}(t)$ of a conservative system are of the form

$$\mathbf{Kw} + \mathbf{M\ddot{w}} = 0 \tag{5.50}$$

where $\mathbf{K}$ is a stiffness operator and $\mathbf{M}$ is an inertia operator. If the system is discrete, $\mathbf{K}$ and $\mathbf{M}$ are symmetric matrices and $\ddot{w} = d^2w/dt^2$. If the system is continuous, $\mathbf{K}$ and $\mathbf{M}$ are differential operators and $\ddot{w} = \partial^2 w/\partial t^2$. Forms of $\mathbf{K}$ and $\mathbf{M}$ for different types of continuous systems, derived using the extended Hamilton's Principle, are specified in Chapter 4.

Using the normal mode solution, the natural frequencies and mode shape vectors for a conservative are determined from

$$\mathbf{Kw} = \omega^2 \mathbf{Mw} \tag{5.51}$$

which is equivalent to

$$\mathbf{M}^{-1}\mathbf{Kw} = \omega^2 \mathbf{w} \tag{5.52}$$

Equation 5.52 shows that the natural frequencies of a linear conservative system are the square roots of the eigenvalues of $\mathbf{M}^{-1}\mathbf{K}$, where $\mathbf{M}$ is the system's inertia operator and $\mathbf{K}$ is the system's stiffness operator. The mode shape vectors are the corresponding eigenvectors. The operators $\mathbf{M}^{-1}\mathbf{K}$ considered in Chapter 4, for discrete and continuous systems, are shown to be self-adjoint and positive definite with respect to the kinetic energy inner product.

The normal mode solution

$$\mathbf{u} = \mathbf{w}e^{i\omega t} \tag{5.53}$$

is so named because the mode shape vectors satisfy an orthogonality condition and are thus normal to each other. Being orthogonal, the mode shape vectors are also linearly independent. The natural frequencies are the square roots of the eigenvalues of $\mathbf{M}^{-1}\mathbf{K}$. The square root of a real and positive value has two values, one positive and one negative. Only the positive value has physical meaning in terms of natural frequency. Indeed, using either the positive or negative value for ω in Equation 5.52 leads to the same mode shape vector. However, the negative value of ω is important mathematically in that if $\mathbf{w}e^{i\omega t}$ is a solution of the differential equations, then so is the linearly independent vector $\mathbf{w}e^{-i\omega t}$.

The following applies to discrete and continuous systems. Upper limits for the summations are not specified with the understanding that the upper limit is n, the dimension of the system for a discrete system, or ∞ if the system is continuous.

The most general solution for a linear homogeneous system is a linear combination of all possible solutions. Thus the general form of the free response is

$$\mathbf{u} = \sum_k \mathbf{w}_k(D_k e^{i\omega_k t} + \bar{D}_k e^{i\omega_k t}) \tag{5.54}$$

Euler's identity is used in Equation 5.54, replacing the exponential functions of complex exponents with real trigonometric functions

$$\mathbf{u} = \sum_k \mathbf{w}_k[C_{k,1}\cos{(\omega_k t)} + C_{k,2}\sin{(\omega_k t)}] \tag{5.55}$$

The constants $C_{k,1}$ and $C_{k,2}$ are determined through application of initial conditions of the form

$$\mathbf{u}(t = 0) = \mathbf{u}_0 \tag{5.56}$$

$$\frac{\partial \mathbf{u}}{\partial t}(t = 0) = \dot{\mathbf{u}}_0 \tag{5.57}$$

where $\mathbf{u}_0$ and $\dot{\mathbf{u}}_0$ are vectors residing in the domain of $\mathbf{K}$. Application of the expansion theorem to the initial condition vectors leads to

$$\mathbf{u}_0 = \sum_k (\mathbf{u}_0,\mathbf{w}_k)_M \mathbf{w}_k \tag{5.58}$$

$$\dot{\mathbf{u}}_0 = \sum_k (\dot{\mathbf{u}}_0,\mathbf{w}_k)_M \mathbf{w}_k \tag{5.59}$$

Application of the initial conditions to Equation 5.55 leads to

$$\mathbf{u}_0 = \sum_k C_{k,1}\mathbf{w}_k \tag{5.60}$$

$$\dot{\mathbf{u}}_0 = \sum_k \omega_k C_{k,2}\mathbf{w}_k \tag{5.61}$$

Since the mode shape vectors are linearly independent, Equation 5.58 through Equation 5.61 imply

$$C_{k,1} = (\mathbf{u}_0,\mathbf{w}_k)_M \tag{5.62}$$

$$C_{k,2} = \frac{1}{\omega_k}(\dot{\mathbf{u}}_0,\mathbf{w}_k)_M \tag{5.63}$$

If $\mathbf{K}$ is positive semi-definite, then $\omega = 0$ is the system's lowest natural frequency. The general form of the free response for such an unconstrained system is

$$\mathbf{u} = \mathbf{w}_1(C_{1,1} + C_{1,2}t) + \sum_{k>1} \mathbf{w}_k[C_{k,1}\cos(\omega_k t) + C_{k,2}\sin{(\omega_k t)}] \tag{5.64}$$

where $\mathbf{w}_1$ is the mode shape vector corresponding to $\omega = 0$. Application of the initial conditions to Equation 5.64 leads to

$$\mathbf{u}_0 = C_{1,1}\mathbf{w}_1 + \sum_{k>1} C_{k,1}\mathbf{w}_k \tag{5.65}$$

$$\dot{\mathbf{u}}_0 = C_{1,2}\mathbf{w}_1 + \sum_{k>1} \omega_k C_{k,2}\mathbf{w}_k \tag{5.66}$$

The evaluation of the constants for $k > 1$ is as in Equation 5.62 and Equation 5.63. Comparison of Equation 5.58, Equation 5.59, Equation 5.65, and Equation 5.66 leads to

$$C_{1,1} = (\mathbf{u}_0, \mathbf{w}_1)_M \tag{5.67}$$

$$C_{1,2} = (\dot{\mathbf{u}}_0, \mathbf{w}_1)_M \tag{5.68}$$

Example 5.6. Determine the time dependent response of the system of Example 5.1 if the initial conditions are

$$\mathbf{u}_0 = \begin{bmatrix} 1 \\ -1 \\ 0 \end{bmatrix} \quad \dot{\mathbf{u}}_0 = \begin{bmatrix} 0 \\ 0 \\ 0 \end{bmatrix} \tag{a}$$

Solution: The natural frequencies and normalized mode shape vectors are determined in Example 5.1 and Example 5.3, as

$$\omega_1 = 0.733\sqrt{\frac{k}{m}} \quad \omega_2 = 1.313\sqrt{\frac{k}{m}} \quad \omega_3 = 1.800\sqrt{\frac{k}{m}} \tag{b}$$

$$\mathbf{w}_1 = \frac{1}{\sqrt{m}}\begin{bmatrix} 0.2469 \\ 0.6081 \\ 0.3158 \end{bmatrix} \quad \mathbf{w}_2 = \frac{1}{\sqrt{m}}\begin{bmatrix} 0.2209 \\ 0.2819 \\ -0.6292 \end{bmatrix} \quad \mathbf{w}_3 = \frac{1}{\sqrt{m}}\begin{bmatrix} -0.9439 \\ 0.2244 \\ -0.0646 \end{bmatrix} \tag{c}$$

The expansion theorem is used in Example 5.4 for the same vector as $\mathbf{u}_0$, resulting in

$$(\mathbf{u}_0, \mathbf{w}_1)_M = -0.9693\sqrt{m} \quad (\mathbf{u}_0, \mathbf{w}_2)_M = -0.3429\sqrt{m}$$

$$(\mathbf{u}_0, \mathbf{w}_3)_M = -1.3927\sqrt{m} \tag{d}$$

Noting that $\dot{\mathbf{u}}_0 = 0$, application of Equation 5.55 through Equation 5.63 leads to

$$
\mathbf{u} = -0.9693 \begin{bmatrix} 0.2469 \\ 0.6081 \\ 0.3158 \end{bmatrix} \cos\left(0.733 \sqrt{\frac{k}{m}} t \right) - 0.3429 \begin{bmatrix} 0.2209 \\ 0.2819 \\ -0.6292 \end{bmatrix}
$$

$$
\times \cos\left(1.313 \sqrt{\frac{k}{m}} t \right) - 1.3927 \begin{bmatrix} -0.9439 \\ 0.2244 \\ -0.0646 \end{bmatrix} \cos\left(1.800 \sqrt{\frac{k}{m}} t \right)
$$

$$
= \begin{bmatrix} -0.2393 \\ -0.5894 \\ -0.3061 \end{bmatrix} \cos\left(0.733 \sqrt{\frac{k}{m}} t \right) + \begin{bmatrix} -0.0757 \\ -0.0967 \\ 0.2158 \end{bmatrix} \cos\left(1.313 \sqrt{\frac{k}{m}} t \right)
$$

$$
+ \begin{bmatrix} 1.3146 \\ -0.3125 \\ 0.0900 \end{bmatrix} \cos\left(1.800 \sqrt{\frac{k}{m}} t \right) \tag{e}
$$

The response of each mass is illustrated in Figure 5.3.

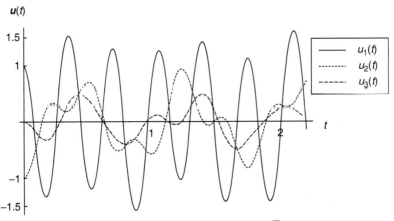

FIGURE 5.3 Response of system of Example 5.6 when $\sqrt{\frac{k}{m}} = 10$.

Example 5.7. Determine the time dependent response of the system of Example 5.2 if

$$u(x,0) = x - \frac{1}{2}x^2 \qquad \frac{\partial u}{\partial t}(x,0) = 0 \tag{a}$$

Solution: The natural frequencies and normalized mode shapes are determined in Example 5.2 and Example 5.3, as

$$\omega_k = (2k-1)\frac{\pi}{2} \tag{b}$$

$$w_k(x) = \sqrt{2}\sin\left((2k-1)\frac{\pi}{2}x\right) \tag{c}$$

The expansion theorem is used in Example 5.4 for the given $u(x, 0)$, leading to

$$(u_0(x),w_k(x))_M = \frac{8}{\pi^3(8k^3 - 12k^2 + 6k - 1)} \tag{d}$$

The free response of the system is obtained using Equation 5.55 as

$$u(x,t) = \frac{16}{\pi^3}$$

$$\times \sum_{k=1}^{\infty} \frac{1}{8k^3 - 12k^2 + 6k - 1}\sin\left((2k-1)\frac{\pi}{2}x\right)\cos\left((2k-1)\frac{\pi}{2}t\right) \tag{e}$$

The time dependent responses of the middle and end of the bar are given in Figure 5.4.

5.5.2 Principal Coordinates

At any instant of time, the free response is a vector in S, the domain of $\mathbf{K}$, and thus has an expansion in terms of the normalized mode shape vectors as

$$\mathbf{u} = \sum_{i=1}^{\infty} c_i \mathbf{w}_i \tag{5.69}$$

Since the free response is a continuous function of time, the coefficients in the expansion must also be continuous functions of time, $c_i = c_i(t)$. Equation 5.69 is substituted into the differential equations governing the free response leading to

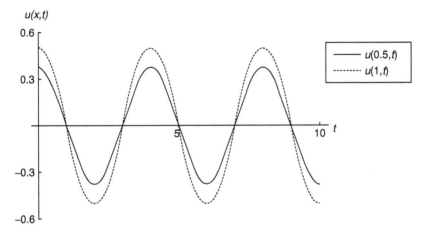

FIGURE 5.4 Response of middle and end of bar of Example 5.7. (The summation of Equation (e) is calculated using 30 terms.)

$$\sum_{i=1} \ddot{c}_i(t)\mathbf{M}\mathbf{w}_i + \sum_{i=1} c_i(t)\mathbf{K}\mathbf{w}_i = 0 \tag{5.70}$$

Taking the standard inner product of Equation 5.70 with $\mathbf{w}_j$ for an arbitrary j leads to

$$\left(\sum_{i=1} \ddot{c}_i\mathbf{M}\mathbf{w}_i, \mathbf{w}_j\right) + \left(\sum_{i=1} c_i\mathbf{K}\mathbf{w}_i, \mathbf{w}_j\right) = 0$$

$$\sum_{i=1} \ddot{c}_i(\mathbf{w}_i, \mathbf{w}_j)_M + \sum_{i=1} c_i(\mathbf{w}_i, \mathbf{w}_j)_K = 0$$

$$\sum_{i=1} \ddot{c}_i \delta_{i,j} + \sum_{i=1} c_i \omega_i^2 \delta_{i,j} = 0$$

$$\ddot{c}_j + \omega_j^2 c_j = 0 \tag{5.71}$$

Equation 5.71 represents a set of uncoupled differential equations to solve for the coefficients in the expansion of $\mathbf{u}(t)$ in terms of the normalized mode shapes. These differential equations can be solved individually and then substituted into Equation 5.69 to obtain the free response in terms of the original generalized coordinates.

Equation 5.71 implies that a set of generalized coordinates exists such that the differential equations, Equation 5.50, are uncoupled using these coordinates. Such a set of coordinates is called a set of principal coordinates. The principal coordinates are related to the chosen generalized coordinates through Equation 5.69.

The use of principal coordinates provides an alternate method for the determination of the free response of a conservative system. Once the natural frequencies and mode shapes are determined, the equation for each principal coordinate can be written in the form of Equation 5.71. The general solution of Equation 5.71 is

$$c_i(t) = A_i \cos(\omega_i t) + B_i \sin(\omega_i t) \tag{5.72}$$

where A_i and B_i are determined through application of initial conditions.

Appropriate initial conditions are of the form of Equation 5.56 and Equation 5.57. Using these in the expansion of Equation 5.69 leads to

$$\mathbf{u}_0 = \sum_{i=1} c_i(0) \mathbf{w}_i \tag{5.73}$$

$$\dot{\mathbf{u}}_0 = \sum_{i=1} \dot{c}_i(0) \mathbf{w}_i \tag{5.74}$$

Taking the kinetic energy inner product of Equation 5.73 and Equation 5.74 with $\mathbf{w}_j$ and using orthonormality of mode shapes leads to

$$c_j(0) = (\mathbf{u}_0, \mathbf{w}_j)_M \tag{5.75}$$

$$\dot{c}_j(0) = (\dot{\mathbf{u}}_0, \mathbf{w}_j)_M \tag{5.76}$$

Application of Equation 5.75 and Equation 5.76 to Equation 5.72 leads to evaluation of the constants of integration as

$$A_j = (\mathbf{u}_0, \mathbf{w}_j)_M \tag{5.77}$$

$$B_j = \frac{1}{\omega_j}(\dot{\mathbf{u}}_0, \mathbf{w}_j)_M \tag{5.78}$$

Substitution of the constants of integration into Equation 5.72, which is then used in Equation 5.69, the general form of the response is

$$\mathbf{u} = \sum_{i=1} \left[(\mathbf{u}_0, \mathbf{w}_i)_M \cos(\omega_i t) + \frac{1}{\mathbf{w}_i}(\dot{\mathbf{u}}_0, \mathbf{w}_i)_M \sin(\omega_i t) \right] \mathbf{w}_i \tag{5.79}$$

5.6 DISCRETE SYSTEMS

The mass and stiffness operators for an n-degree-of-freedom system are nxn symmetric matrices. The natural frequencies and mode shape vectors are determined through solution of a matrix or algebraic eigenvalue problem.

5.6.1 THE MATRIX EIGENVALUE PROBLEM

Let $\mathbf{D}$ be a real nxn matrix. Its eigenvalues are the values of λ such that

$$\mathbf{Du} = \lambda\mathbf{u} \qquad (5.80)$$

The eigenvectors of $\mathbf{D}$ are the corresponding nontrivial solutions. Equation 5.80 can be rewritten as

$$(\mathbf{D} - \lambda\mathbf{I})\mathbf{u} = 0 \qquad (5.81)$$

where $\mathbf{I}$ is the $n \times n$ identity matrix. Equation 5.81 can be formulated as a system of simultaneous algebraic equations to solve for the components of the mode shape vector $\mathbf{u}$. In this formulation, Cramer's rule is used to determine each component of the mode shape vector as

$$u_i = \frac{|\mathbf{C}_i|}{|\mathbf{D} - \lambda\mathbf{I}|} \quad i = 1,2,\ldots,n \qquad (5.82)$$

where $\mathbf{C}_i$ is the nxn matrix obtained by replacing the ith column of $\mathbf{D} - \lambda\mathbf{I}$ with the right hand side vector of Equation 5.81, the zero vector. The determinant of a matrix with a column of zeroes is equal to zero. Thus Equation 5.82 becomes

$$u_i = \frac{0}{|\mathbf{D} - \lambda\mathbf{I}|} \quad i = 1,2,\ldots,n \qquad (5.83)$$

Equation 5.83 shows that $\mathbf{u} = 0$, the trivial solution unless

$$|\mathbf{D} - \lambda\mathbf{I}| = 0 \qquad (5.84)$$

The eigenvalues of $\mathbf{D}$ are calculated as the values of λ, which are solutions of Equation 5.84, which is written as

$$\begin{vmatrix} d_{1,1} - \lambda & d_{1,2} & d_{1,3} & \cdots & d_{1,n} \\ d_{2,1} & d_{2,2} - \lambda & d_{2,3} & \cdots & d_{2,n} \\ d_{3,1} & d_{3,2} & d_{3,3} - \lambda & \cdots & d_{3,n} \\ \vdots & \vdots & \vdots & \ddots & \vdots \\ d_{n,1} & d_{n,2} & d_{n,3} & \cdots & d_{n,n} - \lambda \end{vmatrix} = 0 \qquad (5.85)$$

The expansion of the determinant of Equation 5.85 leads to a polynomial of order n in λ whose roots are the eigenvalues of $\mathbf{D}$. The resulting equation, called the characteristic equation, is of the form

$$\lambda^n + a_{n-1}\lambda^{n-1} + a_{n-2}\lambda^{n-2} + \ldots + a_1\lambda + a_0 = 0 \qquad (5.86)$$

Equation 5.86 has n solutions. Since the elements of $\mathbf{D}$ are real, the coefficients in Equation 5.86 are real; and if complex roots occur, they occur in complex conjugate pairs. Repeated roots of the characteristic equation may occur.

Once an eigenvalue for the matrix has been determined its corresponding eigenvector is determined through solution of Equation 5.80. If $\mathbf{D}$ is an nxn matrix, application of Equation 5.80 leads to at most $n-1$ linearly independent equations to solve for the components of the eigenvector. As noted previously, an eigenvector can never be uniquely determined; it can be determined at best to a multiplicative constant. A solution of Equation 5.80 is often obtained by arbitrarily choosing one component of the eigenvector and then proceeding with the solution of a system of $n-1$ equations for $n-1$ unknowns. This technique is successful unless the element of the eigenvector arbitrarily chosen is actually equal to zero.

The solutions of Equation 5.86 are difficult to obtain analytically. Thus approximate and numerical solutions are often employed to determine eigenvalues and eigenvectors and numerical solutions are often used. Software such as MATLAB MathCad and Mathematica are also useful for eigenvalue and eigenvector determination.

The free response of the initial value problem is illustrated in Example 5.6 of Section 5.3. The response can also be obtained using a method derived from the expansion theorem. Let $\omega_1 \leq \omega_2 \leq \ldots \leq \omega_n$ be the natural frequencies of an nDOF system with corresponding normalized mode shapes $\mathbf{w}_1$, $\mathbf{w}_2$, ...$\mathbf{w}_n$. Normalization of the mode shapes implies $(\mathbf{w}_i, \mathbf{w}_j)_M = \delta_{i,j}$, and $(\mathbf{w}_i, \mathbf{w}_j)_K = \omega_i^2 \delta_{i,j}$.

The principal coordinates for a discrete system are related to the chosen generalized coordinates by

$$\mathbf{x} = \mathbf{Pc} \qquad (5.87)$$

where $\mathbf{c}$ is the vector of principal coordinates and $\mathbf{P}$, called the modal matrix, is a matrix whose columns are the normalized mode shape vectors. The modal matrix is a transformation matrix between the vector of generalized coordinates and the vector of principal coordinates. Since $\mathbf{P}$ is a matrix whose columns are clearly linearly independent, it is nonsingular and hence its inverse exists. Pre-multiplication of Equation 5.1 by $\mathbf{P}^{-1}$ leads to

$$\mathbf{c} = \mathbf{P}^{-1}\mathbf{x} \qquad (5.88)$$

Since the transformation matrix is nonsingular, the principal coordinates exist and the relation between the generalized coordinates and the principal coordinates is unique.

The solution for each principal coordinate is of the form

$$c_i(t) = A_i \cos(\omega_i t) + B_i \sin(\omega_i t) \qquad (5.89)$$

If the system's initial conditions are $\mathbf{x}(0) = \mathbf{x}_0$ and $\dot{\mathbf{x}}(0) = \dot{\mathbf{x}}_0$, Equation 5.88 leads to $\mathbf{c}(0) = \mathbf{P}^{-1}\mathbf{x}_0$ and $\dot{\mathbf{c}}(0) = \mathbf{P}^{-1}\dot{\mathbf{x}}_0$.

The mode shape vector $\mathbf{w}_i$ is a vector of relative displacements of the particles whose motion is described by the generalized coordinates that results when the system vibrates at the natural frequency ω_i. This result is achieved when the initial conditions are such that $c_k(0) = 0$ and $\dot{c}_k(0) = 0$ for all $k \neq i$. The mode shape vectors often have graphical representations, showing the displacement of one particle relative to another. A particle for which the displacement is identically zero for a particular mode is called a node for that mode. Mode shape diagrams for the system of Example 5.1 and Example 5.3 are illustrated in Figure 5.4. The first mode has no nodes, the second mode has one node located in the spring connecting the two rightmost blocks, while the third mode has two nodes, one in each of the springs. As discussed in the next example, the locations of nodes are related to the definition of principal coordinates).

Example 5.8. Figure 5.5 illustrates a two-degree-of-freedom system in which a uniform slender bar of mass m is attached to two springs. Using x and θ as the chosen generalized coordinates, (a) determine the system's natural frequencies and normalized mode shapes, (b) describe the physical meaning of each mode shape, and (c) determine the physical meaning of each of the principal coordinates.

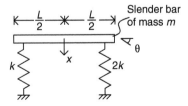

FIGURE 5.5 System of Example 5.8.

Solution (a): The kinetic energy of the system at an arbitrary instant is

$$T = \frac{1}{2}m\dot{x}^2 + \frac{1}{2}\left(\frac{1}{12}mL^2\right)\dot{\theta}^2 \tag{a}$$

The mass matrix is determined from the quadratic form of kinetic energy as

$$\mathbf{M} = \begin{bmatrix} m & 0 \\ 0 & \frac{1}{12}mL^2 \end{bmatrix} \tag{b}$$

The potential energy of the system is

$$V = \frac{1}{2}k\left(x - \frac{L}{2}\theta\right)^2 + \frac{1}{2}2k\left(x + \frac{L}{2}\theta\right)^2 \tag{c}$$

The stiffness matrix is obtained from Equation c as

$$\mathbf{K} = \begin{bmatrix} 3k & \frac{1}{2}kL \\ \frac{1}{2}kL & \frac{3}{4}kL^2 \end{bmatrix} \tag{d}$$

The natural frequencies are the square roots of the eigenvalues of $\mathbf{M}^{-1}\mathbf{K}$. To this end,

$$\mathbf{M}^{-1}\mathbf{K} - \lambda\mathbf{I} = \begin{bmatrix} \frac{1}{m} & 0 \\ 0 & \frac{12}{mL^2} \end{bmatrix}\begin{bmatrix} 3k & \frac{1}{2}kL \\ \frac{1}{2}kL & \frac{3}{4}kL^2 \end{bmatrix} - \lambda\begin{bmatrix} 1 & 0 \\ 0 & 1 \end{bmatrix}$$

$$= \begin{bmatrix} 3\phi - \lambda & \frac{1}{2}L\phi \\ \frac{6\phi}{L} & 9\phi - \lambda \end{bmatrix} \tag{e}$$

where $\phi = (k/m)$. The characteristic equation is obtained as

$$\lambda^2 - 12\phi\lambda + 24\phi^2 = 0 \tag{f}$$

The solutions of Equation f are $\lambda = 2.536\varphi, 9.464\varphi$, which leads to natural frequencies of

$$\omega_1 = \sqrt{\lambda_1} = 1.593\sqrt{\frac{k}{m}} \tag{g}$$

$$\omega_2 = \sqrt{\lambda_2} = 3.076\sqrt{\frac{k}{m}} \tag{h}$$

The mode shapes are obtained by obtaining a nontrivial solution to

$$\begin{bmatrix} 3\phi - \lambda & \frac{1}{2}L\phi \\ \frac{6\phi}{L} & 9\phi - \lambda \end{bmatrix} \begin{bmatrix} a \\ b \end{bmatrix} = \begin{bmatrix} 0 \\ 0 \end{bmatrix} \tag{i}$$

The top row of the system of Equation a is

$$(3\phi - \lambda)a + \frac{1}{2}L\phi b = 0$$

$$b = -\frac{2(3\phi - \lambda)}{L\phi} \tag{j}$$

Substituting $\lambda = 2.546\varphi$ in Equation j and selecting $a = 1$ leads to the first mode shape as

$$\mathbf{X}_1 = C_1 \begin{bmatrix} 1 \\ \dfrac{-0.920}{L} \end{bmatrix} \tag{k}$$

Substituting $\lambda = 9.464\varphi$ in Equation j and selecting $a = 1$ leads to

$$\mathbf{X}_2 = C_2 \begin{bmatrix} 1 \\ \dfrac{12.928}{L} \end{bmatrix} \tag{l}$$

The mode shapes are normalized by requiring $(\mathbf{X}_i, \mathbf{X}_j)_M = 1$. To this end,

$$\mathbf{X}_1^T \mathbf{M} \mathbf{X}_1 = 1$$

$$C_1^2 \begin{bmatrix} 1 & \dfrac{-0.928}{L} \end{bmatrix} \begin{bmatrix} m & 0 \\ 0 & \dfrac{1}{12} mL^2 \end{bmatrix} \begin{bmatrix} 1 \\ \dfrac{-0.928}{L} \end{bmatrix} = 1 \tag{m}$$

$$1.072 m C_1^2 = 1 \Rightarrow C_1 = \frac{0.9660}{\sqrt{m}}$$

$$\mathbf{X}_2^T \mathbf{M} \mathbf{X}_2 = 1$$

$$C_2^2 \begin{bmatrix} 1 & \dfrac{12.928}{L} \end{bmatrix} \begin{bmatrix} m & 0 \\ 0 & \dfrac{1}{12} mL^2 \end{bmatrix} \begin{bmatrix} 1 \\ \dfrac{12.928}{L} \end{bmatrix} = 1 \tag{n}$$

$$14.93 m C_2^2 = 1 \Rightarrow C_2 = \frac{0.2588}{\sqrt{m}}$$

Thus the normalized mode shapes are

$$\mathbf{X}_1 = \frac{1}{\sqrt{m}} \begin{bmatrix} 0.9660 \\ \dfrac{-0.8944}{L} \end{bmatrix} \qquad \mathbf{X}_2 = \frac{1}{\sqrt{m}} \begin{bmatrix} 0.2588 \\ \dfrac{3.3458}{L} \end{bmatrix} \tag{o}$$

(b) A mode shape vector provides relative values of each generalized coordinate when vibrations occur at the frequency corresponding to the mode shape. For example, if initial conditions are set such that the free response occurs only at the lowest natural frequency then, if $x = 0.9660$, $\theta = -(0.8966/L)$. The mode shapes are illustrated in Figure 5.6a.

There are no nodes for the first mode. An extension of the bar shows that it crosses the equilibrium line a distance of $1.103L$ to the right of the mass center. The first mode is a rigid body rotation about an axis through this point. A node for the second mode occurs a distance $0.0774L$ to the left of the mass center. The second mode is a rigid body rotation of the bar about an axis through this point.

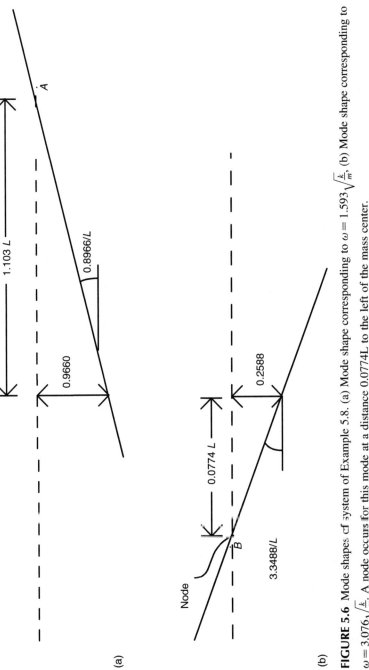

FIGURE 5.6 Mode shapes of system of Example 5.8. (a) Mode shape corresponding to $\omega = 1.593\sqrt{\frac{k}{m}}$, (b) Mode shape corresponding to $\omega = 3.076\sqrt{\frac{k}{m}}$. A node occurs for this mode at a distance 0.0774L to the left of the mass center.

(c) Suppose the generalized coordinates are chosen to be the displacements of the points about the bar rotates for each mode, labeled A and B in Figure 5.6. The displacement of point B is zero for the second mode and the displacement of point A is zero for the first mode. Thus, if the displacements of these points are chosen as generalized coordinates, the differential equations describing the vibrations must be uncoupled. Thus the displacements of A and B are the principal coordinates. Point B is the principal coordinate for the first mode while point A is the principal coordinate for the second mode.

Example 5.9. Three identical bodies are at rest in equilibrium, as illustrated in Figure 5.7, when another identical body moving at a speed of 10 (m/s) engages the system. Determine the subsequent responses of each body.

Solution: The differential equations governing the motion of the system are

$$\begin{bmatrix} m & 0 & 0 & 0 \\ 0 & m & 0 & 0 \\ 0 & 0 & m & 0 \\ 0 & 0 & 0 & m \end{bmatrix} \begin{bmatrix} \ddot{x}_1 \\ \ddot{x}_2 \\ \ddot{x}_3 \\ \ddot{x}_4 \end{bmatrix} + \begin{bmatrix} k & -k & 0 & 0 \\ -k & 2k & -k & 0 \\ 0 & -k & 2k & -k \\ 0 & 0 & -k & k \end{bmatrix} \begin{bmatrix} x_1 \\ x_2 \\ x_3 \\ x_4 \end{bmatrix} = \begin{bmatrix} 0 \\ 0 \\ 0 \\ 0 \end{bmatrix} \quad \text{(a)}$$

The natural frequencies are the square roots of the eigenvalues of $\mathbf{M}^{-1}\mathbf{K}$. To this end,

$$\mathbf{M}^{-1}\mathbf{K} = \begin{bmatrix} \phi & -\phi & 0 & 0 \\ -\phi & 2\phi & -\phi & 0 \\ 0 & -\phi & 2\phi & -\phi \\ 0 & 0 & -\phi & \phi \end{bmatrix} \quad \text{(b)}$$

where $\phi = (k/m) = 3.33 \times 10^4 (1/s^2)$. The natural frequencies are calculated as

$$\omega_1 = 0 \quad \omega_2 = 139.73 \ r/s \quad \omega_3 = 258.20 \ r/s \quad \omega_4 = 337.35 \ r/s \quad \text{(c)}$$

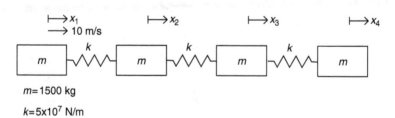

$m = 1500$ kg

$k = 5 \times 10^7$ N/m

FIGURE 5.7 Four-degree-of-freedom system of Example 5.9 is unrestrained.

Since the system is unrestrained, its lowest natural frequency is zero. The normalized mode shape vectors are obtained as

$$\mathbf{X}_1 = \begin{bmatrix} 0.01291 \\ 0.01291 \\ 0.01291 \\ 0.01291 \end{bmatrix} \qquad \mathbf{X}_2 = \begin{bmatrix} -0.01687 \\ -0.00698 \\ 0.00698 \\ 0.01687 \end{bmatrix}$$

$$\mathbf{X}_3 = \begin{bmatrix} 0.01291 \\ -0.01291 \\ -0.01291 \\ 0.01291 \end{bmatrix} \qquad \mathbf{X}_4 = \begin{bmatrix} -0.00698 \\ 0.01687 \\ -0.01687 \\ 0.00698 \end{bmatrix} \qquad \text{(d)}$$

The mode shapes are illustrated in Figure 5.8. The mode corresponding to the natural frequency of zero is a rigid body mode in which all bodies move as if they are rigidly connected. A node exists in the spring between the middle bodies for the second mode. The bodies move back and forth about the node as if the system were periodically expanding and contracting. There is no force in the spring between the middle bodies in the symmetric third mode. The fourth mode also has a node at the midpoint of the middle spring, but also has two additional nodes.

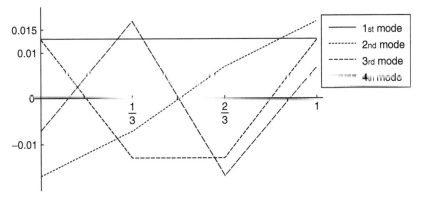

FIGURE 5.8 Normalized mode shapes for system of Example 5.9. The first mode corresponds to rigid body motion of the system.

The initial conditions are

$$\mathbf{x}(0) = \begin{bmatrix} 0 \\ 0 \\ 0 \\ 0 \end{bmatrix} \qquad \dot{\mathbf{x}}(0) = \begin{bmatrix} 10 \\ 0 \\ 0 \\ 0 \end{bmatrix} \tag{e}$$

The general response is of the form

$$\mathbf{x}(t) = \begin{bmatrix} 0.01291 \\ 0.01291 \\ 0.01291 \\ 0.01291 \end{bmatrix} (C_1 + C_2 t) + \begin{bmatrix} -0.01687 \\ -0.00698 \\ 0.00698 \\ 0.01687 \end{bmatrix} [C_3 \cos(139.73t)$$

$$+ C_4 \sin(139.73t)] + \begin{bmatrix} 0.01291 \\ -0.01291 \\ -0.01291 \\ 0.01291 \end{bmatrix} [C_5 \cos(258.8t) + C_6 \sin(258.8t)]$$

$$+ \begin{bmatrix} -0.00698 \\ 0.01687 \\ -0.01687 \\ 0.00698 \end{bmatrix} [C_7 \cos(337.35t) + C_8 \sin(337.35t)] \tag{f}$$

Application of the initial conditions leads to

$$C_1 = C_3 = C_5 = C_7 = 0 \quad C_2 = 143.6 \quad C_4 = -1.81 \quad C_6 = 0.75$$
$$C_8 = -0.3107 \tag{g}$$

Plots for displacement and velocity of each body are given in Figure 5.9 and Figure 5.10. Since the system is conservative and unrestrained, the displacement of each mass away from its initial condition grows indefinitely.

5.7 NATURAL FREQUENCY CALCULATIONS USING FLEXIBILITY MATRIX

The flexibility matrix $\mathbf{A}$ is, when it exists, the inverse of the stiffness matrix $\mathbf{K}$. The flexibility matrix exists when the stiffness matrix is nonsingular.

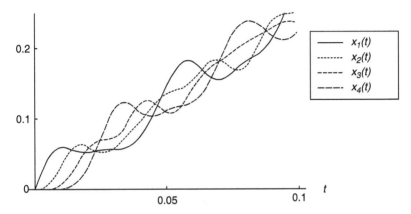

FIGURE 5.9 Displacements of each mass in the system of Example 5.9. Since the system is unrestrained and conservative, a non-zero initial velocity leads to unbounded displacement of each mass.

However, the stiffness matrix is singular only for unrestrained systems. Premultiplying Equation 5.1 by **A**, assuming it exists, leads to

$$\mathbf{A}\mathbf{M}\ddot{\mathbf{x}} + \mathbf{x} = 0 \tag{5.90}$$

Applying the normal mode solution ($\mathbf{x} = \mathbf{v}e^{i\omega t}$) to Equation 5.90 leads to

$$-\omega^2 \mathbf{A}\mathbf{M}\mathbf{v} + \mathbf{v} = 0 \tag{5.91}$$

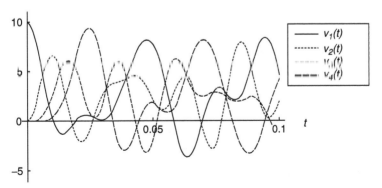

FIGURE 5.10 Velocities of each mass in the system of Example 5.9. Contrary to displacements, velocities are bounded.

Equation 5.91 is rearranged as

$$\mathbf{AMv} = \frac{1}{\omega^2}\mathbf{v} \tag{5.92}$$

Equation 5.92 is in the form of a standard matrix eigenvalue problem form which it can be concluded that the eigenvalue problem. It is deduced from Equation 5.92 that the natural frequencies are the reciprocals of the square roots of the eigenvalues of $\mathbf{AM}$ and the mode shapes are the corresponding eigenvectors.

Since, when it exists, $\mathbf{A}$ is symmetric and positive definite, a flexibility inner product is defined. Mode shape vectors corresponding to distinct natural frequencies are orthogonal with respect to the flexibility inner product.

In problems where it is easier to determine the flexibility matrix than the stiffness matrix, the natural frequencies and mode shape vectors can be determined using the flexibility matrix and the mass matrix.

Example 5.10. Determine the natural frequencies and mode shapes for the system of Figure 5.11. Three machines are equally placed along the span of a simply supported beam. Use a three-degree-of-freedom model ignoring the inertia of the beam.

Solution: The generalized coordinates are taken as the displacements of the machines. The flexibility matrix for this system is determined as Equation h of Example 4.3 and is repeated below.

$$\mathbf{A} = \frac{L^3}{EI}\begin{bmatrix} \dfrac{3}{256} & \dfrac{11}{768} & \dfrac{7}{768} \\[2mm] \dfrac{11}{768} & \dfrac{1}{48} & \dfrac{11}{768} \\[2mm] \dfrac{7}{768} & \dfrac{11}{768} & \dfrac{3}{256} \end{bmatrix} \tag{a}$$

Substituting given values into Equation a leads to

$$\mathbf{A} = 10^{-7}\begin{bmatrix} 2.00 & 2.44 & 1.56 \\ 2.44 & 3.56 & 2.44 \\ 1.56 & 2.44 & 2.00 \end{bmatrix} \tag{b}$$

The kinetic energy of the system is

$$T = \frac{1}{2}m_1\dot{x}_1^2 + \frac{1}{2}m_2\dot{x}_2^2 + \frac{1}{2}m_3\dot{x}_3^2 \tag{c}$$

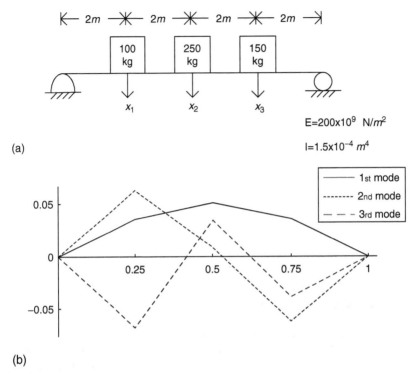

FIGURE 5.11 (a) Natural frequencies for three-degree-of-freedom model of three machines along the span of a simply supported beam are calculated as reciprocals of square roots of eigenvalues of **AM**. (b) Normalized mode shape vectors.

from which the mass matrix is determined as

$$\mathbf{M} = \begin{bmatrix} 100 & 0 & 0 \\ 0 & 250 & 0 \\ 0 & 0 & 150 \end{bmatrix} \qquad \text{(d)}$$

The product of the flexibility matrix and the mass matrix is

$$\mathbf{AM} = \phi \begin{bmatrix} 2 & 6.11 & 2.33 \\ 2.44 & 8.89 & 3.67 \\ 1.56 & 6.11 & 3 \end{bmatrix} \qquad \text{(e)}$$

where $\phi = 10^{-5}$. The characteristic equation is developed using Equation e as

$$-\lambda^3 + 13.89\phi - 9.469\phi^2 + 1.153\phi^3 = 0$$

$$-\left(\frac{\lambda}{\phi}\right)^3 + 13.89\left(\frac{\lambda}{\phi}\right)^2 - 9.469\left(\frac{\lambda}{\phi}\right) + 1.153 = 0 \qquad \text{(f)}$$

The solutions of Equation f are

$$\lambda_1 = 0.1577 \times 10^{-5} \qquad \lambda_2 = 0.554 \times 10^{-5} \qquad \lambda_3 = 13.179 \times 10^{-5} \quad \text{(g)}$$

The natural frequencies are the reciprocals of the square roots of the eigenvalues leading to

$$\omega_1 = 87.1 \ r/s \qquad \omega_2 = 424.8 \ r/s \qquad \omega_3 = 796.1 \ r/s \qquad \text{(h)}$$

The normalized mode shapes are illustrated in Figure 5.11b.

5.8 MATRIX ITERATION

Many numerical methods exist for determination of the eigenvalues and eigenvectors. Meirovitch presents a comprehensive discussion of the most useful methods for application to vibrations problems. Computational software such as MATLAB, MathCad, and Mathematica have reduced the need for such a thorough discussion of these methods. However, these packages do have limitations on the size of matrices that may be used. Efficient computational algorithms are necessary for large-scale finite element simulations. However, many of these methods require more background in linear algebra and are best explained in a study solely focused on numerical methods.

The only numerical method developed here is matrix iteration. Matrix iteration is perfectly suited for approximating the eigenvalues and eigenvectors for vibrations problems when only a few modes are required. However, it is clearly not suitable for obtaining all natural frequencies and mode shapes for a system with a large number of degrees of freedom. Its advantages include:

- A natural frequency and its corresponding mode shape vector are determined simultaneously.
- Distinct natural frequencies and mode shape vectors are determined successively. If some, but not all, of the natural frequencies and mode shapes are needed, only those needed are calculated.
- The method is simple to apply and to program.

Some disadvantages of the method include:

- In order to approximate the lowest natural frequencies and their mode shapes first, the flexibility matrix must be attained. If the stiffness matrix is obtained in the modeling process, it must be inverted to determine the flexibility matrix.
- The method is an iterative process in which the rate of convergence is dependent upon an initial guess.
- The rate of convergence is dependent upon the separation of the natural frequencies—the closer the natural frequencies, the slower the convergence.

Matrix iteration is an iterative method through which eigenvalues and eigenvectors of a self-adjoint matrix are obtained sequentially beginning with the matrix's largest eigenvalue. Thus, in application to a vibrations problem, in order to obtain the lowest natural frequency and mode shape first, matrix iteration is used with $\mathbf{AM}$ as the natural frequencies are the reciprocals of the square roots of the eigenvalues of $\mathbf{AM}$. Consider thus a problem to determine the natural frequencies and mode shape vectors formulated as

$$\mathbf{AMv} = \frac{1}{\omega^2}\mathbf{v} \tag{5.93}$$

Let $\omega_1 \leq \omega_2 \leq, \ldots \leq \omega_n$ be the n natural frequencies with corresponding mode shape vectors $\mathbf{v}_1, \mathbf{v}_2, \ldots, \mathbf{v}_n$ such that $\mathbf{AMv}_i = \frac{1}{\omega_i^2}\mathbf{v}_i$. Let $\mathbf{w}_0$ be an n-dimensional vector. The expansion theorem states that there exist coefficients $\alpha_1, \alpha_2, \ldots, \alpha_n$, such that

$$\mathbf{w}_0 = \sum_{i=1}^{n} \alpha_i \mathbf{v}_i \tag{5.94}$$

Define the sequence of vectors

$$\mathbf{u}_k = \mathbf{AMw}_{k-1} \tag{5.95}$$

$$\mathbf{w}_k = \frac{\mathbf{u}_k}{\|\mathbf{u}_k\|_M} \tag{5.96}$$

Substituting Equation 5.94 into Equation 5.95 and computing the first few terms of the sequence

$$\mathbf{u}_1 = \mathbf{AM}\left(\sum_{i=1}^{n}\alpha_i\mathbf{v}_i\right) = \sum_{i=1}^{n}\alpha_i\mathbf{AMv}_i = \sum_{i=1}^{n}\frac{\alpha_i}{\omega_i^2}\mathbf{v}_i \tag{5.97}$$

$$\|\mathbf{u}_1\|_M = \left[\sum_{i=1}^{n} \left(\frac{\alpha_i}{\omega_i^2}\right)^2\right]^{1/2} \tag{5.98}$$

$$\mathbf{w}_1 = \frac{1}{\|\mathbf{u}_1\|_M} \sum_{i=1}^{n} \frac{\alpha_i}{\omega_i^2} \mathbf{v}_i \tag{5.99}$$

$$\mathbf{u}_2 = \mathbf{AM}\left(\sum_{i=1}^{n} \frac{\alpha_i}{\omega_i^2 \|\mathbf{u}_1\|_M} \mathbf{v}_i\right) = \frac{1}{\|\mathbf{u}_1\|_M} \sum_{i=1}^{n} \frac{\alpha_i}{\omega_i^2} \mathbf{AMv}_i$$

$$= \frac{1}{\|\mathbf{u}_1\|_M} \sum_{i=1}^{n} \frac{\alpha_i}{\omega_i^4} \mathbf{v}_i \tag{5.100}$$

$$\|\mathbf{u}_2\|_M = \frac{1}{\|\mathbf{u}_1\|_M} \left[\sum_{i=1}^{n} \left(\frac{\alpha_i}{\omega_i^4}\right)^2\right]^{1/2} \tag{5.101}$$

$$\mathbf{w}_2 = \frac{1}{\left[\sum_{i=1}^{n} \left(\frac{\alpha_i}{\omega_i^4}\right)^2\right]^{1/2}} \sum_{i=1}^{n} \frac{\alpha_i}{\omega_i^4} \mathbf{v}_i \tag{5.102}$$

By induction, the general form for $\mathbf{w}_k$ is

$$\mathbf{w}_k = \frac{1}{\left[\sum_{i=1}^{n} \left(\frac{\alpha_j}{\omega_j^{2k}}\right)^2\right]^{1/2}} \sum_{i=1}^{n} \frac{\alpha_i}{\omega_i^{2k}} \mathbf{v}_i \tag{5.103}$$

Equation 5.103 is written as

$$\mathbf{w}_k = C_k \sum_{i=1}^{n} \frac{\alpha_i}{\omega_i^{2k}} \mathbf{v}_i \tag{5.104}$$

where

$$C_k = \frac{1}{\left[\sum_{i=1}^{n} \left(\frac{\alpha_i}{\omega_i^{2k}}\right)^2\right]^{1/2}} \tag{5.105}$$

Equation 5.104 is rewritten as

$$\mathbf{w}_k = C_k \left(\frac{\alpha_1}{\omega_1^{2k}} \mathbf{v}_1 + \sum_{i=2}^{n} \frac{\alpha_i}{\omega_i^{2k}} \mathbf{v}_i \right) = \frac{C_k}{\omega_1^{2k}} \left[\alpha_1 \mathbf{v}_1 + \sum_{i=2}^{n} \alpha_k \left(\frac{\omega_1}{\omega_i} \right)^{2k} \mathbf{v}_i \right] \quad (5.106)$$

Noting that each ratio $(\omega_1/\omega_k) < 1$ as k increases, each term in the summation in Equation 5.106 becomes successively smaller than the first such that

$$\lim_{k \to \infty} \mathbf{w}_k = \left(\lim_{k \to \infty} \frac{C_k \alpha_1}{\omega_1^{2k}} \right) \mathbf{v}_1 \quad (5.107)$$

Using Equation 5.105

$$\lim_{k \to \infty} C_k = \lim_{k \to \infty} \frac{1}{\left[\left(\frac{\alpha_1}{\omega_1^{2k}} \right)^2 + \sum_{i=2}^{n} \left(\frac{\alpha_i}{\omega_i^{2k}} \right)^2 \right]^{1/2}} = \frac{\omega_1^{2k}}{\alpha_1} \quad (5.108)$$

Then, from Equation 5.107 and Equation 5.108,

$$\lim_{k \to \infty} \mathbf{w}_k = \mathbf{v}_1 \quad (5.109)$$

Since α_1 is arbitrary, it can be taken to be one, and thus the reciprocal of the kinetic energy inner product norm converges to ω_1^2 and the iteration converges to its corresponding mode shape vector.

The procedure described above will lead to an approximation for the lowest natural frequency and its mode shape vector. It might be suspected that the iteration described above converges to the mode shape vector for ω_2 and leads to the value of ω_2 if an initial guess is made orthogonal to the lowest mode shape. In theory this will work, however, in practice numerical errors will prevent this convergence unless the orthogonality is reinforced on every iteration. Indeed, the Gram–Schmidt procedure may be used to develop an initial guess vector which is orthogonal to the lowest mode shape approximation. The iteration proceeds as follows except that Gram–Schmidt is reinforced on every step.

The above describes a tedious process that can be simplified by redefining the original eigenvalue problem such that orthogonality with the first mode shape is automatically enforced. The matrix

$$(\mathbf{AM})_2 = \mathbf{AM} - \frac{1}{\omega_1^2} \mathbf{v}_1 \mathbf{v}_1^T \mathbf{M} \quad (5.110)$$

is called the deflated matrix and is constructed such that

$$((\mathbf{AM})_2 \mathbf{x}, \mathbf{v}_1)_M = 0 \quad (5.111)$$

for any $\mathbf{x}$. The deflated matrix enforces Gram–Schmidt on each iteration. Matrix iteration used on the deflated matrix converges to the second natural frequency. The process continues with the deflated matrix computed on each step such that

$$(\mathbf{AM})_k = (\mathbf{AM})_{k-1} = \frac{1}{\omega_{k-1}^2} \mathbf{v}_{k-1} \mathbf{v}_{k-1}{}^T \mathbf{M} \qquad (5.112)$$

The deflated matrix enforces the following orthogonality conditions:

$$((\mathbf{AM})_k, \mathbf{v}_\ell)_M = 0 \quad \ell = 1,2,\ldots k-1 \qquad (5.113)$$

Note that an iteration process is not necessary to determine the largest natural frequency and its corresponding mode shape. For an n-degree-of-freedom system, $(\mathbf{AM})_n \mathbf{x}$ is orthogonal to all previous mode shape vectors, implying that $(\mathbf{AM})_n \mathbf{x}$ is an eigenvector and that

$$(\mathbf{AM})_n \mathbf{x} = \frac{1}{\omega_n^2} \mathbf{x} \qquad (5.114)$$

Example 5.11. A discrete mass-spring system is attached to the midspan of the beam of Example 5.10, as illustrated in Figure 5.12. This system is studied as an example of a vibration absorber in Chapter 7. Use matrix iteration to determine the natural frequencies and normalized mode shape vectors.

Solution: The flexibility influence coefficients in the first three rows and columns of the flexibility matrix for the four-degree-of-freedom system are the same as the flexibility influence coefficients of Example 5.10. Consider a

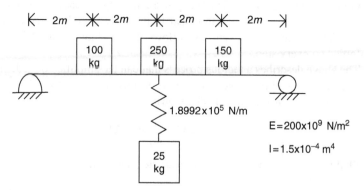

FIGURE 5.12 System of Example 5.11.

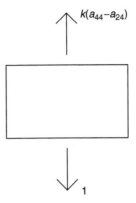

FIGURE 5.13 Free-body diagram used to determine a_{44} in the flexibility matrix of the system of Figure 5.10 and Example 5.11.

unit load applied to the first mass. Since no force is applied to the discrete mass, the force in the spring must be zero. Thus the displacement of the discrete mass must be equal to the displacement of the particle at the midspan of the beam. Thus, $a_{41} = a_{21}$. In a similar fashion, it is shown that $a_{42} = a_{22}$ and $a_{43} = a_{23}$. Finally, consider a unit load applied to the discrete mass as in Figure 5.13. In order for the system to be in equilibrium when the load is applied,

$$1 - k(a_{44} - a_{24}) = 0$$

$$a_{44} = \frac{1}{k} + a_{24} \tag{a}$$

The flexibility matrix for this system is

$$\mathbf{A} = \frac{L^3}{EI} \begin{bmatrix} \dfrac{3}{256} & \dfrac{11}{768} & \dfrac{7}{768} & \dfrac{11}{768} \\[2mm] \dfrac{11}{768} & \dfrac{1}{48} & \dfrac{11}{768} & \dfrac{1}{48} \\[2mm] \dfrac{7}{768} & \dfrac{11}{768} & \dfrac{3}{256} & \dfrac{11}{768} \\[2mm] \dfrac{11}{768} & \dfrac{1}{48} & \dfrac{11}{768} & \dfrac{EI}{kL^3} + \dfrac{1}{48} \end{bmatrix} \tag{b}$$

Substitution of given values in Equation a leads to

$$
\mathbf{A} = 10^{-7}
\begin{bmatrix}
2.00 & 2.44 & 1.56 & 2.44 \\
2.44 & 3.56 & 2.44 & 3.56 \\
1.56 & 2.44 & 2.00 & 2.44 \\
2.44 & 3.56 & 2.44 & 56.21
\end{bmatrix}
\frac{m}{n}
\tag{c}
$$

The mass matrix is determined from the kinetic energy functional as

$$
\mathbf{M} =
\begin{bmatrix}
100 & 0 & 0 & 0 \\
0 & 250 & 0 & 0 \\
0 & 0 & 150 & 0 \\
0 & 0 & 0 & 25
\end{bmatrix}
\text{kg}
\tag{d}
$$

Equation c and Equation d lead to

$$
\mathbf{AM} = 10^{-5}
\begin{bmatrix}
2.00 & 6.11 & 2.33 & 0.611 \\
2.44 & 8.89 & 3.67 & 0.889 \\
1.56 & 6.11 & 3.00 & 0.611 \\
2.44 & 8.89 & 3.67 & 14.05
\end{bmatrix}
s^2
\tag{e}
$$

The initial guess for the mode shape vector for the first mode is

$$
\mathbf{w}_0 =
\begin{bmatrix}
0.359 \\
0.0518 \\
0.0366 \\
1
\end{bmatrix}
\tag{f}
$$

The vectors on the first iteration step are calculated as

$$
\mathbf{u}_1 = 10^{-5}
\begin{bmatrix}
2.00 & 6.11 & 2.33 & 0.611 \\
2.44 & 8.89 & 3.67 & 0.889 \\
1.56 & 6.11 & 3.00 & 0.611 \\
2.44 & 8.89 & 3.67 & 14.05
\end{bmatrix}
\begin{bmatrix}
0.359 \\
0.0518 \\
0.0366 \\
1
\end{bmatrix}
= 10^{-6}
\begin{bmatrix}
2.128 \\
3.081 \\
2.144 \\
28.89
\end{bmatrix}
\tag{g}
$$

$$\frac{1}{(\omega_1^2)_1} = \|\mathbf{u}_1\|_M$$

$$= \left\{ 10^{-12}[2.128 \quad 3.081 \quad 2.144 \quad 28.89] \begin{bmatrix} 100 & 0 & 0 & 0 \\ 0 & 250 & 0 & 0 \\ 0 & 0 & 150 & 0 \\ 0 & 0 & 0 & 25 \end{bmatrix} \right.$$

$$\left. \times \begin{bmatrix} 2.128 \\ 3.081 \\ 2.144 \\ 28.89 \end{bmatrix} \right\} = 1.56 \times 10^{-4} \Rightarrow (\omega_1)_1 = 80.02 \tag{h}$$

$$\mathbf{w}_1 = \frac{1}{1.56 \times 10^{-3}} 10^{-6} \begin{bmatrix} 2.128 \\ 3.081 \\ 2.144 \\ 28.89 \end{bmatrix} = \begin{bmatrix} 0.0136 \\ 0.0197 \\ 0.0137 \\ 0.1850 \end{bmatrix} \tag{i}$$

The second iteration step consists of

$$\mathbf{u}_2 = 10^{-5} \begin{bmatrix} 2.00 & 6.11 & 2.33 & 0.611 \\ 2.44 & 8.89 & 3.67 & 0.889 \\ 1.56 & 6.11 & 3.00 & 0.611 \\ 2.44 & 8.89 & 3.67 & 14.05 \end{bmatrix} \begin{bmatrix} 0.0136 \\ 0.0197 \\ 0.0137 \\ 0.1850 \end{bmatrix} = 10^{-6} \begin{bmatrix} 2.92 \\ 4.23 \\ 2.96 \\ 28.59 \end{bmatrix} \tag{j}$$

$$\frac{1}{(\omega_1^2)_2} = \|\mathbf{u}_2\|_M$$

$$= \left\{ 10^{-12}[2.92 \quad 4.23 \quad 2.96 \quad 28.59] \begin{bmatrix} 100 & 0 & 0 & 0 \\ 0 & 250 & 0 & 0 \\ 0 & 0 & 150 & 0 \\ 0 & 0 & 0 & 25 \end{bmatrix} \right.$$

$$\left. \times \begin{bmatrix} 2.92 \\ 4.23 \\ 2.96 \\ 28.59 \end{bmatrix} \right\} = 1.64 10^{-3} \Rightarrow (\omega_1)_2 = 77.95 \tag{k}$$

$$\mathbf{w}_2 = \frac{1}{\|\mathbf{u}_2\|_M}\mathbf{u}_2 = \frac{1}{1.64 \times 10^{-3}} 10^{-6} \begin{bmatrix} 2.92 \\ 4.23 \\ 2.96 \\ 28.59 \end{bmatrix} = \begin{bmatrix} 0.0178 \\ 0.0257 \\ 0.0179 \\ 0.174 \end{bmatrix} \qquad (1)$$

The iterations continue. The sequence of approximations for the lowest natural frequency is (80.02, 77.95, 77.08, 76.76, 76.64, 78.60, 76.58, 76.58, 76.58, 76.58). The normalized mode shape vector converges after ten iterations to

$$\mathbf{v}_1 = \begin{bmatrix} 0.0236 \\ 0.0341 \\ 0.0239 \\ 0.1506 \end{bmatrix} \qquad (m)$$

The deflation matrix to determine the second mode shape is

$$(\mathbf{AM})_2 = \mathbf{AM} - \frac{1}{\omega_1^2}\mathbf{v}_1\mathbf{v}_1\mathbf{M}$$

$$= 10^{-5} \begin{bmatrix} 2.00 & 6.11 & 2.33 & 0.611 \\ 2.44 & 8.89 & 3.67 & 0.889 \\ 1.56 & 6.11 & 3.00 & 0.611 \\ 2.44 & 8.89 & 3.67 & 14.05 \end{bmatrix} - 1.70 \times 10^{-5} \begin{bmatrix} 0.0236 \\ 0.0341 \\ 0.0239 \\ 0.1506 \end{bmatrix}$$

$$\times [0.0236 \quad 0.0341 \quad 0.0239 \quad 0.1506] \begin{bmatrix} 100 & 0 & 0 & 0 \\ 0 & 250 & 0 & 0 \\ 0 & 0 & 150 & 0 \\ 0 & 0 & 0 & 25 \end{bmatrix}$$

$$= 10^{-5} \begin{bmatrix} 1.047 & 2.672 & 0.8846 & -0.9065 \\ 1.069 & 3.924 & 1.575 & -1.31 \\ 0.5897 & 2.625 & 1.531 & -0.9271 \\ -3.620 & -13.02 & -5.562 & 4.384 \end{bmatrix}$$

$$(n)$$

Using an initial guess of $\mathbf{w}_0 = [0.06357 \quad 0.0086 - 0.0620 - 1]^T$, the sequence of natural frequency approximations is (123.59, 99.1391, 99.1322, 99.1322, 99.1322). The normalized mode shape vector obtained after convergence is

$$\mathbf{v}_2 = \begin{bmatrix} 0.02714 \\ 0.03900 \\ 0.02777 \\ -0.1313 \end{bmatrix} \tag{o}$$

The deflation matrix to determine the third mode is

$$(\mathbf{A}M)_3 = (\mathbf{A}M)_2 - \frac{1}{\omega_2^2}\mathbf{v}_2\mathbf{v}_2^T M$$

$$= 10^{-5} \begin{bmatrix} 1.047 & 2.672 & 0.8846 & -0.9065 \\ 1.069 & 3.924 & 1.575 & -1.31 \\ 0.5897 & 2.625 & 1.531 & -0.9271 \\ -3.620 & -13.02 & -5.562 & 4.384 \end{bmatrix} - 1.018x$$

$$\times 10^{-4} \begin{bmatrix} 0.02714 \\ 0.03900 \\ 0.02777 \\ -0.1313 \end{bmatrix} [0.0217 \quad 0.03900 \quad 0.0277 \quad -0.1313]$$

$$\tag{p}$$

$$\times \begin{bmatrix} 100 & 0 & 0 & 0 \\ 0 & 250 & 0 & 0 \\ 0 & 0 & 150 & 0 \\ 0 & 0 & 0 & 25 \end{bmatrix}$$

$$= 10^{-7} \begin{bmatrix} 29.77 & -1.873 & -20.66 & -0.02164 \\ -0.7491 & 5.959 & -7.680 & -0.00998 \\ -17.71 & 012.80 & 25.46 & 0.00396 \\ -0.00866 & -0.00998 & 0.2376 & .000270 \end{bmatrix}$$

Using an initial guess for the third mode shape of $\mathbf{w}_0 = [0.0683 -0.0353 \; 0.0385 -1]^T$, the sequence of natural frequency approximations for the third mode is (1807.0, 793.76, 768.43, 624.91, 464.16, 428.54, 425.11 424.83, 424.81, 424.80). The converged mode shape is

$$
\mathbf{v}_3 =
\begin{bmatrix}
0.0636 \\
0.0086 \\
-0.062 \\
-0.00038
\end{bmatrix}
\tag{q}
$$

The fourth mode shape vector must be orthogonal to the first three. Thus, assuming $\mathbf{v}_4 = [q_1 \; q_2 \; q_3 \; q_4]^T$, imposition of orthonormality conditions

$$(\mathbf{v}_1,\mathbf{v}_4)_M = 0 \tag{r}$$

$$(\mathbf{v}_2,\mathbf{v}_4)_M = 0 \tag{s}$$

$$(\mathbf{v}_3,\mathbf{v}_4)_M = 0 \tag{t}$$

$$(\mathbf{v}_4,\mathbf{v}_4)_{M=1} \tag{u}$$

Satisfaction of Equation r through Equation u leads to

$$
\mathbf{v}_4 =
\begin{bmatrix}
0.0683 \\
0.0353 \\
0.0384 \\
0.000428
\end{bmatrix}
\tag{v}
$$

The natural frequency for the fourth mode is calculated from

$$\mathbf{AMv}_4 = \frac{1}{\omega_4^2}\mathbf{v}_4$$

$$
10^{-5}
\begin{bmatrix}
2.00 & 6.11 & 2.33 & 0.611 \\
2.44 & 8.89 & 3.67 & 0.889 \\
1.56 & 6.11 & 3.00 & 0.611 \\
2.44 & 8.89 & 3.67 & 14.05
\end{bmatrix}
\begin{bmatrix}
0.0683 \\
0.0353 \\
0.0384 \\
0.000428
\end{bmatrix}
=
\frac{1}{\omega_4^2}
\begin{bmatrix}
0.0683 \\
0.0353 \\
0.0384 \\
0.000428
\end{bmatrix}
\tag{w}
$$

The equation obtained using the top row of Equation w is

$$10^{-5}[(2.00)(0.0683) + (6.11)(0.0353) + (2.33)(0.0384) + (0.611)$$

$$\times (0.000428)]$$

$$= \frac{1}{\omega_4^4}(0.0683) \tag{x}$$

The solution to Equation x is

$$\omega_4 = 796.24 \tag{y}$$

5.9 CONTINUOUS SYSTEMS

The stiffness operator $\mathbf{K}$ for a continuous system is a differential operator or a matrix of differential operators while the inertia operator is a function of spatial variables or a matrix of functions of spatial variables. The domain of the stiffness operator is defined by the boundary conditions that must be satisfied by the system response.

The operators for several systems are considered in Chapter 4. All operators considered are shown to be self-adjoint and positive definite with respect to an appropriately defined inner product. In addition, the operators are shown to be self-adjoint with respect to the kinetic energy inner product and the potential energy inner product and positive definite with respect to the kinetic energy inner product. Thus the results summarized in Section 5.2 are applicable to the continuous systems considered here.

A significant difference between discrete systems and continuous systems is that a discrete system has a finite number of natural frequencies whereas a continuous system has a finite, but countable, number of natural frequencies. The natural frequencies for a continuous system can be ordered by

$$0 \le \omega_1 \le \omega_2 \le \quad \le \omega_{k-1} < \omega_k \le \omega_{k+1} \le \dots \tag{5.115}$$

The corresponding mode shape vectors are $\mathbf{w}_1, \mathbf{w}_2, \dots, \mathbf{w}_{k-1}, \mathbf{w}_k, \mathbf{w}_{k+1}, \dots$ The mode shape vectors are functions of the spatial variables.

The general free response is a linear combination of all possible mode shapes. Applying the results of Section 5.5,

$$\mathbf{u}(\mathbf{r},t) = \sum_{k=1}^{\infty} \mathbf{w}_k(\mathbf{r}) \left[(\mathbf{u}(\mathbf{r},0),\mathbf{w}_k(\mathbf{r}))_M \cos(\omega_k t) + \frac{1}{\omega_k} \left(\frac{\partial \mathbf{u}}{\partial t}(\mathbf{r},0), \mathbf{w}_k(\mathbf{r}) \right)_M \sin(\omega_k t) \right]$$

$$\tag{5.116}$$

The techniques of Section 5.1 through Section 5.3 are applied in subsequent sections to analyze the response of continuous systems. However, physical understanding of the response is sometimes obtained through application of another method or formulation of the response in a different form than that of Equation 5.1.

5.10 SECOND-ORDER PROBLEMS (WAVE EQUATION)

The free transverse vibrations of strings, longitudinal vibrations of elastic bars, and torsional oscillations of shafts are all governed by the wave equation, which in nondimensional form is

$$\frac{\partial}{\partial x}\left[\alpha(x)\frac{\partial u}{\partial x}\right] = \beta(x)\frac{\partial^2 u}{\partial t^2} \tag{5.117}$$

where $u(x, t)$ is the nondimensional displacement, $\alpha(x)$ is a nondimensional function describing the variation of the system's stiffness properties, and $\beta(x)$ is a nondimensional function describing the variation of the system's inertia properties.

Application of the normal-mode solution, Equation 5.2, to Equation 5.117 leads to the eigenvalue problem for the natural frequency and mode shapes as

$$-\frac{d}{dx}\left[\alpha(x)\frac{dw}{dx}\right] = \omega^2 \beta(x)w \tag{5.118}$$

It is shown in Section 4.10 that the inertia operator is self-adjoint and positive definite with respect to the standard inner product on $C^2[0,1]$ while the stiffness operator is self-adjoint and positive definite with respect to the standard inner product on S, the domain of $\mathbf{K}$, a subspace of $C^2[0,1]$ defined by the boundary conditions. The exception occurs when the system is free at both ends, in which case the stiffness operator is positive semi-definite. Thus energy inner products are defined and the operator

$$\mathbf{M}^{-1}\mathbf{K} = -\frac{1}{\beta(x)}\frac{d}{dx}\left[\alpha(x)\frac{dw}{dx}\right] \tag{5.119}$$

is positive definite and self-adjoint with respect to the kinetic energy inner product.

The eigenvalue problem for the second-order system

$$-\frac{1}{\beta(x)}\frac{d}{dx}\left[\alpha(x)\frac{dw}{dx}\right] = \lambda w \tag{5.120}$$

is a form of a mathematical problem called a Sturm–Liouville problem. The Sturm–Liouville Theorem adapted for a nondimensional formulation is presented below.

Theorem 5.8. (*Sturm–Liouville Problems*) *Let* **L** *be a second-order differential operator of the form*

$$\mathbf{L}f = -\frac{1}{\beta(x)}\left\{\frac{d}{dx}\left[\alpha(x)\frac{df}{dx}\right] + q(x)f\right\} \tag{5.121}$$

with $\beta(x) \geq 0$, $\alpha(x) \geq 0$ and $q(x) \leq 0$ for all x, $0 \leq x \leq 1$. If, S, the domain of **L**, is defined such that either both of conditions (a) and (b) are satisfied for all elements of S or (c) is satisfied for all elements of S, then **L** is self-adjoint and at least positive semi-definite with respect to the inner product

$$(f,g) = \int_0^1 f(x)g(x)\beta(x)dx \tag{5.122}$$

where

a. $\alpha(0) = 0$ or $f(0) + \kappa_0(df/dx)(0) = 0$ with $\kappa_0 \leq 0$
$\left(\kappa_0 = -\infty \Rightarrow (df/dx)(0) = 0\right)$

b. $\alpha(1) = 0$ or $f(1) + \kappa_1(df/dx)(1) = 0$ with $\kappa_1 \geq 0$
$\left(\kappa_1 = \infty \Rightarrow (df/dx)(1) = 0\right)$

c. $f(0) = f(1)0$ and $(df/dx)(0) = (df/dx)(1)$

Application of Theorem 5.8 requires that either conditions a and b hold or condition c is true. If a and b hold such that both a(0)=0 and a(1)=0 then L is positive definite. If a(0)=0 then L is positive definite if $\kappa 1$ is finite. If a(1)=0 then L is positive definite if $\kappa 0$ is finite. If neither a(0)=0 or a(1)=0 then L is positive definite only if both $\kappa 0$ and $\kappa 1$ are infinite. The operator L is positive semi-definite in the case that condition c is true.

The system of Example 5.2, in which the natural frequencies and mode shapes of a uniform bar are obtained, is governed by the wave equation, while the eigenvalue problem, obtained through application of the normal mode solution, is a Sturm–Liouville problem.

Example 5.12. Determine the natural frequencies and normalized mode shapes for the transverse vibrations of a string which is fixed at both ends.

Solution: The transverse vibrations of a string with constant tension are governed by the wave equation

$$\frac{\partial^2 u}{\partial x^2} = \frac{\partial^2 u}{\partial t^2} \tag{a}$$

The boundary conditions for the case where the string is fixed at both ends are

$$u(0,t) = 0 \tag{b}$$

$$u(1,t) = 0 \tag{c}$$

Application of the normal-mode solution, $u(x,t) = w(x)e^{i\omega t}$, to Equation a through Equation c leads to

$$-\frac{d^2 w}{dx^2} = \omega^2 w \tag{d}$$

$$w(0) = 0 \tag{e}$$

$$w(1) = 0 \tag{f}$$

Since the stiffness operator for this problem is positive definite only, solutions corresponding to positive values of ω^2 are sought. The general solution of Equation d is thus

$$w(x) = C_1 \cos(\omega x) + C_2 \sin(\omega x) \tag{g}$$

Application of Equation e to Equation g leads to $C_1 = 0$. Application of Equation f to Equation g with $C_1 = 0$ leads to

$$C_2 \sin(\omega) = 0 \tag{h}$$

Non-trivial solutions exist only for values of ω such that $\sin(\omega) = 0$. The nondimensional natural frequencies are

$$\omega_k = k\pi \quad k = 1,2,3,... \tag{i}$$

The mode shape corresponding to a natural frequency of $\omega_k = k\pi$ is

$$w_k(x) = C_k \sin(k\pi x) \tag{j}$$

The mode shapes are normalized by requiring

$$\int_0^1 [w_k(x)]^2 dx = 1$$

$$\int_0^1 C_k^2 \sin^2(k\pi x)dx = 1 \qquad (k)$$

$$\frac{C_k^2}{2} \int_0^1 [1 - \cos(2k\pi x)]dx = 1$$

Evaluation of the integral leads to $C_k = \sqrt{2}$ and the normalized mode shapes are

$$w_k(x) = \sqrt{2} \sin(k\pi x) \qquad (l)$$

Example 5.13. A one-degree-of-freedom approximation is often used to determine the natural frequency of torsional oscillations of a thin disk attached to an elastic shaft. The system is modeled by a disk of mass moment of inertia I attached to the end of a torsional spring of torsional stiffness, $k_t = JG/L$. If all inertia effects of the shaft are neglected, then the natural frequency of the system is approximated as

$$\omega_n = \sqrt{\frac{JG}{IL}} \qquad (a)$$

a. Determine under what conditions the approximation of Equation a within 5% of the true fundamental frequency for this system. A common approximation for inclusion of inertia effects of the shaft when clearly not negligible it to increase the moment of inertia of the disk by $(1/3)\rho JL$. For what values of $\eta = (I/\rho JL)$ does this provide a good approximation to the lowest natural frequency?

b. Now consider the case where $\eta = (I/\rho JL) = 3.13$. The disk is rotated 30° from its equilibrium position, held in this position and then released. Determine the resulting free response of the system

Solution: The torsional vibrations of a uniform bar are governed by the nondimensional wave equation

$$\frac{\partial^2 \theta}{\partial x^2} = \frac{\partial^2 \theta}{\partial t^2} \qquad (b)$$

where $\theta(x,t)$ represent the angular displacement of the shaft in a cross section a distance x from the left end of the shaft. The characteristic time used in the nondimensionalization is obtained from Table 4.1 as

$$t^* = \frac{t}{L\sqrt{\frac{\rho}{G}}} \tag{c}$$

If ω^* is a nondimensional natural frequency, then the corresponding dimensional frequency is

$$\omega = \frac{\omega^*}{L\sqrt{\rho/G}} \tag{d}$$

The boundary condition for the fixed end at $x = 0$ is

$$\theta(x,0) = 0 \tag{e}$$

The appropriate boundary condition at the end where the disk is attached is

$$JG\frac{\partial\theta}{\partial x}(L,t) = -I\frac{\partial^2\theta}{\partial x^2}(L,t) \tag{f}$$

Nondimensional variables listed in Table 4.1 are used to nondimensionalize the boundary condition, Equation f, resulting in

$$\frac{\partial\theta}{\partial x}(1,t) = -\eta\frac{\partial^2\theta}{\partial t^2}(1,t) \tag{g}$$

where

$$\eta = \frac{I}{\rho JL} \tag{h}$$

Substitution of the normal mode solution $\theta(x, t) = w(x)e^{i\omega t}$ into Equation a, Equation e, and Equation f leads to

$$\frac{d^2w}{dx^2} = -\omega^2 w \tag{i}$$

$$w(0) = 0 \tag{j}$$

$$\frac{dw}{dx}(1) = \eta\omega^2 w(1) \tag{k}$$

Because the system is positive definite, only solutions for positive values of ω^2 are considered. Under this assumption, the solution of Equation i is

$$w(x) = C_1 \cos(\omega x) + C_2 \sin(\omega x) \tag{l}$$

Application of Equation j to Equation l leads to $C_1 = 0$. Application of Equation k to Equation j with $C_1 = 0$ leads to

$$\omega \cos(\omega) = \eta \omega^2 \sin(\omega)$$

$$\tan(\omega) = \frac{1}{\eta \omega} \tag{m}$$

Note that the one-degree-of-freedom approximation for the natural frequency can be written as

$$\omega_n = \sqrt{\frac{JG}{IL}} \sqrt{\frac{\rho L}{\rho L}} = \sqrt{\frac{(\rho JL)G}{I(L^2 \rho)}} = \frac{1}{T\sqrt{\eta}} \tag{n}$$

Comparison of Equation d and Equation n shows that the one-degree-of-freedom approximation for the lowest natural frequency is exact if

$$\omega = \frac{1}{\sqrt{\eta}} \tag{o}$$

If Equation o is true, then Equation n becomes

$$\tan\left(\frac{1}{\sqrt{\eta}}\right) = \frac{1}{\sqrt{\eta}} \tag{p}$$

Equation a is approximately true for small arguments of the tangent or for large values of η. Figure 5.14 presents the four lowest natural frequencies as a function of η while Figure 5.15 presents a comparison between the one-degree-of-freedom natural frequency approximation and the exact frequency calculated from Equation p. The percent error in using the

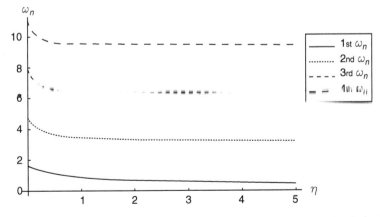

FIGURE 5.14 Four lowest natural frequencies for torsional oscillations of a shaft with an attached disk.

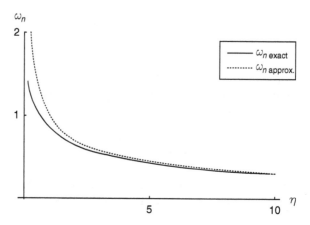

FIGURE 5.15 Comparison between lowest natural frequency of Example 5.13 and one-degree-of-freedom approximation which neglects inertia of shaft.

one-degree-of-freedom approximation is shown in Table 5.1. The error is within 5% for values of $\eta > 3.3$ as illustrated in Figure 5.16.

The inertia effcts of the shaft are often approximated by adding a disk of mass moment of inertia $(1/3)\rho J L$ to the end disk. In this case, the natural

TABLE 5.1
Error in Use of One-Degree-of-Freedom Model to Approximate Lowest Natural Frequency of Shaft with End Disk

η	Exact Natural Frequency	One-Degree-of-freedom Model Without Added Inertia	Error in Approximation $\sqrt{1/\eta}$	One-Degree-of-Freedom Model with Added Mass $\sqrt{3/(3\eta+1)}$	Error in Approximation (%)
0.1	1.43	3.16	121.3%	1.52	6.3
0.2	1.31	2.24	70.2%	1.36	4.5
0.5	1.08	1.41	31.3%	1.30	2.1
1	0.86	1.00	16.2%	0.87	0.7
2	0.65	0.71	8.2%	0.65	0.7
3	0.55	0.58	5.5%	0.55	≈ 0
3.5	0.51	0.54	4.9%	0.51	≈ 0
4	0.48	0.50	4.1%	0.48	≈ 0
6	0.40	0.41	2.8$	0.40	≈ 0
10	0.31	0.32	1.7%	0.31	≈ 0

Error [%]

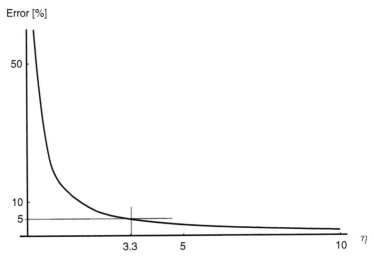

FIGURE 5.16 Percent error in lowest natural frequency of system of example 5.13 when a one-degree-of-freedom approximation is used.

frequency approximation for the one-degree-of-freedom model is

$$\omega = \sqrt{\frac{1}{\eta + (1/3)}} = \sqrt{\frac{3}{3\eta + 1}} \tag{q}$$

The percent error in using the one-degree-of-freedom approximation with and without the approximation for the inertia effects is shown in Table 5.1. Clearly, the added inertia improves the approximation.

(b) The nondimensional natural frequencies are designated as ω_k, $k = 1, 2, 3, \ldots$, where ω_k is the kth solution of

$$\tan(\omega) = \frac{1}{3.13\omega} = \frac{0.319}{\omega} \tag{r}$$

The natural frequencies are illustrated in Figure 5.17 as the points of intersection between the graphs of $\tan(\omega)$ and $(1/\eta\omega)$. For larger values of ω, $1/\eta\omega$ approaches zero. As it gets closer to zero, the natural frequencies approach integer values of π.

$$\lim_{k\to\infty} \omega_k = (k-1)\pi \tag{s}$$

The corresponding mode shapes are of the form

$$w_k(x) = C_k \sin(\omega_k x) \tag{t}$$

Advanced Vibration Analysis

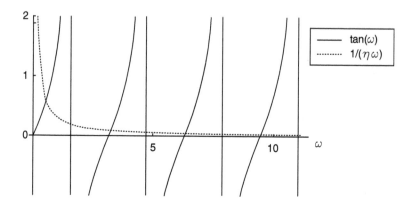

FIGURE 5.17 Illustration of solutions of equation (r) of example 5.13. Natural frequencies correspond to intersections of curves.

The inner product for a system with an attached inertia element is given in Equation 4.160 and used to normalize the mode shapes. To this end,

$$\int_0^1 [w_k(x)]^2 dx + \eta[w_k(1)]^2 = 1$$

$$\int_0^1 C_k^2 \sin^2(\omega_k x)dx + 3.13 C_k^2 \sin^2(\omega_k) = 1$$

$$C_k^2 \left\{ \frac{1}{2}\int_0^1 [1 - \cos(2\omega_k x)]dx + 3.13 \sin^2(\omega_k) \right\} = 1 \qquad (u)$$

$$C_k^2 \left\{ \frac{1}{2}\left[1 - \frac{1}{2\omega_k}\sin(2\omega_k)\right] + 3.13 \sin^2(\omega_k) \right\} = 1$$

$$C_k = \left\{ \frac{1}{2}\left[1 - \frac{1}{2\omega_k}\sin(2\omega_k)\right] + 3.13 \sin^2(\omega_k) \right\}^{-1/2}$$

The expression for C_k, Equation u, may be simplified by using the trigonometric identity $\sin(2\omega_k) = 2\sin(\omega_k)\cos(\omega_k)$ and by noting from Equation r that $(1/\omega_k) = 3.13 \tan(\omega_k)$. Equation u eventually simplifies to

$$C_k = \sqrt{\frac{2}{1 + 3.13 \sin^2(\omega_k)}} \qquad (v)$$

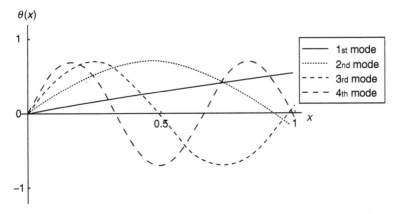

FIGURE 5.18 Mode shapes corresponding to first four natural frequencies for torsional system of example 5.13 with η=3.13.

The first four mode shapes for the bar are illustrated in Figure 5.18. As is the case for discrete systems, a point where the displacement is zero for a mode is called a node. There are no nodes for the first mode, one node for the second mode, two nodes for the third mode, etc.

When the shaft is statically rotated through $30° = 0.524$ rad, the angular displacement of the bar is linear in x. The initial conditions are

$$u(x,0) = 0.524x \qquad \text{(w)}$$

$$\frac{\partial u}{\partial t}(x,0) = 0 \qquad \text{(x)}$$

The free response is given by Equation 5.55, where

$$
\begin{aligned}
(u(x,0),w_k(x)) &= \int_0^1 u(x,0)w_k(x)dx + 3.13u(1,0)w_k(1) \\
&= \int_0^1 (0.514x)C_k \sin(\omega_k x)dx + 3.13(0.524)C_k \sin(\omega_x) \\
&= 0.514C_k \int_0^1 x \sin(\omega_k x)dx + 1.640C_k \sin(\omega_k) \\
&= 0.514C_k \left[\frac{\sin(\omega_k) - \omega_k \cos(\omega_k)}{\omega_k^2} \right] + 1.640C_k \sin(\omega_k) \qquad \text{(y)}
\end{aligned}
$$

Equation r may rearranged as $\cos(\omega_k) = 3.13\omega_k \sin(\omega_k)$, which when used in Equation y leads to

$$(u(x,0),w_k(x)) = 0.514C_k \left[\frac{\sin(\omega_k)}{\omega_k^2} - 3.13 \sin(\omega_k) \right] + 1.609C_k \sin(\omega_k)$$

$$= 0.514C_k \frac{\sin(\omega_k)}{\omega_k^2} \qquad (z)$$

Equation 5.55 leads to the time dependent response as

$$u(x,t) = \sum_{k=1}^{\infty} \left[0.514C_k \frac{\sin(\omega_k)}{\omega_k^2} \right] [C_k \sin(\omega_k x)]\cos(\omega_k t) \qquad (aa)$$

Equation t is used in Equation y leading to

$$u(x,t) = 1.028 \sum_{k=1}^{\infty} \frac{\sin(\omega_k)}{\omega_k^2[1 + 3.13 \sin^2(\omega_k)]} \sin(\omega_k x)\cos(\omega_k t) \qquad (bb)$$

The free response of the end of the shaft is shown in Figure 5.19. Note that the shaft is in its equilibrium position when $t = (2n-1)\pi/2$ for $n = 1, 2, \ldots$ The spatial dependence of the response at several times is plotted in Figure 5.20.

Example 5.14. Consider a bar of nonuniform cross-section, as illus trated in Figure 5.21. The bar has a linear taper and is circular in cross-section with a taper rate of μ. The elastic modulus and mass density are constant.

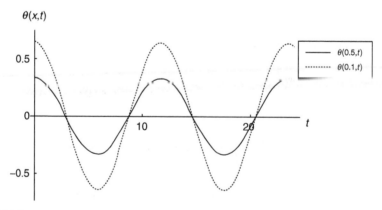

FIGURE 5.19 Time dependence of angular rotation of bar at two locations along its axis.

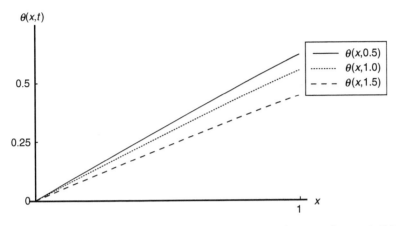

FIGURE 5.20 Spatial dependence of angular displacement of system of example 5.13 when bar has an initial linear angular displacement.

a. Determine the exact values of the system's four lowest natural frequencies and their corresponding mode shapes for $\varepsilon = 0.1$ and $\varepsilon = 0.5$, where $\varepsilon = (\mu L/r_0)$, r_0 is the radius of the bar at $x = 0$ and L is the length of the bar.

b. Use the perturbation method described in Example 5.5 to approximate the natural frequencies and mode shapes for $\varepsilon = 0.1$ and $\varepsilon = 0.5$.

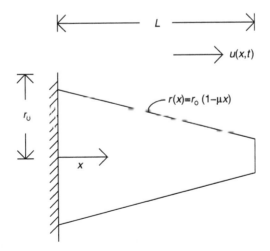

FIGURE 5.21 Tapered bar of Example 5.14.

Solution: (a) The radius of the bar varies as

$$r(x) = r_0 - \mu x \tag{a}$$

where r_0 is the radius at $x = 0$. The cross section area of the circular bar is

$$A(x) = \pi(r_0 - \mu x)^2 \tag{b}$$

Equation b is rewritten using nondimensional variables, $x^* = x/L$ as

$$A(x^*) = \pi r_0^2 (1 - \varepsilon x^*)^2 \tag{c}$$

where

$$\varepsilon = \frac{\mu L}{r_0} \tag{d}$$

Thus the nondimensional functions of Equation 4.65 and Equation 4.66 become

$$\alpha(x^*) = \frac{EA}{E_0 A_0} = \frac{E\pi r_0^2 (1 - \varepsilon x^*)^2}{E(\pi r_0^2)} = (1 - \varepsilon x^*)^2 \tag{e}$$

$$\beta(x^*) = \frac{\rho A}{\rho_0 A_0} = \frac{\rho \pi r_0^2 (1 - \varepsilon x^*)^2}{\rho \pi r_0^2} = (1 - \varepsilon x^*)^2 \tag{f}$$

The bar is fixed at $x = 0$ and free at $x = L$, which leads to boundary conditions of

$$u(0,t) = 0 \tag{g}$$

$$\frac{\partial u}{\partial x}(1) = 0 \tag{h}$$

The nondimensional differential equation governing $u(x,t)$ the longitudinal vibrations of the bar, Equation 5.117 becomes

$$\frac{\partial}{\partial x}\left[(1 - \varepsilon x)^2 \frac{\partial u}{\partial x}\right] = (1 - \varepsilon x)^2 \frac{\partial^2 u}{\partial t^2} \tag{i}$$

When the normal-mode solution, $u(x, t) = w(x)e^{i\omega t}$ is used in Equation g through Equation i, the Sturm–Liouville problem for $w(x)$ is

$$-\frac{1}{(1-\varepsilon x)^2}\frac{d}{dx}\left[(1-\varepsilon x)^2\frac{dw}{dx}\right]=\omega^2 w \qquad \text{(j)}$$

$$w(0)=0 \qquad \text{(k)}$$

$$\frac{dw}{dx}(1)=0 \qquad \text{(l)}$$

The mode shapes are orthogonal with respect to the kinetic energy inner product

$$(f,g)=\int_0^1 f(x)g(x)(1-\varepsilon x)^2\,dx \qquad \text{(m)}$$

The solution of Equation i is obtained through a change of variables. Let $z=1-\varepsilon x$, such that $d/dx = d/dz\,dz/dx = -\varepsilon\,d/dz$. Using z as the dependent variable in Equation j through Equation l leads to

$$\frac{d}{dz}\left(z^2\frac{dw}{dz}\right)+\frac{\omega^2}{\varepsilon^2}z^2 w=0 \qquad \text{(n)}$$

$$w(1)=0 \qquad \text{(o)}$$

$$\frac{dw}{dz}(1-\varepsilon)=0 \qquad \text{(p)}$$

The solution of Equation n is obtained using the template for solution of differential equations whose solutions are written using Bessel functions.

$$w(z)=C_1 z^{-\frac{1}{2}}J_{\frac{1}{2}}\left(\frac{\omega}{\varepsilon}z\right)+C_2 z^{-\frac{1}{2}}Y_{\frac{1}{2}}\left(\frac{\omega}{\varepsilon}z\right) \qquad \text{(q)}$$

Application of Equation o to Equation q gives

$$C_2=-\frac{J_{\frac{1}{2}}\left(\frac{\omega}{\varepsilon}\right)}{Y_{\frac{1}{2}}\left(\frac{\omega}{\varepsilon}\right)} \qquad \text{(r)}$$

Use of Equation p in Equation q and using Equation r leads to the characteristic equation

$$Y_{\frac{1}{2}}\left(\frac{\omega}{\varepsilon}\right)\left\{-\frac{1}{2}(1-\varepsilon)^{-\frac{3}{2}}J_{\frac{1}{2}}\left[\frac{\omega}{\varepsilon}(1-\varepsilon)\right]+(1-\varepsilon)^{-\frac{1}{2}}\left(\frac{\omega}{\varepsilon}\right)J_{\frac{1}{2}}'\left[\frac{\omega}{\varepsilon}(1-\varepsilon)\right]\right\}$$

$$-J_{\frac{1}{2}}\left(\frac{\omega}{\varepsilon}\right)\left\{-\frac{1}{2}(1-\varepsilon)^{-\frac{3}{2}}Y_{\frac{1}{2}}\left[\frac{\omega}{\varepsilon}(1-\varepsilon)\right]+(1-\varepsilon)^{-\frac{1}{2}}\left(\frac{\omega}{\varepsilon}\right)Y_{\frac{1}{2}}'\left[\frac{\omega}{\varepsilon}(1-\varepsilon)\right]\right\}=0$$

(s)

The first four natural frequencies determined using Equation s are plotted as a function of ε in Figure 5.22. The corresponding mode shapes for $\varepsilon = 1.5$ are illustrated in Figure 5.23.

(b) Expansion of the derivative in Equation i leads to

$$-\frac{d^2w}{dx^2}+\frac{2\varepsilon}{1-\varepsilon x}\frac{dw}{dx}=\omega^2 w \tag{t}$$

Use of a binominal expansion for $(1-\varepsilon x)^{-1}$ in Equation t leads to

$$-\frac{d^2w}{dx^2}+2\varepsilon\frac{dw}{dx}+2\varepsilon^2+\ldots=\omega^2 w \tag{u}$$

Equation u is written in the form of Equation 5.b, of example 5.j with

$$Lw=-\frac{d^2w}{dx^2} \tag{v}$$

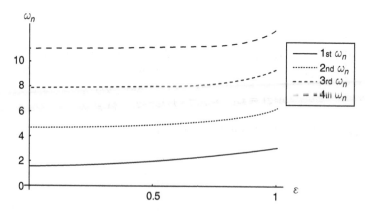

FIGURE 5.22 First four natural frequencies for fixed-free bar with linear taper as function of taper rate.

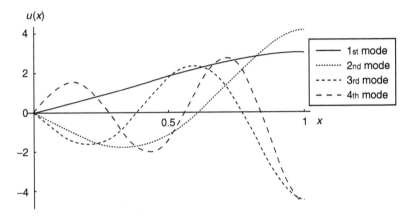

FIGURE 5.23 Mode shapes corresponding to four lowest natural frequencies of tapered fixed-free bar for ε=1.5.

$$L_1 w = 2 \frac{dw}{dx} \tag{w}$$

A perturbation expansion for each mode of the bar's displacement is assumed as

$$W_k(x,t) = W_{k,0}(x,t) + \varepsilon w_{k,1}(x,t) + \dots \tag{x}$$

The perturbations in natural frequencies and mode shapes due to the variation in cross-section of the bar are perturbations from those of a uniform bar fixed at $x = 0$ and free at $x = 1$. The nondimensional natural frequencies of the uniform fixed-free bar are obtained by solving

$$-\frac{d^2 w}{dx^2} = \lambda w \tag{y}$$

subject to boundary conditions of Equation k and Equation l. The natural frequencies and normalized mode shapes for this problem are

$$\omega_k = (2k-1)\frac{\pi}{2} \tag{z}$$

$$w_k = \sqrt{2} \sin\left[(2k-1)\frac{\pi}{2}x\right] \tag{aa}$$

The first-order eigenvalue perturbations are obtained using Equation k of Example 5.5,

$$\lambda_{k,1} = (\mathbf{L}_1 w_k, w_k) \tag{bb}$$

where the inner product is the standard inner product, the inner product for orthogonality of mode shapes for the uniform bar. The eigenvalue perturbations are calculated by

$$\lambda_{k,1} = 2\int_0^1 w_k(x)\frac{dw_k}{dx}dx$$

$$= 2\int_0^1 \left\{\sqrt{2}\sin\left[(2k-1)\frac{\pi}{2}x\right]\right\}\left\{\sqrt{2}(2k-1)\frac{\pi}{2}\cos\left[(2k-1)\frac{\pi}{2}x\right]\right\}dx$$

$$= 2\frac{\pi(2k-1)}{2}\int_0^1 \sin[(2k-1)\pi x]dx$$

$$= \cos[(2k-1)\pi x]\Big|_{x=0}^{x=1} = 2 \tag{cc}$$

The first-order approximations for the eigenvalues of the nonuniform bar are

$$\lambda_k = \left[(2k-1)\frac{\pi}{2}\right]^2 + \frac{1}{2}\varepsilon \tag{dd}$$

The natural frequency approximations are

$$\omega_k = (2k-1)\frac{\pi}{2} + \frac{2}{\pi(2k-1)}\varepsilon + O(\varepsilon^2) \tag{ee}$$

The mode shape perturbations are obtained using Equation o of Example 5.5 by

$$w_{m,1} = \sum_{\substack{k=1 \\ k\neq m}}^{\infty} \frac{(\mathbf{L}_1 w_m, w_k)}{\lambda_m - \lambda_k}w_k \tag{ff}$$

where

$$(\mathbf{L}_1 w_m, w_k) = 2 \int_0^1 w_k(x) \frac{dw_m}{dx} dx$$

$$= 2 \int_0^1 \left\{ \sqrt{2} \sin\left[(2k-1)\frac{\pi}{2} x \right] \right\}$$

$$\times \left\{ \sqrt{2}(2m-1)\frac{\pi}{2} \cos\left[(2m-1)\frac{\pi}{2} x \right] \right\} dx$$

$$= \pi(2m-1) \int_0^1 \left\{ \sin[(k+m-1)\pi x] + \sin[(k-m)\pi x] \right\} dx \quad \text{(gg)}$$

$$= -(2m-1) \left\{ \frac{1}{k+m-1} \cos[(k+m-1)\pi x] \right.$$

$$\left. + \frac{1}{k-m} \cos[(k-m)\pi x] \right\} \Big|_{x=0}^{x=1}$$

$$= \frac{(2m-1)}{(k+m-1)(k-m)} \{(k-m)\cos[(k+m-1)\pi$$

$$+ (k+m-a)\cos[(k-m)\pi - (2k+1)] \}$$

Substitution into Equation ff leads to

$$w_{m,1} = \sum_{\substack{k=1 \\ k \neq m}}^{\infty} \frac{-2\sqrt{2}(2m-1)}{\pi(k+m-1)(k-m)^2} \{(k-m)\cos[(k+m-1)\pi$$

$$+ (k+m-a)\cos[(k-m)\pi - (2k+1)] \} \sin\left[(2k-1)\frac{\pi}{2} x \right] \quad \text{(hh)}$$

The approximation for the mode shape corresponding to the mth mode of the perturbed system is

$$w_m(x) = \sqrt{2} \sin\left[(2k-1)\frac{\pi}{2} x \right] + \varepsilon w_{m,1}(x) \quad \text{(ii)}$$

The error in the approximation is larger for the larger ε, but in both cases the error is smaller in approximation of higher frequencies. Comparisons of the exact mode shapes and the mode shapes obtained using the perturbation method are shown in Figure 5.24 for $\varepsilon = 0.1$ and Figure 5.25 for $\varepsilon = 0.5$.

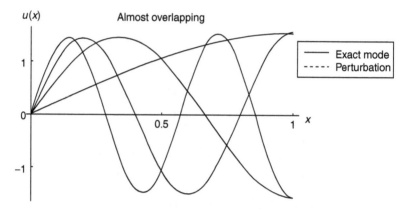

FIGURE 5.24 comparison between exact and approximate mode shapes for system of example 5.14 with ε=0.1.

The agreement between the perturbation solution and exact solution is excellent for e=0.1, but as expected the error increases with increasing and loading to perceptible error for e=0.5.

5.11 EULER–BERNOULLI BEAMS

The nondimensional differential equation for the free transverse vibrations of an Euler–Bernoulli beam is

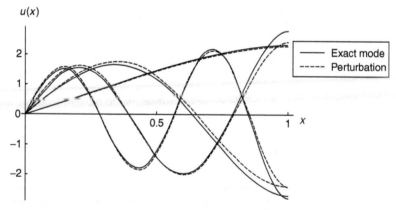

FIGURE 5.25 Comparison between exact and approximate mode shapes for system of example 5.14 with ε=0.5.

$$\frac{\partial^2}{\partial x^2}\left[\alpha(x)\frac{\partial^2 u}{\partial x^2}\right] + \beta(x)\frac{\partial^2 u}{\partial t^2} = 0 \tag{5.123}$$

Application of the normal mode solution to Equation 5.123 leads to

$$\frac{d^2}{dx^2}\left[\alpha(x)\frac{d^2 w}{dx^2}\right] - \omega^2\beta(x)w = 0 \tag{5.124}$$

The kinetic energy inner product for the Euler–Bernoulli beam is

$$(f,g)_M = \int_0^1 f(x)g(x)\beta(x)dx \tag{5.125}$$

It is shown in Section 4.10 that the stiffness operator is self-adjoint with respect to the kinetic energy inner product and the potential energy inner product.

For a uniform beam, $\alpha(x) = \beta(x) = 1$,

$$\frac{d^4 w}{dx^4} - \omega^2 w = 0 \tag{5.126}$$

A solution of Equation 5.126 is assumed as

$$w(x) = e^{\gamma x} \tag{5.127}$$

Substitution of Equation 5.127 into Equation 5.126 leads to

$$\gamma^4 = \omega^2$$
$$\gamma = \pm\sqrt{\omega}, \ \pm i\sqrt{\omega} \tag{5.128}$$

The general solution of Equation 5.126 is

$$w(x) = C_1 \cos\left(\sqrt{\omega}x\right) + C_2 \sin\left(\sqrt{\omega}x\right) + C_3 \cosh\left(\sqrt{\omega}x\right)$$
$$+ C_4 \sinh\left(\sqrt{\omega}x\right) \tag{5.129}$$

The natural frequencies and mode shapes are obtained through application of appropriate boundary conditions as illustrated in the following examples.

Example 5.15. Determine the natural frequencies and normalized mode shapes for a beam that is fixed at each end.

Solution: The boundary conditions for a fixed–fixed beam are

$$u(0,t) = 0 \tag{a}$$

$$\frac{\partial u}{\partial x}(0,t) = 0 \tag{b}$$

$$u(1,t) = 0 \tag{c}$$

$$\frac{\partial u}{\partial x}(1,t) = 0 \tag{d}$$

which after application of the normal mode solution becomes

$$w(0) = 0 \tag{e}$$

$$\frac{dw}{dx}(0) = 0 \tag{f}$$

$$w(1) = 0 \tag{g}$$

$$\frac{dw}{dx}(1) = 0 \tag{h}$$

Sequential application of Equation e through Equation h to Equation 5.129 leads to

$$C_1 + C_3 = 0 \tag{i}$$

$$\sqrt{\omega}C_2 + \sqrt{\omega}C_4 = 0 \tag{j}$$

$$C_1 \cos\left(\sqrt{\omega}\right) + C_2 \sin\left(\sqrt{\omega}\right) + C_3 \cosh\left(\sqrt{\omega}\right) + C_4 \sinh\left(\sqrt{\omega}\right) = 0 \tag{k}$$

$$\begin{aligned}-\sqrt{\omega}C_1 \sin\left(\sqrt{\omega}\right) + \sqrt{\omega}C_2 \cos\left(\sqrt{\omega}\right) + C_3\sqrt{\omega} \sinh\left(\sqrt{\omega}\right) \\ + \sqrt{\omega}C_4 \cosh\left(\sqrt{\omega}\right) = 0\end{aligned} \tag{l}$$

Equation i through Equation l are summarized in matrix form as

$$
\begin{bmatrix}
1 & 0 & 1 & 0 \\
0 & 1 & 0 & 1 \\
\cos(\sqrt{\omega}) & \sin(\sqrt{\omega}) & \cosh(\sqrt{\omega}) & \sinh(\sqrt{\omega}) \\
-\sin(\sqrt{\omega}) & \cos(\sqrt{\omega}) & \sinh(\sqrt{\omega}) & \cosh(\sqrt{\omega})
\end{bmatrix}
\begin{bmatrix}
C_1 \\ C_2 \\ C_3 \\ C_4
\end{bmatrix}
=
\begin{bmatrix}
0 \\ 0 \\ 0 \\ 0
\end{bmatrix}
\qquad (m)
$$

Since the right-hand side vector in Equation m is the zero vector, a nontrivial solution of Equation m exists if and only if the determinant of the coefficient matrix is zero. Such a requirement leads to

$$
\begin{vmatrix}
1 & 0 & 1 & 0 \\
0 & 1 & 0 & 1 \\
\cos(\sqrt{\omega}) & \sin(\sqrt{\omega}) & \cosh(\sqrt{\omega}) & \sinh(\sqrt{\omega}) \\
-\sin(\sqrt{\omega}) & \cos(\sqrt{\omega}) & \sinh(\sqrt{\omega}) & \cosh(\sqrt{\omega})
\end{vmatrix}
= 0 \qquad (n)
$$

There are a number of ways to evaluate the determinant in Equation n. Expansion by the first row leads to

$$
\begin{vmatrix}
1 & 0 & 1 \\
\sin(\sqrt{\omega}) & \cosh(\sqrt{\omega}) & \sinh(\sqrt{\omega}) \\
\cos(\sqrt{\omega}) & \sinh(\sqrt{\omega}) & \cosh(\sqrt{\omega})
\end{vmatrix}
+
\begin{vmatrix}
0 & 1 & 1 \\
\cos(\sqrt{\omega}) & \sin(\sqrt{\omega}) & \sinh(\sqrt{\omega}) \\
-\sin(\sqrt{\omega}) & \cos(\sqrt{\omega}) & \cosh(\sqrt{\omega})
\end{vmatrix}
= 0
$$

$$
\begin{vmatrix}
\cosh(\sqrt{\omega}) & \sinh(\sqrt{\omega}) \\
\sinh(\sqrt{\omega}) & \cosh(\sqrt{\omega})
\end{vmatrix}
+
\begin{vmatrix}
\sin(\sqrt{\omega}) & \cosh(\sqrt{\omega}) \\
\cos(\sqrt{\omega}) & \sinh(\sqrt{\omega})
\end{vmatrix}
-
\begin{vmatrix}
\cos(\sqrt{\omega}) & \sinh(\sqrt{\omega}) \\
-\sin(\sqrt{\omega}) & \cosh(\sqrt{\omega})
\end{vmatrix}
$$

$$
+
\begin{vmatrix}
\cos(\sqrt{\omega}) & \sin(\sqrt{\omega}) \\
-\sin(\sqrt{\omega}) & \cos(\sqrt{\omega})
\end{vmatrix}
= 0
$$

$$
\cosh^2(\sqrt{\omega}) - \sinh^2(\sqrt{\omega}) + \sin(\sqrt{\omega})\sinh(\sqrt{\omega})
$$

$$
-\cos(\sqrt{\omega})\cosh(\sqrt{\omega}) - \cos(\sqrt{\omega})\cosh(\sqrt{\omega})
$$

$$
-\sin(\sqrt{\omega})\sinh(\sqrt{\omega}) + \cos^2(\sqrt{\omega}) + \sin^2(\sqrt{\omega}) = 0
$$

$$
(o)
$$

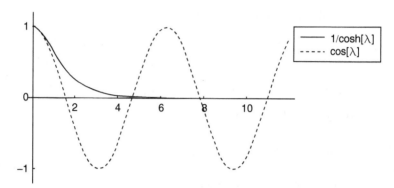

FIGURE 5.26 Natural frequencies of uniform fixed-fixed beam correspond to the squares of the values where the graphs intersect

After simplification, Equation o becomes

$$\cos\left(\sqrt{\omega}\right)\cosh\left(\sqrt{\omega}\right) = 1 \tag{p}$$

The natural frequencies are the values of ω which solve Equation p. The graphs of $\cos(\lambda)$ and $(1/\cosh(\lambda))$ are plotted simultaneously in Figure 5.26. Their points of intersection are the square roots of the natural frequencies. As λ gets large $(1/\cosh(\lambda))$ approaches zero and the points where the two graphs intersect approach an odd multiple of $\pi/2$,

$$\lim_{k\to\infty} \omega_k = \left[(2k+1)\frac{\pi}{2}\right]^2 \tag{q}$$

The first four natural frequencies obtained by solving Equation p are $\omega_1 = 22.37$, $\omega_2 = 61.67$, $\omega_3 = 120.9$ and $\omega_4 = 199.9$. These are nondimensional frequencies. The dimensional frequencies are obtained by multiplying the nondimensional frequencies by $\sqrt{EI/\rho AL^4}$.

Mode shapes are obtained using Equation i through Equation l. Equation i implies that $C_3 = -C_1$, while Equation j implies that $C_4 = -C_2$. Substitution of these results into Equation k leads to

$$C_2 = \left[\frac{\cos\left(\sqrt{\omega}\right) - \cosh\left(\sqrt{\omega}\right)}{\sinh\left(\sqrt{\omega}\right) - \sin\left(\sqrt{\omega}\right)}\right] C_1 \tag{r}$$

Substitution of Equation r into Equation 5.129 leads to

$$w_k(x) = C_k \{ \cos(\sqrt{\omega_k}x) - \cosh(\sqrt{\omega_k}x) + \alpha_k [\sin(\sqrt{\omega_k}x) - \sinh(\sqrt{\omega_k}x)] \}$$

(s)

where

$$\alpha_k = \frac{\cosh(\sqrt{\omega_k}) - \cos(\sqrt{\omega_k})}{\sinh(\sqrt{\omega_k}) - \sin(\sqrt{\omega_k})}$$

(t)

The mode shape is normalized by requiring

$$\int_0^1 [w_k(x)]^2 \, dx = 1$$

$$\int_0^1 C_k^2 \{ \cos(\sqrt{\omega_k}x) - \cosh(\sqrt{\omega_k}x) + \alpha_k [\sin(\sqrt{\omega_k}x) - \sinh(\sqrt{\omega_k}x)] \}^2 \, dx = 1$$

(u)

Normalized mode shapes corresponding to the four lowest natural frequencies are illustrated in Figure 5.27.

A similar technique is used to derive the natural frequencies, mode shapes, and normalized mode shapes for beams with other end conditions. The mode shape equations and the four lowest natural frequencies for each case are

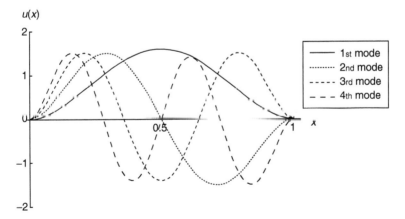

FIGURE 5.27 Normalized mode shapes corresponding to the first four natural frequencies of a fixed-fixed beam

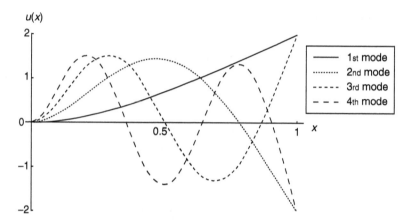

FIGURE 5.28 Mode shapes corresponding to the first four natural frequencies of a uniform fixed-free beam.

given. Mode shape plots for fixed-free and fixed-pinned conditions are given in Figure 5.28 and Figure 5.29 respectively.

Table 5.2 shows that the characteristic equation for the free–free beam is the same as that for a fixed–fixed beam, however the lowest frequency for the free–free beam is listed as $\omega = 0$ while the lowest frequency for the fixed–fixed beam is listed as $\omega = 22.37$. Whereas the characteristic equations are the same for both beams, the boundary conditions and the four equations,

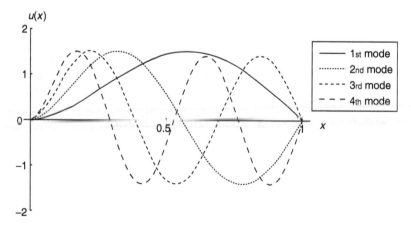

FIGURE 5.29 Mode shapes corresponding to the first four natural frequencies of a uniform fixed-pinned beam.

TABLE 5.2

Characteristic Equations for Euler-Bernoulli Beams with a Variety of End Conditions

End Conditions	Characteristic Equation	Lowest Natural Frequencies
Fixed-fixed	$\cos(\sqrt{\omega})\cosh(\sqrt{\omega}) = 1$	$\omega_1 = 22.37$
		$\omega_2 = 61.66$
		$\omega_3 = 120.9$
		$\omega_4 = 199.9$
		$\omega_5 = 298.6$
Pinned-pinned	$\sin(\sqrt{\omega}) = 1$	$\omega_1 = 9.870$
		$\omega_2 = 39.48$
		$\omega_3 = 88.83$
		$\omega_4 = 157.9$
		$\omega_5 = 246.7$
Fixed-free	$\cos(\sqrt{\omega})\cosh(\sqrt{\omega}) = -1$	$\omega_1 = 3.514$
		$\omega_2 = 22.03$
		$\omega_3 = 61.70$
		$\omega_4 = 120.9$
		$\omega_5 = 199.9$
Free-free	$\cos(\sqrt{\omega})\cosh(\sqrt{\omega}) = 1$	$\omega_1 = 0$
		$\omega_2 = 22.37$
		$\omega_3 = 61.66$
		$\omega_4 = 120.9$
		$\omega_5 = 199.9$
Fixed-pinned	$\tan(\sqrt{\omega}) = \tanh(\sqrt{\omega})$	$\omega_1 = 15.42$
		$\omega_2 = 49.96$
		$\omega_3 = 104.2$
		$\omega_4 = 178.3$
		$\omega_5 = 272.0$
Pinned-free	$\tan(\sqrt{\omega}) = \tanh(\sqrt{\omega})$	$\omega_1 = 0$
		$\omega_2 = 15.42$
		$\omega_3 = 49.96$
		$\omega_4 = 104.2$
		$\omega_5 = 178.3$

similar to those of Equation i through Equation l of Example 5.15, are different. Indeed $\omega = 0$ is a solution of Equation p of Example 5.15, however its use in Equation i through Equation l leads to the trivial solution as the only possible solution.

Application of the boundary conditions for a uniform free–free beam, $\partial^2 \mu/\partial x^2(0,t) = 0, \partial^3 \mu/\partial x^3(0,t) = 0, \partial^2 \mu/\partial x^2(1,t) = 0$, and $\partial^3 \mu/\partial x^3(1,t) = 0$ to

Equation 5.129 leads to

$$
\begin{bmatrix}
-\omega & 0 & \omega & 0 \\
0 & -\omega^{\frac{3}{2}} & 0 & \omega^{\frac{3}{2}} \\
-\omega\cos\left(\sqrt{\omega}\right) & -\omega\sin\left(\sqrt{\omega}\right) & \omega\cosh\left(\sqrt{\omega}\right) & \omega\sinh\left(\sqrt{\omega}\right) \\
\omega^{\frac{3}{2}}\sin\left(\sqrt{\omega}\right) & -\omega^{\frac{3}{2}}\cos\left(\sqrt{\omega}\right) & \omega^{\frac{3}{2}}\sinh\left(\sqrt{\omega}\right) & \omega^{\frac{3}{2}}\cosh\left(\sqrt{\omega}\right)
\end{bmatrix}
\begin{bmatrix} C_1 \\ C_2 \\ C_3 \\ C_4 \end{bmatrix}
$$

$$
= \begin{bmatrix} 0 \\ 0 \\ 0 \\ 0 \end{bmatrix}
$$

$$(5.130)$$

If $\omega \neq 0$, evaluation of the determinant of the matrix in Equation 5.130 leads to the same characteristic equation from which the natural frequencies of a fixed–fixed beam are determined. Clearly, if $\omega = 0$, Equation 5.130 cannot be solved for the arbitrary constants.

The general form of the solution of Equation 5.126 corresponding to $\omega = 0$ is

$$w(x) = C_1 + C_2 x + C_3 x^2 + C_4 x^3 \qquad (5.131)$$

For a fixed-fixed beam, $w(0) = 0 \Rightarrow C_1 = 0$, $(dw/dx)(0) = 0 \Rightarrow C_2 = 0$, and $(d^3 w/dx^3)(1) = 0 \Rightarrow C_4 = 0$ and then $(d^2 w/dx^2)(1) = 0 \Rightarrow C_3 = 0$. Thus, for a fixed-fixed beam, $\omega = 0$ leads to only the trivial solution and thus is not a natural frequency.

When applying the boundary conditions for a free–free beam to Equation 5.131, $(d^3 w/dx^3)(1) = 0 \Rightarrow C_4 = 0$, and then $(d^2 w/dx^2)(1) = 0 \Rightarrow C_3 = 0$. Application of $w(0) = 0$ and $(dw/dx)(0) = 0$ both lead to the identity $0 = 0$. Thus, arbitrary choice of both C_1 and C_2 lead to nontrivial solutions. The general mode shape is of the form

$$w(x) = C_1 + C_2 x \qquad (5.132)$$

The free-free beam is unrestrained and $\lambda = 0$ is an eigenvalue of multiplicity 2. This implies there are two rigid body modes. One corresponds to a rigid body translation of the beam while the second corresponds to a rigid-body rotation of the beam.

The two mode shapes suggested by Equation 5.132 are $\hat{w}_{0,1} = 1$ and $\hat{w}_{0,2} = x$. It is desirable to render these orthonormal in order to use in applications of the expansion theorem. Thus the Gram–Schmidt process of Section 3.12 is employed. Since the beam is uniform, the kinetic energy inner product is the same as the standard inner product on $C^4[0,1]$. It can be shown that $\|\hat{w}_{0,1}\| = (\hat{w}_{0,1}, \hat{w}_{0,1})^{1/2} = 1$. Thus

$$w_{0,1} = 1 \tag{5.133}$$

The Gram–Schmidt theorem is used to obtain a function orthogonal to $\hat{w}_{0,1}$, as

$$\tilde{w}_{0,2} = \hat{w}_{0,2} - (\hat{w}_{0,2}, w_{0,1}) w_{0,1}$$

$$= x - \left(\int_0^1 x(1)dx \right)(1) \tag{5.134}$$

$$= x - \frac{1}{2}$$

The function in Equation 5.134 is normalized by

$$w_{0,2} = \frac{\tilde{w}_{0,2}}{\|\tilde{w}_{0,2}\|} \tag{5.135}$$

where

$$\|\tilde{w}_{0,2}\| = \left[\int_0^1 \left(x - \frac{1}{2} \right)^2 dx \right]^{1/2} = \frac{1}{2\sqrt{3}} \tag{5.136}$$

Thus the second mode shape corresponding to $\omega = 0$, which is orthonormal to the $w_{0,1} = 1$, is

$$w_{0,2}(x) = \sqrt{3}(2x - 1) \tag{5.137}$$

These mode shapes are illustrated in Figure 5.30.

The lowest natural frequencies for a beam with an attached mass are plotted in Figure 5.31 as a function of η the ratio of the end mass to the mass of the

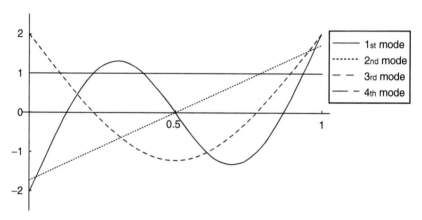

FIGURE 5.30 Mode shapes for free-free uniform beam.

beam. As η increases, the natural frequencies decrease as the inertia of the beam becomes larger. The effect is more pronounced for the higher modes.

Example 5.16. The differential equation governing the transverse displacement, $w(x, t)$, of a beam on an elastic foundation, as illustrated in Figure 5.32, is

$$EI\frac{\partial^4 w}{\partial x^4} + kw + \rho A\frac{\partial^2 w}{\partial t^2} = 0 \qquad (a)$$

Determine the natural frequencies and mode shapes for a simply supported beam on an elastic foundation.

Solution: Use of the nondimensional variables of Equation 4.112 in Equation a leads to the following nondimensional form of the partial differential equation, as

$$\frac{\partial^4 w}{\partial x^4} \mid \eta w \mid \frac{\partial^2 w}{\partial t^2} - 0 \qquad (b)$$

where

$$\eta = \frac{kL^3}{EI} \qquad (c)$$

A normal mode solution, $w(x)e^{i\omega t}$, when substituted into Equation b leads to

$$\frac{d^4 w}{dx^4} + \eta w - \omega^2 w = 0 \qquad (d)$$

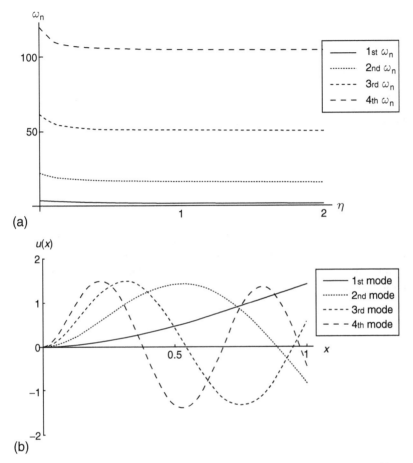

FIGURE 5.31 (a) Natural frequencies of a uniform beam fixed at $x = 0$ with an attached mass at $x = 1$. (b) Mode shapes for fixed-fixed uniform beam with end mass for $\eta = 0.25$.

Assuming a solution of the form of Equation 5.127 for Equation d leads to

$$\gamma^4 - (\omega^2 - \eta) = 0$$
$$\lambda^4 - \nu^2 = 0$$

(e)

where

$$\nu = \sqrt{\omega^2 - \eta}$$

(f)

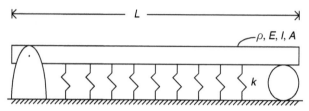

FIGURE 5.32 Beam on an elastic foundation.

The solutions of Equation f are of the form $\lambda = \pm\sqrt{\nu}, \pm i\sqrt{\nu}$ and the general solution to Equation d is

$$w(x) = C_1 \cos(\sqrt{\nu}x) + C_2 \sin(\sqrt{\nu}x) + C_3 \cosh(\sqrt{\nu}x) + C_4 \sinh(\sqrt{\nu}x) \quad \text{(g)}$$

Equation g is of the form of Equation 5.129 with ω replaced by ν. Thus the solutions for comparable boundary conditions are comparable. From Table 5.2, the natural frequencies of a pinned–pinned beam on an elastic foundation are

$$\nu_k = (k\pi)^2$$
$$\omega_k^2 - \eta = (k\pi)^4 \quad \text{(h)}$$
$$\omega_k = \sqrt{(k\pi)^4 + \eta}$$

The normalized mode shapes are

$$w_k(x) = \sqrt{2}\sin(k\pi x) \quad \text{(i)}$$

Equation h shows that the effect of the elastic foundation on the natural frequencies is small for small η.

Example 5.17. The nondimensional differential equation for a uniform beam with a constant axial load is derived in Example 4.14, as

$$\varepsilon\frac{\partial^4 w}{\partial x^4} - \frac{\partial^2 w}{\partial x^2} + \frac{\partial^2 w}{\partial t^2} = 0 \quad \text{(a)}$$

Determine the natural frequencies and normalized mode shapes for a uniform beam with a constant axial load for (a) a fixed-fixed beam (b) a fixed-free beam.

Solution: The normal mode solution when applied to Equation a leads to

$$\varepsilon \frac{d^4w}{dx^4} - \frac{d^2w}{dx^2} - \omega^2 w = 0 \tag{b}$$

A solution for Equation b is sought of the form $w(x) = e^{rx}$, which when substituted into Equation b leads to

$$\varepsilon \gamma^4 - \gamma^2 - \omega^2 = 0 \tag{c}$$

The quadratic formula is used to solve Equation c leading to

$$\gamma = \pm \left[\frac{1}{2\varepsilon} \left(1 \pm \sqrt{1 + 4\varepsilon\omega^2} \right) \right]^{1/2} \tag{d}$$

Evaluation of Equation d shows that there are two real solutions of Equation c and two imaginary solutions. The real solutions are

$$\gamma = \pm \left[\frac{1}{2\varepsilon} \left(1 + \sqrt{1 + 4\varepsilon\omega^2} \right) \right]^{1/2} = \pm \mu \tag{e}$$

while the imaginary solutions are

$$\gamma = \pm i \left[\frac{1}{2\varepsilon} \left(1 - \sqrt{1 + 4\varepsilon\omega^2} \right) \right]^{1/2} = \pm i\nu \tag{f}$$

The general solution of Equation b can be written as

$$w(x) = C_1 \cosh(\mu x) + C_2 \sinh(\mu x) + C_3 \cos(\nu x) + C_4 \sin(\nu x) \tag{g}$$

4. An axial load in a fixed-free beam may be caused by thermal expansion. The boundary conditions for a fixed–fixed beam are

$$w(0) = 0 \tag{h}$$

$$\frac{dw}{dx}(0) = 0 \tag{i}$$

$$w(1) = 0 \tag{j}$$

$$\frac{dw}{dx}(1) = 0 \qquad \text{(k)}$$

Application of Equation h through Equation k to Equation g leads to

$$\begin{bmatrix} 1 & 0 & 1 & 0 \\ 0 & \mu & 0 & \nu \\ \cosh(\mu) & \sinh(\mu) & \cos(\nu) & \sin(\nu) \\ \mu\sinh(\mu) & \mu\cosh(\mu) & -\nu\sin(\nu) & \nu\cos(\nu) \end{bmatrix} \begin{bmatrix} C_1 \\ C_2 \\ C_3 \\ C_4 \end{bmatrix} = \begin{bmatrix} 0 \\ 0 \\ 0 \\ 0 \end{bmatrix} \qquad \text{(l)}$$

Setting the determinant of the coefficient matrix in Equation l to zero leads to

$$2\mu\nu - 2\mu\nu\cos(\nu)\cosh(\mu) + (\mu^2 - \nu^2)\sin(\nu)\sinh(\mu) = 0 \qquad \text{(m)}$$

Plots of the first five frequencies for various values ε are illustrated in 26. Figure 5.32. For small ε the effects of axial stress is larger than bending stress. The natural frequencies are close to those for a fixed-fixed bar. As ε increases the effects of bending stresses are more pronounced, but the axial stress still has a significant effect on the natural frequencies.

b. The boundary conditions for a fixed-free beam with an axial load are

$$w(0) = 0 \qquad \text{(n)}$$

$$\frac{dw}{dx}(0) = 0 \qquad \text{(o)}$$

$$\frac{d^2w}{dx^2}(1) = 0 \qquad \text{(p)}$$

$$\varepsilon\frac{d^3w}{dx^3}(1) = \frac{dw}{dx}(1) \qquad \text{(q)}$$

Application of Equation n through Equation q to Equation g leads to

$$
\begin{bmatrix}
1 & 0 & 1 & 0 \\
0 & \mu & 0 & \nu \\
\mu^2\cosh(\mu) & \mu^2\sinh(\mu) & -\nu^2\cos(\nu) & -\nu^2\sin(\nu) \\
(\varepsilon\mu^3-\mu)\sinh(\mu) & (\varepsilon\mu^3-\mu)\cosh(\mu) & (\varepsilon\nu^3+\nu)\sin(\nu) & -(\varepsilon\nu^3+\nu)\cos(\nu)
\end{bmatrix}
$$

$$
\times
\begin{bmatrix}
C_1 \\
C_2 \\
C_3 \\
C_4
\end{bmatrix}
=
\begin{bmatrix}
0 \\
0 \\
0 \\
0
\end{bmatrix}
\tag{r}
$$

Setting the determinant of the coefficient matrix of Equation r to zero leads to

$$
-\mu^2+\varepsilon\mu^4+\nu^2+\varepsilon\nu^2[-\nu^2+\mu^2(1+2\varepsilon\nu^2)]\ \cos(\nu)\ \cosh(\mu)-\mu\nu(-2
$$
$$
+\varepsilon\mu^2-\varepsilon\nu^2)\ \sin(\nu)\ \sinh(\mu)
\tag{s}
$$

The first five natural frequencies for several values of ε are given in Figure 5.33b.

5.12 REPEATED STRUCTURES

Vibrations of a set of beams connected by flexible elastic layers leads to the set of coupled partial differential equations of Equation 4.115. A normal-mode

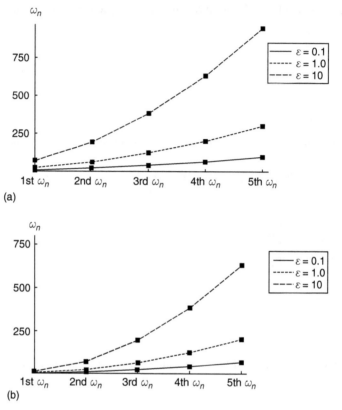

FIGURE 5.33 (a) Natural frequencies of stretched fixed-fixed beam for various values of ε. (b) Natural frequencies of stretched fixed-free beam for various values of ε.

solution is assumed as

$$\mathbf{w}(x,t) = \mathbf{f}(x)e^{i\omega t} = \begin{bmatrix} f_1(x) \\ f_2(x) \\ f_3(x) \\ \vdots \\ f_{n-1}(x) \\ f_n(x) \end{bmatrix} e^{i\omega t} \qquad (5.138)$$

Substitution of the normal mode solution into Equation 4.115 leads to

$$
\begin{bmatrix}
\dfrac{d^4}{dx^4} + \kappa_0 + \kappa_1 & -\kappa_1 & 0 & 0 & \cdots & 0 \\[2ex]
-\kappa_1 & \phi_2\dfrac{d^4}{dx^4} + \kappa_1 + \kappa_2 & -\kappa_2 & 0 & \cdots & 0 \\[2ex]
0 & -\kappa_2 & \phi_3\dfrac{d^4}{dx^4} + \kappa_2 + \kappa_3 & -\kappa_3 & \cdots & 0 \\[2ex]
\vdots & \vdots & \vdots & \vdots & \ddots & \vdots \\[2ex]
0 & 0 & 0 & 0 & \cdots & -\kappa_{n-1} \\[2ex]
0 & 0 & 0 & 0 & \cdots & \phi_n\dfrac{d^4}{dx^4} + \kappa_{n-1} + \kappa_n
\end{bmatrix}
$$

$$
\times
\begin{bmatrix} f_1 \\ f_2 \\ f_3 \\ \vdots \\ f_{n-1} \\ f_6 \end{bmatrix}
+
\begin{bmatrix}
-\omega^2 & 0 & 0 & 0 & \cdots & 0 \\
0 & -\mu_2\omega^2 & 0 & 0 & \cdots & 0 \\
0 & 0 & -\mu_3\omega^2 & 0 & \cdots & 0 \\
0 & 0 & 0 & -\mu_4\omega^2 & \cdots & 0 \\
\vdots & \vdots & \vdots & \vdots & \ddots & \vdots \\
0 & 0 & 0 & 0 & \cdots & -\mu_n\omega^2
\end{bmatrix}
\begin{bmatrix} f_1 \\ f_2 \\ f_3 \\ \vdots \\ f_{n-1} \\ f_6 \end{bmatrix}
=
\begin{bmatrix} 0 \\ 0 \\ 0 \\ 0 \\ 0 \\ 0 \end{bmatrix}
$$

$$(5.139)$$

Let ϕ_k be the kth mode shape for a single uniform beam with the same type of supports as the periodic structure and let $\tilde{\omega}_k$ be the nondimensional natural frequency corresponding to the kth mode. The mode shapes and several natural frequencies are given in Table 5.2 The differential equation for which ϕ_k is a nontrivial solution is

$$\frac{d^4\phi_k}{dx^4} = \tilde{\omega}_k^2\phi_k \qquad (5.140)$$

The solution of Equation 5.139 is assumed as

$$
\mathbf{f} = \mathbf{a}_k \phi_k =
\begin{bmatrix}
a_1 \\
a_2 \\
a_3 \\
a_4 \\
\vdots \\
a_n
\end{bmatrix}
\phi_k(x)
\tag{5.141}
$$

where $\mathbf{a}_k$ is a vector of constants. Substitution of Equation 5.141 into Equation 5.139 and using Equation 5.140 leads to

$$
\begin{bmatrix}
\tilde{\omega}_k^2 + \kappa_0 + \kappa_1 & -\kappa_1 & 0 & 0 & \cdots & 0 \\
-\kappa_1 & \phi_2\tilde{\omega}_k^2 + \kappa_1 + \kappa_2 & -\kappa_2 & 0 & \cdots & 0 \\
0 & -\kappa_2 & \phi_3\tilde{\omega}_k^2 + \kappa_2 + \kappa_3 & -\kappa_3 & \cdots & 0 \\
\vdots & \vdots & \vdots & \vdots & \ddots & \vdots \\
0 & 0 & 0 & 0 & \cdots & -\kappa_{n-1} \\
0 & 0 & 0 & 0 & \cdots & \phi_n\tilde{\omega}_k^2 + \kappa_{n-1} + \kappa_n
\end{bmatrix}
$$

$$
\times
\begin{bmatrix}
a_1 \\
a_2 \\
a_3 \\
\vdots \\
a_{n-1} \\
a_n
\end{bmatrix}
= \omega^2
\begin{bmatrix}
1 & 0 & 0 & 0 & \cdots & 0 \\
0 & \mu_2 & 0 & 0 & \cdots & 0 \\
0 & 0 & \mu_3 & 0 & \cdots & 0 \\
\vdots & \vdots & \vdots & \vdots & \ddots & \vdots \\
0 & 0 & 0 & 0 & \ddots & 0 \\
0 & 0 & 0 & 0 & \cdots & \mu_n
\end{bmatrix}
\begin{bmatrix}
a_1 \\
a_2 \\
a_3 \\
\vdots \\
a_{n-1} \\
a_n
\end{bmatrix}
$$

$$
\tag{5.142}
$$

A general matrix formulation of Equation 5.142 is

$$
\mathbf{K}_k \mathbf{a} = \omega^2 \mathbf{M} \mathbf{a}
\tag{5.143}
$$

where

$$
\mathbf{K}_k =
\begin{bmatrix}
\tilde{\omega}_k^2 + \kappa_0 + \kappa_1 & -\kappa_1 & 0 & 0 & \cdots & 0 \\
-\kappa_1 & \phi_2 \tilde{\omega}_k^2 + \kappa_1 + \kappa_2 & -\kappa_2 & 0 & \cdots & 0 \\
0 & -\kappa_2 & \phi_3 \tilde{\omega}_k^2 + \kappa_2 + \kappa_3 & -\kappa_3 & \cdots & 0 \\
\vdots & \vdots & \vdots & \vdots & \ddots & \vdots \\
0 & 0 & 0 & 0 & \cdots & -\kappa_{n-1} \\
0 & 0 & 0 & 0 & \cdots & \phi_n \tilde{\omega}_k^2 + \kappa_{n-1} + \kappa_n
\end{bmatrix}
$$

$$(5.144)$$

$$
\mathbf{M} =
\begin{bmatrix}
1 & 0 & 0 & 0 & \cdots & 0 \\
0 & \mu_2 & 0 & 0 & \cdots & 0 \\
0 & 0 & \mu_3 & 0 & \cdots & 0 \\
\vdots & \vdots & \vdots & \vdots & \ddots & \vdots \\
0 & 0 & 0 & 0 & \ddots & 0 \\
0 & 0 & 0 & 0 & \cdots & \mu_n
\end{bmatrix}
$$

$$(5.145)$$

Equation 5.143 may be written in the form of a matrix eigenvalue problem as

$$\mathbf{M}^{-1}\mathbf{K}_k \mathbf{a}_k = \omega^2 \mathbf{a}_k \qquad (5.146)$$

The above discussion leads to the following conclusions:

- The spatial mode shapes $\phi_k(x)$ of the periodic set of Euler–Bernoulli beams are the same as a single Euler–Bernoulli beam.
- For each spatial mode, there are n natural frequencies, $\omega_{k,1} \le \omega_{k,2} \le \ldots \le \omega_{k,n}$.
- The mode shape for each natural frequency is of the form $\mathbf{a}_k \phi_k(x)$, where $\mathbf{a}_k$ is a $n \times 1$ transverse mode shape vector.
- The natural frequencies for each spatial mode are the square roots of the eigenvalues of $\mathbf{M}^{-1}\mathbf{K}_k$ and the transverse mode shape vectors are the corresponding eigenvectors.
- Since $\mathbf{M}$ and $\mathbf{K}_k$ are symmetric matrices, the eigenvalues of $\mathbf{M}^{-1}\mathbf{K}_k$ are all real. It is easy to show that $\mathbf{M}^{-1}\mathbf{K}_k$ is positive definite with respect to a kinetic energy inner product defined on R^n. The mode shape vectors for a specific k are mutually orthogonal with respect to a kinetic energy product, $(\mathbf{a}_{k,i}, \mathbf{a}_{k,j})_M = 0$ for $i \ne j$ where $(\mathbf{f}, \mathbf{g})_M = \mathbf{g}^T \mathbf{M} \mathbf{f}$.
- Each spatial mode corresponds to n distinct natural frequencies. The kth mode of an Euler–Bernoulli beam has $(k-1)$ nodes along the span of the beam. The above implies that there are n frequencies

which lead to modes with $(k-1)$ nodes. That is, the system has n natural frequencies which lead to mode shapes that have the properties of a kth node.

The case when all beams are identical, $\mu_2 = \mu_3 = \ldots = \mu_n = 1$ and $\phi_2 = \phi_3 = \ldots = \phi_n = 1$, is a special case. M is the identity matrix and

$$\mathbf{K}_k = \tilde{\omega}_k^2 \mathbf{I} + \mathbf{K} \tag{5.147}$$

where

$$\mathbf{K} = \begin{bmatrix} \kappa_0 + \kappa_1 & -\kappa_1 & 0 & 0 & \cdots & 0 \\ -\kappa_1 & \kappa_1 + \kappa_2 & -\kappa_2 & 0 & \cdots & 0 \\ 0 & -\kappa_2 & \kappa_2 + \kappa_3 & -\kappa_3 & \cdots & 0 \\ \vdots & \vdots & \vdots & \vdots & \ddots & \vdots \\ 0 & 0 & 0 & 0 & \cdots & -\kappa_{n-1} \\ 0 & 0 & 0 & 0 & \cdots & +\kappa_{n-1} + \kappa_n \end{bmatrix} \tag{5.148}$$

Equation 5.146 is written as

$$(\tilde{\omega}_k^2 \mathbf{I} + \mathbf{K})\mathbf{a}_k = \omega^2 \mathbf{a}_k$$
$$\mathbf{K}\mathbf{a}_k = (\omega^2 - \tilde{\omega}_k^2)\mathbf{a}_k \tag{5.149}$$

Consider the eigenvalue problem

$$\mathbf{K}\mathbf{b} = \lambda\mathbf{b} \tag{5.150}$$

Comparison between Equation 5.138 and Equation 5.139 shows that if $\lambda_1, \lambda_2, \ldots \lambda_n$ are the eigenvalues of $\mathbf{K}$, then the natural frequencies of the beams are

$$\omega_{k,j} = \sqrt{\lambda_j + \tilde{\omega}_k^2} \quad k = 1,2,\ldots \quad j = 1,2,\ldots,n \tag{5.151}$$

If $\mathbf{b}_j$ is an eigenvector of $\mathbf{K}$ corresponding to its eigenvalue λ_j, then $\mathbf{b}_j$ is a discrete mode shape vector corresponding to the natural frequencies $\omega_{k,j}$ $k = 2,\ldots$.

If $\kappa_0 = \kappa_n = 0$, $\mathbf{K}$ corresponds to that of an unrestrained discrete system. Its lowest eigenvalue is zero and thus the natural frequencies of a single beam are the natural frequencies corresponding to $j = 0$ and correspond to a rigid body motion of the beams.

Example 5.18. Consider a set of two fixed–fixed Euler–Bernoulli beams connected by an elastic layer. The spatial mode shapes for a fixed–fixed Euler–Bernoulli beam are

$$\phi_k(x) = \cosh\left(\sqrt{\tilde{\omega}_k}x\right) - \cos\left(\sqrt{\tilde{\omega}_k}x\right)$$

$$-\left(\frac{\cosh\left(\sqrt{\tilde{\omega}_k}\right) - \cos\left(\sqrt{\tilde{\omega}_k}\right)}{\sinh\left(\sqrt{\tilde{\omega}_k}\right) - \sin\left(\sqrt{\tilde{\omega}_k}\right)}\right)\left(\sinh\left(\sqrt{\tilde{\omega}_k}x\right) - \sin\left(\sqrt{\tilde{\omega}_k}x\right)\right) \quad \text{(a)}$$

The characteristic equation for the beam is

$$\cos\left(\sqrt{\tilde{\omega}_k}\right)\cosh\left(\sqrt{\tilde{\omega}_k}\right) = 1 \quad \text{(b)}$$

(a) Determine the natural frequencies for the first five modes and their corresponding mode shapes for the first five spatial modes if both beams are identical, (b) Repeat part (a) for a simply supported beam.

Solution: For two beams, Equation 5.142 is reduced to

$$\begin{bmatrix} \tilde{\omega}_k^2 + \kappa & -\kappa \\ -\kappa & \phi_2\tilde{\omega}_k^2 + \kappa \end{bmatrix}\begin{bmatrix} a_1 \\ a_2 \end{bmatrix} = \omega^2 \begin{bmatrix} 1 & 0 \\ 0 & \mu_2 \end{bmatrix}\begin{bmatrix} a_1 \\ a_2 \end{bmatrix} \quad \text{(c)}$$

The natural frequencies are determined by solving

$$\begin{vmatrix} \tilde{\omega}_k^2 + \kappa - \omega^2 & -\kappa \\ -\kappa & \phi_2\tilde{\omega}_k^2 + \kappa - \mu_2\omega^2 \end{vmatrix} = 0$$

$$(\tilde{\omega}_k^2 + \kappa - \omega^2)(\phi_2\tilde{\omega}_k^2 + \kappa - \mu_2\omega^2) - \kappa^2 = 0$$

$$\mu_2\omega^4 - [\mu_2(\tilde{\omega}_k^2 + \kappa) + \phi_2\tilde{\omega}_k^2 + \kappa]\omega^2 + \phi_2\tilde{\omega}_k^4 + \phi_2\kappa\tilde{\omega}_k^2 + \kappa\tilde{\omega}_k^2 = 0 \quad \text{(d)}$$

The natural frequencies for each mode can be determined by finding the roots of Equation d using the quadratic formula. For each k there are two natural frequencies $\omega_{k,1} \leq \omega_{k,2}$. The mode shape vector corresponding to a natural frequency is

TABLE 5.3
Natural Frequencies and Mode Shapes for Dual Fixed-Fixed Beams with
(a) $\mu = 1$ $\phi = 1$ $\kappa = 1.5$. (b) $\mu = 1.2$ $\phi = 1.1$ $\kappa = 5000$

(a)

K	1	2	3	4	5
$\tilde{\omega}_k$	22.37	61.67	120.90	199.9	298.6
$\omega_{k,1}$	22.37	61.67	120.90	199.9	298.6
$\omega_{k,2}$	22.44	61.70	120.92	199.9	299.6
$X_{k,1}$	$\begin{bmatrix} 1 \\ 1 \end{bmatrix}$	$\begin{bmatrix} 1 \\ 1 \end{bmatrix}$	$\begin{bmatrix} 1 \\ 1 \end{bmatrix}$	$\begin{bmatrix} 1 \\ 1 \end{bmatrix}$	$\begin{bmatrix} 1 \\ 1 \end{bmatrix}$
$X_{k,2}$	$\begin{bmatrix} 1 \\ -1 \end{bmatrix}$	$\begin{bmatrix} 1 \\ -1 \end{bmatrix}$	$\begin{bmatrix} 1 \\ -1 \end{bmatrix}$	$\begin{bmatrix} 1 \\ -1 \end{bmatrix}$	$\begin{bmatrix} 1 \\ -1 \end{bmatrix}$

(b)

K	1	2	3	4	5
$\tilde{\omega}_k$	22.37	61.66	120.90	199.9	298.6
$\omega_{k,1}$	22.86	60.23	117.96	194.52	289.57
$\omega_{k,2}$	98.22	113.26	152.55	218.81	310.10
$X_{k,1}$	$\begin{bmatrix} 1 \\ 1.005 \end{bmatrix}$	$\begin{bmatrix} 1 \\ 1.035 \end{bmatrix}$	$\begin{bmatrix} 1 \\ 1.141 \end{bmatrix}$	$\begin{bmatrix} 1 \\ 1.419 \end{bmatrix}$	$\begin{bmatrix} 1 \\ 2.057 \end{bmatrix}$
$X_{k,2}$	$\begin{bmatrix} 1 \\ -0.8296 \end{bmatrix}$	$\begin{bmatrix} 1 \\ -0.8051 \end{bmatrix}$	$\begin{bmatrix} 1 \\ -0.730 \end{bmatrix}$	$\begin{bmatrix} 1 \\ -0.587 \end{bmatrix}$	$\begin{bmatrix} 1 \\ -0.405 \end{bmatrix}$

$$\begin{bmatrix} f_{1,k,n}(x) \\ f_{2,k,n}(x) \end{bmatrix} = \begin{bmatrix} 1 \\ \dfrac{\tilde{\omega}_k^2 + \kappa - \omega_{k,n}^2}{\kappa} \end{bmatrix} \psi_k(x) \qquad \text{(e)}$$

Tables 5.3 gives the fundamental frequency and the intermodal frequency for the first five spatial modes of the fixed–fixed beam as well as the mode shape vector for two sets of parameters. Note that for the special case in which the beams are identical, $\mu_2 = \phi_2 = 1$, the sets of intermodal mode shape vectors are the same for each mode. The lowest mode shape corresponds to a mode in which the springs between the beams are unstretched, whereas the higher mode corresponds to a mode shape in which the beams move opposite to one another. When the beams are not identical, the intermodal mode shapes are different for each mode.

Mode shapes are presented in Figure 5.34 for several cases.

(b) The mode shapes and natural frequencies for a simply supported beam are

$$\phi_k(x) = \sin(k\pi x) \tag{f}$$

$$\tilde{\omega}_k = (k\pi)^2 \tag{g}$$

The analysis for simply supported beams is the same as for fixed–fixed except for the numerical values of $\tilde{\omega}_k$. Results for a pair of simply supported beams are shown in Table 5.4.

Example 5.19. Consider a series of concentric shafts, illustrated in Figure 5.35, each connected to the next through an elastic layer which is modeled as a layer of torsional springs. Let $k_{t,i}$ be the torsional stiffness per unit length of the layer connecting the ith and $(i+1)$st shafts. (a) Derive a general theory for the natural frequencies and mode shapes if one end of each shaft is free and the other fixed. (b) Apply the theory to a set of three shafts of equal thickness.

Solution: (a) The governing equations for the free response of a set of concentric shafts connected by layers of torsional springs are

$$G_1 J_1 \frac{\partial^2 \theta_1}{\partial x^2} - k_{t,1}(\theta_1 - \theta_2) = \rho_1 J_1 \frac{\partial^2 \theta_1}{\partial t^2}$$

$$G_2 J_2 \frac{\partial^2 \theta_2}{\partial x^2} + k_{t,1}(\theta_1 - \theta_2) - k_{t,2}(\theta_2 - \theta_3) = \rho_2 J_2 \frac{\partial^2 \theta_2}{\partial t^2}$$

$$\vdots$$

$$G_j J_j \frac{\partial^2 \theta_j}{\partial x^2} + k_{t,j-1}(\theta_{j-1} - \theta_j) - k_{t,j}(\theta_j - \theta_{j+1}) = \rho_j J_j \frac{\partial^2 \theta_j}{\partial t^2}$$

$$\vdots$$

$$G_n J_n \frac{\partial^2 \theta_n}{\partial x^2} + k_{t,n-1}(\theta_{n-1} - \theta_n) = \rho_n J_n \frac{\partial^2 \theta_n}{\partial t^2} \tag{a}$$

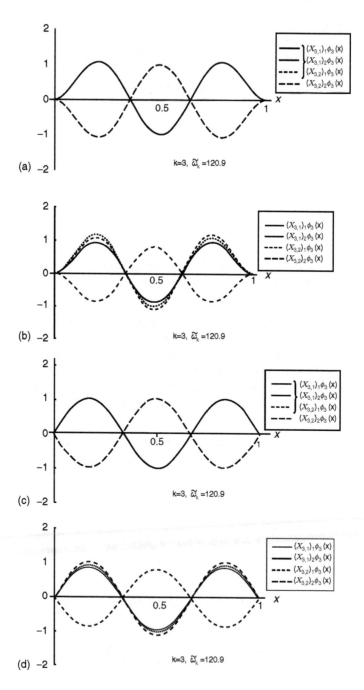

TABLE 5.4
Natural Frequencies and Mode Shapes for Dual Pinned-Pinned Beams with (a) $\mu = 1$ $\phi = 1$ $\kappa = 1000$, (b) $\mu = 1.2$ $\phi = 1.1$ $\kappa = 5000$

(a)

K	1	2	3	4	5
$\tilde{\omega}_k$	9.87	39.47	88.82	157.91	246.74
$\omega_{k,1}$	9.87	39.47	88.82	157.91	246.74
$\omega_{k,2}$	45.80	59.65	99.45	164.12	250.76
$X_{k,1}$	$\begin{bmatrix} 1 \\ 1 \end{bmatrix}$	$\begin{bmatrix} 1 \\ 1 \end{bmatrix}$	$\begin{bmatrix} 1 \\ 1 \end{bmatrix}$	$\begin{bmatrix} 1 \\ 1 \end{bmatrix}$	$\begin{bmatrix} 1 \\ 1 \end{bmatrix}$
$X_{k,2}$	$\begin{bmatrix} 1 \\ -1 \end{bmatrix}$	$\begin{bmatrix} 1 \\ -1 \end{bmatrix}$	$\begin{bmatrix} 1 \\ -1 \end{bmatrix}$	$\begin{bmatrix} 1 \\ -1 \end{bmatrix}$	$\begin{bmatrix} 1 \\ -1 \end{bmatrix}$

(b)

K	1	2	3	4	5
$\tilde{\omega}_k$	9.87	39.47	88.82	157.91	246.74
$\omega_{k,1}$	9.64	38.55	86.72	153.92	239.77
$\omega_{k,2}$	96.23	103.28	129.50	182.41	261.47
$X_{k,1}$	$\begin{bmatrix} 1 \\ 1.000 \end{bmatrix}$	$\begin{bmatrix} 1 \\ 1.014 \end{bmatrix}$	$\begin{bmatrix} 1 \\ 1.074 \end{bmatrix}$	$\begin{bmatrix} 1 \\ 1.249 \end{bmatrix}$	$\begin{bmatrix} 1 \\ 1.678 \end{bmatrix}$
$X_{k,2}$	$\begin{bmatrix} 1 \\ -0.832 \end{bmatrix}$	$\begin{bmatrix} 1 \\ -0.822 \end{bmatrix}$	$\begin{bmatrix} 1 \\ -0.776 \end{bmatrix}$	$\begin{bmatrix} 1 \\ -0.667 \end{bmatrix}$	$\begin{bmatrix} 1 \\ -0.497 \end{bmatrix}$

The equations are nondimensionalized using

$$x^* = \frac{x}{L} \tag{b}$$

$$t^* = \frac{t}{L\sqrt{\rho_1/G_1}} \tag{c}$$

FIGURE 5.34 (a) Mode shapes for two Euler-Bernoulli beams connected by an elastic layer: Fixed-fixed case ($\mu = 1$, $\Phi = 4$, $\kappa = 1.5$). (b) Mode shapes for two Euler-Bernoulli beams connected by an elastic layer: Fixed-fixed case ($\mu = 1.2$, $\Phi = 1.1$, $\kappa = 5000$). (c) Mode shapes for two Euler-Bernoulli beams connected by an elastic layer: Fixed-fixed case ($\mu = 1$, $\Phi = 1$, $\kappa = 1.5$). (d) Mode shapes for ($\mu = 1.2$, $\Phi = 1.1$, $\kappa = 5000$).

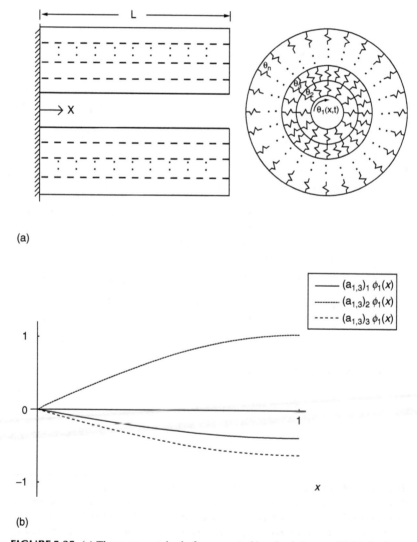

(a)

(b)

FIGURE 5.35 (a) Three concentric shafts connected by elastic layers. (b) Mode shapes for k = 1, m = 3.

The resulting nondimensional equations are

$$\frac{\partial^2 \theta_1}{\partial x^2} - \kappa_1(\theta_1 - \theta_2) = \frac{\partial^2 \theta_1}{\partial t^2}$$

$$\delta_2 \frac{\partial^2 \theta_2}{\partial x^2} + \kappa_1(\theta_1 - \theta_2) - \kappa_2(\theta_2 - \theta_3) = \mu_2 \frac{\partial^2 \theta_2}{\partial t^2}$$

$$\vdots$$

$$\delta_j \frac{\partial^2 \theta_j}{\partial x^2} + \kappa_{j-1}(\theta_{j-1} - \theta_j) - \kappa_j(\theta_j - \theta_{j+1}) = \mu_j \frac{\partial^2 \theta_j}{\partial t^2}$$

$$\vdots$$

$$\delta_n \frac{\partial^2 \theta_n}{\partial x^2} + \kappa_{n-1}(\theta_{n-1} - \theta_n) = \mu_n \frac{\partial^2 \theta_n}{\partial t^2} \qquad \text{(d)}$$

where

$$\delta_j = \frac{G_j J_j}{G_1 J_1} \qquad \text{(e)}$$

$$\mu_j = \frac{\rho_j J_j}{\rho_1 J_1} \qquad \text{(f)}$$

$$\kappa_{t,j} = \frac{\kappa_{t,j} L^2}{G_1 J_1} \qquad \text{(g)}$$

A normal mode solution to Equation d is assumed as

$$\begin{bmatrix} \theta_1 \\ \theta_2 \\ \vdots \\ \theta_n \end{bmatrix} = \begin{bmatrix} f_1(x) \\ f_2(x) \\ \vdots \\ f_n(x) \end{bmatrix} e^{i\omega t} \qquad \text{(h)}$$

which when substituted into Equation d leads to

$$\begin{bmatrix} \dfrac{d^2}{dx^2}-\kappa_1 & \kappa_1 & 0 & 0 & \cdots & 0 \\[2ex] \kappa_1 & \delta_2\dfrac{d^2}{dx^2}-\kappa_1-\kappa_2 & \kappa_2 & 0 & \cdots & 0 \\[2ex] 0 & \kappa_2 & \delta_3\dfrac{d^2}{dx^2}-\kappa_2-\kappa_3 & \kappa_3 & \cdots & 0 \\[2ex] \vdots & \vdots & \vdots & \vdots & \ddots & \vdots \\[2ex] 0 & 0 & 0 & 0 & \cdots & \kappa_{n-1} \\[2ex] 0 & 0 & 0 & 0 & \cdots & \delta_n\dfrac{d^2}{dx^2}-\kappa_{n-1} \end{bmatrix}$$

$$\times \begin{bmatrix} f_1 \\ f_2 \\ f_3 \\ \vdots \\ f_{n-1} \\ f_n \end{bmatrix} = -\omega^2 \begin{bmatrix} 1 & 0 & 0 & 0 & \cdots & 0 \\ 0 & \mu_2 & 0 & 0 & \cdots & 0 \\ 0 & 0 & \mu_3 & 0 & \cdots & 0 \\ \vdots & \vdots & \vdots & \ddots & \vdots & \vdots \\ 0 & 0 & 0 & 0 & \cdots & 0 \\ 0 & 0 & 0 & 0 & \cdots & \mu_n \end{bmatrix} \begin{bmatrix} f_1 \\ f_2 \\ f_3 \\ \vdots \\ f_{n-1} \\ f_n \end{bmatrix}$$

(i)

The natural frequencies and mode shapes for a fixed-free torsional shaft are

$$\tilde{\omega}_k = (2k-1)\frac{\pi}{2} \tag{j}$$

$$\phi_k(x) = \sin\left[(2k-1)\frac{\pi}{2}x\right] \tag{k}$$

It is noted that $\phi_k(x)$ is a nontrivial solution of $d^2\phi_k/dx^2 = -\tilde{\omega}_k^2\phi_k$. A solution of Equation i is assumed of the form

$$\mathbf{f} = \mathbf{a}_k\phi_k(x) \tag{l}$$

Which when substituted into Equation i leads to

$$
\begin{bmatrix}
\tilde{\omega}_k^2 + \kappa_1 & -\kappa_1 & 0 & 0 & \cdots & 0 \\
-\kappa_1 & \delta_2\tilde{\omega}_k^2 + \kappa_1 + \kappa_2 & -\kappa_2 & 0 & \cdots & 0 \\
0 & -\kappa_2 & \delta_3\tilde{\omega}_k^2 + \kappa_2 + \kappa_3 & -\kappa_3 & \cdots & 0 \\
\vdots & \vdots & \vdots & \vdots & \ddots & \vdots \\
0 & 0 & 0 & 0 & \cdots & -\kappa_{n-1} \\
0 & 0 & 0 & 0 & \cdots & \delta_n\tilde{\omega}_k^2 + \kappa_{n-1}
\end{bmatrix}
$$

$$
\times
\begin{bmatrix}
a_1 \\ a_2 \\ a_3 \\ \vdots \\ a_{n-1} \\ a_n
\end{bmatrix}
= \omega^2
\begin{bmatrix}
1 & 0 & 0 & 0 & \cdots & 0 \\
0 & \mu_2 & 0 & 0 & \cdots & 0 \\
0 & 0 & \mu_3 & 0 & \cdots & 0 \\
\vdots & \vdots & \vdots & \ddots & \vdots & \vdots \\
0 & 0 & 0 & 0 & \cdots & 0 \\
0 & 0 & 0 & 0 & \cdots & \mu_n
\end{bmatrix}
\begin{bmatrix}
a_1 \\ a_2 \\ a_3 \\ \vdots \\ a_{n-1} \\ a_n
\end{bmatrix}
$$

(m)

Equation m is a matrix eigenvalue problem of the form of Equation 5.16. Thus, for each spatial mode there are n radial mode shapes. The radial mode shapes are the eigenvectors corresponding to each natural frequency. The mode shapes are of the form

$$
\mathbf{w}_{k,m}(x) = \mathbf{a}_{k,m}\phi_k(x) \quad k = 1, 2, \ldots, \infty \quad m = 1, 2, \ldots, n \tag{n}
$$

where $\phi_k(x)$ is given in Equation k and $\mathbf{a}_{k,m}$ is the eigenvector of $\mathbf{M}^{-1}\mathbf{K}_k$, which corresponds to the natural frequency $\omega_{k,m}$. The condition for mode shape orthonormality is

$$
\int_0^1 \phi_k(x)\phi_\ell(x)\mathbf{a}^T_{l,i}\mathbf{M}\mathbf{a}_{k,i}dx = \delta_{k,\ell}\delta_{j,i} \tag{o}
$$

(b) Consider a system of three concentric annular shafts with $J_2 = (11/9)J_1$, $J_3 = (13/9)J_1$. If the mass densities and shear moduli of each shaft are the same, then $\delta_2 = \mu_2 = (11/9)$ and $\delta_3 = \mu_3 = (13/9)$. Assuming $\kappa_1 = 0.5$ and $\kappa_2 = 1.0$, Equation m becomes

$$\begin{bmatrix} \tilde{\omega}_k^2 + 0.5 & -0.5 & 0 \\ -0.5 & \dfrac{11}{9}\tilde{\omega}_k^2 + 1.5 & -1 \\ 0 & -1 & \dfrac{13}{9}\tilde{\omega}_k^2 + 1 \end{bmatrix} \begin{bmatrix} a_1 \\ a_2 \\ a_3 \end{bmatrix} = \omega^2 \begin{bmatrix} 1 & 0 & 0 \\ 0 & \dfrac{11}{9} & 0 \\ 0 & 0 & \dfrac{13}{9} \end{bmatrix} \begin{bmatrix} a_1 \\ a_2 \\ a_3 \end{bmatrix}$$

(p)

For a fixed-free shaft,

$$\tilde{\omega}_k = (2k-1)\frac{\pi}{2}$$

(q)

For each value of (k), Equation p is a matrix eigenvalue problem with three eigenvalues and three mode shape vectors. The set of mode shape vectors is the same for each mode. The natural frequencies for $k = 1,2,\ldots,4$ are given in Table 5.5. For each k there is a fundamental mode and two inter-radial modes. The normalized mode shape vectors are given in Table 5.6.

The mode shapes are plotted in Figure 5.35(b). As the mode number increases, the inter-radial frequencies approach the fundamental frequency for the mode.

Example 5.20. Determine the natural frequencies and mode shapes for two stretched fixed-fixed Euler–Bernoulli beams connected by an elastic layer The nondimensional equations governing the transverse displacements of the beams are

TABLE 5.5
Torsional Natural Frequencies of Three Concentric Shafts Connected by Layers of Torsional Springs.

	$k=1$	$k=$	$k=3$	$k=4$
$n=1$	1.5708	4.71239	7.85398	10.99561
$n=2$	1.7393	4.7712	7.88941	11.0209
$n=3$	2.08068	4.90596	7.97163	11.0799

TABLE 5.6
Normalized Mode Shape Vectors for Torsional
Oscillations of Three Concentric Shaftes Connected
by Layers of Torsional Springs, K = 1

$n = 1$	$n = 2$	$n = 3$
0.739	−1.144	−0.380
0.739	0.132	1.036
0.739	0.680	−0.613

$$\varepsilon \frac{d^4 w_1}{dx^4} - \frac{d^2 w_1}{dx^2} + \nu(w_1 - w_2) + \frac{\partial^2 w_1}{\partial t^2} = 0 \tag{a}$$

$$\varepsilon\delta \frac{d^4 w_2}{dx^4} - \frac{d^2 w_2}{dx^2} - \nu(w_1 - w_2) + \mu \frac{\partial^2 w_2}{\partial t^2} = 0 \tag{b}$$

where

$$\varepsilon = \frac{E_1 I_1}{PL^2} \tag{c}$$

$$\nu = \frac{kL^2}{P} \tag{d}$$

$$\delta = \frac{E_2 I_2}{E_1 I_1} \tag{e}$$

$$\mu = \frac{\rho_2 A_2}{\rho_1 A_1} \tag{f}$$

Substitution of the normal mode solution

$$\begin{bmatrix} w_1(x,t) \\ w_2(x,t) \end{bmatrix} = \begin{bmatrix} f_1(x) \\ f_2(x) \end{bmatrix} e^{i\omega t} \tag{g}$$

into Equation a and Equation b leads to

$$\varepsilon \frac{d^4 f_1}{dx^4} - \frac{d^2 f_1}{dx^2} + \nu(f_1 - f_2) - \omega^2 f_1 = 0 \tag{h}$$

$$\varepsilon\delta\frac{d^4 f_2}{dx^4} - \frac{d^2 f_2}{dx^2} - \nu(f_1 - f_2) - \omega^2 \mu f_2 = 0 \tag{i}$$

Due to the presence of the second derivative terms in Equation h and Equation i, a solution of the form obtained in Example 5.18 and Example 5.19 is not, in general, possible. An exception occurs when the beams are identical, $\delta = \mu = 1$. In this case, assume

$$\begin{bmatrix} f_1(x) \\ f_2(x) \end{bmatrix} = \begin{bmatrix} a_1 \\ a_2 \end{bmatrix} \phi_k(x) \tag{j}$$

where $\phi_k(x)$ is a nontrivial solution of

$$\varepsilon\frac{d^4 \phi_k}{dx^4} - \frac{d^2 \phi_k}{dx^2} - \tilde{\omega}_k^2 \phi_k = 0 \tag{k}$$

corresponding to the natural frequency $\tilde{\omega}_k$ of the fixed-fixed stretched beam with a stretching parameter ε. The natural frequencies and mode shapes are determined in Example 4.14 and Example 5.17.

Substitution of Equation j and Equation k into Equation h and Equation i using Equation k for the case of identical beams leads to

$$\begin{bmatrix} \tilde{\omega}_k^2 - \omega^2 + \nu & -\nu \\ -\nu & \tilde{\omega}_k^2 - \omega^2 + \nu \end{bmatrix} \begin{bmatrix} a_1 \\ a_2 \end{bmatrix} = \begin{bmatrix} 0 \\ 0 \end{bmatrix} \tag{l}$$

The natural frequencies $\omega_{k,1}$ and $\omega_{k,2}$ are determined by setting the determinant of the matrix in Equation l to zero and solving for ω. The results are

$$\omega_{k,1} = \tilde{\omega}_k \tag{m}$$

$$\omega_{k,2} = \sqrt{\tilde{\omega}_k + 2\nu} \tag{n}$$

The transverse mode shapes are determined as

$$\mathbf{a}_{k,1} = \begin{bmatrix} 1 \\ 1 \end{bmatrix} \qquad \mathbf{a}_{k,2} = \begin{bmatrix} 1 \\ -1 \end{bmatrix} \tag{o}$$

In the case where the beams are identical, the lowest transverse mode for each spatial mode is similar to a rigid body mode in which corresponding points on each beam have the same displacements and the elastic layer is undeformed. The higher frequency mode is one in which the beams are completely out of phase. The transverse motion is similar to that of an unrestrained two-degree-of-freedom system.

If the beams are not identical, it is not possible to obtain a solution of the form of the solutions obtained for the unstreched Euler–Bernoulli beams. Instead, a general solution is assumed as

$$\begin{bmatrix} f_1(x) \\ f_2(x) \end{bmatrix} = \begin{bmatrix} a_1 \\ a_2 \end{bmatrix} e^{\gamma x} \tag{p}$$

Substitution of Equation p into Equation h and Equation i leads to

$$\begin{bmatrix} \varepsilon\gamma^4 - \gamma^2 + \nu - \omega^2 & -\nu \\ -\nu & \varepsilon\delta\gamma^4 - \gamma^2 + \nu - \mu\omega^2 \end{bmatrix} \begin{bmatrix} a_1 \\ a_2 \end{bmatrix} = \begin{bmatrix} 0 \\ 0 \end{bmatrix} \tag{q}$$

Setting the determinant of the coefficient matrix to zero leads to

$$\varepsilon^2\delta\gamma^8 - \varepsilon(1+\delta)\gamma^6 + [\varepsilon(\nu - \mu\omega^2 + \nu\delta - \delta\omega^2) + 1]\gamma^4 +$$
$$[\omega^2(1+\mu) - 2\nu]\gamma^2 + \omega^2(\omega^2\mu - \nu - 2\mu) = 0 \tag{r}$$

Equation r has 8 solutions, but since it contains only even powers of γ, if b is a solution then $-b$ must also be a solution. The general solution is of the form

$$\begin{bmatrix} f_1(x) \\ f_2(x) \end{bmatrix} = \sum_{i=1}^{4} \begin{bmatrix} 1 \\ a_{2,i} \end{bmatrix} [C_{k,i1}e^{\gamma_i x} + C_{k,i2}e^{-\gamma_{k,i}x}] \tag{s}$$

Boundary conditions are applied to Equation s leading to 8 homogeneous equations for the eight constants of integration. The determinant of the resulting coefficient matrix is set equal to zero. However, the roots of Equation s are dependent upon the value of the natural frequency which is unknown. A closed form solution is not possible, thus an iterative procedure is employed to determine natural frequencies.

- An initial guess for a natural frequency is made. Since the system is self-adjoint, the natural frequencies are real and positive.
- The roots of Equation s are determined using the initial guess.
- The solution of the form of Equation s is developed. Some roots may be real while others may be complex. Complex roots lead to trigonometric solutions.

- Boundary conditions are applied to the appropriate form of Equation s leading to a system of eight equations for eight unknowns.
- The determinant of the coefficient matrix is calculated. If the guessed value is a natural frequency, the determinant will be zero.
- If the determinant is not zero, a new guess may be made and the procedure repeated.
- There are many methods to determine a new guess. The determinant of the matrix is a continuous function of ω. Thus any method for finding solutions of $F(\omega)$ can be employed.
- There are an infinite number of natural frequencies. An iterative method that may be employed is to make guesses for ω incremen- tally. That is, $\omega^{(i)} = \omega^{(i-1)} + \Delta\omega$, where $\Delta\omega$ is a fixed value. The function $F(\omega)$ is calculated until a sign change occurs. When this occurs, the search interval is refined until the desired accuracy is obtained. A half-interval search or a false position search may be employed for the refinement.
- Higher natural frequencies are obtained by reinitiating the search again looking for a value of ω for which $F(\omega)$ changes sign.
- There are, of course, numerical problems with any search. The Euler–Bernoulli beam example shows that natural frequencies corresponding to distinct modes may be close to one another. If the search interval is too large, two natural frequencies may be inadvertently missed. Large values of ω lead to evaluation of exponentials with large arguments which often causes numerical difficulties.

The addition of an axial load leads to significant differences in the identification of mode shapes. For an unstretched Euler–Bernoulli beam, a kth spatial mode corresponds to n distinct natural frequencies; there are n natural frequencies that have mode shapes with $(k-1)$ nodes located along thelength of the beam. This is not necessarily the case for the stretched beam.

Mirrors used in MFMS systems are often modeled as a chain of bars connected by linear springs, as illustrated in Figure 5.36. For simplicity, assume the chain consists of n bars of equal length L and identical properties, mass density ρ, elastic modulus E, and cross sectional area A. The stiffness of the spring connecting bar i and bar $i+1$ is k_i. Let x_i be a local spatial

FIGURE 5.36 Elastic bars connected by elastic springs.

coordinate, measured along bar i, $0 \leq x_i \leq L$. The longitudinal displacement of the ith bar is $w_i(x_i, t)$. Nondimensional variables are introduced as

$$x_i^* = \frac{x_i}{L} \tag{5.152}$$

$$w_i^* = \frac{w_i}{L} \tag{5.153}$$

$$t^* = \frac{t}{L\sqrt{\rho/G}} \tag{5.154}$$

The longitudinal displacement of each bar is governed by the nondimensional wave equation

$$\frac{\partial^2 w_i}{\partial x_i^2} = \frac{\partial^2 w_i}{\partial t^2} \tag{5.155}$$

Non-dimensional spring stiffnesses are defined by

$$\kappa_i = \frac{k_i L}{EA} \tag{5.156}$$

The boundary conditions for the first bar

$$\frac{\partial w_1}{\partial x_1}(0) = \kappa_0 w_1(0) \tag{5.157}$$

$$\frac{\partial w_1}{\partial x_1}(1) = -\kappa_1 [w_1(1) - w_2(0)] \tag{5.158}$$

The boundary conditions for bar i, $i = 2,3,\ldots n-1$ are

$$\frac{\partial w_i}{\partial x_i}(0) = \kappa_{i-1}[w_i(0) - w_{i-1}(1)] \tag{5.159}$$

$$\frac{\partial w_i}{\partial x_i}(1) = -\kappa_i[w_i(1) - w_{i+1}(0)] \tag{5.160}$$

The boundary conditions for the nth bar are

$$\frac{\partial w_n}{\partial x_n}(0) = \kappa_{n-1}[w_n(0) - w_{n-1}(1)] \tag{5.161}$$

$$\frac{\partial w_n}{\partial x_n}(1) + \kappa_n w_n(1) = 0 \tag{5.162}$$

Substitution of the normal mode solution

$$w_i(x,t) = f_i(x)e^{i\omega t} \tag{5.163}$$

in the equations represented by Equation 5.155 leads to

$$-\frac{d^2 f_i}{dx_i^2} = \omega_i^2 f_i \tag{5.164}$$

The solutions of Equation 5.164 are

$$f_i(x_i) = C_{1,i}\cos(\omega x_i) + C_{2,i}\sin(\omega x_i) \tag{5.165}$$

Application of boundary conditions leads to equations of the form

$$(\kappa_{i-1}\cos\omega)C_{1,i-1} + (\kappa_{i-1}\sin\omega)C_{2,i-1} - \kappa_{i-1}C_{1,i} + \omega C_{2,i} = 0$$
$$(-\omega\sin\omega + \kappa_i\cos\omega)C_{1,i} + (\omega\cos\omega + \kappa_i\sin\omega)C_{2,i} - \kappa_i C_{1,i+1} = 0 \tag{5.166}$$

The matrix formulation of Equation 5.166 is

$$
\begin{bmatrix}
-\kappa_0 & \omega & 0 & 0 & 0 & \cdots & 0 & 0 \\
-\omega\sin\omega + \kappa_1\cos\omega & \omega\cos\omega + \kappa_1\sin\omega & -\kappa_1 & 0 & 0 & \cdots & 0 & 0 \\
\kappa_1\cos\omega & \kappa_1\sin\omega & -\kappa_1 & \omega & 0 & \cdots & 0 & 0 \\
0 & 0 & -\omega\sin\omega + \kappa_2\cos\omega & \omega\cos\omega + \kappa_2\sin\omega & \kappa_2 & \cdots & 0 & 0 \\
0 & 0 & \kappa_2\cos\omega & \kappa_2\sin\omega & -\kappa_2 & \cdots & 0 & 0 \\
\vdots & \vdots & \vdots & \vdots & \vdots & \ddots & \vdots & \vdots \\
0 & 0 & 0 & 0 & 0 & \cdots & -\kappa_{n-1} & \omega \\
0 & 0 & 0 & 0 & 0 & \cdots & -\omega\sin\omega + \kappa_n\cos\omega & \omega\cos\omega + \kappa_n\sin\omega
\end{bmatrix}
$$

$$
\times
\begin{bmatrix}
C_{1,1} \\
C_{2,1} \\
C_{1,2} \\
C_{2,2} \\
C_{1,3} \\
\vdots \\
C_{1,n} \\
C_{2,n}
\end{bmatrix}
=
\begin{bmatrix}
0 \\
0 \\
0 \\
0 \\
0 \\
\vdots \\
0 \\
0
\end{bmatrix}
\tag{5.167}
$$

The natural frequencies are determined be setting the determinant of the coefficient matrix to zero.

Example 5.21. Consider a structure modeled as two bars connected by three springs as illustrated in Figure 5.37. The bars are identical as are the springs. (a) Determine the characteristic equation for the system and the first several natural frequencies and mode shapes.

Solution: Application of Equation 5.167 to the two bar system leads to

$$
\begin{bmatrix}
-\kappa & \omega & 0 & 0 \\
-\omega\sin\omega + \kappa\cos\omega & \omega\cos\omega + \kappa\sin\omega & -\kappa & 0 \\
\kappa\cos\omega & \kappa\sin\omega & -\kappa & \omega \\
0 & 0 & -\omega\sin\omega + \kappa\cos\omega & \omega\cos\omega + \kappa\sin\omega
\end{bmatrix}
$$

$$
\times
\begin{bmatrix}
C_{1,1} \\
C_{2,1} \\
C_{1,2} \\
C_{2,2}
\end{bmatrix}
=
\begin{bmatrix}
0 \\
0 \\
0 \\
0
\end{bmatrix}
$$

(a)

A nontrivial solution of Equation a exists if and only if the determinant of the coefficient matrix is zero:

$$
\begin{vmatrix}
-\kappa & \omega & 0 & 0 \\
-\omega\sin\omega + \kappa\cos\omega & \omega\cos\omega + \kappa\sin\omega & -\kappa & 0 \\
\kappa\cos\omega & \kappa\sin\omega & -\kappa & \omega \\
0 & 0 & -\omega\sin\omega + \kappa\cos\omega & \omega\cos\omega + \kappa\sin\omega
\end{vmatrix}
= 0 \quad \text{(b)}
$$

The determinant is expanded leading to the characteristic equation

$$
\omega[\omega(\omega^2 - 2\kappa^2)\sin^4(\omega) \quad 3\kappa^2\omega\cos^2(\omega) + 2k(\kappa^2 - 2\omega^2)\sin(\omega)\cos(\omega)] = 0 \quad \text{(c)}
$$

While $\omega = 0$ satisfies Equation c, it only leads to the trivial solution. For $\kappa = 1.5$, Equation c becomes

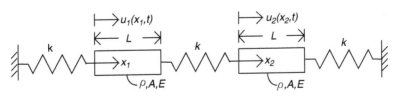

FIGURE 5.37 Chain of two bars and three springs of example 5.21.

$$6.75\omega \cos^2(\omega) + \omega(\omega^2 - 4.5)\sin^2(\omega) + (6.75 - 6\omega^2)\sin(\omega)\cos(\omega) = 0 \quad (d)$$

The five smallest solutions of Equation d are

$$\omega_1 = 0.9882, \qquad \omega_2 = 1.7509, \qquad \omega_3 = 3.5342, \qquad \omega_4 = 4.1201,$$

$$\omega_5 = 6.5097 \tag{e}$$

The mode shape vectors are obtained using Equation a as

$$f_{n,1}(x) = C_n \left[\cos(\omega_n x) + \frac{\kappa}{\omega_n}\sin(\omega_n x) \right]$$

$$f_{n,2}(x) = C_n \left[\left(-2\kappa + \frac{\omega_n^2}{\kappa} \right)\cos(\omega_n) + 3\omega_n \sin(\omega_n) - \frac{\omega_n(\kappa^2 + \omega_n^2)}{\kappa\omega_n\cos(\omega_n) + \kappa\sin(\omega_n)} \right]$$

$$\times \left[\cos(\omega_n x) + \frac{\sin(\omega_n x)(-\kappa \cos(\omega_n) + \omega_n\sin(\omega_n))}{\omega_n \cos(\omega_n) + \kappa \sin(\omega_n)} \right]$$

$$\tag{f}$$

The mode shapes can be normalized by requiring

$$\int_0^1 [(f_{n,1}(x))^2 + (f_{n,2}(x))^2]dx = 1 \tag{g}$$

The mode shapes for the two lowest natural frequencies are illustrated in Figure 5.38.

5.13 TIMOSHENKO BEAMS

The nondimensional equations governing the motion of a Timoshenko beam are those of Equation 4.85 and Equation 4.86, repeated below.

$$\frac{\partial^2 w}{\partial x^2} - \frac{\partial \psi}{\alpha x} = \frac{\partial^2 w}{\partial t^2} \tag{5.168}$$

$$\frac{\partial w}{\partial x} + \eta_1 \frac{\partial^2 \psi}{\partial x^2} - \psi = \eta_2 \frac{\partial^2 \psi}{\partial t^2} \tag{5.169}$$

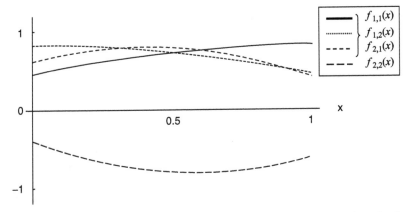

FIGURE 5.38 Mode shapes for system with two elastic bars and three springs for the lower two ω_n ($\omega_1 = 0.1998$, $\omega_2 = 1.7509$).

The normal mode solution for Equation 5.168 and Equation 5.169 is

$$\begin{bmatrix} w(x) \\ \psi(x) \end{bmatrix} = \begin{bmatrix} f_1(x) \\ f_2(x) \end{bmatrix} e^{i\omega t} \tag{5.170}$$

Substitution of Equation 5.170 into Equation 5.168 and Equation 5.169 leads to

$$\frac{d^2 f_1}{dx^2} - \frac{df_2}{dx} = -\omega^2 f_1 \tag{5.171}$$

$$\frac{df_1}{dx} + \eta_1 \frac{d^2 f_2}{dx^2} - f_2 = -\eta_2 \omega^2 f_2 \tag{5.172}$$

Since Equation 5.171 and Equation 5.172 are homogeneous, ordinary differential equations with constant coefficients a solution is assumed as

$$\begin{bmatrix} f_1(x) \\ f_2(x) \end{bmatrix} = \begin{bmatrix} a \\ b \end{bmatrix} e^{\alpha} \tag{5.173}$$

Substitution of Equation 5.173 into Equation 5.171 and Equation 5.172 leads to

$$\alpha^2 a - \alpha b = -\omega^2 a \tag{5.174}$$

$$\alpha a + \eta_1 \alpha^2 b - b = -\eta_2 \omega^2 b \qquad (5.175)$$

Equation 5.174 and Equation 5.175 are written in a matrix form as

$$\begin{bmatrix} \alpha^2 + \omega^2 & -\alpha \\ \alpha & \eta_1 \alpha^2 + \eta_2 \omega^2 - 1 \end{bmatrix} \begin{bmatrix} a \\ b \end{bmatrix} = \begin{bmatrix} 0 \\ 0 \end{bmatrix} \qquad (5.176)$$

A nontrivial solution of Equation 5.176 exists if and only if the determinant of the coefficient matrix is zero. Expansion of the determinant leads to

$$\eta_1 \alpha^4 + (\eta_1 + \eta_2)\omega^2 \alpha^2 + \eta_2 \omega^4 - \omega^2 = 0 \qquad (5.177)$$

For a given ω, Equation 5.177 is viewed as a quadratic equation to solve for α in which case the quadratic formula leads to

$$\begin{aligned} \alpha^2 &= \frac{1}{2\eta_1} \left[-(\eta_1 + \eta_2)\omega^2 \pm \sqrt{(\eta_1 + \eta_2)^2 \omega^4 - 4\eta_1(\eta_2 \omega^4 - \omega^2)} \right] \\ &= \frac{1}{2\eta_1} \left[-(\eta_1 + \eta_2)\omega^2 \pm \sqrt{(\eta_1 - \eta_2)^2 \omega^4 + 4\eta_1 \omega^2} \right] \end{aligned} \qquad (5.178)$$

An algebraic exercise shows that if $\eta_2 < 1$, then $(\eta_1 + \eta_2)\omega^2 < \sqrt{(\eta_1 - \eta_2)^2 \omega^4 + 4\eta_1 \omega^2}$. In this case, the use of the plus $(+)$ sign in Equation 5.178 gives a positive value of α^2 while the use of the minus $(-)$ sign leads to a negative value of α^2. Then the four values of α are of the form

$$\alpha = \pm\mu, \quad \mu = \frac{1}{\sqrt{2\eta_1}} \left[-(\eta_1 + \eta_2)\omega^2 + \sqrt{(\eta_1 - \eta_2)^2 \omega^4 + 4\eta_1 \omega^2} \right]^{1/2} \qquad (5.179)$$

$$\alpha = \pm i\nu, \quad \nu = \frac{1}{\sqrt{2\eta_1}} \left[(\eta_1 + \eta_2)\omega^2 + \sqrt{(\eta_1 - \eta_2)^2 \omega^4 + 4\eta_1 \omega^2} \right]^{1/2} \qquad (5.180)$$

Recall that $\eta_2 = I/AL^2$, which can be rearranged to $\eta_2 = (r/L)^2$, where r is the radius of gyration of the beam's cross section. Thus, η_2 is the reciprocal of the square of the beam's slenderness ratio which is clearly less than one.

The general solution of Equation 5.171 and Equation 5.172 is

$$
\begin{bmatrix} f_1(x) \\ f_2(x) \end{bmatrix} = C_1 \begin{bmatrix} 1 \\ \dfrac{\omega^2 + \mu^2}{\mu} \end{bmatrix} e^{\mu x} + C_2 \begin{bmatrix} 1 \\ \dfrac{\omega^2 + \mu^2}{-\mu} \end{bmatrix} e^{-\mu x}
$$

$$
+ C_3 \begin{bmatrix} 1 \\ \dfrac{\omega^2 - \nu^2}{i\nu} \end{bmatrix} e^{i\nu x} + C_4 \begin{bmatrix} 1 \\ \dfrac{\omega^2 - \nu^2}{i\nu} \end{bmatrix} e^{-i\nu x} \tag{5.181}
$$

Using Euler's identity, the definition of hyperbolic functions, Equation 5.181 is rewritten as

$$
\begin{bmatrix} f_1(x) \\ f_2(x) \end{bmatrix} = D_1 \left\{ \begin{bmatrix} 1 \\ 0 \end{bmatrix} \cosh(\mu x) + \begin{bmatrix} 0 \\ \dfrac{\omega^2 + \mu^2}{\mu} \end{bmatrix} \sinh(\mu x) \right\}
$$

$$
+ D_2 \left\{ \begin{bmatrix} 0 \\ \dfrac{\omega^2 + \mu^2}{\mu} \end{bmatrix} \cosh(\mu x) + \begin{bmatrix} 1 \\ 0 \end{bmatrix} \sinh(\mu x) \right\}
$$

$$
+ D_3 \left\{ \begin{bmatrix} 1 \\ 0 \end{bmatrix} \cos(\nu x) + \begin{bmatrix} 0 \\ \dfrac{\omega^2 - \nu^2}{\nu} \end{bmatrix} \sin(\nu x) \right\}
$$

$$
+ D_4 \left\{ \begin{bmatrix} 0 \\ \dfrac{\omega^2 - \nu^2}{\nu} \end{bmatrix} \cos(\mu x) + \begin{bmatrix} 1 \\ 0 \end{bmatrix} \sin(\nu x) \right\}
\tag{5.182}
$$

Application of the boundary conditions leads to a system of equations of the form

$$
\mathbf{QD} = 0 \tag{5.183}
$$

where $\mathbf{Q}$ is a matrix derived through application of the boundary conditions to Equation 5.182 and whose elements are dependent upon μ, ν and ω and $\mathbf{D} = [D_1\ D_2\ D_3\ D_4]^T$. A solution to this homogeneous system of Equations exists only if

$$q(\mu,\nu,\omega) = |\mathbf{Q}| = 0 \qquad (5.184)$$

The values of ω are the nondimensional natural frequencies of the system. It is shown in Chapter 4 that the stiffness operator is self-adjoint and thus there are an infinite, but countable, number of natural frequencies. The parameters μ and ν are functions of ω through Equation 5.179 and Equation 5.180. Thus Equation 5.184 can be viewed as the equation to solve for ω. A closed form solution of Equation 5.184 is clearly not possible, thus an iterative or numerical solution is necessary. One possible algorithm is to guess a value of ω, calculate μ and ν from Equation 5.179 and Equation 5.180, then evaluate Equation 5.184. If the left-hand side is not zero, then a rational approach can be employed to improve the guess. The procedure is complicated by the fact that there are an infinite number of natural frequencies and that such an iterative procedure may miss some solutions. A second complication is that as the magnitude of the frequency increases evaluation of the hyperbolic trigonometric functions lead to large numerical values. The slope, $\frac{d|Q|}{d\omega}$ is large in the vicinity of a natural frequency.

The modes of a second-order system and an Euler–Bernoulli beam are countable in the sense that the mode shape for the nth mode contains $n-1$ nodes and that there is only one frequency corresponding to an nth mode. The number of nodes is determined by the argument of the trigonometric functions in the mode shape. The argument in the trigonometric function for the Timoshenko beam is. Recalling that $\alpha = i\nu$, Equation 5.177 can be rewritten as

$$
\begin{aligned}
\eta_1 \nu^4 - (\eta_1 + \eta_2)\nu^2\omega^2 + \eta_2\omega^4 - \omega^2 = 0 \\
\eta_2\omega^4 - [(\eta_1 + \eta_2)\nu^2 + 1]\omega^2 + \eta_1\nu^4 = 0
\end{aligned}
\qquad (5.185)
$$

Equation 5.180 is viewed as a quadratic equation that given ν is used to solve for ω. To this end,

$$\omega^2 = \frac{1}{2\eta_2}\left[(\eta_1 + \eta_2)\nu^2 + 1 \pm \sqrt{[(\eta_1 + \eta_2)\nu^2 + 1]^2 - 4\eta_1\eta_2\nu^4}\right] \qquad (5.186)$$

Clearly, Equation 5.186 gives two real values of ω, leading to the following conclusions:

- If a natural frequency ω is calculated, the value of ν calculated using Equation 5.180 then another natural frequency is calculated from Equation 5.186.

- The mode number of the mode shape is determined by the value of v. For the Timoshenko beam, two frequencies correspond to each mode.
- The two frequencies for mode k can be called the lower modal frequency and labeled $\omega_{k,l}$ and the higher modal frequency and labeled $\omega_{k,u}$.
- Even though the mode number are the same, the mode shapes corresponding to the lower and higher modal frequencies are different.
- Since the mode shapes are different they satisfy the orthonormality condition

$$\int_0^1 [f_{1,k,\ell}(x)f_{1,k,u}(x) + \eta_2 f_{2,k,\ell}(x)f_{2,k,u}(x)]dx = \delta_{k,u} \tag{5.187}$$

Example 5.22. Determine the natural frequencies and mode shapes for a simply-supported Timoshenko beam if $\eta_1 = 1.6 \times 10^{-3}$ and $\eta_2 = 6.25 \times 10^{-4}$

Solution: The boundary conditions for a pinned–pinned beam are

$$w(0,t) = 0 \tag{a}$$

$$\frac{\partial \psi}{\partial x}(0,t) = 0 \tag{b}$$

$$w(1,t) = 0 \tag{c}$$

$$\frac{\partial \psi}{\partial x}(1,t) = 0 \tag{d}$$

Application of Equation a through Equation d to Equation 5.102 leads to

$$f_1(0) = 0 = D_1 + D_3 \tag{e}$$

$$\frac{df_2}{dx}(0) = 0 = (\omega^2 + \mu^2)D_1 + (\omega^2 - \nu^2)D_3 \tag{f}$$

$$f_1(1) = 0 = D_1 \cosh(\mu) + D_2 \sinh(\mu) + D_3 \cos(\nu) + D_4 \sin(\nu) \tag{g}$$

$$\frac{df_2}{dx}(1) = 0 = (\omega^2 + \mu^2)\cosh(\mu)D_1 + (\omega^2 + \mu^2)\sinh(\mu)D_2$$

$$+D_3(\omega^2 - \nu^2)\cos(\nu) - (\omega^2 - \nu^2)D_4\sin(\nu) \tag{h}$$

Equation e and Equation f imply that $D_1 = D_3 = 0$. Setting the determinant of the matrix generated by the remaining two equations leads to

Evaluation of the determinant of the matrix generated from Equation f through Equation h leads to

$$(\nu^2 + \mu^2)\sinh(\mu)\sin(\nu) = 0 \tag{i}$$

The solutions of Equation i are

$$\nu = k\pi \quad k = 1,2,\dots \tag{j}$$

Substitution of Equation j into Equation g leads to the conclusion of $D_2 = 0$.

The natural frequencies corresponding to this value of ν are calculated using Equation 5.186, which for $k = 1$ leads to

$$\omega_{1,\ell} = 0.1243 \qquad \omega_{1,u} = 20.44 \tag{k}$$

Substitution of Equation j through Equation n in Equation 5.182 leads to the mode shape for the lower frequency for the first mode as

$$\begin{bmatrix} f_{1,1,\ell}(x) \\ f_{2,1,\ell}(x) \end{bmatrix} = D_4 \left\{ \begin{bmatrix} 0 \\ -3.167 \end{bmatrix} \cos(\pi x) + \begin{bmatrix} 1 \\ 0 \end{bmatrix} \sin(\pi x) \right\} \tag{l}$$

The mode shape for the upper frequency of the first mode leads to

$$\begin{bmatrix} f_{1,1,u}(x) \\ f_{2,1,u}(x) \end{bmatrix} = D_4 \left\{ \begin{bmatrix} 1 \\ 129.85 \end{bmatrix} \cos(\pi x) + \begin{bmatrix} 1 \\ 0 \end{bmatrix} \sin(\pi x) \right\} \tag{m}$$

The normalized mode shapes of Equation l and Equation m are plotted in Figure 5.39 and Figure 5.40.

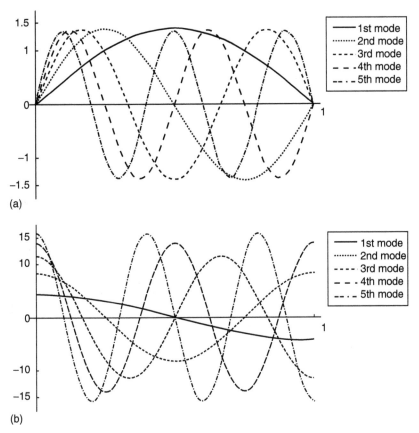

FIGURE 5.39 Mode shapes for lowest natural frequencies of Timoshenko beam of example 5.22 (a) $\omega(x)$ (b) $\psi(x)$.

The mode shapes are normalized by requiring

$$\langle \mathbf{w}_i, \mathbf{w}_i \rangle_M = 1$$

$$\int_0^1 D_{4,k}^2 \left[\sin^2(k\pi x) + \eta_2 \left(\frac{\omega^2 - \nu^2}{\nu} \right)^2 \cos^2(k\pi x) \right] dx = 1 \tag{n}$$

$$D_{4,k} = \nu \sqrt{\frac{2}{\nu^2 + \eta_2(\omega^2 - \nu^2)^2}}$$

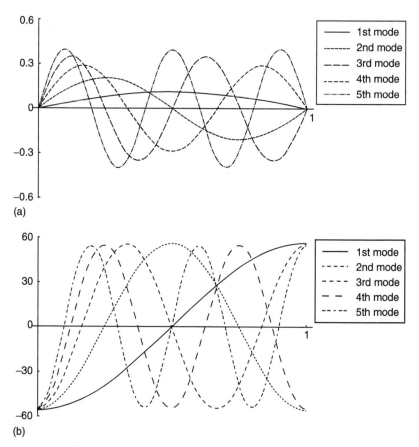

FIGURE 5.40 Mode shapes for higher natural frequencies of ezch mode for Timoshenko beam of example 5.22 (a) $\omega(x)$ (b) $\psi(x)$.

Example 5.23. Determine the natural frequencies and mode shapes for a fixed-free Timoshenko beam.

Solution: The boundary conditions for a fixed-free Timoshenko beam are

$$w(0,t) = 0 \tag{a}$$

$$\psi(0,t) = 0 \tag{b}$$

$$\frac{\partial w}{\partial x}(1,t) - \psi(1,t) = 0 \tag{c}$$

$$\frac{\partial \psi}{\partial x}(1,t) = 0 \tag{d}$$

Application of Equation a through Equation d to the general form of the mode shape, Equation 5.182 leads to

$$D_1 + D_3 = 0 \tag{e}$$

$$\left(\frac{\omega^2 + \mu^2}{\mu}\right)D_2 + \left(\frac{\omega^2 - \nu^2}{\nu}\right)D_4 = 0 \tag{f}$$

$$\left[\mu \sinh\mu - \left(\frac{\omega^2 + \mu^2}{\mu}\right)\sinh\mu\right]D_1 + \left[\mu \cosh\mu - \left(\frac{\omega^2 + \mu^2}{\mu}\right)\cosh\mu\right]D_2 +$$

$$\left[-\nu \sin \nu - \left(\frac{\omega^2 - \nu^2}{\nu}\right)\sin \nu\right]D_3 + \left[\nu \cos \nu - \left(\frac{\omega^2 - \nu^2}{\nu}\right)\cos \nu\right]D_4 = 0 \tag{g}$$

$$(\omega^2 + \mu^2)\cosh \mu D_1 + (\omega^2 + \mu^2)\sinh \mu D_2 + (\omega^2 - \nu^2)\cos \nu D_3$$
$$- (\omega^2 - \nu^2)D_4 = 0 \tag{h}$$

The determinant of the matrix obtained from Equation e and Equation f is set equal to zero. A trail and error or numerical solution is used to solve for values of ν. The lower and upper natural frequencies for this mode are obtained from Equation 5.186. The corresponding values of μ are determined using Equation 5.179. The mode shape vectors are then determined from Equation e through Equation h. Assuming an arbitrary value for D_1, Equation e, Equation f, and Equation h lead to

$$D_2 = \frac{(\omega^2 + \mu^2)\cosh \mu - (\omega^2 - \nu^2)\cos \nu}{(\omega^2 + \mu^2)\left(\frac{\nu}{\mu}\cos \nu - \sinh \mu\right)} D_1 \tag{i}$$

$$D_3 = -D_1 \tag{j}$$

$$D_4 = -\frac{\nu}{\mu} \frac{(\omega^2 + \mu^2)\cosh \mu - (\omega^2 - \nu^2)\cos \nu}{(\omega^2 - \nu^2)\left(\frac{\nu}{\mu}\cos \nu - \sinh \mu\right)} D_1 \tag{k}$$

Table 5.7 gives the first five modes, the corresponding upper and lower frequencies, the values of μ, and the values of the coefficients in the mode shape.

Mode shapes for the lower modes of the fixed–fixed Timoshenko beam are illustrated in Figure 5.41. Figure 5.42 and Figure 5.43 illustrate the mode shapes for fixed-free and fixed-pinned Timoshenko beams.

TABLE 5.7
Lower (a) and Upper (b) Natural Frequencies for Fixed-Free Timoshenko Beam

(a)

ν	1.8740	4.6682	7.7929	10.88	13.9587
Mode number	1	2	3	4	5
$\omega_{k,l}$	0.1383	0.8417	2.2586	4.1886	6.5100
μ	0.1948	1.1625	2.9947	5.2412	7.6568
D_2/D_1	27.40	−0.8044	−1.0459	−0.9748	−1.006
D_4/D_1	0.9303	0.6223	−0.80427	115.70	1777

(b)

ν	1.8740	4.6682	7.7929	10.88	13.9587
Mode number	1	2	3	4	5
$\omega_{k,u}$	40.223	41.335	43.898	47.358	51.667
μ	−0.2988	−1.8811	−5.3892	−10.883	−18.640
D_2/D_1	−1610	1649	1.003	1.00	1.00
D_4/D_1	0.6172	1.0261	−152.69	26750	−5.12×10^7

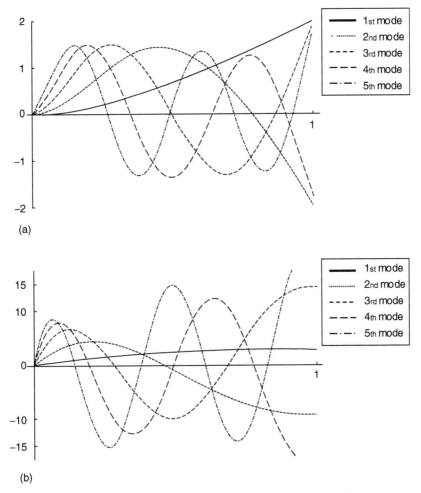

FIGURE 5.41 Mode shapes corresponding to lower mode frequencies for up pinned-pinned Timoshenko beam. (a) $w(x)$ (b) $\psi(x)$.

5.14 COMBINED CONTINUOUS AND DISCRETE SYSTEMS

Consider the system of Figure 5.44, a discrete mass-spring system attached to the end of a cantilever beam. The differential equations and boundary conditions for the system is derived in Example 2.17, as

$$\rho A \frac{\partial^2 w}{\partial t^2} + \frac{\partial^2}{\partial x^2}\left(EI \frac{\partial^2 w}{\partial x^2} \right) = 0 \tag{5.188}$$

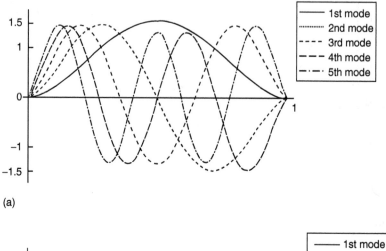

(a)

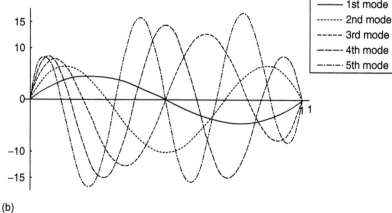

(b)

FIGURE 5.42 Mode shapes corresponding to lower mode frequencies of a fixed-free Timoshenko beam (a) $w(x)$ (b) $\psi(x)$.

$$m\ddot{y} + k(y - w(L,t)) = 0 \tag{5.189}$$

$$\frac{\partial}{\partial x}\left(EI\frac{\partial^2 w}{\partial x^2}\right)(0,t) = 0 \quad \text{or} \quad w(0,t) = 0 \tag{5.190}$$

$$EI\frac{\partial^2 w}{\partial x^2}(0,t) = 0 \quad \text{or} \quad \frac{\partial w}{\partial x}(0,t) = 0 \tag{5.191}$$

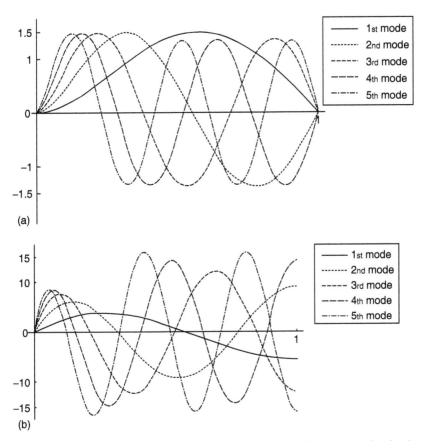

FIGURE 5.43 Mode shapes corresponding to lower mode frequencies of a fixed-pinned Timoshenko beam (a) $w(x)$ (b) $\psi(x)$.

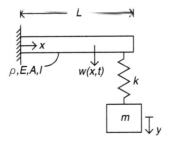

FIGURE 5.44 Discrete mass-spring system attached to end of cantilever beam.

$$\frac{\partial}{\partial x}\left(EI\frac{\partial^2 w}{\partial x^2}(L,t)\right) + k[y - w(L,t)] = 0 \quad \text{or} \quad w(L,t) = 0 \qquad (5.192)$$

$$I\frac{\partial^2 w}{\partial x^2}(L,t) = 0 \quad \text{or} \quad \frac{\partial w}{\partial x}(L,t) = 0 \qquad (5.193)$$

Use of the nondimensional variables defined in Equation 4.151 through Equation 4.154 for a uniform beam leads to

$$\frac{\partial^4 w}{\partial x^4} + \frac{\partial^2 w}{\partial t^2} = 0 \qquad (5.194)$$

$$\kappa[y - w(1,t)] + \mu\frac{d^2 y}{dt^2} = 0 \qquad (5.195)$$

The boundary conditions for a uniform fixed-free beam are

$$w(0,t) = 0 \qquad (5.196)$$

$$\frac{\partial w}{\partial x}(0,t) = 0 \qquad (5.197)$$

$$\frac{\partial^3 w}{\partial x^3} + \kappa[y - w(1,t)] = 0 \qquad (5.198)$$

$$EI\frac{\partial^2 w}{\partial x^2}(1,t) = 0 \qquad (5.199)$$

Substitution of the normal mode solution

$$w(x,t) = w(x)e^{i\omega t} \qquad (5.200)$$

$$y(t) = Ye^{i\omega t} \qquad (5.201)$$

in Equation 5.194 and Equation 5.195 leads to

$$\frac{d^4 w}{dx^4} - \omega^2 w = 0 \qquad (5.202)$$

$$-\mu\omega^2 Y + \kappa[Y - w(1)] = 0 \qquad (5.203)$$

The boundary conditions become

$$w(0) = 0 \qquad (5.204)$$

$$\frac{dw}{dx}(0) = 0 \tag{5.205}$$

$$\frac{d^3w}{dx^3} + \kappa[Y - w(1)] = 0 \tag{5.206}$$

$$\frac{d^2w}{dx^2}(1) = 0 \tag{5.207}$$

The general solution of Equation 5.202 is Equation 5.129

$$w(x) = C_1 \cosh(\sqrt{\omega}x) + C_2 \sinh(\sqrt{\omega}x) + C_3 \cos(\sqrt{\omega}x)$$
$$+ C_4 \sin(\sqrt{\omega}x) \tag{5.208}$$

Substitution of the boundary conditions into Equation 5.208 and use of Equation 5.203 leads to

$$\begin{bmatrix} 1 & 0 & 1 & 0 & 0 \\ 0 & \sqrt{\omega} & 0 & \sqrt{\omega} & 0 \\ \omega\cosh(\sqrt{\omega}) & \omega\sinh(\sqrt{\omega}) & -\omega\cos(\sqrt{\omega}) & -\omega\sin(\sqrt{\omega}) & 0 \\ \omega^{\frac{3}{2}}\sinh(\sqrt{\omega})-\cosh(\sqrt{\omega}) & \omega^{\frac{3}{2}}\cosh(\sqrt{\omega})-\kappa\sinh(\sqrt{\omega}) & \omega^{\frac{3}{2}}\sin(\sqrt{\omega})-\cos(\sqrt{\omega}) & -\omega^{\frac{3}{2}}\cos(\sqrt{\omega})-\kappa\sin(\sqrt{\omega}) & \kappa \\ -\kappa\cosh(\sqrt{\omega}) & -\kappa\sinh(\sqrt{\omega}) & -\kappa\cos(\sqrt{\omega}) & \kappa\sin(\sqrt{\omega}) & \kappa-\mu\omega^2 \end{bmatrix} \begin{bmatrix} C_1 \\ C_2 \\ C_3 \\ C_4 \\ Y \end{bmatrix}$$
$$= \begin{bmatrix} 0 \\ 0 \\ 0 \\ 0 \\ 0 \end{bmatrix} \tag{5.209}$$

A nontrivial solution is obtained only if the determinant of the coefficient matrix is zero.

Example 5.25. Determine the natural frequencies and normalized mode shapes of the system of Figure 5.44 if $\kappa = 2$ and $\mu = 0.5$.

Solution: Setting the determinant of the matrix in Equation 5.209 to zero leads to

$$\kappa - \mu\omega^2 - \cosh(\sqrt{\omega})\{(-\kappa + \mu\omega^2)\cos(\sqrt{\omega}) + \kappa\mu\sqrt{\omega}\sin(\sqrt{\omega})\}$$
$$+ \kappa\mu\sqrt{\omega}\cos(\sqrt{\omega})\sinh(\sqrt{\omega}) = 0 \tag{a}$$

Substituting $\kappa = 2$ and $\mu = 0.5$ in Equation a and solving for ω leads to

$$\omega_1 = 1.48749 \quad \omega_2 = 4.68096 \quad \omega_3 = 22.2192 \quad \omega_4 = 61.7623$$
$$\omega_5 = 120.935$$

(b)

The normalized mode shape vectors for the first two modes are

$$\begin{bmatrix} w_1(x) \\ Y_1 \end{bmatrix}$$
$$= \begin{bmatrix} 0.6279[\cos(1.219x) - \cosh(1.299x)] + 0.5511[\sinh(1.219x) - \sin(1.219x)] \\ -1.3542 \end{bmatrix}$$

(c)

$$\begin{bmatrix} w_2(x) \\ Y_2 \end{bmatrix}$$
$$= \begin{bmatrix} 0.78877[\cos(2.1635x) - \cosh(2.1635x)] + 0.59275[\sinh(2.1635x) - \sin(2.1635x)] \\ 0.41639 \end{bmatrix}$$

(d)

The mode shape orthonormality is stated as

$$\int_0^1 w_i(x)w_j(x)\,dx + \mu Y_i Y_j = \delta_{i,j}$$

(e)

The normalized mode shapes for the beam are plotted in Figure 5.45.

5.15 MEMBRANES

The nondimensional equation governing the free response of a membrane is

$$\nabla^2 W = \frac{\partial^2 W}{\partial t^2}$$

(5.210)

where $W(\mathbf{r},t)$, the transverse displacement, is a function of two spatial variables and time. It has been shown that the operator $-\nabla^2$ is positive definite and self-adjoint with respect to the standard inner product on S, the two dimensional region on which W is defined.

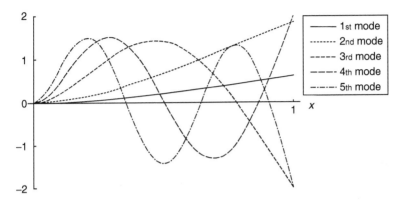

FIGURE 5.45 Normalized mode shapes for beam fixed at x=0 with a discrete mass-spring system at x=1, $\mu = 0.5$, $\kappa = 2$.

Substitution of the normal mode solution

$$W(\mathbf{r},t) = w(\mathbf{r})e^{i\omega t} \tag{5.211}$$

into Equation 5.210 leads to

$$-\nabla^2 w = \omega^2 w \tag{5.212}$$

The natural frequencies are the square roots of the eigenvalues of $-\nabla^2 w$, with specified boundary conditions.

Consider the vibrations of a rectangular membrane, as illustrated in Figure 5.46. Let x and y be spatial variables whose axes are parallel to the sides of the membrane, $0 \le x \le a$, $0 \le y \le b$. Nondimensional spatial variables

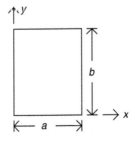

FIGURE 5.46 Rectangular membrane.

are defined as

$$x^* = \frac{x}{a} \tag{5.213}$$

$$y^* = \frac{y}{a} \tag{5.214}$$

$$t^* = \frac{t}{a\sqrt{\rho/T}} \tag{5.215}$$

The region, in terms of nondimensional variables, becomes $0 \leq x \leq 1$, $0 \leq y \leq \eta$, where $\eta = b/a$. Using these nondimensional variables, the eigenvalue problem is written as

$$-\left(\frac{\partial^2 w}{\partial x^2} + \frac{\partial^2 w}{\partial y^2}\right) = \omega^2 w \tag{5.216}$$

The method of separation of variables is used to determine the natural frequencies and mode shapes. A product solution is assumed as

$$w(x,y) = X(x)Y(y) \tag{5.217}$$

where $X(x)$ is a function of x only and $Y(y)$ is a function of y only. Substitution of Equation 5.217 into Equation 5.216 leads to

$$-\frac{d^2 X}{dx^2} Y(y) - X(x)\frac{d^2 Y}{dy^2} = \omega^2 X(x)Y(y) \tag{5.218}$$

Dividing Equation 5.218 by $X(x)Y(y)$ leads to

$$-\frac{1}{X(x)}\frac{d^2 X}{dx^2} = \frac{1}{Y(y)}\frac{d^2 Y}{dy^2} + \omega^2 \tag{5.219}$$

The left-hand side of Equation 5.219 is a function of x only while the right-hand side of Equation 5.219 is a function of y only. However, x and y are independent spatial variables. The only case in which a function of x can be equal to a function of y for all x and y is if both functions are equal to the same constant. This separation argument implies that for a product solution to exist, both sides of Equation 5.219 must be equal to a constant (the separation constant); call it λ.

$$\frac{d^2 X}{dx^2} = -\lambda X \tag{5.220}$$

$$\frac{d^2 Y}{dy^2} + \omega^2 Y = \lambda Y \tag{5.221}$$

To proceed further, boundary conditions must be specified. Consider a rectangular membrane that is simply supported over its entire boundary. The appropriate boundary conditions are

$$W(0,y,t) = 0 \tag{5.222}$$

$$W(1,y,t) = 0 \tag{5.223}$$

$$W(x,0,t) = 0 \tag{5.224}$$

$$W(x,\eta,t) = 0 \tag{5.225}$$

After application of the normal mode solution of Equation 5.211 and the product solution of Equation 5.217, the first boundary condition, Equation 5.222 is written as

$$X(0)Y(y) = 0 \tag{5.226}$$

Equation 5.226 can be satisfied for all y if and only if

$$X(0) = 0 \tag{5.227}$$

Similar analysis of the remaining boundary conditions leads to

$$X(1) = 0 \tag{5.228}$$

$$Y(0) = 0 \tag{5.229}$$

$$Y(\eta) = 0 \tag{5.230}$$

Equation 5.220 subject to boundary conditions, Equation 5.227 and Equation 5.228 is a differential eigenvalue problem. The separation constants, λ, are the eigenvalues of $Lf = -d^2 f/dx^2$ where the domain of L is the subspace S of $C^2[0,1]$ such that if f is in S, then $f(0) = 0$ and $f(1) = 0$. The problem for $X(x)$ and the separation constants is a Sturm–Liouville problem.

The solution of Equation 5.220 is

$$X(x) = C_1 \cos(\sqrt{\lambda}) + C_2 \sin(\sqrt{\lambda}x) \tag{5.231}$$

Application of Equation 5.227 Then, application of Equation 5.228 leads to

$$C_2 \sin(\sqrt{\lambda}) = 0 \tag{5.232}$$

A nontrivial solution for $X(x)$ is obtained for values of λ such that $\sin(\sqrt{\lambda}) = 0$, which are of the form

$$\lambda_k = (k\pi)^2 \quad k = 1,2,\dots \tag{5.233}$$

The eigenvalues of the Sturm–Liouville problem for $X(x)$ are given in Equation 5.233. The corresponding eigenvectors are

$$X_k(x) = C_k \sin(k\pi x) \tag{5.234}$$

For each $k = 1,2,\dots$, a corresponding solution in the y direction is obtained by solving

$$-\frac{d^2 Y_k}{dx^2} + (k\pi)^2 Y_k = \omega^2 Y_k \tag{5.235}$$

$$Y_k(0) = 0 \tag{5.236}$$

$$Y_k(\eta) = 0 \tag{5.237}$$

The natural frequencies are the square roots of the eigenvalues of the Sturm–Liouville problem of Equation 5.235 through Equation 5.237.

The general solution of Equation 5.235 is

$$Y_k(x) = C_3 \cos(\sqrt{\omega^2 - k^2 \pi^2} y) + C_4 \sin(\sqrt{\omega^2 - k^2 \pi^2} y) \tag{5.238}$$

Application of Equation 5.236 to Equation 5.238 leads to $C_3 = 0$. Application of Equation 5.237 then leads to

$$\sin\left[(\sqrt{\omega^2 - k^2 \pi^2})\eta\right] = 0 \tag{5.239}$$

Equation 5.239 is satisfied by values of ω such that

$$\left(\sqrt{\omega^2 - k^2 \pi^2}\right)\eta = m\pi \quad m = 1,2,\dots \tag{5.240}$$

The natural frequencies are thus obtained as

$$\omega_{k,m} = \sqrt{(k\pi)^2 + \left(\frac{m\pi}{\eta}\right)^2} \tag{5.241}$$

The eigenvector of the Sturm–Liouville problem in the y direction is

$$Y_m(y) = C_m \sin\left(\frac{m\pi}{\eta}y\right) \tag{5.242}$$

The mode shape corresponding to a natural frequency of Equation 5.241 is

$$w_{k,m}(x,y) = X_k(x)Y_m(y) = C_{k,m}\sin(k\pi x)\sin\left(\frac{m\pi}{\eta}y\right) \tag{5.243}$$

Since L is self-adjoint the mode shapes satisfy an orthogonality condition of

$$(w_{k,m},w_{\ell,n}) = 0 \quad k \neq \ell \text{ and } m \neq n$$

$$\int_0^1\int_0^\eta \sin(k\pi x)\sin\left(\frac{m\pi}{\eta}y\right)\sin(\ell\pi x)\sin\left(\frac{n\pi}{\eta}y\right)dydx = 0 \tag{5.244}$$

The mode shapes are normalized by requiring

$$(w_{k,m},w_{k,m}) = 1$$

$$(C_{k,m})^2\int_0^1\int_0^\eta\left[\sin(k\pi x)\sin\left(\frac{m\pi}{\eta}y\right)\right]^2 dydx = 1 \tag{5.245}$$

$$C_{k,m} = 2$$

Thus the normalized mode shapes are

$$w_{k,m} = 2\sin(k\pi x)\sin\left(\frac{m\pi}{\eta}y\right) \tag{5.246}$$

Mode shapes for several natural frequencies of a rectangular membrane with $\eta = 2$ are illustrated in Figure 5.47. Nodal lines are lines for each mode shape for which the displacement is zero. For example, consider

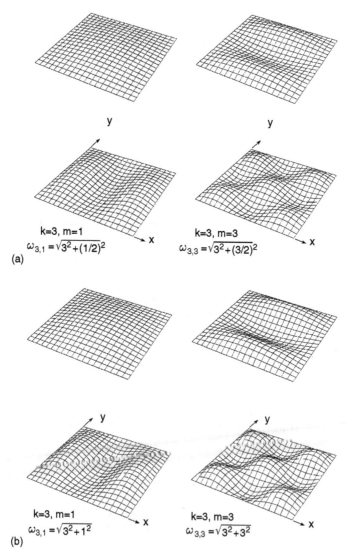

k=3, m=1
$\omega_{3,1} = \sqrt{3^2 + (1/2)^2}$

k=3, m=3
$\omega_{3,3} = \sqrt{3^2 + (3/2)^2}$

(a)

k=3, m=1
$\omega_{3,1} = \sqrt{3^2 + 1^2}$

k=3, m=3
$\omega_{3,3} = \sqrt{3^2 + 3^2}$

(b)

FIGURE 5.47. Mode shapes for (a) Rectangular membrane with η=2. (b) Square membrane with individual mode shapes for repeated frequencies. (c) Combined mode shapes for repeated frequencies.

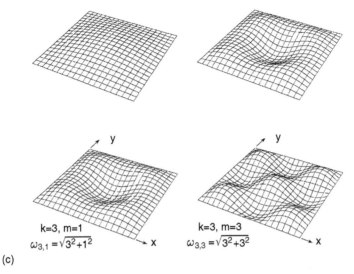

$$k=3, m=1$$
$$\omega_{3,1} = \sqrt{3^2+1^2}$$

$$k=3, m=3$$
$$\omega_{3,3} = \sqrt{3^2+3^2}$$

(c)

Figure 5.47   (*continued*)

$$w_{2,3}(x, y) = 2 \sin(2\pi x)\sin\left(\frac{3\pi}{\eta}y\right) \tag{5.247}$$

Modal lines for this mode are $x = 1/2$, $y = 1/3\eta$, and $y = 2/3\eta$.

The square membrane ($\eta = 1$) is a special case. The natural frequency corresponding to the mode shape $w_{k,m}$ is the same as the natural frequency corresponding to $w_{m,k}$ for $k \neq m$; $\omega_{m,k} = \omega_{k,m} = \pi\sqrt{k^2 + m^2}$. While $w_{k,m}$ and $w_{m,k}$ are orthogonal, any linear combination of these mode shapes is also a mode shape corresponding to this natural frequency. Thus the frequency $\omega_{m,k}$ has two linearly independent mode shapes,

$$w_{m,k} = 2 \sin(m\pi x)\sin(k\pi y) \tag{5.248}$$
$$w_{k,m} = 2 \sin(k\pi x)\sin(m\pi y) \tag{5.249}$$

which are illustrated in Figure 5.47b. However, a linear combination of these mode shapes is also a mode shape. The combined mode shapes are illustrated in Figure 5.47c.

Polar coordinates are used in the analysis of the circular membrane of Figure 5.48. When the nondimensional variables

$$r^* = \frac{r}{R} \tag{5.250}$$

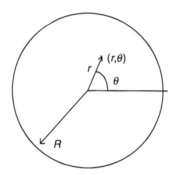

FIGURE 5.48 Polar coordinates are used to analyze vibrations of circular membrane.

$$w^* = \frac{w}{R} \tag{5.251}$$

$$t^* = \frac{t}{R\sqrt{\rho/T}} \tag{5.252}$$

are introduced, Equation 5.222 becomes

$$-\left(\frac{\partial^2 w}{\partial r^2} + \frac{1}{r}\frac{\partial w}{\partial r} + \frac{1}{r^2}\frac{\partial^2 w}{\partial \theta^2}\right) = \omega^2 w \tag{5.253}$$

A product solution to Equation 5.253 is assumed as

$$w(r, \theta) = R(r)\Theta(\theta) \tag{5.254}$$

Substitution of Equation 5.254 into Equation 5.253 leads to

$$\frac{d^2 R}{dr^2}\Theta(\theta) + \frac{1}{r}\frac{dR}{dr}\Theta(\theta) + \frac{1}{r^2}R(r)\frac{d^2\Theta}{d\theta^2} = -\omega^2 R(r)\Theta(\theta) \tag{5.255}$$

Multiplying Equation 5.243 by r^2, dividing by $R(r)\Theta(\theta)$ and rearranging leads to

$$\frac{1}{R}\left(r^2\frac{d^2 R}{dr^2} + r\frac{dR}{dr}\right) + \omega^2 r^2 = -\frac{1}{\Theta}\frac{d^2\Theta}{d\theta^2} \tag{5.256}$$

The left-hand side of Equation 5.256 is a function of r only while the right hand side is a function of θ only. Since r and θ are independent variables, the equation is satisfied for all r and θ only if both sides are equal to the same constant (the separation argument); call it λ

$$\frac{d^2\Theta}{d\theta^2} + \lambda\Theta = 0 \qquad (5.257)$$

$$r^2\frac{d^2R}{dr^2} + r\frac{dR}{dr} + (\omega^2 r^2 - \lambda)R = 0 \qquad (5.258)$$

If the membrane is simply supported over its entire boundary, $W(1,\theta,t) = 0$, which after substitution of the normal mode solution and the product solution leads to the boundary condition

$$R(1) = 0 \qquad (5.259)$$

The boundary is defined in polar coordinates by the circle, $r = 1$. Thus there are no specific boundary conditions $\Theta(\theta)$ must satisfy. The polar coordinate points, (r,θ) and $(r,\theta+2\pi)$ for $r<1$, both represent the same point of the surface of the membrane. However, the displacement must be single valued; $W(r, \theta, t) = W(r, \theta+2\pi, t)$ for all r and t. This condition requires $\Theta(\theta)$ to be periodic of period 2π. That is

$$\Theta(\theta) = \Theta(\theta + 2\pi) \qquad (5.260)$$

The problem to determine $\Theta(\theta)$ is described by Equation 5.257 and Equation 5.260. This is a Sturm–Liouville problem in which condition (c) of the Sturm–Liouville Theorem, Theorem 5.8, applies. The problem is self-adjoint, but only positive semi definite in that $\lambda = 0$ is an eigenvalue with an eigenfunction

$$\Theta_0(\theta) = C_0 \qquad (5.261)$$

where C_0 is a constant. For $\lambda>0$, the general solution of Equation 5.257 is

$$\Theta(\theta) = C_1 \cos(\sqrt{\lambda}\theta) + C_2 \sin(\sqrt{\lambda}\theta) \qquad (5.262)$$

The periodicity condition, Equation 5.260, is satisfied when $\lambda = n^2$, $n = 1,2,\ldots$. However, no relationship exists between C_1 and C_2. Hence two independent eigenvectors correspond to the eigenvalue $\lambda = n^2$

$$\Theta_{n,1} = C_{n,1} \cos(n\theta) \qquad (5.263)$$
$$\Theta_{n,2} = C_{n,2} \sin(n\theta) \qquad (5.264)$$

Use of the eigenvectors, determined from Sturm–Liouville problem defined by Equation 5.257 and Equation 5.260, in the Expansion Theorem lead to the trigonometric Fourier series representation for a periodic function.

The separation constants are $\lambda_n = n^2$, $n = 0,1,2,\ldots..$ Equation 5.258 becomes

$$r^2 \frac{d^2 R_n}{dr^2} + r \frac{dR_n}{dr} + (\omega^2 r^2 - n^2)R_n = 0 \tag{5.265}$$

The general solution of Equation 5.265 is obtained in terms of Bessel functions, as

$$R_n(r) = D_{n,1} J_n(\omega r) + D_{n,2} Y_n(\omega r) \tag{5.266}$$

The solution for $R_n(r)$ is obtained for $0 \le r \le 1$ and must be finite within this range. However, the Bessel function of the second kind $Y_n(r)$ is singular at $r = 0$. This singularity is eliminated from Equation 5.266 only by requiring $D_{n,2} = 0$. Application of Equation 5.259 then leads to $J_n(\omega) = 0$. Thus the natural frequencies are obtained as the zeroes of the Bessel function of the first kind. Each Bessel function has an infinite but countable number of zeroes. Thus the natural frequencies are obtained through solution of

$$J_n(\omega_{n,m}) = 0 \tag{5.267}$$

The corresponding eigenvectors are

$$R_{n,m}(r) = D_{n,m} J_n(\omega_{n,m} r) \tag{5.268}$$

The problem described by Equation 5.258 and Equation 5.259 is a Sturm–Liouville problem in which condition (b) of the Sturm–Liouville Theorem applies for $r = 0$. Equation 5.258 may be rewritten in the form of Equation 5.265 with $p(r) = r$.

The mode shapes are of the form

$$w_{0,m} = C_{0,m} J_0(\omega_{0,m} r) \tag{5.269}$$

$$w_{1,m,n} = C_{1,m,n} J_n(\omega_{n,m} r)\cos(n\theta) \tag{5.270}$$

$$w_{2,m,n} = C_{2,m,n} J_n(\omega_{n,m} r)\sin(n\theta) \tag{5.271}$$

Mode shapes corresponding to distinct natural frequencies are orthogonal with respect to the inner product

$$(f(r,\theta), g(r,\theta)) = \int_0^{2\pi} \int_0^1 f(r,\theta)g(r,\theta)r\,dr\,d\theta \tag{5.272}$$

Specifically stated, the orthogonality conditions are

$$(w_{0,m}, w_{0,k}) = 0 \quad k \neq m \tag{5.273}$$

$$(w_{0,m}, w_{1,\ell,n}) = 0 \tag{5.274}$$

$$(w_{0,m}, w_{2,\ell,n}) = 0 \tag{5.275}$$

$$(w_{1,m,n}, w_{1,\ell,k}) = 0 \quad m \neq \ell \text{ and } n \neq k \tag{5.276}$$

$$(w_{1,m,n}, w_{2,\ell,k}) = 0 \tag{5.277}$$

$$(w_{2,m,m}, w_{2,\ell,k}) = 0 \quad m \neq \ell \text{ and } n \neq k \tag{5.278}$$

Plots of mode shapes for several modes are presented in Figure 5.49. These plots illustrate that the mode shapes have nodal lines and nodal circles. The mode $w_{2,m,n}$ has nodal lines corresponding to values of θ, such that $\sin(n\theta) = 0$ or $\theta = 0, \pi/n, 2\pi/n, \ldots, (n-1)\pi/n, \pi, (n+1)\pi/n, \ldots, (2n-1)\pi/n$ as well as nodal circles at all values of r such that $J_n(\omega_{n,m}r) = 0$.

Example 5.25. Determine the natural frequencies and mode shapes for a semi-circular simply supported membrane as illustrated in Figure 5.50.

Solution: The method of separation of variables is used with an assumed product solution of the form of Equation 5.254. The solution remains the same as for the fully circular membrane through Equation 5.258. The region in which a solution is sought is described by $0 \leq r \leq 1$, with $0 \leq \theta \leq \pi$. Boundary conditions for the membrane are

$$R(1) = 0 \tag{a}$$

$$\Theta(0) = 0 \tag{b}$$

$$\Theta(\pi) = 0 \tag{c}$$

The solution for $\Theta(\theta)$ is as in Equation 5.263. Application of the boundary conditions leads to

$$\Theta(0) = 0 \Rightarrow C_1 = 0 \tag{d}$$

$$\Theta(\pi) = 0 \Rightarrow C_2 \sin(\sqrt{\lambda}\pi) = 0 \tag{e}$$

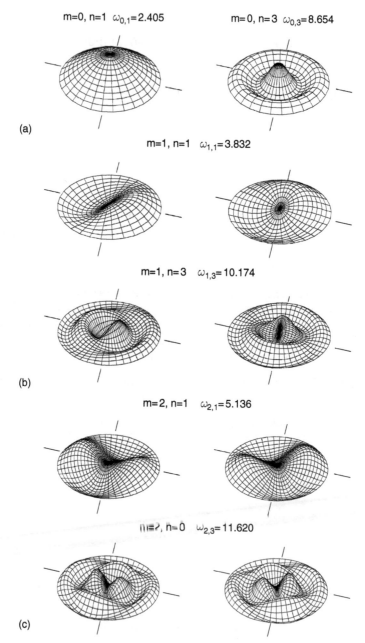

FIGURE 5.49 Mode shapes for circular membrane. Radial nodes are circles corresponding to solutions of $Jm(\omega m,n, r)$. Circumferential nodes are lines corresponding to $\sin[m\theta]=0$ or $\cos[m\theta] = 0$. (a) $m = 0$ (b) $m = 1$ (c) $m = 2$.

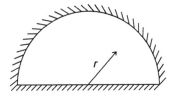

FIGURE 5.50 Simply supported semi-circular membrane of Example 5.25.

Equation e is satisfied if

$$\sqrt{\lambda}\pi = n\pi \quad n = 1,2,\dots$$
$$\lambda_n = n^2 \quad n = 1,2,\dots \tag{f}$$

The general solution for $\Theta(\theta)$ is

$$\Theta(\theta) = C_n \sin(n\theta) \tag{g}$$

The eigenvalues in θ are the same as those for the fully circular plate. However, the eigenvectors only include the sine functions for the semi-circular plate.

Since the eigenvalues are the same as those for the fully circular plate, the equation for $R(r)$ is the same as Equation 5.265, except the equation corresponding to $n = 0$ is not part of the free response. The solution for $R(r)$ is given by Equation 5.268 and the natural frequencies are obtained by solving Equation 5.267.

The natural frequencies for the semi-circular membrane are the same as those for the fully circular membrane, except for solutions of $J_0(\omega) = 0$. There is only one mode shape for each natural frequency of the semi-circular plate whereas the circular plate has two independent mode shapes for each natural frequency. The inner product for orthogonality and normalization is

$$(f,g) = \int_0^\pi \int_0^1 f(r,\theta), g(r,\theta) r \, dr \, d\theta \tag{h}$$

Example 5.26. The plate of Figure 5.51 is nearly circular. Its boundary is described by

$$r = 1 + \varepsilon f(\theta) \tag{a}$$

where ε is a small dimensionless parameter. Use the perturbation method to

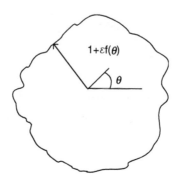

FIGURE 5.51 Nearly circular membrane of Example 5.26.

determine approximations to the natural frequencies of a simply supported nearly circular plate.

Solution: Let $w_{1,n,m}(r,\theta)$ be a mode shape of a circular plate corresponding to a natural frequency $\omega_{n,m}$ as determined from Equation 5.270. First-order approximations to the natural frequency and mode shape of the nearly circular plate are

$$\hat{\omega}_{n,m} = \omega_{n,m} + \varepsilon\phi \tag{b}$$

$$\hat{w}_{1,n,m}(r,\theta) = w_{1,n,m}(r,\theta) + \varepsilon g(r,\theta) \tag{c}$$

The vibrations of the membrane for this mode are governed by

$$-\nabla^2 \hat{w}_{1,n,m} = \hat{\omega}_{n,m}^2 \hat{w}_{1,n,m} \tag{d}$$

$$\hat{w}_{1,n,m}(1 + \varepsilon f, \theta) = 0 \tag{e}$$

Substitution of Equation d and Equation e into Equation b gives

$$-\nabla^2(w_{1,n,m}(r,\theta) + \varepsilon g(r,\theta)) = (\omega_{n,m} + \varepsilon\phi)^2[w_{1,n,m}(r,\theta) + \varepsilon g(r,\theta)] \tag{f}$$

Expanding and collecting terms of like powers of ε in Equation f leads to

$$0 = \nabla^2 w_{1,n,m} + \omega_{n,m}^2 w_{1,n,m} + \varepsilon[\nabla^2 g + \omega_{n,m}^2 g + 2\phi w_{1,n,m}] + O(\varepsilon^2) \tag{g}$$

Substitution of Equation c into the boundary condition, Equation e results in

$$w_{1,n,m}(1 + \varepsilon f, \theta) + \varepsilon g(1 + \varepsilon f, \theta) = 0 \tag{h}$$

Use of a Taylor series expansion about $r = 1$ in Equation h gives

$$w_{1,n,m}(1,\theta) + \varepsilon\left[g(1,\theta) + f(\theta)\frac{\partial w_{1,n,m}}{\partial r}(1,\theta)\right] + O(\varepsilon^2) = 0 \qquad \text{(i)}$$

Setting coefficients of powers of ε to zero independently leads to the following problems for $g(r,\theta)$

$$\nabla^2 g + \omega_{m,n}^2 g = -2\phi w_{1,m,n} \qquad \text{(j)}$$

$$g(1,\theta) = -f(\theta)\frac{\partial w_{1,m,n}}{\partial\theta}(1,\theta) \qquad \text{(k)}$$

The system of Equation j and Equation k has both a nonhomogeneous term in the differential equation as well as a nonhomogeneous term in the boundary conditions. The corresponding homogeneous system has a nontrivial solution, $w_{1,m,n}(r,\theta)$. Thus a solution of the nonhomogeneous problem exists only if a solvability condition is met. In order to impose the solvability condition, the nonhomogeneity in the boundary condition is transferred to the differential equation. To this end, define

$$\hat{g}(r,\theta) = g(r,\theta) + f(\theta)\frac{\partial w_{1,m,n}}{\partial\theta}(1,\theta) \qquad \text{(l)}$$

Rewriting Equation j and Equation k using $\hat{g}(r,\theta)$ as the dependent variable leads to

$$\nabla^2\hat{g} + \omega_{1,m,n}^2\hat{g} = -2\phi w_{1,m,n} + \nabla^2\left[f(\theta)\frac{\partial w_{1,m,n}}{\partial\theta}(1,\theta)\right]$$

$$- -2\phi w_{1,m,n} + \frac{1}{r^2}\frac{d^2}{d\theta^2}\left[f(\theta)\frac{\partial w_{1,m,n}}{\partial\theta}(1,\theta)\right] \qquad \text{(m)}$$

$$\hat{g}(1,\theta) = 0 \qquad \text{(n)}$$

The solvability condition requires that for a solution to exist, the nonhomogeneous term must be orthogonal to $w_{1,m,n}(r,\theta)$,

$$\int_0^{2\pi}\int_0^1\left\{-2\phi w_{1,m,n1} + \frac{1}{r^2}\frac{d^2}{d\theta^2}\left[f(\theta)\frac{\partial w_{1,m,n}}{\partial\theta}(1,\theta)\right]\right\}w_{1,m,n}r\,dr\,d\theta = 0 \qquad \text{(o)}$$

Equation o is solved for the natural frequency perturbation as

$$\phi = \frac{\int\limits_{0}^{2\pi}\int\limits_{0}^{1} \frac{w_{1,m,n}}{r} \frac{d^2}{d\theta^2}\left[f(\theta)\frac{\partial w_{1,m,n}}{\partial\theta}(1,\theta)\right]drd\theta}{2\int\limits_{0}^{2\pi}\int\limits_{0}^{1}(w_{1,m,n})^2 rdrd\theta} \tag{p}$$

5.16 GREEN'S FUNCTIONS

The uniform cantilever beam of Figure 5.52 is subject to a concentrated unit load a distance ξ from its fixed end. Elementary beam theory is used to determine the displacement of a particle along the neutral axis of the beam a distance x from the fixed support, as

$$EIw(x) = \begin{cases} -\dfrac{1}{6}x^3 + \dfrac{1}{2}\xi x^2 & x \le \xi \\[2ex] \dfrac{1}{6}(x-\xi)^3 - \dfrac{1}{6}x^3 + \dfrac{1}{2}\xi x^2 & x > \xi \end{cases} \tag{5.279}$$

The function defined in Equation 5.279 is denoted by $K(x,\xi)$ and is often called the flexibility influence function. The influence function is a continuous version of the flexibility influence coefficient for a discrete system.

The cantilever beam of Figure 5.53 is subject to a distributed load over the length of the beam. The superposition principle indicates that the deflection at any point due to a set of concentrated loads is equal to the sum of the deflections at that point due to each individual load. In addition, the deflection is directly proportional to the magnitude of the concentrated load. The resultant of the distributed load applied over a differential element of length dx, located a distance ξ from the left end of the beam, is $f(\xi)d\xi$. The deflection

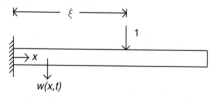

FIGURE 5.52 Uniform cantilever beam with unit concentrated load located a distance ξ from left support.

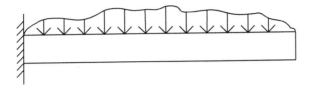

FIGURE 5.53 Cantilever beam with distributed load.

of the beam a distance x from the left end due to the load on this differential element is

$$dw = K(x, \xi)f(\xi)d\xi \tag{5.280}$$

The superposition principle is applied to determine the deflection of the beam at x due to the distributed load as

$$w(x) = \int_0^L K(x, \xi)f(\xi)d\xi \tag{5.281}$$

Equation 5.281 is used to determine the deflection of the beam for any particle along the neutral axis of the beam. The work done by the external forces during loading is

$$W = \frac{1}{2}\int_0^L w(x)f(x)dx = \frac{1}{2}\int_0^L \left(\int_0^L K(x, \xi)f(\xi)d\xi\right)f(x)dx$$

$$= \frac{1}{2}\int_0^L\int_0^L K(x, \xi)f(x)f(\xi)d\xi dx \tag{5.282}$$

Suppose the deflection is calculated at a point along the neutral axis a distance ξ from the left end of the beam due to the same distributed load

$$w(\xi) = \int_0^L f(x)K(\xi, x)dx \tag{5.283}$$

The work done by the loading is

$$W = \frac{1}{2}\int_0^L w(\xi)f(\xi)d\xi = \int_0^L\int_0^L K(\xi, x)f(\xi)f(x)dxd\xi \tag{5.284}$$

Requiring the work to be the same for the same loading, Equation 5.282 and Equation 5.284 are equal if and only if

$$K(x, \xi) = K(\xi, x) \qquad (5.285)$$

Equation 5.285 is a statement of the symmetry of the flexibility influence function.

The function $K(x,\xi)$ is of a class of functions called Green's functions. A Green's function is defined for a homogeneous boundary-value problem, a differential equation and its boundary conditions. Assume the operator has been nondimensionalized such that it has a coefficient of one on its highest derivative and the range of the independent variable, x, is $x, \leq x \leq 1$. A Green's function, $G(x,\xi)$ is required to satisfy four properties:

- The Green's function satisfies the differential equation for all x and ξ such that $0 \leq x \leq \xi$ and $\xi < x \leq 1$. The Green's function satisfies the differential equation for all x except at $x = \xi$.
- The Green's function must satisfy all boundary conditions.
- $G(x,\xi)$ is continuous for all $0 \leq x \leq 1$. If the operator is fourth-order, then $\partial G/\partial x(x,\xi)$ is also continuous for all $x, 0 \leq x \leq 1$.
- For a second-order operator, $\partial G/\partial x(x,\xi)$ has a jump discontinuity of -1 at $x = \xi$. For a fourth-order operator, $\partial^3 G/\partial x^3$ has a jump discontinuity of -1 at $x = \xi$. This property is a result of the application of a concentrated load at $x = \xi$.

If $K(x,\xi)$ is the Green's function for the system, then the solution of the nonhomogeneous system with $f(x)$ as the nonhomogeneity is

$$w(x) = \int_0^L f(\xi)K(x, \xi)d\xi \qquad (5.286)$$

The force applied to a beam undergoing free vibrations is the inertia force developed due to the motion of the beam. The inertia force applied to a differential element is

$$f(\xi)d\xi = -\rho A \frac{\partial^2 w}{\partial t^2}d\xi \qquad (5.287)$$

Thus the deflection of the beam at a location described by the coordinate x is

$$w(x, t) = -\int_0^L \rho A(\xi)\frac{\partial^2 w}{\partial t^2} K(x, \xi)d\xi \qquad (5.288)$$

Equation 5.288 is an integral equation in that the beam displacement, the dependent variable, is part of the integrand of an integral.

The standard nondimensional variables for beam problems are introduced into Equation 5.288. The influence function, having dimensions of length per force, is also nondimensionalized such that

$$K^*(x^*, \xi^*) = K(x, \xi) \frac{L^2}{E_0 I_0} \tag{5.289}$$

The resulting nondimensional form of Equation 5.288 is

$$w(x, t) = -\int_0^1 \alpha(\xi) \frac{\partial^2 w}{\partial t^2} K(x, \xi) d\xi \tag{5.290}$$

Equation 5.290 is an example of a Fredholm integral equation of the second kind and is an alternate formulation of the problem to determine the time dependent displacement of a structure.

A normal mode solution, $w(x, t) = \hat{w}(x) e^{i\omega t}$, is applied to Equation 5.290 leading to

$$\hat{w}(x) = \omega^2 \int_0^1 \alpha(\xi) K(x, \xi) w(\xi) d\xi \tag{5.291}$$

Define the linear operator **L** by

$$\mathbf{L}w = \int_0^1 \alpha(\xi) K(x, \xi) w(\xi) d\xi \tag{5.292}$$

Equation 5.292 can be written using the operator **L** as

$$\mathbf{L}\hat{w} = \frac{1}{\omega^2} \hat{w} \tag{5.293}$$

Equation 5.293 is in the form of a standard eigenvalue problem. The natural frequencies are the reciprocals of the square roots of the eigenvalues of **L**.

The alternate formulation of the eigenvalue problem for a continuous system, Equation 5.293 is analogous to the formulation of the eigenvalue problem for discrete systems in which the natural frequencies are obtained as the eigenvalues of $\mathbf{AM}$. Indeed, The operator $\mathbf{L}$ is analogous to the matrix $\mathbf{AM}$ in that the mass operator for the continuous system is $\alpha(x)$ and $K(x,\xi)$ is the flexibility influence function. For a discrete system, $\mathbf{AM} = (\mathbf{M}^{-1}\mathbf{K})^{-1}$. A similar analogy can be made that $\mathbf{L} = (\mathbf{M}^{-1}\mathbf{K})^{-1}$ for a continuous system.

Since the eigenvalues of $\mathbf{L}$ are the reciprocals of the eigenvalues of the differential operator, it should be possible to show that the eigenvalues and eigenvectors of $\mathbf{L}$ satisfy the same properties. The operator $\mathbf{M}^{-1}\mathbf{K}$ is self-adjoint with respect to the kinetic energy inner product. The self-adjointness of $\mathbf{L}$ with respect to this inner product is tested by examining

$$(\mathbf{L}f,g)_{\mathbf{M}} = \int_0^1 \left[\int_0^1 \alpha(\xi)K(x,\xi)f(\xi)d\xi \right] \alpha(x)g(x)dx$$

$$= \int_0^1 \int_0^1 \alpha(\xi)\alpha(x)K(x,\xi)f(\xi)g(x)d\xi dx \qquad (5.294)$$

Interchanging the names of the variables in Equation 5.294 followed by interchanging the order of integration leads to

$$(\mathbf{L}f,g)_{\mathbf{M}} = \int_0^1 \int_0^1 \alpha(x)\alpha(\xi)K(\xi,x)f(x)g(\xi)dxd\xi$$

$$= \int_0^1 \left(\int_0^1 \alpha(\xi)K(\xi,x)g(\xi)d\xi \right) \alpha(x)f(x)dx \qquad (5.295)$$

Using the symmetry of K leads to

$$(\mathbf{L}f,g)_{\mathbf{M}} = \int_0^1 \left(\int_0^1 \alpha(\xi)K(x,\xi)g(\xi)d\xi \right) \alpha(x)f(x)dx = (f,\mathbf{L}g)_{\mathbf{M}} \qquad (5.296)$$

Thus $\mathbf{L}$ is self-adjoint with respect to the kinetic energy inner product defined for all functions in the domain of the stiffness operator. The flexibility influence function is always positive leading to the conclusion that $\mathbf{L}$ is positive definite.

Green's functions are mostly of theoretical value and lend little value to the exact determination of natural frequencies and mode shapes. However, it does lend itself to an iterative method to approximate atural frequencies and mode shapes. An iteration method similar to the matrix iteration method used to approximate the natural frequencies and mode shapes of a continuous system. The method is developed along the same lines as the matrix iteration.

6 Non-Self-Adjoint Systems

6.1 NON-SELF-ADJOINT OPERATORS

When viscous damping and/or gyroscopic effects are included, it is often not possible to formulate the problem for the system's free response in the form of an eigenvalue problem for a self-adjoint operator. Formulation of the problem for the free response may be formulated as an eigenvalue problem for an operator that is not self-adjoint.

Eigenvalues for self-adjoint operators are real. Thus, it has not been necessary to work in complex vector spaces. A major difference between real and complex vector spaces is the commutative property required for defining a valid inner product. For a real vector space, the requirement is simply that the inner product operation is commutative, whereas, for a complex vector space, one property required for the definition of an inner product is

$$(\mathbf{u}, \mathbf{v}) = (\overline{\mathbf{v}, \mathbf{u}}) = (\bar{\mathbf{v}}, \bar{\mathbf{u}}) \tag{6.1}$$

As a result of the required property of inner products that $(\lambda\mathbf{u},\mathbf{v}) = \lambda(\mathbf{u},\mathbf{v})$

$$
\begin{aligned}
(\mathbf{u}, \mu\mathbf{v}) \quad &= (\overline{\mu\mathbf{v}, \mathbf{u}}) \\
&= (\overline{\mu}\bar{\mathbf{v}}, \bar{\mathbf{u}}) \\
&= \bar{\mu}(\bar{\mathbf{v}}, \bar{\mathbf{u}}) \\
&= \bar{\mu}(\mathbf{u}, \mathbf{v}) \tag{6.2}
\end{aligned}
$$

The inner product generated norm for a complex vector space is defined by

$$\|\mathbf{u}\| = (\mathbf{u}, \bar{\mathbf{u}})^{1/2} = (\bar{\mathbf{u}}, \mathbf{u})^{1/2} \tag{6.3}$$

Let $\mathbf{N}$ be an operator whose domain is a vector space S and whose range is P. An inner product is defined for all elements of S and P. The adjoint of $\mathbf{N}$

(denoted by $\mathbf{N}^*$) is an operator whose domain is P and range is S such that for all $\mathbf{u}$ in S and $\mathbf{v}$ in P

$$(\mathbf{Nu}, \mathbf{v}) = (\mathbf{u}, \mathbf{N}^*\mathbf{v}) \tag{6.4}$$

Let λ be an eigenvalue of $\mathbf{N}$ with a corresponding eigenvector $\mathbf{u}$. Consider

$$\begin{aligned} (\mathbf{Nu}, \bar{\mathbf{u}}) &= (\lambda\mathbf{u}, \bar{\mathbf{u}}) \\ &= \lambda(\mathbf{u}, \bar{\mathbf{u}}) \\ &= \lambda\|\mathbf{u}\|^2 \end{aligned} \tag{6.5}$$

Using the definition of the adjoint, Equation 6.5 can also be written as

$$(\mathbf{Nu}, \bar{\mathbf{u}}) = (\mathbf{u}, \mathbf{N}^*\bar{\mathbf{u}}) \tag{6.6}$$

In order for Equation 6.5 to reconcile with Equation 6.6, there must exist a μ such that

$$\mathbf{N}^*\bar{\mathbf{u}} = \mu\bar{\mathbf{u}} \tag{6.7}$$

Equation 6.7 implies that $\bar{\mathbf{u}}$ is an eigenvector of $\mathbf{N}^*$ corresponding to the eigenvalue μ. Using Equation 6.7 in Equation 6.6 leads to

$$\begin{aligned} (\mathbf{Nu}, \bar{\mathbf{u}}) &= (\mathbf{u}, \mu\bar{\mathbf{u}}) \\ &= \bar{\mu}(\mathbf{u}, \bar{\mathbf{u}}) \\ &= \bar{\mu}\|\mathbf{u}\|^2 \end{aligned} \tag{6.8}$$

Comparison of Equation 6.5 and Equation 6.8 shows that $\bar{\mu} = \lambda$ and thus the following theorem is proved.

Theorem 6.1. *The eigenvalues of an operator are the complex conjugates of the eigenvalues of its adjoint.*

Assume λ is an eigenvlaue of $\mathbf{N}$ with a corresponding eigenvector $\mathbf{u}$ and $\mu \neq \bar{\lambda}$ is an eigenvalue of $\mathbf{N}^*$ with a corresponding eigenvector $\mathbf{v}$. Consider the following inner products

$$(\mathbf{Nu}, \mathbf{v}) = (\lambda\mathbf{u}, \mathbf{v}) \tag{6.9}$$

and

$$(\mathbf{u}, \mathbf{N}^*\mathbf{v}) = (\mathbf{u}, \mu\mathbf{v}) \tag{6.10}$$

Subtracting Equation 6.10 from Equation 6.9 gives

$$(\mathbf{Nu}, \mathbf{v}) - (\mathbf{u}, \mathbf{N}^*\mathbf{v}) = (\lambda\mathbf{u}, \mathbf{v}) - (\mathbf{u}, \mu\mathbf{v}) \tag{6.11}$$

The left-hand side of Equation 6.11 is identically zero because of the definition of the adjoint operator. Thus,

$$\begin{aligned} 0 &= (\lambda\mathbf{u}, \mathbf{v}) - (\mathbf{u}, \mu\mathbf{v}) \\ &= \lambda(\mathbf{u}, \mathbf{v}) - \bar{\mu}(\mathbf{u}, \mathbf{v}) \\ &= (\lambda - \bar{\mu})(\mathbf{u}, \mathbf{v}) \end{aligned} \tag{6.12}$$

By hypothesis $\mu \neq \bar{\lambda}$, Equation 6.12 implies

$$(\mathbf{u}, \mathbf{v}) = 0 \tag{6.13}$$

Theorem 6.2. *(Bi-orthogonality) Eigenvectors corresponding to eigenvalues of an operator and its adjoint, which are not complex conjugates, are orthogonal, Equation 6.13.*

Eigenvectors of a non-self-adjoint operator are normalized by dividing by their inner product generated norms. If $\lambda_1, \lambda_2, \ldots$ are eigenvalues of an operator $\mathbf{N}$ with corresponding eigenvectors $\hat{\mathbf{u}}_1, \hat{\mathbf{u}}_2, \ldots$, then a set of normalized eigenvectors are obtained by

$$\mathbf{u_i} = \frac{\hat{\mathbf{u}}_\mathbf{i}}{\|\hat{\mathbf{u}}_\mathbf{i}\|} \tag{6.14}$$

Using the normalization of Equation 6.14, the eigenvectors of the adjoint operator also become normalized as they are the complex conjugates of the eigenvectors of $\mathbf{N}$.

If the eigenvectors have been normalized, then Theorem 6.2 can be extended to the principle of bi-orthonormality such that

$$(\mathbf{u_i}, \mathbf{v_j}) = \delta_{i,j} \tag{6.15}$$

Let λ_i be an eigenvalue of $\mathbf{N}$ with a corresponding eigenvector $\mathbf{u_i}$, then

$$\mathbf{Nu_i} = \lambda_i\mathbf{u_i} \tag{6.16}$$

Taking the inner product of Equation 6.16 with $\mathbf{v_j}$, an eigenvector of $\mathbf{N}^*$,

$$
\begin{aligned}
(\mathbf{Nu_i}, \mathbf{v_j}) &= (\lambda_i \mathbf{u_i}, \mathbf{v_j}) \\
&= \lambda_i(\mathbf{u_i}, \mathbf{v_j}) \\
&= \lambda_i \delta_{i,j}
\end{aligned}
\tag{6.17}
$$

Equation 6.17 is an orthogonality condition satisfied by the operator $\mathbf{N}$.

Both discrete and continuous systems are considered. Whereas the basic theory is the same, considerations exist for continuous systems that do not exist for discrete systems. The eigenvalues of a matrix $\mathbf{A}$, whether it is self-adjoint or not, are calculated by determining the values of λ such that $|\mathbf{A} - \lambda \mathbf{I}| = 0$. Some computational methods that are employed to numerically determine eigenvalues and eigenvectors of a matrix do require the matrix to be symmetric. However, the characteristic equation is an nth order polynomial in λ, which has n roots. If the matrix has only real components, then when complex roots occur they occur as complex conjugate pairs. The resulting normalized eigenvectors are linearly independent, and while they are not orthogonal they do satisfy the bi-orthonormality condition, Equation 6.13.

The n linearly independent eigenvectors of a matrix $\mathbf{A}$ span R^n. Thus, any n-dimensional vector, $\mathbf{w}$, has a representation in terms of the eigenvectors of $\mathbf{A}$ as

$$
\mathbf{w} = \sum_{i=1}^{n} \alpha_i \mathbf{u_i}
\tag{6.18}
$$

Taking the inner product of Equation 6.16 with $\mathbf{v_j}$, an eigenvector of the adjoint of $\mathbf{A}$ leads to

$$
\begin{aligned}
(\mathbf{w}, \mathbf{v_j}) &= \left(\sum_{i=1}^{n} \alpha_i \mathbf{u_i}, \mathbf{v_j} \right) \\
&= \sum_{i=1}^{n} \alpha_i (\mathbf{u_i}, \mathbf{v_j}) \\
&= \sum_{i=1}^{n} \alpha_i \delta_{i,j} \\
&= \alpha_j
\end{aligned}
\tag{6.19}
$$

In light of Equation 6.17, the expansion theorem for the non-self-adjoint operator becomes

$$
\mathbf{w} = \sum_{i=1}^{n} (\mathbf{w}, \mathbf{v}_i) \mathbf{u_i}
\tag{6.20}
$$

Questions of convergence and completeness should be addressed when deriving the expansion theorem for non-self-adjoint operators. In any case, it is assumed that the expansion theorem, Equation 6.18, is also valid for non-self-adjoint operators.

6.2 DISCRETE SYSTEMS WITH PROPORTIONAL DAMPING

The free response of a linear discrete system with viscous damping is governed by the set of equations represented by

$$\mathbf{M}\ddot{\mathbf{x}} + \mathbf{C}\dot{\mathbf{x}} + \mathbf{K}\mathbf{x} = 0 \tag{6.21}$$

where $\mathbf{M}$ is the symmetric $n \times n$ mass matrix, $\mathbf{C}$ is the symmetric $n \times n$ damping matrix, and $\mathbf{K}$ is the symmetric $n \times n$ stiffness matrix. The damping matrix is determined using the dissipation function, as described in Chapter 2.

Even though the damping matrix is symmetric, it should be suspected that it is not possible to formulate the response of a system with viscous damping using a self-adjoint eigenvalue problem. A normal mode assumption for a damped single-degree-of-freedom system leads to complex values of the natural frequency, contrary to what may be obtained with a self-adjoint formulation.

The free response of an undamped system is obtained by determining the system's natural frequencies and normalized mode shapes and using the expansion theorem to develop a set of uncoupled equations in terms of the system's principal coordinates. Let $\omega_1 \leq \omega_2 \leq \cdots \leq \omega_n$ be the undamped natural frequencies of the system with corresponding normalized mode shape vectors $\mathbf{w}_1, \mathbf{w}_2, \ldots, \mathbf{w}_n$. The expansion theorem suggests that $\mathbf{x}(t)$ can be expanded in terms of the normalized mode shapes as

$$\mathbf{x}(t) = \sum_{i=1}^{n} c_i(t)\mathbf{w_i} \tag{6.22}$$

Substitution of Equation 6.22 into Equation 6.21 leads to

$$\sum_{i=1}^{n} \ddot{c}_i \mathbf{M}\mathbf{w_i} + \sum_{i=1}^{n} \dot{c}_i \mathbf{C}\mathbf{w_i} + \sum_{i=1}^{n} c_i \mathbf{K}\mathbf{w_i} = 0 \tag{6.23}$$

Taking the standard inner product of Equation 6.23 with $\mathbf{w_j}$ for an arbitrary j leads to

$$\sum_{i=1}^{n} \ddot{c}_i(\mathbf{Mw_i}, \mathbf{w_j}) + \sum_{i=1}^{n} \dot{c}_i(\mathbf{Cw_i}, \mathbf{w_j}) + \sum_{i=1}^{n} c_i(\mathbf{Kw_i}, \mathbf{w_j}) = 0 \qquad (6.24)$$

Using mode shape orthonormality, Equation 6.24 becomes

$$\ddot{c}_j + \sum_{i=1}^{n} \dot{c}_i(\mathbf{Cw_i}, \mathbf{w_j}) + \omega_j^2 c_j = 0 \qquad (6.25)$$

No orthogonality condition exists that collapses the sum involving the damping inner product to a single term. Equation 6.25 shows that, in general, the principal coordinates of the undamped system cannot be used to uncouple the differential equations for a damped system.

A special case is that of proportional damping in which the damping matrix is a linear combination of the mass and stiffness matrices,

$$\mathbf{C} = \alpha\mathbf{K} + \beta\mathbf{M} \qquad (6.26)$$

While this may seem an obscure case, a proportional assumption may be used when no other information exists as to the nature of the damping matrix. A designer may choose to use proportional damping in the design of a discrete system using discrete viscous dampers, as it makes the analysis of the response much easier to determine and thus vibration control easier to achieve.

If the system has proportional damping, the middle term on the left-hand side of Equation 6.24 becomes

$$\sum_{i=1}^{n} \dot{c}_i(\mathbf{Cw_i}, \mathbf{w_j}) = \sum_{i=1}^{n} \dot{c}_i[\alpha(\mathbf{Kw_i}, \mathbf{w_j}) + \beta(\mathbf{Mw_i}, \mathbf{w_j})]$$

$$= \sum_{i=1}^{n} \dot{c}_i(\alpha\omega_j^2\delta_{i,j} + \beta\delta_{i,j})$$

$$= \dot{c}_j(\alpha\omega_j^2 + \beta) \qquad (6.27)$$

Substitution of Equation 6.27 in Equation 6.24 leads to

$$c_j + (\alpha\omega_j^2 + \beta)\dot{c}_j + \omega_j^2 c_j = 0 \qquad (6.28)$$

The standard form of the differential equation governing the free response of a one-degree-of-freedom system of viscous damping ratio ζ and natural frequency ω_n is

$$\ddot{x} + 2\zeta\omega_n\dot{x} + \omega_n^2 x = 0 \qquad (6.29)$$

By analogy with Equation 6.29, Equation 6.28 is rewritten in the form

$$\ddot{c}_j + 2\zeta_j\omega_j\dot{c}_j + \omega_j^2 c_j = 0 \tag{6.30}$$

where the modal damping ratios are defined as

$$\zeta_j = \frac{1}{2}\alpha\omega_j + \frac{1}{2\omega_j}\beta \tag{6.31}$$

The general solution of Equation 6.30 depends upon the value of ζ_j. The mode is underdamped if $0 < \zeta_j < 1$ and the response is

$$c_j(t) = A_j e^{-\zeta_j\omega_j t}\sin\left(\omega_i\sqrt{1-\zeta_j^2}\,t + \phi_j\right) \tag{6.32}$$

The mode is critically damped if $\zeta_j = 1$ and the general response is

$$c_j(t) = e^{-\omega_j t}(A_j + B_j t) \tag{6.33}$$

The mode is overdamped if $\zeta_j > 1$ and the general response is

$$c_j(t) = e^{-\zeta_j\omega_j t}\left(A_j e^{-\omega_j \sqrt{\zeta_j^2-1}\,t} + B_j e^{\omega_j \sqrt{\zeta_j^2-1}\,t}\right) \tag{6.34}$$

For a system with $\beta = 0$, the modal damping ratio increases with increasing modal frequency. Some or all modes may be overdamped.

Example 6.1. Determine the free response for the system of Figure 6.1 if the 20 kg block is displaced 1 cm while the 10 kg block is held in the system's equilibrium position and then the system is released.

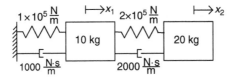

FIGURE 6.1 System of Example 6.1 is proportionally damped.

Solution: The differential equations governing the motion of the system of Figure 6.1 are

$$\begin{bmatrix} 10 & 0 \\ 0 & 20 \end{bmatrix} \begin{bmatrix} \ddot{x}_1 \\ \ddot{x}_2 \end{bmatrix} + \begin{bmatrix} 3000 & -2000 \\ -2000 & 2000 \end{bmatrix} \begin{bmatrix} \dot{x}_1 \\ \dot{x}_2 \end{bmatrix}$$

$$+ \begin{bmatrix} 3 \times 10^5 & -2 \times 10^5 \\ -2 \times 10^5 & 2 \times 10^5 \end{bmatrix} \begin{bmatrix} x_1 \\ x_2 \end{bmatrix} = \begin{bmatrix} 0 \\ 0 \end{bmatrix} \tag{a}$$

The damping matrix **C** of Equation a is proportional to the stiffness matrix **K** with a constant of proportionality of $\alpha = 0.01$.

The response of a discrete system with proportional damping is built from the free undamped response. The natural frequencies of the system are the square roots of the eigenvalues of $\mathbf{M}^{-1}\mathbf{K}$, and the mode shape vectors are the corresponding normalized eigenvectors. The natural frequencies and eigenvectors are determined as

$$\omega_1 = 80.57 \text{r/s} \qquad \mathbf{X}_1 = \begin{bmatrix} 0.1630 \\ 0.1916 \end{bmatrix} \tag{b}$$

$$\omega_2 = 196.23 \text{r/s} \qquad \mathbf{X}_2 = \begin{bmatrix} -0.2710 \\ 0.1153 \end{bmatrix} \tag{c}$$

The modal damping ratios are obtained using Equation 6.31 as

$$\zeta_1 = \frac{1}{2}\alpha\omega_1 = \frac{1}{2}(0.01)(80.57) = 0.403 \tag{d}$$

$$\zeta_2 = \frac{1}{2}\alpha\omega_2 = \frac{1}{2}(0.01)(196.23) = 0.981 \tag{e}$$

Thus, both modes are underdamped. The differential equations for each mode are

$$\ddot{c}_1 + 64.92\dot{c}_1 + 6.50 \times 10^3 c_1 = 0 \tag{f}$$

$$\ddot{c}_2 + 385.1\dot{c}_1 + 3.85 \times 10^4 c_1 = 0 \tag{g}$$

The initial conditions are

$$\begin{bmatrix} x_1(0) \\ x_2(0) \end{bmatrix} = \begin{bmatrix} 0 \\ 0.01 \end{bmatrix} \tag{h}$$

$$\begin{bmatrix} \dot{x}_1(0) \\ \dot{x}_2(0) \end{bmatrix} = \begin{bmatrix} 0 \\ 0 \end{bmatrix} \tag{i}$$

The relation between the generalized coordinates and the principal coordinates is that of Equation 6.22. Applying Equation 6.22 at $t = 0$ leads to

$$\begin{bmatrix} 0 \\ 0.01 \end{bmatrix} = c_1(0) \begin{bmatrix} 0.1630 \\ 0.1916 \end{bmatrix} + c_2(0) \begin{bmatrix} -0.2910 \\ -0.1153 \end{bmatrix} \tag{j}$$

Solution of the initial values of the principal coordinates from Equation j gives $c_1(0) = 0.0382$, $c_2(0) = 0.0231$. The initial velocities are $\dot{c}_1(0) = \dot{c}_2(0) = 0$.

The free response of an underdamped one-degree-of-freedom system is given in Equation 1.160 with parameters defined in Equation 1.161 through Equation 1.163. Adapting these equations for the response of the principal coordinates leads to

$$c_1(t) = A_1 e^{-\zeta_1 \omega_1 t} \sin(\omega_{d1} t + \phi_1) \tag{k}$$

where

$$\omega_{d1} = \omega_1 \sqrt{1 - \zeta_1^2} = 73.75 \frac{r}{s} \tag{l}$$

$$A_1 = \sqrt{(c_1(0))^2 + \left(\frac{\dot{c}_1(0) + \zeta_1 \omega_1 c_1(0)}{\omega_{d1}} \right)^2} = 0.0419 \tag{m}$$

$$\phi_1 = \tan^{-1} \left(\frac{c_1(0)\omega_{d1}}{\dot{c}_1(0) + \zeta_1 \omega_1 c_1(0)} \right) = 1.156 \text{ rad} \tag{n}$$

Substitution of Equation l through Equation n in Equation k leads to

$$c_1(t) = 0.0419 e^{-32.46t} \sin(73.75t + 1.156) \tag{o}$$

The time-dependent form of $c_2(t)$ is obtained in a similar fashion leading to

$$c_2(t) = 0.1194e^{-192.5t}\sin(37.90t + 0.1944) \qquad \text{(p)}$$

The response, in terms of the original generalized coordinates, is obtained using Equation 6.22 as

$$\begin{bmatrix} x_1(t) \\ x_2(t) \end{bmatrix} = \begin{bmatrix} 0.1630 \\ 0.1916 \end{bmatrix} c_1(t) + \begin{bmatrix} -0.2710 \\ 0.1153 \end{bmatrix} c_2(t) \qquad \text{(q)}$$

6.3 DISCRETE SYSTEMS WITH GENERAL DAMPING

The differential equations governing the free response of a discrete system with a general damping matrix cannot be uncoupled using the principal coordinates of the undamped system. The response of such a system is determined using the state-space method. Using this method, a set of n second-order differential equations with the generalized coordinates as dependent variables is recast as a set of $2n$ first-order differential equations, with the generalized coordinates and their first time derivatives as generalized coordinates.

Application of Lagrange's equations to an nDOF system with viscous damping leads to a set of n second-order differential equations in terms of $\mathbf{x}$, the vector of generalized coordinates as

$$\mathbf{M\ddot{x}} + \mathbf{C\dot{x}} + \mathbf{Kx} = 0 \qquad (6.35)$$

Define the vectors

$$\mathbf{y}_1 = \mathbf{x} \qquad (6.36)$$

$$\mathbf{y}_2 = \mathbf{\dot{x}} \qquad (6.37)$$

By definition

$$\mathbf{\dot{y}}_1 = \mathbf{y}_2 \qquad (6.38)$$

Rewriting Equation 6.35 in terms of $\mathbf{y}_1$ and $\mathbf{y}_2$

$$\mathbf{M\dot{y}}_2 + \mathbf{Cy}_2 + \mathbf{Ky}_1 = 0 \qquad (6.39)$$

$$\mathbf{\dot{y}}_2 = -\mathbf{M}^{-1}\mathbf{Ky}_1 - \mathbf{M}^{-1}\mathbf{Cy}_2 \qquad (6.40)$$

Equation 6.39 and Equation 6.40 are recast in matrix form using a single matrix as

$$
\begin{bmatrix} \dot{\mathbf{y}}_1 \\ - \\ \dot{\mathbf{y}}_2 \end{bmatrix} = \begin{bmatrix} \mathbf{0} & | & \mathbf{I} \\ - & | & - \\ -\mathbf{M}^{-1}\mathbf{K} & | & -\mathbf{M}^{-1}\mathbf{C} \end{bmatrix} \begin{bmatrix} \mathbf{y}_1 \\ - \\ \mathbf{y}_2 \end{bmatrix} \tag{6.41}
$$

Each of the two vectors in Equation 6.41 are of size $2n \times 1$ while the matrix is of size $2n \times 2n$. Defining $\mathbf{z} = [\mathbf{y}_1 | \mathbf{y}_2]^T$, Equation 6.41 is rewritten as

$$
\dot{\mathbf{z}} = \mathbf{Q}\mathbf{z} \tag{6.42}
$$

where $\mathbf{Q}$ is the $2n \times 2n$ matrix of Equation 6.41.

A solution of Equation 6.42 is sought in the form of

$$
\mathbf{z} = \mathbf{\Phi} e^{\gamma t} \tag{6.43}
$$

where $\mathbf{\Phi}$ is a $2n \times 1$ vector of constants. Substitution of Equation 6.43 into Equation 6.59 leads to

$$
\mathbf{Q}\mathbf{\Phi} = \gamma\mathbf{\Phi} \tag{6.44}
$$

Equation 6.44 is in the form of a matrix eigenvalue problem where the eigenvalues are the values of the exponent γ and the non-trivial forms of $\mathbf{\Phi}$ as the corresponding eigenvectors.

The matrix $\mathbf{Q}$ is not symmetric and thus not self-adjoint with respect to the standard inner product on R^{2n}. There is no inner product for which $\mathbf{Q}$ is self-adjoint. Its adjoint with respect to the standard inner product is $\mathbf{Q}^T$. The eigenvalues of $\mathbf{Q}^T$ are the complex conjugates of the eigenvalues of $\mathbf{Q}$.

Let γ_i be an eigenvalue of $\mathbf{Q}$ with corresponding eigenvector $\mathbf{\Phi}_i$. Then by definition

$$
\mathbf{Q}\mathbf{\Phi}_i = \gamma_i\mathbf{\Phi}_i \tag{6.45}
$$

Let $\hat{\mathbf{\Phi}}_i$ be the eigenvector of $\mathbf{Q}^T$ corresponding to its eigenvector $\bar{\gamma}_i$. Then

$$
\mathbf{Q}^T\hat{\mathbf{\Phi}}_i = \bar{\gamma}_i\hat{\mathbf{\Phi}}_i \tag{6.46}
$$

Recalling that $(\mathbf{AB})^T = \mathbf{B}^T\mathbf{A}^T$ for two matrices for which the product $\mathbf{AB}$ can be formed, transposing both sides of Equation 6.63 leads to

$$\hat{\boldsymbol{\Phi}}_i^T \mathbf{Q} = \bar{\gamma}_i \hat{\boldsymbol{\Phi}}_i^T \qquad (6.47)$$

Equation 6.47 implies that $\hat{\boldsymbol{\Phi}}_i$ is a left eigenvector of $\mathbf{Q}$.
Eigenvectors are normalized by requiring

$$(\boldsymbol{\Phi}_i, \hat{\boldsymbol{\Phi}}_i) = 1 \qquad (6.48)$$

The bi-orthogonality condition for the normalized eigenvectors is

$$(\boldsymbol{\Phi}_i, \hat{\boldsymbol{\Phi}}_j) = \delta_{i,j} \qquad (6.49)$$

Taking the inner product of Equation 6.46 with $\hat{\boldsymbol{\Phi}}_j$ gives

$$(\mathbf{Q}\boldsymbol{\Phi}_i, \hat{\boldsymbol{\Phi}}_j) = (\gamma_i \boldsymbol{\Phi}_i, \hat{\boldsymbol{\Phi}}_j) = \gamma_i (\boldsymbol{\Phi}_i, \hat{\boldsymbol{\Phi}}_j) = \gamma_i \delta_{i,j} \qquad (6.50)$$

The general solution is a linear combination of all possible solutions

$$\mathbf{z} = \sum_{j=1}^{2n} C_j \boldsymbol{\Phi}_j e^{\gamma_j t} \qquad (6.51)$$

Since the elements of $\mathbf{Q}$ are real and thus the coefficients of its characteristic equation are real, complex eigenvalues occur in complex conjugate pairs. However, since the characteristic equation includes odd powers of γ if $\gamma = a$ is an eigenvalue of $\mathbf{Q}$, then it is not necessarily the case that $\gamma = -a$ is also an eigenvalue of $\mathbf{Q}$. Complex eigenvlaues correspond to underdamped modes whereas real eigenvalues correspond to overdamped modes. However, over-damped modes do not occur in pairs as they do for a system with proportional damping; there are no values of ζ and ω for which two eigenvalues can be written, such as in the overdamped response of Equation 1.167. All eigenvalues of $\mathbf{Q}$ have negative real parts if the system is stable (as all considered in this study are).

Example 6.2. The two-degree-of-freedom system of Figure 6.2 is not proportionally damped. (a) Write the governing differential equations using the state-space formulation of Equation 6.41. (b) Determine the eigenvalues and normalized eigenvectors of $\mathbf{Q}$. (c) Determine the free response if the 30 kg mass is given a displacement of 0.01 m while the 10 kg mass is held in its equilibrium position and then the system is released from the rest.

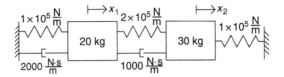

FIGURE 6.2 System of Example 6.2 has general damping.

Solution: The differential equations governing the motion of the system are

$$
\begin{bmatrix} 20 & 0 \\ 0 & 30 \end{bmatrix} \begin{bmatrix} \ddot{x}_1 \\ \ddot{x}_2 \end{bmatrix} + \begin{bmatrix} 3000 & -1000 \\ -1000 & 1000 \end{bmatrix} \begin{bmatrix} \dot{x}_1 \\ \dot{x}_2 \end{bmatrix}
$$
$$
+ \begin{bmatrix} 3 \times 10^5 & -2 \times 10^5 \\ -2 \times 10^5 & 3 \times 10^5 \end{bmatrix} \begin{bmatrix} x_1 \\ x_2 \end{bmatrix} = \begin{bmatrix} 0 \\ 0 \end{bmatrix} \tag{a}
$$

Define the vectors

$$
\mathbf{y}_1 = \begin{bmatrix} x_1 \\ x_2 \end{bmatrix} \qquad \mathbf{y}_2 = \begin{bmatrix} \dot{x}_1 \\ \dot{x}_2 \end{bmatrix} \tag{b}
$$

Calculations lead to

$$
\mathbf{M}^{-1}\mathbf{C} = \begin{bmatrix} 150 & -50 \\ -33.33 & 33.33 \end{bmatrix} \qquad \mathbf{M}^{-1}\mathbf{K} = \begin{bmatrix} 15000 & -10000 \\ -6666.67 & 10000 \end{bmatrix} \tag{c}
$$

The differential equations are rewritten in the form of Equation 6.41 as

$$
\begin{bmatrix} \dot{y}_{1,1} \\ \dot{y}_{1,2} \\ \dot{y}_{2,1} \\ \dot{y}_{2,2} \end{bmatrix} = \begin{bmatrix} 0 & 0 & 1 & 0 \\ 0 & 0 & 0 & 1 \\ -15000 & 10000 & -150 & 50 \\ 6666.67 & -10000 & 33.33 & -33.33 \end{bmatrix} \begin{bmatrix} y_{1,1} \\ y_{1,2} \\ y_{2,1} \\ y_{2,2} \end{bmatrix} \tag{d}
$$

The eigenvalues of the matrix $\mathbf{Q}$ are obtained from

$$
\begin{vmatrix}
-\lambda & 0 & 1 & 0 \\
0 & -\lambda & 0 & 1 \\
-15000 & 10000 & -150-\lambda & 50 \\
6666.67 & -10000 & 33.33 & -33.33-\lambda
\end{vmatrix} = 0
\tag{e}
$$

The eigenvalues of Q and their corresponding eigenvectors are

$$
\lambda_1 = -73.54 + 114.4i \qquad \mathbf{Y}_1 =
\begin{bmatrix}
-0.00364 - 0.00567i \\
0.00290 + 0.000510i \\
0.9164 \\
-0.2715 + 0.249i
\end{bmatrix}
\tag{f}
$$

$$
\lambda_2 = -73.54 - 114.4i \qquad \mathbf{Y}_2 =
\begin{bmatrix}
-0.00364 + 0.00567i \\
0.00290 - 0.000510i \\
0.9164 \\
-0.2715 - 0.249i
\end{bmatrix}
\tag{g}
$$

$$
\lambda_3 = -18.13 + 64.63i \qquad \mathbf{Y}_3 =
\begin{bmatrix}
0.00588 + 0.00775i \\
0.00304 + 0.01086i \\
-0.6076 + 0.2396i \\
-0.7571
\end{bmatrix}
\tag{h}
$$

$$
\lambda_4 = -18.13 - 64.63i \qquad \mathbf{Y}_4 =
\begin{bmatrix}
0.00588 - 0.00775i \\
0.00304 - 0.01086i \\
-0.6075 - 0.2396i \\
-0.7571
\end{bmatrix}
\tag{i}
$$

The general solution is assumed as

$$
\mathbf{y} = \sum_{i=1}^{4} C_i \mathbf{Y}_i e^{\lambda_i t}
\tag{j}
$$

The constants of integration are obtained through application of the initial conditions

$$
\begin{bmatrix} 0 \\ 0.01 \\ 0 \\ 0 \end{bmatrix} = C_1 \begin{bmatrix} -0.00364 - 0.00567i \\ 0.00290 + 0.000510i \\ 0.9164 \\ -0.2715 + 0.249i \end{bmatrix} + C_2 \begin{bmatrix} -0.00364 + 0.00567i \\ 0.00290 - 0.000510i \\ 0.9164 \\ -0.2715 - 0.249i \end{bmatrix}
$$

$$
+ C_3 \begin{bmatrix} 0.00588 + 0.00775i \\ 0.00304 + 0.01086i \\ -0.6076 + 0.2396i \\ -0.7571 \end{bmatrix} + C_4 \begin{bmatrix} 0.00588 - 0.00775i \\ 0.00304 - 0.01086i \\ -0.6076 - 0.2396i \\ -0.7571 \end{bmatrix} \quad \text{(k)}
$$

The constants of integration are calculated by solving a system of simultaneous algebraic equations generated from Equation k leading to

$$
C_1 = 0.0673 - 0.6737i \quad \text{(l)}
$$

$$
C_2 = 0.0673 + 0.6737i \quad \text{(m)}
$$

$$
C_3 = 0.2375 - 0.3441i \quad \text{(n)}
$$

$$
C_4 = 0.2375 + 0.3441i \quad \text{(o)}
$$

Equation l through Equation o are substituted into Equation j. The solutions for the original generalized coordinates are

$$
x_1(t) = e^{-73.54t}[-0.004064 \cos(114.4t) - 0.002072 \sin(114.4t)]
$$

$$
+ e^{-18.13t}[0.004064 \cos(64.6t) + 0.0001827 \sin(64.6t)] \quad \text{(p)}
$$

$$
x_2(t) = e^{-73.54t}[0.0005392 \cos(114.4t) - 0.001908 \sin(114.4t)]
$$

$$
+ e^{-18.13t}[0.004461 \cos(64.6t) + 0.001531 \sin(64.6t)] \quad \text{(q)}
$$

The time-dependent response is illustrated in Figure 6.3.

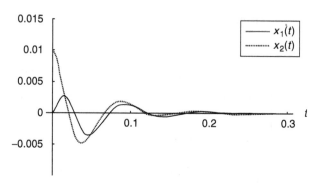

FIGURE 6.3 Time dependent response of system of Example 6.2.

6.4 DISCRETE GYROSCOPIC SYSTEMS

The general form of the differential equations for the free response of a discrete gyroscopic system is

$$\mathbf{M}\ddot{\mathbf{x}} + \mathbf{G}\dot{\mathbf{x}} + \mathbf{K}\mathbf{x} = 0 \tag{6.52}$$

where $\mathbf{M}$ and $\mathbf{K}$ are the symmetric $n \times n$ mass and stiffness matrices respectively and $\mathbf{G}$ is the $n \times n$ skew-symmetric gyroscopic matrix. Application of the normal mode solution, $\mathbf{x}(t) = \mathbf{X}e^{i\omega t}$, into Equation 6.52 leads to

$$(-\omega^2 \mathbf{M} + i\omega \mathbf{G} + \mathbf{K})\mathbf{X} = 0 \tag{6.53}$$

A non-trivial solution to Equation 6.53 exists if and only if

$$|-\omega^2 \mathbf{M} + i\omega \mathbf{G} + \mathbf{K}| = 0 \tag{6.54}$$

Define the matrix

$$\mathbf{U}(\omega) = -\omega^2 \mathbf{M} + i\omega \mathbf{G} + \mathbf{K} \tag{6.55}$$

Equation 6.54 and Equation 6.55 show that $|\mathbf{U}| = 0$. The transpose of $\mathbf{U}$ is

$$\mathbf{U}^{\mathrm{T}}(\omega) = -\omega^2 \mathbf{M}^{\mathrm{T}} + i\omega \mathbf{G}^{\mathrm{T}} + \mathbf{K}^{\mathrm{T}} \tag{6.56}$$

However, $\mathbf{M}$ and $\mathbf{K}$ are symmetric so that $\mathbf{M}^T = \mathbf{M}$ and $\mathbf{K}^T = \mathbf{K}$. The gyroscopic matrix is skew symmetric, $\mathbf{G}^T = -\mathbf{G}$. Thus

$$\mathbf{U}^T(\omega) = -\omega^2 \mathbf{M} - i\omega \mathbf{G} + \mathbf{K} \qquad (6.57)$$

The determinant of the transpose of a matrix is equal to the determinant of the matrix. Since $|\mathbf{U}(\omega)| = 0$, $|\mathbf{U}^T(\omega)| = 0$ and vice versa. Suppose ω is a natural frequency of the system and thus $|\mathbf{U}(\omega)| = 0$. Consider

$$\mathbf{U}^T(-\omega) = -\omega^2 \mathbf{M} + i\omega \mathbf{G} + \mathbf{K} = \mathbf{U}(\omega) \qquad (6.58)$$

Equation 6.58 implies that $-\omega$ is also a solution to Equation 6.54. Since complex solutions occur in complex conjugate pairs, the pairing of ω and $-\omega$ as solutions implies that only purely real or purely imaginary solutions to Equation 6.54 exist.

A pair of imaginary values, $\pm i\omega$, lead to normal mode responses of $e^{\omega t}$ and $e^{-\omega t}$. If such values are obtained, the former response grows without bound and the system is unstable. Thus, for a stable system, the solutions of $|\mathbf{U}(\omega)| = 0$ are pairs of real numbers.

Although the natural frequencies are real, the mode shapes are complex. Equation 6.53 is not cast in the form of a traditional eigenvalue problem. This is accomplished by using a state-space formulation of the equations in a fashion similar to that for systems with viscous damping. Define

$$\mathbf{y}_1 = \mathbf{x} \qquad (6.59)$$

$$\mathbf{y}_2 = \dot{\mathbf{x}} \qquad (6.60)$$

The state-space formulation of Equation 6.52 is

$$\begin{bmatrix} \dot{\mathbf{y}}_1 \\ - \\ \dot{\mathbf{y}}_2 \end{bmatrix} = \begin{bmatrix} \mathbf{0} & | & \mathbf{I} \\ - & | & - \\ -\mathbf{M}^{-1}\mathbf{K} & | & -\mathbf{M}^{-1}\mathbf{G} \end{bmatrix} \begin{bmatrix} \mathbf{y}_1 \\ - \\ \mathbf{y}_2 \end{bmatrix} \qquad (6.61)$$

A solution of Equation 6.61 is assumed as

$$\mathbf{y} = \mathbf{Y}e^{\lambda t} \qquad (6.62)$$

that, when substituted into Equation 6.62, leads to

$$\mathbf{AY} = \lambda \mathbf{Y} \tag{6.63}$$

where the $2n \times 2n$ matrix in Equation 6.61 is designated as $\mathbf{A}$. From the previous discussion, the eigenvalues of $\mathbf{A}$ are purely imaginary, although its eigenvectors are complex.

Example 6.3. The differential equations governing the motion of the system of Figure 6.4 are derived in Example 2.13 as

$$\begin{bmatrix} m & 0 \\ 0 & m \end{bmatrix} \begin{bmatrix} \ddot{x} \\ \ddot{y} \end{bmatrix} + \begin{bmatrix} 0 & -2m\omega \\ 2m\omega & 0 \end{bmatrix} \begin{bmatrix} \dot{x} \\ \dot{y} \end{bmatrix}$$

$$+ \begin{bmatrix} k_1 - m\omega^2 & 0 \\ 0 & k_2 - m\omega^2 \end{bmatrix} \begin{bmatrix} x \\ y \end{bmatrix} = \begin{bmatrix} 0 \\ 0 \end{bmatrix} \tag{a}$$

Determine the natural frequencies and mode shape vectors of the system and the system response if the particle is initially displaced such that

$$\begin{bmatrix} x_1(0) \\ x_2(0) \end{bmatrix} = \begin{bmatrix} 0.01 \\ 0.02 \end{bmatrix} \tag{b}$$

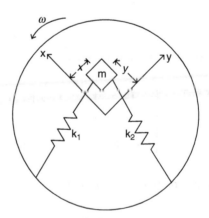

FIGURE 6.4 Gyroscopic system of Example 6.3.

with initial velocities of zero. Use values of

$$m = 4 \text{ kg}, \qquad k_1 = 400 \text{ N/m}, \qquad k_2 = 500 \text{ N/m} \qquad \text{and} \qquad \omega = 8 \text{ r/s}.$$

Solution: Substituting the given values into Equation a leads to

$$\begin{bmatrix} 4 & 0 \\ 0 & 4 \end{bmatrix} \begin{bmatrix} \ddot{x} \\ \ddot{y} \end{bmatrix} + \begin{bmatrix} 0 & -64 \\ 64 & 0 \end{bmatrix} \begin{bmatrix} \dot{x} \\ \dot{y} \end{bmatrix} + \begin{bmatrix} 144 & 0 \\ 0 & 244 \end{bmatrix} \begin{bmatrix} x \\ y \end{bmatrix} = \begin{bmatrix} 0 \\ 0 \end{bmatrix} \qquad \text{(c)}$$

Equation 6.61 is formulated as

$$A = \begin{bmatrix} 0 & 0 & 1 & 0 \\ 0 & 0 & 0 & 1 \\ -36 & 0 & 0 & 16 \\ 0 & -61 & -16 & 0 \end{bmatrix} \qquad \text{(d)}$$

The eigenvalues and eigenvectors of **A** are calculated as

$$\lambda_1 = 18.619i \qquad \Phi_1 = \begin{bmatrix} -0.0371 \\ -0.387i \\ -0.6911i \\ 0.7207 \end{bmatrix} \qquad \text{(e)}$$

$$\lambda_2 = 18.019i \qquad \Phi_2 = \begin{bmatrix} -0.0371 \\ 0.387i \\ 0.6911i \\ 0.7207 \end{bmatrix} \qquad \text{(f)}$$

$$\lambda_3 = 2.5169i \qquad \Phi_3 = \begin{bmatrix} 0.2973i \\ 0.2190 \\ -0.7482 \\ 0.5512i \end{bmatrix} \qquad \text{(g)}$$

$$\lambda_3 = -2.5169i \qquad \Phi_4 = \begin{bmatrix} -0.2973i \\ 0.2190 \\ -0.7482 \\ -0.5512i \end{bmatrix} \qquad \text{(h)}$$

Equation d through Equation h show that the natural frequencies are 18.61 r/s and 2.519 r/s. The free response is of the form

$$\mathbf{Y} = \sum_{i=1}^{4} c_i \Phi_i e^{\lambda_i t} = c_1 \Phi_1 e^{i\omega_1 t} + c_2 \bar{\Phi}_1 e^{-i\omega_1 t} + c_3 \Phi_3 e^{i\omega_2 t} + c_4 \bar{\Phi}_3 e^{-i\omega_2 t} \qquad \text{(i)}$$

Application of initial conditions to Equation i leads to

$$\begin{bmatrix} 0.01 \\ 0.02 \\ 0 \\ 0 \end{bmatrix} = c_1 \begin{bmatrix} -0.0371 \\ -0.387i \\ -0.6911i \\ 0.7207 \end{bmatrix} + c_2 \begin{bmatrix} -0.0371 \\ 0.387i \\ 0.6911i \\ 0.7207 \end{bmatrix} + c_3 \begin{bmatrix} 0.2973i \\ 0.2190 \\ -0.7482 \\ 0.5512i \end{bmatrix}$$

$$+ c_4 \begin{bmatrix} -0.2973i \\ 0.2190 \\ -0.7482 \\ -0.5512i \end{bmatrix} \qquad \text{(j)}$$

Simultaneous solution of the equations developed in Equation j leads to

$$\begin{aligned} c_1 &= -0.117 + 0.0415i \\ c_2 &= -0.117 - 0.0415i \\ c_3 &= 0.383 - 0.0154i \\ c_4 &= 0.383 + 0.154i \end{aligned} \qquad \text{(k)}$$

Substitution of Equation k into Equation i leads to

$$
\begin{bmatrix} x_1 \\ x_2 \\ \dot{x}_1 \\ \dot{x}_2 \end{bmatrix} = (0.117 + 0.0415i) \begin{bmatrix} -0.0371 \\ -0.387i \\ -0.6911i \\ 0.7207 \end{bmatrix} e^{18.619it}
$$

$$
+ (-0.117 - 0.415i) \begin{bmatrix} -0.0371 \\ 0.387i \\ 0.6911i \\ 0.7207 \end{bmatrix} e^{-18.619it}
$$

$$
+ (0.383 - 0.154i) \begin{bmatrix} 0.2973i \\ 0.2190 \\ -0.7482 \\ 0.5512i \end{bmatrix} e^{2.5169it}
$$

$$
+ (0.383 + 0.154i) \begin{bmatrix} -0.2973i \\ 0.2190 \\ -0.7482 \\ -0.5512i \end{bmatrix} e^{-2.5169it} \tag{l}
$$

Note that the first two terms are complex conjugates as are the last two terms. The sum of a complex number and its complex conjugate is twice the real part of the number. Using this analysis, the solution for the generalized coordinates are obtained as

$$
x_1(t) = -0.00434 \cos(18.619t) + 0.00154 \sin(18.619t)
$$

$$
+ 0.0458 \cos(2.5169t) - 0.119 \sin(2.5169) \tag{m}
$$

$$
x_2(t) = 0.0161 \cos(18.619t) + 0.0453 \sin(18.619t)
$$

$$
+ 0.0839 \cos(2.5169t) + 0.0337 \sin(2.5169) \tag{n}
$$

6.5 CONTINUOUS SYSTEMS WITH VISCOUS DAMPING

The general form of the differential equations governing the free response of a continuous system with viscous damping is

$$\mathbf{K}\mathbf{u} + \mathbf{C}\frac{\partial \mathbf{u}}{\partial t} + \mathbf{M}\frac{\partial^2 \mathbf{u}}{\partial t^2} = 0 \qquad (6.64)$$

where $\mathbf{K}$ is the stiffness operator, $\mathbf{C}$ is the viscous damping operator, and $\mathbf{M}$ is the inertia operator.

Assume that $\mathbf{M}$ and $\mathbf{K}$ are self-adjoint with respect to the standard inner product and that $\mathbf{M}^{-1}\mathbf{K}$ is self-adjoint with respect to the kinetic energy inner product. Let $\omega_1 \leq \omega_2 \leq \cdots \leq \omega_k \leq \cdots$ be the system's natural frequencies with corresponding normalized mode shape vectors $\mathbf{w}_1, \mathbf{w}_2, \ldots, \mathbf{w}_k$. Assume $\mathbf{u}$ can be expanded in terms of the principal coordinates, $c_1(t), c_2(t), \ldots c_k(t)$, of the system when $\mathbf{C} = \mathbf{0}$,

$$\mathbf{u} = \sum_{i=1}^{\infty} c_i(t)\mathbf{w}_i \qquad (6.65)$$

Substituting Equation 6.65 into Equation 6.64, taking the inner product with $\mathbf{w}_j$, and using mode shape orthonormality leads to

$$\ddot{c}_j(t) + \sum_{i=1}^{\infty} \dot{c}_i(t)(\mathbf{C}\mathbf{w}_i, \mathbf{w}_j) + \omega_j^2 c_j(t) = 0 \qquad (6.66)$$

Thus, as in the case for discrete systems, the use of the principal coordinates for the undamped system does not uncouple the differential equations for a continuous system with viscous damping.

The system of Equation 6.64 can be rewritten using a state-space formulation. To this end, define

$$y_1 = \mathbf{u} \qquad (6.67)$$

$$y_2 = \frac{\partial \mathbf{u}}{\partial t} \qquad (6.68)$$

From the definitions in Equation 6.67 and Equation 6.68,

$$\frac{\partial y_1}{\partial t} = y_2 \qquad (6.69)$$

Substitution of Equation 6.66 and Equation 6.68 into Equation 6.61 leads to

$$\mathbf{K}\mathbf{y}_1 + \mathbf{C}\mathbf{y}_2 + \mathbf{M}\frac{\partial \mathbf{y}_2}{\partial t} = 0 \tag{6.70}$$

$$\frac{\partial \mathbf{y}_2}{\partial t} = -\mathbf{M}^{-1}\mathbf{K}\mathbf{y}_1 - \mathbf{M}^{-1}\mathbf{C}\mathbf{y}_1 \tag{6.71}$$

Equation 6.70 and Equation 6.71 are summarized in a matrix form as

$$\begin{bmatrix} \dfrac{\partial \mathbf{y}_1}{\partial t} \\[2mm] \dfrac{\partial \mathbf{y}_2}{\partial t} \end{bmatrix} = \begin{bmatrix} 0 & \mathbf{I} \\[2mm] -\mathbf{M}^{-1}\mathbf{K} & -\mathbf{M}^{-1}\mathbf{C} \end{bmatrix} \begin{bmatrix} \mathbf{y}_1 \\[2mm] \mathbf{y}_2 \end{bmatrix} \tag{6.72}$$

A solution of Equation 6.72 is assumed as

$$\mathbf{y} = \Phi(x)e^{\gamma t} \tag{6.73}$$

that when substituted into Equation 6.72 leads to

$$\gamma\Phi = \mathbf{Q}\Phi \tag{6.74}$$

where

$$\mathbf{Q} = \begin{bmatrix} 0 & \mathbf{I} \\ -\mathbf{M}^{-1}\mathbf{K} & -\mathbf{M}^{-1}\mathbf{C} \end{bmatrix} \tag{6.75}$$

The values of γ are the eigenvalues of $\mathbf{Q}$ with corresponding eigenvectors $\Phi(x)$. Since $\mathbf{Q}$ is not symmetric, it is not self-adjoint with respect to the standard inner product on the appropriate vector space.

Separation of variables is an alternate method used to determine the free response of a continuous system with viscous damping when $\mathbf{L}$, $\mathbf{M}$, and $\mathbf{C}$ are of appropriate forms. The method is illustrated in Example 6.4.

Example 6.4. A fixed-free shaft rotates in a viscous liquid such that it is subject to a resisting moment per length of the form $c(\partial\theta/\partial t)$ over the entire length of the shaft. The equation governing the free response of the shaft is

$$\rho J\frac{\partial^2\theta}{\partial t^2} + c\frac{\partial\theta}{\partial t} - JG\frac{\partial^2\theta}{\partial x^2} = 0 \tag{a}$$

(a) Nondimensionalize Equation a using the nondimensional variable.

$$t^* = \frac{t}{\sqrt{\frac{\rho L^2}{G}}} \tag{b}$$

$$x^* = \frac{x}{L} \tag{c}$$

(b) Determine the free response when a torque is applied to the free end such that it is rotated through an angle θ_0 such that $\theta(x,0) = \theta_0 x$ and $(\partial\theta/\partial t)(x, 0) = 0$.

Solution: (a) Substitution of Equation b into Equation a leads to

$$\frac{\partial^2 \theta}{\partial t^2} + 2\zeta \frac{\partial\theta}{\partial t} - \frac{\partial^2 \theta}{\partial x^2} = 0 \tag{d}$$

where

$$\zeta = \frac{cL}{2J\sqrt{\rho G}} \tag{e}$$

The boundary and initial conditions are

$$\theta(0, t) = 0 \tag{f}$$

$$\frac{\partial\theta}{\partial x}(1, t) = 0 \tag{g}$$

$$\theta(x, 0) = \theta_0 x \tag{h}$$

$$\frac{\partial\theta}{\partial t}(x, 0) = 0 \tag{i}$$

A separation of variables solution is assumed as

$$\theta(x, t) = f(x)g(t) \tag{j}$$

Substitution of Equation j into Equation d leads to

$$\frac{d^2 g}{dt^2} f(x) + 2\zeta \frac{dg}{dt} f(x) - \frac{d^2 f}{dx^2} g(t) = 0 \tag{k}$$

Dividing Equation k by $f(x)g(t)$ and rearranging leads to

$$\frac{1}{g(t)}\left(\frac{d^2g}{dt^2} + 2\zeta\frac{dg}{dt}\right) = \frac{1}{f}\frac{d^2f}{dx^2} \tag{1}$$

The usual separation argument is used: The left-hand side of Equation 1 is a function of t only, while the right-hand side of Equation 1 is a function of x only. Since t and x are independent variables, this can be true only if both sides are equal to the same constant, call it $-\lambda$. Equation 1 is then separated into

$$\frac{d^2f}{dx^2} = -\lambda f \tag{m}$$

$$\frac{d^2g}{dt^2} + 2\zeta\frac{dg}{dt} = -\lambda g \tag{n}$$

Application of the separation of variables to the boundary conditions, Equation f and Equation g, leads to

$$f(0) = 0 \tag{o}$$

$$\frac{df}{dx}(1) = 0 \tag{p}$$

Equation m, Equation o, and Equation p constitute a Sturm–Liouville problem as defined in Chapter 5. Since it is a proper Sturm–Liouville problem, the values of λ must be positive. Thus, the general solution of Equation m is

$$f(x) = C_1\sin\left(\sqrt{\lambda}x\right) + C_2\cos\left(\sqrt{\lambda}x\right) \tag{q}$$

Application of Equation o to Equation q leads to $C_2 = 0$. Subsequent application of Equation p leads to

$$C_1\sqrt{\lambda}\cos\left(\sqrt{\lambda}\right) = 0 \tag{r}$$

The eigenvalues obtained from Equation r are

$$\lambda = \left[(2n-1)\frac{\pi}{2}\right]^2 \tag{s}$$

The solutions for $f(x)$ are

$$f_n(x) = \sqrt{2} \sin\left[(2n-1)\frac{\pi}{2}x\right] \tag{t}$$

Equation n becomes

$$\frac{d^2g}{dt^2} + 2\zeta\frac{dg}{dt} + \left[(2n-1)\frac{\pi}{2}\right]^2 g = 0 \tag{u}$$

Equation u is analogous to the equation governing the free response of a one-degree-of-freedom system. The mathematical form of the response depends upon the value of ζ. For values of n such that

$$4\zeta^2 - [(2n-1)\pi]^2 < 0 \tag{v}$$

The solution of Equation u is

$$g_n(t) = e^{-\zeta t}[A_n \cos(\omega_n t) + B_n \sin(\omega_n t)] \tag{w}$$

where

$$\omega_n = \frac{1}{2}\sqrt{[(2n-1)\pi]^2 - 4\zeta^2} \tag{x}$$

Application of the product solution to the initial condition, Equation i, leads to $(dg/dt)(0) = 0$, which when applied to Equation w leads to $B_n = 0$. Thus, for values of n such that Equation w is satisfied

$$g_n(t) = A_n e^{-\zeta t}\cos(\omega_n t) \tag{y}$$

For values of n such that Equation v is not satisfied, the solution that satisfies the initial condition $(dg/dt)(0) = 0$ is

$$g_n(t) = A_n e^{-\zeta t}\cosh\left[\frac{1}{2}\sqrt{[(2n-1)\pi]^2 - 4\zeta^2}t\right] \tag{z}$$

There are an infinite number of solutions of the form of Equation j that satisfy Equation a, two boundary conditions and one initial condition.

The most general solution is a linear combination of all solutions

$$\theta(x,t) = \sum_{n=1}^{\infty} f_n(x)g_n(t) \tag{aa}$$

Application of the initial condition of Equation h through Equation aa leads to

$$\theta_0(x) = \sum_{n=1}^{\infty} f_n(x)g_n(0)$$
$$= \sum_{n=1}^{\infty} A_n\sqrt{2}\sin\left[(2n-1)\frac{\pi}{2}x\right] \tag{bb}$$

Since the eigenvalue problem for $f(x)$ is a Sturm–Liouville problem, the expansion theorem is used as

$$\theta_0(x) = \theta_0 \sum_{n=1}^{\infty} (x, f_n(x)) f_n(x) \tag{cc}$$

where

$$(x, f_n) = \sqrt{2}\int_0^1 x\sin\left[(2n-1)\frac{\pi}{2}x\right]dx$$
$$= \frac{(-1)^n 4\sqrt{2}}{[(2n-1)\pi]^2} \tag{dd}$$

Use of Equation cc in Equation bb leads to $\sqrt{2}A_n = 4\sqrt{2}\theta_0/[(2n-1)\pi]^2$ and thus

$$\theta(x,t) = \frac{4\sqrt{2}\theta_0}{\pi^2}\sum_{n=1}^{\infty}\frac{(-1)^n}{[(2n-1)\pi]^2}\hat{g}_n(t)\sin\left[(2n-1)\frac{\pi}{2}x\right] \tag{ee}$$

where $\hat{g}_n(t) = (1/A_n)g_n(t)$.

The damping in the system of Example 6.4 is analogous to proportional damping defined for discrete systems. The first few modes in the expansion of Equation ee may be underdamped. Higher modes are overdamped.

7 Forced Response

The forced response of a system is the response due to an external nonconservative force. The study of forced response of one-degree-of-freedom systems is distinguished between the steady-state response due to harmonic excitations and the transient response due to more general excitations. The same is true for multi-degree-of-freedom systems and continuous systems. The steady-state response due to harmonic excitations is first considered, first for discrete systems and then for continuous systems.

The Laplace transform method is applied to both discrete and continuous systems to determine forced response, but the algebra is often intractable. However, its advantage is that it may be applied to undamped, damped, and gyroscopic systems. Modal analysis is a more general method that is applicable to any type of excitation. Successful application of modal analysis depends upon orthogonality of the normal modes. The method is applied to both discrete and continuous systems. However, since the normal modes for damped and gyroscopic systems satisfy different orthogonality conditions than undamped systems, the methods of analysis are different.

7.1 RESPONSE OF DISCRETE SYSTEMS FOR HARMONIC EXCITATIONS

7.1.1 GENERAL THEORY

The general form of the equations governing the response of a linear discrete system with viscous damping due to a single frequency harmonic excitation is

$$\mathbf{M}\ddot{x} + \mathbf{C}\dot{x} + \mathbf{K}x = \mathbf{F}\sin(\omega t) + \mathbf{G}\cos(\omega t) \tag{7.1}$$

where $\mathbf{F}$ and $\mathbf{G}$ are vectors of amplitudes, and ω is the frequency of the excitation.

If the system is subject to a multi-frequency excitation, then the response is obtained for each frequency and linear superposition is used to determine the total response.

465

A steady-state response for Equation 7.1 is assumed of the form

$$\mathbf{x} = \mathbf{A} \sin(\omega t) + \mathbf{B} \cos(\omega t) \tag{7.2}$$

Substitution of Equation 7.2 into Equation 7.1 leads to

$$(-\omega^2 \mathbf{MA} - \omega \mathbf{CB} + \mathbf{KA})\sin(\omega t) + (-\omega^2 \mathbf{MB} + \omega \mathbf{CA} + \mathbf{KB})\cos(\omega t)$$

$$= \mathbf{F} \sin(\omega t) + \mathbf{G} \cos(\omega t) \tag{7.3}$$

Setting coefficients of $\sin(\omega t)$ and $\cos(\omega t)$ to zero independently leads to

$$-\omega^2 \mathbf{MA} - \omega \mathbf{CB} + \mathbf{KA} = \mathbf{F} \tag{7.4}$$

$$-\omega^2 \mathbf{MB} + \omega \mathbf{CA} + \mathbf{KB} = \mathbf{G} \tag{7.5}$$

Equation 7.4 and Equation 7.5 represent $2n$ simultaneous algebraic equations to solve for the components of $\mathbf{A}$ and $\mathbf{B}$.

If the system is undamped, $\mathbf{C} = 0$, then Equation 7.4 and Equation 7.5 are uncoupled. The solutions for $\mathbf{A}$ and $\mathbf{B}$ are

$$\mathbf{A} = (\mathbf{K} - \omega^2 \mathbf{M})^{-1} \mathbf{F} \tag{7.6}$$

$$\mathbf{B} = (\mathbf{K} - \omega^2 \mathbf{M})^{-1} \mathbf{G} \tag{7.7}$$

Solutions of Equation 7.6 and Equation 7.7 exist unless $|\mathbf{K} - \omega^2 \mathbf{M}| = 0$. Such is the case when the excitation frequency coincides with any of the system's natural frequencies. The result is a resonance condition like that for a one-degree-of-freedom system. The response when resonance occurs is considered using modal analysis later in this chapter.

If the system is damped, then Equation 7.4 and Equation 7.5 are solved simultaneously. Equation 7.4 is rearranged as

$$\mathbf{A} = (\mathbf{K} - \omega^2 \mathbf{M})^{-1} (\mathbf{F} + \omega \mathbf{CB}) \tag{7.8}$$

Substitution of Equation 7.8 into Equation 7.5 leads to

$$[\mathbf{K} - \omega^2 \mathbf{M} + \omega^2 \mathbf{C}(\mathbf{K} - \omega^2 \mathbf{M})^{-1} \mathbf{C}]\mathbf{B} = \mathbf{G} - \omega \mathbf{C}(\mathbf{K} - \omega^2 \mathbf{M})^{-1} \mathbf{F} \tag{7.9}$$

Equation 7.9 may be solved for $\mathbf{B}$ and then $\mathbf{A}$ is obtained from Equation 7.8. Resonance conditions do not exist for damped systems. If $|(\mathbf{K} - \omega^2 \mathbf{M})|$, then an alternate path to the solution is obtained by rearranging Equation 7.4 for $\mathbf{B}$ in terms of $\mathbf{A}$ and $\mathbf{F}$.

The above method works well for analysis of existing systems in which all parameters have numerical values. However, the algebra required to perform a frequency response analysis may be overwhelming unless a symbolic solver such as MATLAB, Maple, or Mathematica is used.

Example 7.1. Three machines are placed along the span of a simply supported beam as illustrated in Figure 7.1. During operation, the leftmost machine is subject to a harmonic force of amplitude of 20,000 N at a frequency of 84.5 r/s. Determine the steady-state response of each of the machines.

Solution: The flexibility matrix is calculated for this system in Example 5.10, as

$$\mathbf{A} = 10^{-7} \begin{bmatrix} 2.00 & 2.44 & 1.56 \\ 2.44 & 3.56 & 2.44 \\ 1.56 & 2.44 & 2.00 \end{bmatrix} \tag{a}$$

The mass matrix is determined using the kinetic energy functional, as

$$\mathbf{M} = \begin{bmatrix} 100 & 0 & 0 \\ 0 & 250 & 0 \\ 0 & 0 & 150 \end{bmatrix} \tag{b}$$

The force vector for this problem is

$$\mathbf{F} = \begin{bmatrix} 20,000\ \sin(84.5t) \\ 0 \\ 0 \end{bmatrix} \tag{c}$$

Since the flexibility matrix is known rather than the stiffness matrix, it is more

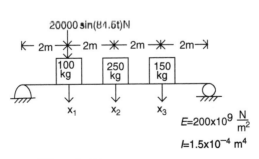

FIGURE 7.1 System of Example 7.1.

convenient to formulate the differential equations in the form

$$\mathbf{AM\ddot{x} + x = AF} \tag{d}$$

Substitution of Equation a through Equation c into Equation d leads to

$$\mathbf{AM} = 10^{-5} \begin{bmatrix} 2 & 6.11 & 2.33 \\ 2.44 & 8.89 & 3.67 \\ 1.56 & 6.11 & 3 \end{bmatrix} \begin{bmatrix} \ddot{x}_1 \\ \ddot{x}_2 \\ \ddot{x}_3 \end{bmatrix} + \begin{bmatrix} x_1 \\ x_2 \\ x_3 \end{bmatrix} = \begin{bmatrix} 0.0400 \\ 0.0488 \\ 0.0312 \end{bmatrix} \sin(84.5t) \tag{e}$$

Since the system is not damped and the force vector does not contain $\cos(84.5t)$ terms, a steady-state solution is assumed as

$$\mathbf{x} = \begin{bmatrix} X_1 \\ X_2 \\ X_3 \end{bmatrix} \sin(84.5t) \tag{f}$$

Substitution of Equation f into Equation e leads to

$$(-\omega^2 \mathbf{AM} + \mathbf{I})\mathbf{X} = \mathbf{AF}$$

$$\begin{bmatrix} 0.8572 & -0.4364 & -0.1666 \\ -0.1745 & 0.3653 & -0.2618 \\ -0.1111 & -0.4364 & 0.7858 \end{bmatrix} \begin{bmatrix} X_1 \\ X_2 \\ X_3 \end{bmatrix} = \begin{bmatrix} 0.0400 \\ 0.0488 \\ 0.0312 \end{bmatrix} \tag{g}$$

Solution of the system of Equation g leads to a steady-state response of

$$\mathbf{x} = \begin{bmatrix} 0.5819 \\ 0.8292 \\ 0.5823 \end{bmatrix} \sin(84.5t) \tag{h}$$

Example 7.2. The differential equations governing the two-degree-of-freedom system of Figure 7.2 are

$$\begin{bmatrix} 1 & 0 \\ 0 & 2 \end{bmatrix} \begin{bmatrix} \ddot{x}_1 \\ \ddot{x}_2 \end{bmatrix} + \begin{bmatrix} 10 & 0 \\ 0 & 0 \end{bmatrix} \begin{bmatrix} \dot{x}_1 \\ \dot{x}_2 \end{bmatrix} + \begin{bmatrix} 100 & -50 \\ -50 & 150 \end{bmatrix} \begin{bmatrix} x_1 \\ x_2 \end{bmatrix}$$

$$= \begin{bmatrix} 100 \\ 0 \end{bmatrix} \sin(20t) + \begin{bmatrix} 0 \\ 100 \end{bmatrix} \cos(20t) \tag{a}$$

Determine the system's steady-state response.

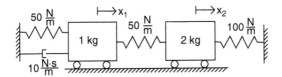

FIGURE 7.2 System of Example 7.2.

Solution: The appropriate forms of Equation 7.4 and Equation 7.5 obtained from Equation a are

$$\begin{bmatrix} -300 & -50 \\ -50 & -650 \end{bmatrix} \begin{bmatrix} A_1 \\ A_2 \end{bmatrix} - \begin{bmatrix} 200 & 0 \\ 0 & 0 \end{bmatrix} \begin{bmatrix} B_1 \\ B_2 \end{bmatrix} = \begin{bmatrix} 100 \\ 0 \end{bmatrix} \tag{b}$$

$$\begin{bmatrix} 200 & 0 \\ 0 & 0 \end{bmatrix} \begin{bmatrix} A_1 \\ A_2 \end{bmatrix} + \begin{bmatrix} -300 & -50 \\ -50 & -650 \end{bmatrix} \begin{bmatrix} B_1 \\ B_2 \end{bmatrix} = \begin{bmatrix} 0 \\ 100 \end{bmatrix} \tag{c}$$

Equation 7.8 is formulated as

$$\begin{bmatrix} A_1 \\ A_2 \end{bmatrix} = \begin{bmatrix} -0.3377 \\ 0.0260 \end{bmatrix} + \begin{bmatrix} -0.6753 & 0 \\ 0.0520 & 0 \end{bmatrix} \begin{bmatrix} B_1 \\ B_2 \end{bmatrix} \tag{d}$$

Substituting Equation d into Equation b leads to

$$\begin{bmatrix} 200 & 0 \\ 0 & 0 \end{bmatrix} \left\{ \begin{bmatrix} -0.3377 \\ 0.0260 \end{bmatrix} + \begin{bmatrix} -0.6753 & 0 \\ 0.0520 & 0 \end{bmatrix} \begin{bmatrix} B_1 \\ B_2 \end{bmatrix} \right\}$$

$$+ \begin{bmatrix} -300 & -50 \\ 50 & -350 \end{bmatrix} \begin{bmatrix} B_1 \\ n_i \end{bmatrix} = \begin{bmatrix} 0 \\ 100 \end{bmatrix}$$

$$\begin{bmatrix} -435.1 & -50 \\ -50 & -650 \end{bmatrix} \begin{bmatrix} B_1 \\ B_2 \end{bmatrix} = \begin{bmatrix} 67.53 \\ 100 \end{bmatrix} \tag{e}$$

The solution of Equation e is

$$\begin{bmatrix} B_1 \\ B_2 \end{bmatrix} = \begin{bmatrix} -0.1388 \\ -0.1432 \end{bmatrix} \tag{f}$$

Substitution of Equation f into Equation d leads to

$$\begin{bmatrix} A_1 \\ A_2 \end{bmatrix} = \begin{bmatrix} -0.2440 \\ 0.0188 \end{bmatrix} \tag{g}$$

Substitution of Equation f and Equation g into Equation 7.2 leads to

$$\begin{bmatrix} x_1(t) \\ x_2(t) \end{bmatrix} = \begin{bmatrix} -0.2440 \\ 0.0188 \end{bmatrix} \sin(20t) + \begin{bmatrix} -0.1388 \\ -0.1432 \end{bmatrix} \cos(20t) \tag{h}$$

An alternate form of Equation h is

$$\begin{bmatrix} x_1(t) \\ x_2(t) \end{bmatrix} = \begin{bmatrix} X_1 \sin(20t - \phi_1) \\ X_2 \sin(20t - \phi_2) \end{bmatrix} \tag{i}$$

The steady-state amplitudes are calculated as

$$X_1 = [(-0.2440)^2 + (-0.1388)^2]^{1/2} = 0.2807 \text{ m} \tag{j}$$

$$X_2 = [(0.0188)^2 + (-0.1432)^2]^{1/2} = 0.1444 \text{ m} \tag{k}$$

The steady-state phases are

$$\phi_1 = \tan^{-1}\left(\frac{-0.1388}{-0.2440}\right) = 2.088 \text{ rad} \tag{l}$$

$$\phi_2 = \tan^{-1}\left(\frac{-0.1432}{0.0188}\right) = -0.1303 \text{ rad} \tag{m}$$

7.1.2 VIBRATION ABSORBERS

The differential equations governing the two-degree-of-freedom system of Figure 7.3 are

$$\begin{bmatrix} m_1 & 0 \\ 0 & m_2 \end{bmatrix} \begin{bmatrix} \ddot{x}_1 \\ \ddot{x}_2 \end{bmatrix} + \begin{bmatrix} k_1 + k_2 & -k_2 \\ -k_2 & k_2 \end{bmatrix} \begin{bmatrix} x_1 \\ x_2 \end{bmatrix} = \begin{bmatrix} F_0 \\ 0 \end{bmatrix} \sin(\omega t) \tag{7.10}$$

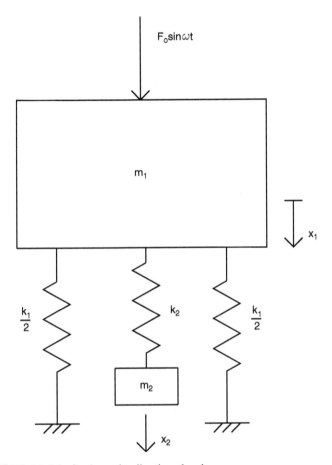

FIGURE 7.3 Model of a dynamic vibration absorber.

The steady-state response for the system of Figure 7.3 is obtained by assuming

$$\begin{bmatrix} x_1(t) \\ x_2(t) \end{bmatrix} = \begin{bmatrix} X_1 \\ X_2 \end{bmatrix} \sin(\omega t) \qquad (7.11)$$

Substituting Equation 7.11 into Equation 7.10 leads to the frequency response equations

$$X_1 = \frac{F_0}{k_1} \left| \frac{1 - r_2^2}{r_1^2 r_2^2 - r_2^2 - (1 + \mu)r_1^2 + 1} \right| \qquad (7.12)$$

$$X_2 = \frac{F_0}{k_1} \left| \frac{1}{r_1^2 r_2^2 - r_2^2 - (1 + \mu)r_1^2 + 1} \right| \tag{7.13}$$

where the following nondimensional parameters are defined

$$\mu = \frac{m_2}{m_1} \tag{7.14}$$

$$r_1 = \frac{\omega}{\omega_{11}} \tag{7.15}$$

$$r_2 = \frac{\omega}{\omega_{22}} \tag{7.16}$$

The quantities ω_{11} and ω_{22} are defined by

$$\omega_{11} = \sqrt{\frac{k_1}{m_1}} \tag{7.17}$$

$$\omega_{22} = \sqrt{\frac{k_2}{m_2}} \tag{7.18}$$

The frequency response curve showing $(k_1 X_1 / F_0)$ plotted against r_2 for selected values of the other parameters is illustrated in Figure 7.4. The frequency response curve shows that the steady-state amplitude is zero for $r = r_2$.

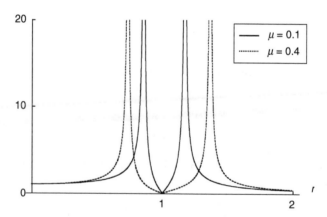

FIGURE 7.4 Frequency response curves for steady-state amplitude X_1 for system of Figure 7.3.

Consider a machine of mass m attached to an undamped foundation of stiffness k as illustrated in Figure 7.5. During operation, the machine is subject to a harmonic force of the form $F_0 \sin(\omega t)$. The frequency response curve of an undamped one-degree-of-freedom system shows that when ω is near ω_n, the steady-state amplitude is very large; and that when $\omega = \omega_n$, resonance occurs and the steady-state response is unbounded. The frequency response curve of the two-degree-of-freedom system suggests a remedy to this dilemma. Add a second mass-spring system such that the resulting system is configured as in Figure 7.3 and

$$\omega_{22} = \omega \qquad (7.19)$$

If this remedy is taken, then the steady-state amplitude of the machine is zero.

The machine in the above discussion is called the primary system. The natural frequency of the primary system, by itself, is ω_{11}. The mass-spring system attached to the primary system is called an auxiliary system. If grounded, the natural frequency of the auxiliary system is ω_{22}. When the auxiliary system is designed such that $\omega_{22} = \omega$, the steady-state amplitude of the primary system is zero. Then the auxiliary system is called a dynamic vibration absorber. The absorber is said to be "tuned" to the excitation frequency. When the absorber is tuned to the excitation frequency, Equation 7.13 leads to

$$X_2 = \frac{F_0}{k_2} \qquad (7.20)$$

The absorber response is 180°out of phase with the excitation.

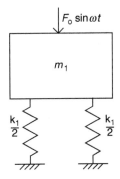

FIGURE 7.5 One-degree-of-freedom system. When $\omega \approx \sqrt{k/m}$ the machine has a large steady-state amplitude.

The addition of the absorber to the primary system adds one degree of freedom to the system. The primary system and the auxiliary system is a two-degree-of-freedom system. The natural frequencies of the system are obtained as

$$
\omega_1 = \frac{\omega_{11}}{\sqrt{2}} \left\{ 1 + \left(\frac{\omega_{22}}{\omega_{11}} \right)^2 (1 + \mu) \right.
$$

$$
\left. - \left[\left(\frac{\omega_{22}}{\omega_{11}} \right)^4 (1 + \mu)^2 + 2(\mu - 1) \left(\frac{\omega_{22}}{\omega_{11}} \right)^2 + 1 \right]^{1/2} \right\}^{1/2}
\tag{7.21}
$$

$$
\omega_2 = \frac{\omega_{11}}{\sqrt{2}} \left\{ 1 + \left(\frac{\omega_{22}}{\omega_{11}} \right)^2 (1 + \mu) \right.
$$

$$
\left. + \left[\left(\frac{\omega_{22}}{\omega_{11}} \right)^4 (1 + \mu)^2 + 2(\mu - 1) \left(\frac{\omega_{22}}{\omega_{11}} \right)^2 + 1 \right]^{1/2} \right\}^{1/2}
\tag{7.22}
$$

Analysis shows that ω_1 is less than the tuned frequency while ω_2 is greater than the tuned frequency. The separation between the natural frequencies is larger for larger values of μ.

The steady-state amplitude of the primary system is zero only when the excitation frequency is equal to the frequency to which the absorber is tuned. If the excitation frequency is different than the tuned frequency, the primary system has a nonzero steady-state amplitude. If μ, the ratio of the absorber mass to the primary mass, is small, the frequency response curve is steep near the frequency to which the absorber is tuned. A small change in excitation frequency may lead to a large change in the steady-state amplitude of the primary system.

Example 7.3. Reconsider the system of Example 7.1. The natural frequencies of the system are calculated in Example 5.10 as 87.1 r/s, 424.8 r/s, and 796.1 r/s. The frequency of the force to which the machine is subject is 87.1 r/s, which is 95% of the system's lowest natural frequency, Thus large amplitude steady-state responses are expected. Design a vibration absorber of mass 25 kg such that when suspended from the midspan of the beam, as illustrated in Figure 7.6 the steady-state amplitude of the midspan is zero. Determine the natural frequencies of the system with the absorber in place, the steady-state response of each machine, and the absorber when the excitation frequency is the same as the tuned frequency.

Solution: To tune the absorber, its grounded natural frequency is equal to the excitation frequency.

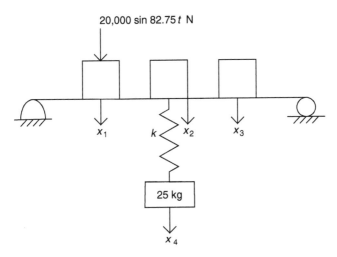

20,000 sin 82.75 t N

x_1 k x_2 x_3

25 kg

x_4

FIGURE 7.6 When vibration absorber is tuned to 82.75 r/s, the steady-state vibrations of the midspan of the beam are eliminated.

$$\sqrt{\frac{k}{m}} = \omega$$

$$k = m\omega^2$$

(a)

$$= (25 \text{ kg}) \left(82.75 \frac{r}{s} \right)^2$$

$$= 1.712 \times 10^5 \text{N/m}$$

The system with the absorber added is a four-degree-of-freedom system. The chosen generalized coordinates are the displacements of the machines, x_1, x_2, and x_3, and the displacement of the absorber, x_4. The flexibility matrix for this system is determined in Example 5.11, as

$$A = 10^{-7} \begin{bmatrix} 2.00 & 2.44 & 1.56 & 2.44 \\ 2.44 & 3.56 & 2.44 & 3.56 \\ 1.56 & 2.44 & 2.00 & 2.44 \\ 2.44 & 3.56 & 2.44 & 61.98 \end{bmatrix}$$

(b)

The mass matrix is obtained from the kinetic energy functional, as

$$
\mathbf{M} = \begin{bmatrix} 100 & 0 & 0 & 0 \\ 0 & 250 & 0 & 0 \\ 0 & 0 & 150 & 0 \\ 0 & 0 & 0 & 25 \end{bmatrix} \tag{c}
$$

The force vector is

$$
\mathbf{F} = \begin{bmatrix} 20,000 \sin(82.75t) \\ 0 \\ 0 \\ 0 \end{bmatrix} \tag{d}
$$

The differential equations are formulated, as

$$
10^{-5} \begin{bmatrix} 2.00 & 6.11 & 2.33 & 0.611 \\ 2.44 & 8.89 & 3.67 & 0.889 \\ 1.56 & 6.11 & 3.00 & 0.611 \\ 2.44 & 8.89 & 3.67 & 15.50 \end{bmatrix} \begin{bmatrix} \ddot{x}_1 \\ \ddot{x}_2 \\ \ddot{x}_3 \\ \ddot{x}_4 \end{bmatrix} + \begin{bmatrix} x_1 \\ x_2 \\ x_3 \\ x_4 \end{bmatrix}
$$

$$
= \begin{bmatrix} 0.0400 \\ 0.0489 \\ 0.0316 \\ 0.0489 \end{bmatrix} \sin(82.75t) \tag{e}
$$

The steady-state solution of Equation e is obtained as

$$
\mathbf{x} = 10^{-3} \begin{bmatrix} 6.566 \\ 0 \\ -2.643 \\ -810.5 \end{bmatrix} \sin(82.75t) \tag{f}
$$

Equation f shows that the steady-state amplitude of the midspan of the beam, where the absorber is attached is zero and the amplitudes of the other machines are significantly less than before the absorber is attached. Since the steady-state amplitude of the machine is zero, the displacements of the other two masses are out of phase, when one is positive the other is negative (Figure 7.7).

The frequency response of the midspan of the beam when the absorber is attached similar to that of Figure 7.7. The band of frequencies over which the amplitude is small is narrow. A slight change in input frequency can lead to a large change in steady-state amplitude.

The natural frequencies of the system when the absorber is added are determined as the reciprocals of the square roots of the eigenvalues of **AM**. The natural frequencies are calculated as 76.6 r/s, 99.1 r/s, 424.8 r/s, and 796.2 r/s. The larger natural frequencies are unaffected by the addition of the absorber. The lowest natural frequency of the system without the absorber is between the two lowest frequencies when the absorber is attached.

Example 7.4. Two problems with the use of a damped vibration absorber are that the narrow band over which the steady-state amplitude is small, and that during start-up, the lowest natural frequency must be passed through. A potential solution is to add damping to the absorber, resulting in the system of Figure 7.8. Determine the steady-state response of the primary system when the damped absorber is added.

Solution: The differential equations of the two-degree-of-freedom system with the damped absorber are

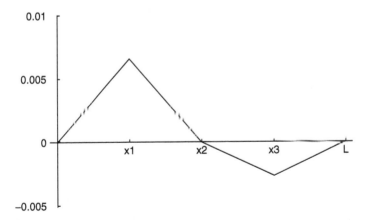

FIGURE 7.7 Graphical representation of machine displacements when absorber is tuned to excitation frequency. Amplitude of machine at midspan is zero. The responses of the other machines are out of phase with one another.

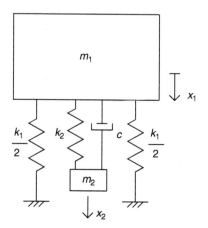

FIGURE 7.8 Model of a damped vibration absorber.

$$\begin{bmatrix} m_1 & 0 \\ 0 & m_2 \end{bmatrix} \begin{bmatrix} \ddot{x}_1 \\ \ddot{x}_2 \end{bmatrix} + \begin{bmatrix} c & -c \\ -c & c \end{bmatrix} \begin{bmatrix} \dot{x}_1 \\ \dot{x}_2 \end{bmatrix} + \begin{bmatrix} k_1 + k_2 & -k_2 \\ -k_2 & k_2 \end{bmatrix} \begin{bmatrix} x_1 \\ x_2 \end{bmatrix}$$

$$= \begin{bmatrix} F_0 \sin(\omega t) \\ 0 \end{bmatrix} \tag{a}$$

A steady-state solution of Equation a is assumed in the form of Equation 7.2.

$$\begin{bmatrix} x_1(t) \\ x_2(t) \end{bmatrix} = \begin{bmatrix} a_1 \\ a_2 \end{bmatrix} \sin(\omega t) + \begin{bmatrix} b_1 \\ b_2 \end{bmatrix} \cos(\omega t) \tag{b}$$

Substitution of Equation a and Equation b into Equation 7.4 and Equation 7.5 leads to

$$\begin{bmatrix} k_1 + k_2 - m_1\omega^2 & -k_2 \\ -k_2 & k_2 - m_2\omega^2 \end{bmatrix} \begin{bmatrix} a_1 \\ a_2 \end{bmatrix} + \begin{bmatrix} -\omega c & \omega c \\ \omega c & -\omega c \end{bmatrix} \begin{bmatrix} b_1 \\ b_2 \end{bmatrix} = \begin{bmatrix} F_0 \\ 0 \end{bmatrix} \tag{c}$$

$$\begin{bmatrix} \omega c & -\omega c \\ -\omega c & \omega c \end{bmatrix} \begin{bmatrix} a_1 \\ a_2 \end{bmatrix} + \begin{bmatrix} k_1 + k_2 - m_1\omega^2 & -k_2 \\ -k_2 & k_2 - m_2\omega^2 \end{bmatrix} \begin{bmatrix} b_1 \\ b_2 \end{bmatrix} = \begin{bmatrix} 0 \\ 0 \end{bmatrix} \tag{d}$$

The algebra is formidable, but Equation c and Equation d can be solved. The solution is of the form of Equation b, but can be rewritten in terms of amplitudes and phases as

$$\begin{bmatrix} x_1 \\ x_2 \end{bmatrix} = \begin{bmatrix} X_1 \sin(\omega t + \phi_1) \\ X_2 \sin(\omega t + \phi_2) \end{bmatrix} \tag{e}$$

where

$$X_1 = \sqrt{(a_1)^2 + (b_1)^2} \tag{f}$$

$$X_2 = \sqrt{(a_2)^2 + (b_2)^2} \tag{g}$$

$$\phi_1 = \tan^{-1}\left(\frac{b_1}{a_1}\right) \tag{h}$$

$$\phi_2 = \tan^{-1}\left(\frac{b_2}{a_2}\right) \tag{i}$$

The algebra eventually leads to

$$X_1 = \frac{F_0}{k_1} \sqrt{\frac{(2\zeta r_1 q)^2 + (r_1^2 - q^2)^2}{\{r_1^4 - [1 + (1 + \mu)q^2]r_1^2 + q^2\}^2 + (2\zeta r_1 q)^2[1 - r_1^2(1 + \mu)]^2}} \tag{j}$$

$$X_2 = \frac{F_0}{k_1} \sqrt{\frac{q^4 + (2\zeta q)^2}{\{r_1^4 - [1 + (1 + \mu)q^2]r_1^2 + q^2\}^2 + (2\zeta r_1 q)^2[1 - r_1^2(1 + \mu)]^2}} \tag{k}$$

where r_1, r_2, and μ are as defined in Equation 7.14 through Equation 7.16.

$$q = \frac{r_2}{r_1} \tag{l}$$

$$\zeta = \frac{c}{2\sqrt{k_2 m_2}} \tag{m}$$

Frequency response plots for the primary system are shown in Figure 7.9 through Figure 7.11. When damping is added to the absorber, it is not possible to tune the absorber such that the steady-state amplitude of the primary system is zero, but it is possible to design the absorber to widen the band over which the steady-state amplitude is small and reduces the amplitude during a startup.

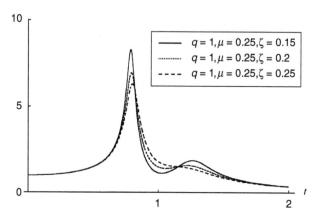

FIGURE 7.9 Frequency response curves for primary mass when damped vibration absorber is added. When the steady-state amplitude is a minimum, a small change in frequency leads to a large change in amplitude.

7.2 HARMONIC EXCITATION OF CONTINUOUS SYSTEMS

The general form of the differential equation for the harmonic excitation of a continuous system is

$$\mathbf{K}\mathbf{w} + \mathbf{C}\frac{\partial \mathbf{w}}{\partial t} + \mathbf{M}\frac{\partial^2 \mathbf{w}}{\partial t^2} = \mathbf{F}\sin(\omega t) + \mathbf{G}\cos(\omega t) \qquad (7.23)$$

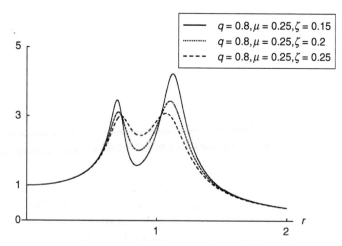

FIGURE 7.10 Frequency response curves of primary mass when damped absorber is added.

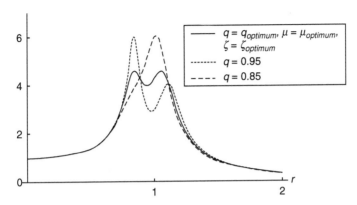

FIGURE 7.11 Use of optimally chosen parameters leads to frequency response curve in which peaks are approximately equal and amplitude varies little peaks.

A solution of Equation 7.23 is assumed in the form of

$$\mathbf{w} = \mathbf{A}(x)\sin(\omega t) + \mathbf{B}(x)\cos(\omega t) \tag{7.24}$$

Substituting Equation 7.24 into Equation 7.23 and equating coefficients of sin (ωt) and cos (ωt) leads to

$$(\mathbf{K} - \omega^2\mathbf{M})\mathbf{A} - \omega\mathbf{C}\mathbf{B} = \mathbf{F} \tag{7.25}$$

$$\omega\mathbf{C}\mathbf{A} + (\mathbf{K} - \omega^2\mathbf{M})\mathbf{A} = \mathbf{G} \tag{7.26}$$

Equation 7.25 and Equation 7.26 are ordinary differential equations to solve for **A** and **B**.

Example 7.5. Determine the steady-state response of the system of Figure 7.12, a thin disk attached to the end of a circular shaft subject to a harmonic torque $T_0 \sin(\omega t)$.

Solution: The nondimensional differential equation governing the angular displacement of the shaft is

$$\frac{\partial^2\theta}{\partial t^2} = \frac{\partial^2\theta}{\partial x^2} \tag{a}$$

The boundary conditions are

$$\theta(x,0) = 0 \tag{b}$$

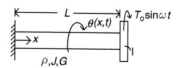

FIGURE 7.12 System of Example 7.5, a thin disk attached to the end of a circular shaft is subjected to a harmonic torque.

$$\frac{\partial \theta}{\partial x}(1,t) = -\beta \frac{\partial^2 \theta}{\partial t^2} + \lambda \sin(\hat{\omega} t) \tag{c}$$

where

$$\beta = \frac{I}{\rho J L} \tag{d}$$

$$\lambda = \frac{T_0 L}{JG} \tag{e}$$

$$\hat{\omega} = \frac{1}{L}\sqrt{\frac{G}{\rho}}\omega \tag{f}$$

The steady-state response is assumed as

$$\theta(x,t) = \Theta(x)\sin(\hat{\omega} t) \tag{g}$$

Substituting Equation g into Equation a through Equation c leads to the following problem for $\Theta(x)$:

$$\frac{d^2 \Theta}{dx^2} + \hat{\omega}^2 \Theta = 0 \tag{h}$$

$$\Theta(0) = 0 \tag{i}$$

$$\frac{d\Theta(1)}{dx} - \beta \hat{\omega}^2 \Theta(1) = \lambda \tag{j}$$

The solution of Equation h through Equation j is

$$\Theta(x) = \frac{\lambda}{[\hat{\omega}\cos(\hat{\omega}) - \beta\hat{\omega}^2\sin(\hat{\omega})]}\sin(\hat{\omega}x) \tag{k}$$

Then, steady-state amplitude of the disk at $x = 1$ is

$$\Theta(1) = \frac{\lambda\sin(\hat{\omega})}{\hat{\omega}\cos(\hat{\omega}) - \beta\hat{\omega}^2\sin(\hat{\omega})} \tag{l}$$

Example 7.6. Determine the steady-state response of the system of Figure 7.13.

Solution: The differential equations governing the motion of the system are

$$EI\frac{\partial^4 w}{\partial x^2} + \rho A\frac{\partial^2 w}{\partial t^2} = 0 \tag{a}$$

$$m\frac{d^2 y}{dt^2} + c\left(\frac{dy}{dt} - \frac{\partial w}{\partial t}(L,t)\right) + k(y - w(L,t)) = F_0\sin(\omega t) \tag{b}$$

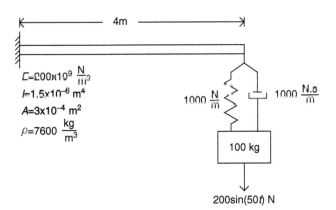

FIGURE 7.13 System of Example 7.6.

The boundary conditions are formulated as

$$w(0,t) = 0 \tag{c}$$

$$\frac{\partial w}{\partial x}(0,t) = 0 \tag{d}$$

$$\frac{\partial^2 w}{\partial x^2}(L,t) = 0 \tag{e}$$

$$EI \frac{\partial^3 w}{\partial x^3}(L,t) - c\left(\frac{dy}{dt} - \frac{\partial w}{\partial t}(L,t)\right) - k(y - w(L,t)) = 0 \tag{f}$$

The following nondimensional variables are introduced

$$x^* = \frac{x}{L} \tag{g}$$

$$y^* = \frac{y}{L} \tag{h}$$

$$w^* = \frac{w}{L} \tag{i}$$

$$t^* = t\sqrt{\frac{EI}{\rho A L^4}} \tag{j}$$

Substitution of Equation g through Equation j into Equation a through Equation f leads to

$$\frac{\partial^4 w}{\partial x^4} + \frac{\partial^2 w}{\partial t^2} = 0 \tag{k}$$

$$\frac{d^2 y}{dt^2} + 2\zeta\phi\left(\frac{dy}{dt} - \frac{\partial w}{\partial t}(1,t)\right) + \kappa(y - w(1,t)) = \lambda \sin(\omega^* t) \tag{l}$$

$$w(0,t) = 0 \tag{m}$$

$$\frac{\partial w}{\partial x}(0,t) = 0 \tag{n}$$

$$\frac{\partial^2 w}{\partial x^2}(1,t) = 0 \tag{o}$$

$$\frac{\partial^3 w}{\partial x^3}(1,t) + \eta\left(\frac{dy}{dt} - \frac{\partial w}{\partial t}(1,t)\right) + \kappa(y - w(1,t)) = 0 \tag{p}$$

where

$$\phi = \sqrt{\left(\frac{k}{m}\right)\left(\frac{\rho AL^4}{EI}\right)} = 0.2416 \tag{q}$$

$$\zeta = \frac{c\phi}{2\sqrt{km}} = 0.1208 \tag{r}$$

$$\eta = \frac{cL^3}{EI\phi}\sqrt{\frac{k}{m}} = 0.0265 \tag{s}$$

$$\kappa = \frac{kL^3}{EI} = 0.640 \tag{t}$$

$$\omega^* = \omega\sqrt{\frac{\rho AL^4}{EI}} = 1.208 \tag{u}$$

$$\lambda = \frac{F_0 L^3}{EI} = 0.128 \tag{v}$$

A solution of Equation k through Equation p is assumed as

$$\begin{bmatrix} w(x,t) \\ y(t) \end{bmatrix} = \begin{bmatrix} f(x) \\ A \end{bmatrix}\sin(\omega t) + \begin{bmatrix} g(x) \\ B \end{bmatrix}\cos(\omega t) \tag{w}$$

Substitution of Equation w into Equation k and Equation m

$$\left(\frac{d^4 f}{dx^4} - \omega^2 f\right)\sin(\omega t) + \left(\frac{d^4 g}{dx^4} - \omega^2 g\right)\cos(\omega t) = 0 \tag{x}$$

$$\{[\kappa(A - f(1)) - \omega^2 A] - 2\zeta\phi\omega(B - g(1))\}\sin(\omega t) + \{2\zeta\phi\omega(A - f(1)) \\ + \kappa(B - g(1)) - \omega^2 B\}\cos(\omega t) = \lambda \sin(\omega t) \tag{y}$$

Coefficients of $\sin(\omega t)$ and $\cos(\omega t)$ are independently set equal to zero leading to

$$\frac{d^4 f}{dx^4} - \omega^2 f = 0 \tag{z}$$

$$\frac{d^4 g}{dx^4} - \omega^2 g = 0 \tag{aa}$$

$$[\kappa(A - f(1)) - \omega^2 A] - 2\zeta\phi\omega(B - g(1)) = \lambda \tag{bb}$$

$$2\zeta\phi\omega(A - f(1)) + \kappa(B - g(1)) - \omega^2 B = 0 \tag{cc}$$

Substitution of Equation w into the boundary conditions, Equation m through Equation p leads to

$$f(0) = 0 \tag{dd}$$

$$g(0) = 0 \tag{ee}$$

$$\frac{df}{dx}(0) = 0 \tag{ff}$$

$$\frac{dg}{dx}(0) = 0 \tag{gg}$$

$$\frac{d^2 f}{dx^2}(1) = 0 \tag{hh}$$

$$\frac{d^2 g}{dx^2}(1) = 0 \tag{ii}$$

$$\frac{d^3 f}{dx^3}(1) + \kappa(A - f(1)) - \eta\omega(B - g(1)) = 0 \tag{jj}$$

$$\frac{d^3 g}{dx^3}(1) + \kappa(B - g(1)) + \eta\omega(A - f(1)) = 0 \tag{kk}$$

The general solutions of Equation z and Equation aa are

$$f(x) = C_1\cosh(\sqrt{\omega}x) + C_2\sinh(\sqrt{\omega}x) + C_3\cos(\sqrt{\omega}x) + C_4\sin(\sqrt{\omega}x) \tag{ll}$$

$$g(x) = C_5\cosh(\sqrt{\omega}x) + C_6\sinh(\sqrt{\omega}x) + C_7\cos(\sqrt{\omega}x) + C_8\sin(\sqrt{\omega}x) \tag{mm}$$

Substitution of Equation ll and Equation mm into Equation dd through Equation kk and Equation bb and Equation cc, respectively, lead to the following ten equations for ten unknowns:

$$C_1 + C_3 = 0 \tag{nn}$$

$$C_5 + C_7 = 0 \tag{oo}$$

$$\sqrt{\omega}C_2 + \sqrt{\omega}C_4 = 0 \tag{pp}$$

$$\sqrt{\omega}C_6 + \sqrt{\omega}C_8 = 0 \qquad \text{(qq)}$$

$$\omega C_1\cosh(\sqrt{\omega}) + \omega C_2\sinh(\sqrt{\omega}) - \omega C_3\cos(\sqrt{\omega}) - \omega C_4\sin(\sqrt{\omega}) = 0 \qquad \text{(rr)}$$

$$\omega C_5\cosh(\sqrt{\omega}) + \omega C_6\sinh(\sqrt{\omega}) - \omega C_7\cos(\sqrt{\omega}) - \omega C_8\sin(\sqrt{\omega}) = 0 \qquad \text{(ss)}$$

$$\begin{aligned}
&(\omega^{3/2}\sinh(\sqrt{\omega}) - \kappa\cosh(\sqrt{\omega}))C_1 + (\omega^{3/2}\cosh(\sqrt{\omega}) - \kappa\sinh(\sqrt{\omega}))C_2 \\
&+ (\omega^{3/2}\sin(\sqrt{\omega}) - \kappa\cos(\sqrt{\omega}))C_3 + (-\omega^{3/2}\cos(\sqrt{\omega}) - \kappa\sin(\sqrt{\omega}))C_4 \\
&+ \eta\omega\cosh(\sqrt{\omega})C_5 + \eta\omega\sinh(\sqrt{\omega})C_6 + \eta\omega\cos(\sqrt{\omega})C_7 \\
&+ \eta\omega\sin(\sqrt{\omega})C_8 + \kappa A - \eta\omega B = 0
\end{aligned} \qquad \text{(tt)}$$

$$\begin{aligned}
&-\eta\omega\cosh(\sqrt{\omega})C_1 - \eta\omega\sinh(\sqrt{\omega})C_2 - \eta\omega\cos(\sqrt{\omega})C_3 \\
&- \eta\omega\sin(\sqrt{\omega})C_4 + (\omega^{3/2}\sinh(\sqrt{\omega}) - \kappa\cosh(\sqrt{\omega}))C_5 + (\omega^{3/2}\cosh(\sqrt{\omega}) \\
&- \kappa\sinh(\sqrt{\omega}))C_6 + (\omega^{3/2}\sin(\sqrt{\omega}) - \kappa\cos(\sqrt{\omega}))C_7 + (-\omega^{3/2}\cos(\sqrt{\omega}) \\
&- \kappa\sin(\sqrt{\omega}))C_8 + \eta\omega A + \kappa B = 0
\end{aligned} \qquad \text{(uu)}$$

$$\begin{aligned}
&-\kappa\cos h(\sqrt{\omega})C_1 - \kappa\sinh(\sqrt{\omega})C_2 - \kappa\cos(\sqrt{\omega})C_3 - \kappa\sin(\sqrt{\omega})C_4 \\
&+ 2\zeta\phi\omega\cos h(\sqrt{\omega})C_5 + 2\zeta\phi\omega\sin h(\sqrt{\omega})C_6 + 2\zeta\phi\omega\cos(\sqrt{\omega})C_7 \\
&+ 2\zeta\phi\omega\sin(\sqrt{\omega})C_8 + (\kappa - \omega^2)A - 2\zeta\omega\phi B = \lambda
\end{aligned} \qquad \text{(vv)}$$

$$\begin{aligned}
&-2\zeta\phi\omega\cosh(\sqrt{\omega})C_1 - 2\zeta\phi\omega\sinh(\sqrt{\omega})C_2 - 2\zeta\phi\omega\cos(\sqrt{\omega})C_3 \\
&- 2\zeta\phi\omega\sin(\sqrt{\omega})C_4 - \kappa\cosh(\sqrt{\omega})C_5 - \kappa\sinh(\sqrt{\omega})C_6 \\
&- \kappa\cos(\sqrt{\omega})C_7 - \kappa\sin(\sqrt{\omega})C_8 + 2\zeta\omega\phi A + (\kappa - \omega^2)B = 0
\end{aligned} \qquad \text{(ww)}$$

Equation nn through Equation ww are a set of ten equations for ten unknowns. Substituting numerical values for the parameters and solving simultaneously leads to

$$\begin{aligned}
&C_1 = -0.03321 \quad C_2 = 0.03166 \quad C_3 = 0.03321 \quad C_4 = -0.03166 \quad C_5 = -0.003168 \\
&C_6 = 0.003022 \quad C_7 = 0.003168 \quad C_8 = -0.003022 \quad A = -0.1353 \quad B = -0.007432
\end{aligned} \qquad \text{(xx)}$$

The time-dependent response of the system is

$$w(x,t) = \cos(1.208t)[0.003166 \cos(1.099x) - 0.003168 \cos h(1.099x)$$
$$-0.003022 \sin(1.099t) + 0.003022 \sin h(1.099x)]$$
$$+ \sin(1.208t)[0.03321 \cos(1.099t) - 0.03321 \cos h(1.099t)$$
$$-0.03166 \sin(1.099t) + 0.03166 \sin h(1.099t)]$$
(yy)

$$y(t) = -0.00743 \cos(1.208t) - 0.1353 \sin(1.208t)$$
(zz)

The time-dependent response of the system is plotted in Figure 7.14.

Example 7.7. The nondimensional differential equation governing the forced response of the system of Figure 7.15 is

$$\frac{\partial^4 w}{\partial x^4} + \frac{\partial^2 w}{\partial t^2} = \lambda \sin(\omega t)\delta\left(x - \frac{1}{2}\right)$$
(a)

where $\lambda = (F_0 L^3/EI)$. Determine the time-dependent response of the beam.

Solution: A solution to Equation a is assumed of the form

$$w(x,t) = f(x)\sin(\omega t)$$
(b)

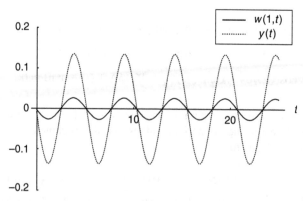

FIGURE 7.14 Steady-state response of system of Example 7.6.

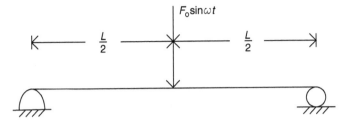

FIGURE 7.15 Simply supported beam with harmonic load at midspan.

Substitution of Equation b into Equation a leads to

$$\frac{d^4 f}{dx^4} - \omega^2 w = \lambda \delta\left(x - \frac{1}{2}\right) \tag{c}$$

The general solution of Equation c is obtained using the method of variation of parameters as

$$f(x) = \frac{\lambda}{2\omega^3}\left\{ \sin\left[\omega\left(x - \frac{1}{2}\right)\right] - \sinh\left[\omega\left(x - \frac{1}{2}\right)\right] u\left(x - \frac{1}{2}\right) \right\} + \tag{d}$$

$$C_1\cos(\omega x) + C_2\sin(\omega x) + C_3\cosh(\omega x) + C_4\sinh(\omega x)$$

The boundary conditions for the imply supported beam are

$$f(0) = 0 \tag{e}$$

$$\frac{d^2 f}{dx^2}(0) = 0 \tag{f}$$

$$f(1) = 0 \tag{g}$$

$$\frac{d^2 f}{dx^2}(1) = 0 \tag{h}$$

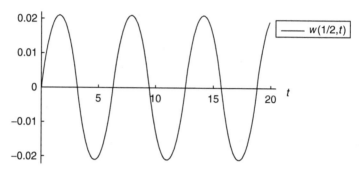

FIGURE 7.16 Forced response of pinned-pinned beam with a concentrated load at midspan.

Application of Equation e through Equation h to Equation d leads to

$$f(x) = \frac{\lambda \left\{ \begin{array}{l} \left[\sin(\omega) + \sinh\left(\dfrac{\omega}{2}\right) \right] [\cos(\omega x) - \cosh(\omega x)] \\[2ex] + \cos(\omega)\sin\left(\dfrac{\omega}{2}\right) [\sin(\omega x)] + \sinh(\omega x)] \end{array} \right\}}{2\omega^3 [\cos(\omega)\sinh(\omega) + \sin(\omega)\cosh(\omega)]}$$

$$+ \frac{\lambda}{2\omega^3} \left\{ \sin\left[\omega\left(x - \frac{1}{2}\right)\right] - \sinh\left[\omega\left(x - \frac{1}{2}\right)\right] u\left(x - \frac{1}{2}\right) \right\}$$ (i)

Figure 7.16 illustrates $f(x)$ which is continuous for all x. Figure 7.17 illustrates (d^3f/dx^3), which is proportional to the shear force. The shear force is discontinuous at $x = 1/2$ due to the application of the concentrated load at the midspan of the beam.

7.3 LAPLACE TRANSFORM SOLUTIONS

The Laplace transform method is often convenient for the determination of the forced response of discrete and continuous systems. Let w be the displacement vector for a linear system. The Laplace transform of the displacement vector is defined as

$$W(s) = \int_0^\infty w(t)e^{-st}\, dt$$ (7.27)

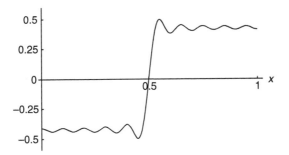

FIGURE 7.17 Shear force of simply supported beam subject to a concentrated load at mid-span.

The Laplace transform of a function of time is a function of a complex variable s. The Laplace transform has properties which allow a set of differential equations to be transformed to a set of algebraic equations and a set of partial differential equations to be transformed into a set of ordinary differential equations. The algebraic equations are solved leading to the transforms of the dependent variables. An inversion procedure, which often relies upon transform tables and properties, is used to obtain the time-dependent form of the dependent variables. The procedure is illustrated through several examples.

7.3.1 DISCRETE SYSTEMS

Taking the Laplace transform of Equation 7.1, using the linearity of the transform and the properties of transform of the first and second derivatives and assuming all initial conditions are zero gives

$$(s^2\mathbf{M} + s\mathbf{C} + \mathbf{K})\mathbf{X}(s) = \mathbf{F}(s) \tag{7.28}$$

where $X(s) = L\{x(t)\}$ and $F(s) - L\{F(t)\}$. The impedance matrix is defined as

$$\mathbf{Z}(s) = s^2\mathbf{M} + s\mathbf{C} + \mathbf{K} \tag{7.29}$$

Equation 7.29 is rearranged as

$$\mathbf{X}(s) = Z^{-1}(s)\mathbf{F}(s) \tag{7.30}$$

Each element of the vector $X(s)$ can be inverted to obtain $x(t)$.

Example 7.8. The two-degree-of-freedom system of Figure 7.18a is subject to the time-dependent force of Figure 7.18b. Use the Laplace transform method to derive the time-dependent response for $x_2(t)$, assuming that both masses are at rest in equilibrium when the force is applied.

Solution: The differential equations governing the motion of the system are

$$\begin{bmatrix} 100 & 0 \\ 0 & 100 \end{bmatrix} \begin{bmatrix} \ddot{x}_1 \\ \ddot{x}_2 \end{bmatrix} + \begin{bmatrix} 3 \times 10^4 & -2 \times 10^4 \\ -2 \times 10^4 & 2 \times 10^4 \end{bmatrix} \begin{bmatrix} x_1 \\ x_2 \end{bmatrix} = \begin{bmatrix} 0 \\ F(t) \end{bmatrix} \qquad (a)$$

The mathematical form of $F(t)$ is

$$F(t) = 1000[u(t) - u(t - 0.5)] \qquad (b)$$

The Laplace transform of $F(t)$ is obtained using the second shifting theorem as

$$F(s) = \frac{1000}{s}(1 - e^{-0.5s}) \qquad (c)$$

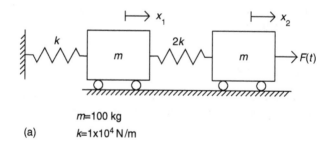

m=100 kg
k=1x10⁴ N/m

(a)

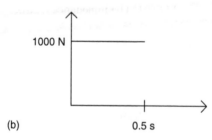

(b)   0.5 s

FIGURE 7.18 (a) System of Example 7.18. (b) Step input.

Define $X_1(s) = \mathcal{L}\{x_1(t)\}$ and $X_2(s) = \mathcal{L}\{x_2(st)\}$. Taking the Laplace transform of Equation a using the properties of linearity of the transform and the transform of derivatives leads to

$$\begin{bmatrix} 100s^2 + 3 \times 10^4 & -2 \times 10^4 \\ -2 \times 10^4 & 100s^2 + 2 \times 10^4 \end{bmatrix} \begin{bmatrix} X_1(s) \\ X_2(s) \end{bmatrix} = \begin{bmatrix} 0 \\ \dfrac{1000}{s}(1 - e^{-0.5s}) \end{bmatrix} \qquad \text{(d)}$$

Multiplying by the inverse of the impedance matrix leads to

$$\begin{bmatrix} X_1(s) \\ X_2(s) \end{bmatrix} = \frac{1}{100} \begin{bmatrix} \dfrac{s^2 + 200}{s^4 + 500s^2 + 20000} & \dfrac{200}{s^4 + 500s^2 + 20000} \\[2mm] \dfrac{200}{s^4 + 500s^2 + 20000} & \dfrac{s^2 + 300}{s^4 + 500s^2 + 20000} \end{bmatrix} \begin{bmatrix} 0 \\ \dfrac{1000}{s}(1 - e^{-0.5s}) \end{bmatrix}$$

$$\text{(e)}$$

$$= \frac{10}{s}(1 - e^{-0.5s}) \begin{bmatrix} \dfrac{200}{s^4 + 500s^2 + 20000} \\[2mm] \dfrac{s^2 + 300}{s^4 + 500s^2 + 20000} \end{bmatrix}$$

A partial fraction decomposition leads to

$$X_2(s) = \frac{10}{s}(1 - e^{-0.5s}) \frac{s^2 + 300}{(s^2 + 43.85)(s^2 + 456.54)}$$

$$\text{(f)}$$

$$= \left[\frac{0.150}{s} - \frac{0.142s}{s^2 + 43.85} - \frac{0.0830}{s^2 + 456.54} \right] (1 - e^{-0.5s})$$

Equation f is inverted, leading

$$x_2(t) = [0.150 - 0.0142 \cos(6.62t) - 0.0830 \cos(21.36t)] -$$
$$\{0.150 - 0.0142 \cos[6.62(t - 0.5)] - 0.0830 \cos[21.36(t - 0.5)]\}u(t - 0.5)$$

$$\text{(g)}$$

Example 7.9. Use the Laplace transform method to show that the application of a properly tuned dynamic vibration absorber to any point in a system leads to a steady-state amplitude of zero at the point where the absorber is attached.

Solution: Consider and undamped n-degree-of-freedom system with a mass matrix $\mathbf{M}$ and a stiffness matrix $\mathbf{K}$. When an absorber is added to the system, the resulting system has $(n+1)$ degrees of freedom. As a model, consider the system of Figure 7.19 where an absorber is added to the particle on the beam whose displacement is x_1. Without loss of generality, it is assumed that the absorber is attached to the particle whose displacement is x_1. The mass and stiffness matrices for the resulting $(n+1)$-degree-of-freedom system are

$$
\mathbf{M_a} = \begin{bmatrix}
m_{1,1} & m_{1,2} & \cdots & m_{1,n} & 0 \\
m_{2,1} & m_{2,2} & \cdots & m_{2,n} & 0 \\
\vdots & \vdots & \ddots & \vdots & \vdots \\
m_{n,1} & m_{n,2} & \cdots & m_{n,n} & 0 \\
0 & 0 & \cdots & 0 & m_a
\end{bmatrix}
\tag{a}
$$

$$
\mathbf{K_a} = \begin{bmatrix}
k_{1,1} & k_{1,2} & \cdots & k_{1,n} & -k_a \\
k_{2,1} & k_{2,2} & \cdots & k_{2,n} & 0 \\
\vdots & \vdots & \ddots & \vdots & \vdots \\
k_{n,1} & k_{n,2} & \cdots & k_{n,n} & 0 \\
-k_a & 0 & 0 & 0 & k_a
\end{bmatrix}
\tag{b}
$$

The force vector for the $(n+1)$-degree-of-freedom system is

$$
\mathbf{F_a} = \begin{bmatrix}
f_1 \\
f_2 \\
\vdots \\
f_n \\
0
\end{bmatrix} \sin(\omega t)
\tag{c}
$$

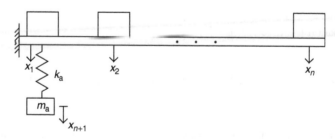

FIGURE 7.19 Vibration absorber is added to particle whose displacement is x. Steady-state amplitude for x_1 is zero when absorber is tuned to excitation frequency.

Application of the Laplace transform method to determine the response of the $(n+1)$-degree-of-freedom system leads to

$$
\begin{bmatrix}
m_{1,1}s^2 + k_{1,1} & m_{1,2}s^2 + k_{1,2} & \cdots & m_{1,n}s^2 + k_{1,n} & -k_a \\
m_{2,1}s^2 + k_{2,1} & m_{2,2}s^2 + k_{2,2} & \ddots & m_{2,n}s^2 + k_{2,n} & 0 \\
\vdots & \vdots & \ddots & \vdots & \vdots \\
m_{n,1}s^2 + k_{n,1} & m_{2,n}s^2 + k_{2,n} & \cdots & m_{n,n}s^2 + k_{n,n} & 0 \\
-k_a & 0 & \cdots & 0 & m_a s^2
\end{bmatrix}
\begin{bmatrix}
X_1(s) \\
X_2(s) \\
\vdots \\
X_n(s) \\
X_a(s)
\end{bmatrix}
$$

$$
=
\begin{bmatrix}
f_1 \\
f_2 \\
\vdots \\
f_n \\
0
\end{bmatrix}
\frac{\omega}{s^2 + \omega^2}
\tag{d}
$$

Cramer's rule may be used to solve for $X_1(s)$, as

$$
X_1(s) = \frac{
\begin{vmatrix}
f_1 & m_{1,2}s^2 + k_{1,2} & \cdots & m_{1,n}s^2 + k_{1,n} & -k_a \\
f_2 & m_{2,2}s^2 + k_{2,2} & \cdots & m_{2,n}s^2 + k_{2,n} & 0 \\
\vdots & \vdots & \ddots & \vdots & \vdots \\
f_n & m_{n,2}s^2 + k_{n,2} & \cdots & m_{n,n}s^2 + k_{n,n} & 0 \\
0 & 0 & 0 & 0 & k_a + m_a s^2
\end{vmatrix}
}{|\mathbf{Z}(s)|}
\left(\frac{\omega}{s^2 + \omega}\right)
\tag{e}
$$

The determinant of the impedance matrix may be expanded and written as the product of n linear factors in s,

$$
|\mathbf{Z}(s)| = C(s - s_1)(s - s_2)\ldots(s - s_n)
\tag{f}
$$

where C is a constant and $s_1, s_2, \ldots s_n$, are the roots of $|\mathbf{Z}(s)|$. A partial fraction decomposition may be used to expand $X_1(s)$ as

$$
X_1(s) = \frac{A_1}{s + i\omega} + \frac{A_2}{s - i\omega} + \sum_{k=1}^{n} \frac{B_k}{s - s_k}
\tag{g}
$$

The terms in the summation of the partial fraction decomposition of Equation g correspond to the free response. The steady-state response for $X_1(s)$ is obtained from the first two terms in Equation g. The coefficients in the partial fraction decomposition are obtained using residues

$$A_1 = \lim_{s \to -i\omega} (s + i\omega)X_1(s)$$

$$= \frac{\omega}{-2j\omega|\mathbf{Z}(-i\omega)|} \begin{vmatrix} f_1 & k_{1,2} - m_{1,2}\omega^2 & \cdots & k_{1,n} - m_{1,n}\omega^2 & -k_a \\ f_2 & k_{2,2} - m_{2,2}\omega^2 & \cdots & k_{2,n} - m_{2,n}\omega^2 & 0 \\ \vdots & \vdots & \ddots & \vdots & \vdots \\ f_n & k_{n,2} - m_{n,2}\omega^2 & \cdots & k_{n,n} - m_{n,n}\omega^2 & 0 \\ 0 & 0 & \cdots & 0 & k_a - m_a\omega^2 \end{vmatrix} \quad \text{(h)}$$

The determinant in Equation h is expanded by row expansion using the $(n+1)$th row, leading to

$$A_1 = -\frac{1}{2|\mathbf{Z}(-i\omega)|}(k_a - m_a\omega^2)|D_{n+1,n+1}| \quad \text{(i)}$$

where $|D_{n+1,n+1}|$ is the determinant of the $n \times n$ matrix obtained by eliminating the $(n+1)$th row and $(n+1)$th column. A similar analysis leads to

$$A_2 = \frac{1}{2i|Z(i\omega)|}(k_a - m_a\omega^2)|D_{n+1,n+1}| \quad \text{(j)}$$

The steady-sate response for $x_1(t)$ is obtained by inverting

$$X_{1ss}(s) = \frac{(k_a - m_a\omega^2)|D_{n+1,n+1}|}{|Z(i\omega)|(2i)}\left(\frac{1}{s + i\omega} - \frac{1}{s - i\omega}\right) \quad \text{(k)}$$

The inverse transform of Equation k is

$$x_{1ss}(t) = \frac{(k_a - m_a\omega^2)|D_{n+1,n+1}|}{|Z(i\omega)|(2i)}(e^{-i\omega t} - e^{i\omega t})$$

$$= -\frac{(k_a - m_a\omega^2)|D_{n+1,n+1}|}{|Z(i\omega)|}\sin(\omega t)$$

$$\text{(l)}$$

The absorber is tuned to the excitation frequency when $(k_a/m_a = \omega^2)$. Equation 1 shows that when the absorber is tuned to the excitation frequency, $x_1(t) = 0$.

The above shows that the steady-state amplitude of the particle in the system to which a vibration absorber is attached has a steady-state amplitude of zero when the excitation is at a frequency equal to the frequency to which the absorber is tuned.

7.3.2 CONTINUOUS SYSTEMS

The Laplace transform of a function of a spatial variable and time is a function of the spatial variable as well as the transform variable.

$$W(x,s) = \int_0^\infty w(x,t)e^{-st}\,dt \tag{7.31}$$

The transform has the property

$$\mathcal{L}\left\{\frac{\partial w}{\partial x}\right\} = \frac{\partial W}{\partial x}(x, s) \tag{7.32}$$

Equation 7.32 implies that

$$\mathcal{L}\{Kw\} = KW(x, s) \tag{7.33}$$

The general form of the differential equations for the forced response of a linear continuous system is

$$\mathbf{K}w \mid \mathbf{C}\frac{\partial \mathbf{w}}{\partial t} + \mathbf{M}\frac{\partial^2 \mathbf{w}}{\partial t^2} = \mathbf{F}(x,t) \tag{7.34}$$

Application of the Laplace transform to Equation 7.34, assuming all initial conditions are zero, leads to

$$\mathbf{K}W(x,s) + s\mathbf{C}W(x,s) + s^2\mathbf{M}W(x,s) = \mathbf{F}(x,s) \tag{7.35}$$

Equation 7.35 is a differential equation which must be supplemented by appropriate boundary conditions. The boundary value problem is solved for $\mathbf{W}(x,s)$, which is subsequently inverted to obtain $w(x,t)$.

Example 7.10. Determine the response of the shaft of Figure 7.20 when it is subject to a time-dependent nondimensional moment $M(t)$.

Solution: The partial differential equation governing the angular oscillations of the shaft is

$$\frac{\partial^2 \theta}{\partial t^2} = \frac{\partial^2 \theta}{\partial x^2} \tag{a}$$

The boundary conditions for a shaft fixed at $x = 0$ and subject to an applied torque at $x = 1$ are

$$\theta(0,t) = 0 \tag{b}$$

$$\frac{\partial \theta}{\partial x}(1,t) = M(t) \tag{c}$$

Define $\Theta(x,s) = L\{\theta(x,t)\}$ and $\bar{M}(s) = L\{M(t)\}$. Taking the Laplace transform of Equation a leads to

$$\frac{d^2 \Theta}{dx^2} - s^2 \Theta(x,s) = 0 \tag{d}$$

Taking the Laplace transform of the boundary conditions leads to

$$\Theta(0,s) = 0 \tag{e}$$

$$\frac{d\Theta}{dx}(1,s) = \bar{M}(s) \tag{f}$$

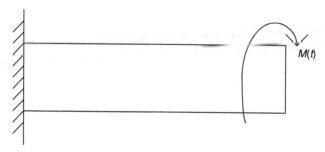

FIGURE 7.20 The Laplace transform methods is used to determine the response of torsional shaft due to time dependent moment applied to end of shaft.

The general solution to Equation d is

$$\Theta(s) = C_1(s)e^{sx} + C_2(s)e^{-sx} \tag{g}$$

Application of the boundary conditions, Equation e and Equation f to Equation g leads to

$$C_1(s) + C_2(s) = 0 \tag{h}$$

$$sC_1(s)e^s - sC_2e^{-s}(s) = \bar{M}(s) \tag{i}$$

Equation h and Equation i are solved simultaneously and the results substituted into Equation g, resulting in

$$\Theta(s) = \frac{\bar{M}(s)}{s}\left(\frac{e^{sx} + e^{-sx}}{e^s + e^{-s}}\right) \tag{j}$$

The Convolution Theorem is used to invert Equation j

$$\theta(x,s) = \int_0^t f(\tau)M(t-\tau)d\tau \tag{k}$$

where

$$f(x,t) = \mathcal{L}^{-1}\left\{\frac{1}{s}\left(\frac{e^{sx} + e^{-sx}}{e^s + e^{-s}}\right)\right\} \tag{l}$$

To invert Equation l, note the following:

$$\frac{1}{e^s + e^{-s}} = \frac{1}{e^s(1 - e^{-2s})}$$

$$= e^{-s}(1 - e^{-2s})^{-1} \tag{m}$$

$$= e^{-s}(1 - e^{-2s} + e^{-4s} - e^{-6s} + e^{-8s} + \cdots)$$

Use of Equation m leads to

$$\frac{1}{s}\left(\frac{e^{sx}+e^{-sx}}{e^{s}+e^{-s}}\right) = e^{-s}(e^{sx}+e^{-sx})(1-e^{-2s}+e^{-4s}-e^{-6s}+\cdots)$$

$$= \frac{1}{s}\{[e^{s(x-1)}-e^{s(x-3)}+e^{s(x-5)}-\cdots]$$

$$+[e^{-s(x+1)}-e^{-s(x+3)}+e^{-s(x+5)}-\cdots]\}$$

(n)

The inverse of the transform in Equation n is

$$f(x,t) = \{u[t-(1-x)]-u[t-(3-x)]+u[t-(5-x)]-\cdots\}$$
$$+\{u[t-(x+1)]-u(t-(x+3)]-u[t-(x+5)-]\cdots\}$$

(o)

Substitution of Equation o into Equation k leads to.

$$\theta(x,t) = \int_0^t \Big[\{u[\tau-(1-x)-u[\tau-(3-x)+u[\tau-(5-x)]-\cdots\}$$

$$+\{u[\tau-(x+1)]-u[(\tau-(x+3)-u[\tau-(x+5)]-\cdots\}\Big]$$
$$\times M(t-\tau)d\tau$$

(p)

Note that

$$\int_0^t u(\tau-a)f(t-\tau)d\tau = u(t-a)\int_a^t f(t-\tau)d\tau$$

(q)

Use of Equation q in Equation p leads to

$$\theta(x,t) = \left\{ u[t-(1-x)]\int_{1-x}^t M(t-\tau)d\tau - u[t-(3-x)]\int_{3-x}^t M(t-\tau)d\tau \right.$$

$$\left.+u[t-(5-x)]\int_{5-x}^t M(t-\tau)d\tau+\cdots\right\} + \left\{ u[t-(x+1)]\int_{x+1}^t M(t-\tau)d\tau \right.$$

$$\left.-u[t-(x+3)]\int_{x+3}^t M(t-\tau)d\tau+u[t-(x+5)]\int_{x+5}^t M(t-\tau)d\tau+\cdots\right\} (r)$$

Equation r is the traveling wave solution for the wave equation. If $M(t) = 1$, Equation r becomes

$$\theta(x,t) = \{[t-(1-x)]u[(t-(1-x)]-[t-(3-x)]u[t-(3-x)]$$
$$+[t-(5-x)u[t-(5-x)]+\cdots\}+\{[t-(x+1)]u[t-(x+1) \qquad (s)$$
$$-[t-(x+3)]u[t-(x+3)]+[t-(x+5)]u[t-(x+5)]+\cdots\}$$

Consider the function $[t-(1-x)]u[t-(1-x)]$. Due to the unit step function for a given value of x, this is zero until $t = 1-x$. For $x = 1$, this occurs at $t = 0$. For $x = 1/2$, this occurs for $t = 1/2$. This function represents a pulse that begins at $t = 0$ at $x = 1$ and travels left along the bar. When the pulse reaches the fixed end, it is reflected into the pulse $[t-(x+1)u(x+1)]$. This pulse does not begin until t is larger than $x+1$, which occurs first at $x = 0$ when $t = 1$. This pulse then propagates right along the bar. The pulse reaches the end of the bar at $t = 2$, at which time the pulse $[t-(3-x)]u[t-(3-x)]$ is initiated, but has an opposite sign than the previous pulse. This process continues indefinitely.

7.4   MODAL ANALYSIS FOR UNDAMPED DISCRETE SYSTEMS

Consider an undamped n-degree-of-freedom discrete system with a general forced excitation. The governing differential equations are of the form

$$\mathbf{M\ddot{x} + Kx = F} \qquad (7.36)$$

Let $\omega_1 \leq \omega_2 \leq \ldots \leq \omega_n$ be the system's natural frequencies with corresponding normalized mode shape vectors $\mathbf{X_1, X_2, \ldots, X_n}$. Mode shape orthonormality implies that $(\mathbf{X_i, X_j})_M = \delta_{ij}$ and $(\mathbf{X_i, X_j})_K = \omega_i^2 \delta_{i,j}$.

At any instant of time, $\mathbf{x}$, the solution of Equation 7.36, is in R^n and hence may be expanded in a series of normalized mode shapes

$$\mathbf{x}(t) = \sum_{i=1}^{n} (\mathbf{x}, X_i)_M X_i \qquad (7.37)$$

Since $\mathbf{x}$ is not known, but is a function of time, Equation 7.37 can be written using unknown time-dependent coefficients in the expansion

$$\mathbf{x}(t) = \sum_{i=1}^{n} c_i(t)\mathbf{X}_i \tag{7.38}$$

Substitution of Equation 7.38 into Equation 7.36 leads to

$$\mathbf{M}\left(\sum_{i=1}^{n} \ddot{c}_i(t)\mathbf{X}_i\right) + \mathbf{K}\left(\sum_{i=1}^{n} c_i(t)\mathbf{X}_i\right) = \mathbf{F} \tag{7.39}$$

Using linearity of $\mathbf{M}$ and $\mathbf{K}$ taking the standard inner product of both sides with $\mathbf{X}_j$ for any j between 1 and n leads to

$$\sum_{i=1}^{n} \ddot{c}_i(t)(\mathbf{MX}_i,\mathbf{X}_j) + \sum_{i=1}^{n} c_i(t)(\mathbf{KX}_i,\mathbf{X}_j) = (\mathbf{F},\mathbf{X}_j) \tag{7.40}$$

Equation 7.40 is rewritten using energy inner product notation as

$$\sum_{i=1}^{n} \ddot{c}_i(t)(\mathbf{X}_i,\mathbf{X}_j)_{\mathbf{M}} + \sum_{i=1}^{n} c_i(t)(\mathbf{X}_i,\mathbf{X}_j)_K = (\mathbf{F},\mathbf{X}_j) \tag{7.41}$$

Application of mode shape orthonormality in Equation 7.41 leads to

$$\ddot{c}_j + \omega_j^2 c_j = g_j(t) \quad j = 1,2,\dots,n \tag{7.42}$$

where

$$g_j(t) = (\mathbf{F},\mathbf{X}_j) = \mathbf{X}_j^T \mathbf{F} \tag{7.43}$$

The differential equations of Equation 7.42 are uncoupled. Each may be solved independently for $c_j(t)$. The response in terms of the original generalized coordinates is then obtained from Equation 7.38.

The general solution of Equation 7.42 is obtained using the convolution integral as

$$c_j(t) = \frac{1}{\omega_j} \int_0^t g_j(\tau)\sin[\omega_n(t-\tau)]d\tau \tag{7.44}$$

The relationship between the chosen generalized coordinates and the coefficients in which the differential equations are uncoupled is Equation 7.38, which may be summarized in matrix form as

$$\mathbf{x} = \mathbf{Pc} \tag{7.45}$$

where c is the vector of coefficients used in the expansion theorem and $\mathbf{P}$, the modal matrix, is the matrix whose columns are the normalized mode shape vectors. Since the columns of $\mathbf{P}$ are orthogonal, they are clearly linearly independent and thus $\mathbf{P}$ is nonsingular. Equation 7.45 can then be viewed as a one-to-one linear transformation between two sets of coordinates. The elements of the vector c are called the principal coordinates. While it is difficult to ascribe physical meaning to principal coordinates, their mathematical usage is clear. They are the set of coordinates in which the system is both statically and dynamically uncoupled. The principal coordinates are often called modal coordinates because each principal coordinate is associated with only one mode of vibration.

Example 7.11. Use modal analysis to determine the forced response of the system of Figure 7.18.

Solution: The differential equations governing the motion of the system are derived in Example 7.8 as

$$\begin{bmatrix} 100 & 0 \\ 0 & 100 \end{bmatrix} \begin{bmatrix} \ddot{x}_1 \\ \ddot{x}_2 \end{bmatrix} + \begin{bmatrix} 3\times 10^4 & -2\times 10^4 \\ -2\times 10^4 & 2\times 10^4 \end{bmatrix} \begin{bmatrix} x_1 \\ x_2 \end{bmatrix} = \begin{bmatrix} 0 \\ F(t) \end{bmatrix} \tag{a}$$

The natural frequencies and normalized mode shape vectors are determined as

$$\omega_1 = 6.6215\frac{r}{s} \qquad \mathbf{X}_1 = \begin{bmatrix} 0.0615 \\ 0.0788 \end{bmatrix} \tag{b}$$

$$\omega_2 = 21.36\frac{r}{s} \qquad \mathbf{X}_2 = \begin{bmatrix} 0.0788 \\ -0.0615 \end{bmatrix} \tag{c}$$

Equation 7.49 is used to obtain

$$g_1(t) = \mathbf{X}_1^T \mathbf{F} = [\,0.0615 \quad 0.0788\,] \begin{bmatrix} 0 \\ F(t) \end{bmatrix} = 0.0788F(t) \tag{d}$$

$$g_2(t) = \mathbf{X}_2^T \mathbf{F} = [\,0.0788 \quad 0.0615\,] \begin{bmatrix} 0 \\ F(t) \end{bmatrix} = -0.0615F(t) \tag{e}$$

where

$$F(t) = 1000[1 - u(t - 0.5)] \tag{f}$$

The modal equations obtained from Equation 7.42 are

$$\ddot{c}_1 + (6.622)^2 c_1 = 78.8[1 - u(t - 0.50)] \tag{g}$$

$$\ddot{c}_2 + (21.36)^2 c_2 = -61.5[1 - u(t - 0.50)] \tag{h}$$

The convolution integral solutions of Equation g and Equation h are

$$c_1(t) = \frac{1}{6.622} \int_0^t 78.8 \sin[6.622(t - \tau)][1 - u(\tau - 0.5)]d\tau \tag{i}$$

$$= 1.797\{1 - \cos(6.622t) - [1 - \cos(6.622t - 3.108)]u(t - 0.5)\}$$

$$c_2(t) = -\frac{1}{21.36} \int_0^t 61.5 \sin[21.36(t - \tau)][1 - u(\tau - 0.5)]d\tau \tag{j}$$

$$= -0.13481\{1 - \cos(21.36t) - [1 - \cos(21.36t - 10.68)]u(t - 0.5)\}$$

The solution for the original generalized coordinates is

$$\begin{bmatrix} x_1(t) \\ x_2(t) \end{bmatrix} = c_1(t) \begin{bmatrix} 0.0615 \\ 0.0788 \end{bmatrix} + c_2(t) \begin{bmatrix} 0.0788 \\ -0.0615 \end{bmatrix} \tag{k}$$

7.5 MODAL ANALYSIS OF UNDAMPED CONTINUOUS SYSTEMS

The differential equations governing the response of an undamped continuous system are written in the form

$$\mathbf{Kw} + \mathbf{M\ddot{w}} = \mathbf{F} \tag{7.46}$$

where $\mathbf{K}$ is the stiffness operator, $\mathbf{M}$ is the mass operator, and $\mathbf{F}$ is the generalized force vector. A self-adjoint and positive definite system has an infinite, but countable, number of natural frequencies $\omega_1 \leq \omega_2 \leq \omega_3 \leq \ldots \omega_k \leq \ldots$ with corresponding mode shape vectors $w_1, w_2, \ldots w_k, \ldots$. The mode shape vectors satisfy the orthonormality conditions with respect to defined energy inner products, as $(\mathbf{w}_i, \mathbf{w}_j)_M = \delta_{i,j}$ and $(\mathbf{w}_i, \mathbf{w}_j)_k = \omega_i^2 \delta_{i,j}$.

At any instant of time, the solution $w(x,t)$ to Equation 7.46 and appropriately specified boundary conditions is in the domain of $\mathbf{K}$. Thus, it may be expanded in an eigenvector expansion of the form

$$\mathbf{w}(x,t) = \sum_{i=1}^{\infty} (\mathbf{w}, \mathbf{w}_i)_M \mathbf{w}_i \tag{7.47}$$

where for a bar or beam with discrete masses

$$(\mathbf{w}, \mathbf{w}_i)_M = \int_0^1 \mathbf{w}(x,t)\mathbf{w}_i(x)\rho A(x)\mathrm{d}x + \sum_k m_k \mathbf{w}(x_k,t)\mathbf{w}_i(x_k) \tag{7.48}$$

Equation 7.48 indicates that the inner products are functions of time which are not explicitly known unless the response is known. Equation 7.47 may be rewritten as

$$\mathbf{w}(x,t) = \sum_{i=1}^{\infty} c_i(t)\mathbf{w}_i(x) \tag{7.49}$$

Substitution of Equation 7.49 into Equation 7.46 leads to

$$\mathbf{K}\left(\sum_{i=1}^{\infty} c_i(t)\mathbf{w}_i(x) \right) + \mathbf{M}\left(\sum_{i=1}^{\infty} \ddot{c}_i(t)\mathbf{w}_i(x) \right) = \mathbf{F}(x,t) \tag{7.50}$$

The linear operators may be interchanged as long as the infinite series converges. The completeness of the eigenvectors in the domain of $\mathbf{K}$ is discussed in Chapter 5. Since $\mathbf{w}(x,t)$ is in the domain of $\mathbf{K}$, for every t there must be a convergent expansion of the form of Equation 7.32, which converges pointwise to $\mathbf{w}$. Thus, the series is convergent and the order of operations may be interchanged. Thus, Equation 7.50 becomes

$$\sum_{i=1}^{\infty} c_i(t)\mathbf{K}\mathbf{w}_i(x) + \sum_{i=1}^{\infty} \ddot{c}_i(t)\mathbf{M}\mathbf{w}_i(x) = \mathbf{F}(x,t) \tag{7.51}$$

Taking the standard inner product of both sides of Equation 7.51 with $w_j(x)$ for any j leads to

$$\left(\sum_{i=1}^{\infty} c_i(t)\mathbf{K}\mathbf{w}_i(x),\mathbf{w}_j(x)\right) + \left(\sum_{i=1}^{\infty} \ddot{c}_i(t)\mathbf{M}\mathbf{w}_i(x),\mathbf{w}_j(x)\right)$$

$$= (\mathbf{F}(x,t),\mathbf{w}_j(x)) \tag{7.52}$$

The summation may be interchanged with the inner product operation because the series is convergent. Using the energy inner product definitions in Equation 7.52 leads to

$$\sum_{i=1}^{\infty} c_i(t)(\mathbf{w}_i,\mathbf{w}_j)_K + \sum_{i=1}^{\infty} \ddot{c}_i(t)(\mathbf{w}_i,\mathbf{w}_j)_M = (\mathbf{F},\mathbf{w}_j) \tag{7.53}$$

Mode shape orthonormality is used in Equation 7.53, reducing it to

$$\ddot{c}_j(t) + \omega_j^2 c_j(t) = g_j(t) \quad j = 1,2,\ldots \tag{7.54}$$

where

$$g_j(t) = (\mathbf{F},\mathbf{w}_j) \tag{7.55}$$

Equation 7.54 represents uncoupled differential equations, which may be solved individually for each $c_j(t)$. The convolution integral is applied leading to

$$c_j(t) = \frac{1}{\omega_j} \int_0^t g_j(\tau)\sin[\omega_j(t-\tau)]d\tau \tag{7.56}$$

The procedure used to solve the nonhomogeneous forced vibrations problem for a continuous system is called modal analysis. The equations for modal analysis for continuous systems and discrete systems are the same. The number of mode shapes and inner product definitions are, of course, different. Considerations exist for continuous systems which do not exist for discrete systems. The question of convergence and the interchange of order of operations has been discussed.

Example 7.12. The simply supported beam of Figure 7.21a is subject to the uniform loading whose time dependence is shown in Figure 7.21b. Determine the forced response of the system.

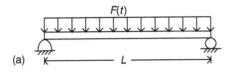

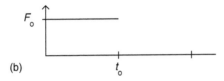

FIGURE 7.21 (a) Simply supported beam with uniform load for Example 7.12. (b) Time dependence of load F(t).

Solution: The governing problem for the forced response of the system is

$$EI\frac{\partial^4 w}{\partial x^4} + \rho A\frac{\partial^2 w}{\partial t^2} = F(x,t) \tag{a}$$

The partial differential equation is nondimensionalized using the variables introduced in Equation 4.64 through Equation 4.68, leading to

$$\frac{\partial^4 w}{\partial x^4} + \frac{\partial^2 w}{\partial t^2} = \lambda F(x,t) \tag{b}$$

where all variables including the force is nondimensional and

$$\lambda = \frac{F_0 L^3}{EI} \tag{c}$$

where F_0 is a characteristic parameter of $F(x,t)$, often taken to be the maximum of the force per unit length.

The nondimensional forced response is governed by Equation c along with the boundary conditions

$$w(0,t) = 0 \qquad \frac{\partial^2 w}{\partial x^2}(0,t) = 0 \qquad w(1,t) = 0 \qquad \frac{\partial^2 w}{\partial x^2}(1,t) = 0 \tag{d}$$

The natural frequencies and normalized mode shapes that satisfy Equation c and Equation d are obtained from Table 5.3 as

$$\omega_k = (k\pi)^2 \quad k = 1,2,3,\ldots \tag{e}$$

$$w_k(x) = \sqrt{2}\sin(k\pi x) \tag{f}$$

The nondimensional load is

$$F(x,t) = [u(t) - u(t - t_0^*)] \tag{g}$$

where

$$t_0^* = t_0\sqrt{\frac{EI}{\rho AL^4}} \tag{h}$$

The use of the normal modes in the expansion theorem leads to uncoupled differential equations of the form of Equation 7.54. The forces appearing in the equations are of the form

$$g_i(t) = (F, w_i)$$

$$= \int_0^1 F(x,t)w_i(x)dx$$

$$\tag{i}$$

$$= \int_0^1 [u(t) - u(t - t_0^*)]\sqrt{2}\sin(i\pi x)dx$$

$$= \frac{\sqrt{2}}{i\pi}[1 - (-1)^i][u(t) - u(t - t_0^*)]$$

Equation i is simplified to

$$g_i(t) = q_i[u(t) - u(t - t_0^*)] \tag{j}$$

where

$$q_i = \begin{cases} 0 & i = 2,4,6,\ldots \\[2mm] \dfrac{2\sqrt{2}}{\pi i} & i = 1,3,5,\ldots \end{cases} \tag{k}$$

The modal equations are of the form

$$\ddot{c}_{k,n} + \omega_{k,n}^2 c_{k,n} = A_{k,n} e^{-\alpha t} \tag{1}$$

The solution of Equation 1, assuming $c_{k,n}(0) = \dot{c}_{k,n}(0) = 0$, is

$$c_{k,n}(t) = \frac{A_{k,n}}{\omega_{k,n}^2 + \alpha^2} \left[\frac{\alpha}{\omega_{k,n}} \sin(\omega_{k,n}t) - \cos(\omega_{k,n}t) + e^{-\alpha t} \right] \tag{m}$$

Substitution of Equation c through Equation e and Equation i through Equation l in Equation f leads to

$$
\begin{bmatrix} \theta_1(x,t) \\ \theta_2(x,t) \\ \theta_3(x,t) \end{bmatrix} = \sum_{k=1}^{\infty} \left\{ \begin{bmatrix} 0.7385 \\ 0.7385 \\ 0.7385 \end{bmatrix} \frac{(0.7385)\sin\left[(2k-1)\frac{\pi}{4}\right]T_0}{\alpha^2 + \omega_{k,1}^2} \right.
$$

$$
\times \left[\frac{\alpha}{\omega_{k,1}} \sin(\omega_{k,1}t) - \cos(\omega_{k,1}t) + e^{-\alpha t} \right]
$$

$$
+ \begin{bmatrix} -1.1445 \\ 0.1322 \\ 0.6804 \end{bmatrix} \frac{(0.6804)\sin\left[(2k-1)\frac{\pi}{4}\right]T_0}{\alpha^2 + \omega_{k,2}^2}
$$

$$
\times \left[\frac{\alpha}{\omega_{k,2}} \sin(\omega_{k,2}t) - \cos(\omega_{k,2}t) + e^{-\alpha t} \right] \tag{n}
$$

$$
+ \begin{bmatrix} -0.3804 \\ 1.0361 \\ -0.6133 \end{bmatrix} \frac{(-0.6133)\sin\left[(2k-1)\frac{\pi}{4}\right]T_0}{\alpha^2 + \omega_{k,3}^2}
$$

$$
\left. \times \left[\frac{\alpha}{\omega_{k,3}} \sin(\omega_{k,3}t) - \cos(\omega_{k,3}t) + e^{-\alpha t} \right] \right\} \sqrt{2}\sin\left[(2k-1)\frac{\pi}{2}r\right]
$$

Equation n is evaluated using the first four sets of natural frequencies from Table 5.6. Using $T_0 = 1$ and $\alpha = 0.1$, the response for each shaft is illustrated in Figure 7.23.

Example 7.15. Use modal analysis to determine the time-dependent response of a circular membrane when it is subject to a force $F(t) = F_0 \sin(\omega t)$ applied on the surface of the membrane at the point corresponding to $r = (1/2)$ and $\theta = (\pi/2)$.

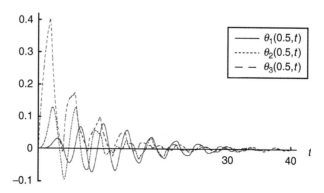

FIGURE 7.23 Forced response of three concentric shafts connected by layers of torsional springs due to applied torque at midspan of outer shaft.

Solution: The equations for the natural frequencies and normalized mode shapes for a circular membrane are determined in Section 5.15. The appropriate modal expansion for the time-dependent response of the membrane is

$$w(r,\theta,t) = \sum_{m=0}^{\infty} \left\{ c_{0,m}(t)w_{0,m}(r,\theta) + \sum_{n=1}^{\infty} [c_{1,m,n}(t)w_{1,m,n}(r,\theta) + c_{2,m,n}(t)w_{2,m,n}(r,\theta)] \right\} \tag{a}$$

where

$$w_{0,m} = D_{0,m}J_0(\omega_{0,m}r) \tag{b}$$

$$w_{1,m,n} = D_{1,m,n}J_n(\omega_{n,m}r)\cos(n\theta) \tag{c}$$

$$w_{2,m,n} = D_{2,m,n}J_n(\omega_{n,m}r)\sin(n\theta) \tag{d}$$

The constants are determined to render the mode shapes orthonormal. For example,

$$D_{1,m,n} = \left\{ \int_0^{2\pi} \int_0^1 [J_n(\omega_{n,m}r)\cos(n\theta)]^2 r \, dr \, d\theta \right\}^{-1/2} \tag{e}$$

The natural frequency $\omega_{n,m}$ is the mth root of $J_n(r)$.

The force is represented as

$$f(r,\theta,t) = F_0 \sin(\omega t)\delta\left(r - \frac{1}{2}\right)\delta\left(\theta - \frac{\pi}{2}\right) \tag{f}$$

The uncoupled modal equations are of the form

$$\ddot{c}_{0,m} + \omega_{0,m}^2 c_{0,m} = g_{0,m}(t) \tag{g}$$

$$\ddot{c}_{1,m,n} + \omega_{n,m}^2 c_{1,m,n} = g_{1,m,n}(t) \tag{h}$$

$$\ddot{c}_{2,m,n} + \omega_{n,m}^2 c_{2,m,n} = g_{2,m,n}(t) \tag{i}$$

where

$$g_{0,m}(t) = (f(r,\theta,t), w_{0,m}(r,\theta))$$

$$= \int_0^{2\pi}\int_0^1 F_0\sin(\omega t)\delta\left(r - \frac{1}{2}\right)\delta\left(\theta - \frac{\pi}{2}\right)D_{0,m}J_0(\omega_{0,m}r)r\,dr\,d\theta$$

$$= \frac{F_0}{2}D_{0,m}J_0\left(\frac{\omega_{0,m}}{2}\right)\sin(\omega t) \tag{j}$$

$$g_{1,m,n}(t) = \int_0^{2\pi}\int_0^1 F_0\sin(\omega t)\delta\left(r - \frac{1}{2}\right)\delta\left(\theta - \frac{\pi}{2}\right)D_{1,m,n}J_n(\omega_{m,m}r)\cos(n\theta)r\,dr\,d\theta$$

$$= \frac{F_0}{2}D_{1,m,n}J_n\left(\frac{\omega_{n,m}}{2}\right)\sin(\omega t)\cos\left(n\frac{\pi}{2}\right) \tag{k}$$

$$= \frac{F_0}{2}D_{1,m,n}J_n\left(\frac{\omega_{n,m}}{2}\right)\sin(\omega t)\begin{cases} 0 & n = 1,3,5,\ldots \\ -1 & n = 2,6,10,\ldots \\ 1 & n = 4,8,14,\ldots \end{cases}$$

$$
\begin{aligned}
g_{2,m,n}(t) &= \int_0^{2\pi}\int_0^1 F_0\sin(\omega t)\delta\left(r-\frac{1}{2}\right)\delta\left(\theta-\frac{\pi}{2}\right)D_{2,m,n}J_n(\omega_{m,m}r)\sin(n\theta)r\,dr\,d\theta \\[2mm]
&= \frac{F_0}{2}D_{2,m,n}J_n\left(\frac{\omega_{n,m}}{2}\right)\sin(\omega t)\sin\left(n\frac{\pi}{2}\right) \qquad\qquad\qquad (1) \\[2mm]
&= \frac{F_0}{2}D_{1,m,n}J_n\left(\frac{\omega_{n,m}}{2}\right)\sin(\omega t)
\begin{cases}
0 & n = 2,4,6,\ldots \\
1 & n = 1,5,9,\ldots \\
-1 & n = 3,7,11,\ldots
\end{cases}
\end{aligned}
$$

The steady-state response of Equation g is

$$
c_{0,m} = \frac{F_0 D_{0,m}J_0\left(\frac{\omega_{0,m}}{2}\right)}{2(\omega_{0,m}^2 - \omega^2)}\sin(\omega t) \qquad\qquad (m)
$$

Similar equations are obtained for the steady-state response of Equation h and Equation i, which are substituted into Equation a to determine the membrane's response.

7.6 DISCRETE SYSTEMS WITH DAMPING

The free response of a discrete system with proportional damping is obtained in Chapter 6 of Section 2, and the free response of a system with general viscous damping is determined in Chapter 6 of Section 3.

7.6.1 PROPORTIONAL DAMPING

A system is proportionally damped when the viscous damping matrix is a linear combination of the mass and stiffness matrices as defined in Equation 6.26. The forced response of a system with proportional damping is determined by solving

$$
\mathbf{M\ddot{x}} + (\alpha\mathbf{K} + \beta\mathbf{M})\mathbf{\dot{x}} + \mathbf{Kx} = \mathbf{F} \qquad\qquad (7.57)
$$

The principal coordinates of the undamped system may be used to decouple the equations represented by Equation 7.57. Let $\omega_1 \leq \omega_2 \leq \ldots \leq \omega_n$ be the undamped natural frequencies of the system with corresponding normalized mode shape vectors $w_1, w_2, \ldots, w_n$. The expansion theorem suggests that $\mathbf{x}(t)$ can be expanded in terms of the normalized mode shapes as

$$\mathbf{x}(t) = \sum_{i=1}^{n} c_i(t)\mathbf{w}_i \qquad (7.58)$$

Substituting Equation 7.58 into Equation 7.57 and following the steps used in Section 6.2 leads to

$$\ddot{c}_1 + 2\zeta_i\omega_i\dot{c}_i + \omega_i^2 c_i = g_i(t) \qquad (7.59)$$

where

$$g_i(t) = (\mathbf{F}, \mathbf{w}_i) \qquad (7.60)$$

The convolution integral solution of Equation 7.59 for an underdamped mode is

$$c_i(t) = \frac{1}{\omega_i\sqrt{1-\zeta_i^2}} \int_0^t g_i(\tau)e^{-\zeta_i\omega_i(t-\tau)}\sin\left[\omega_i\sqrt{1-\zeta_i^2}(t-\tau)\right]d\tau \qquad (7.61)$$

7.6.2 GENERAL VISCOUS DAMPING

The differential equations for the forced response of a discrete system with general viscous damping are

$$\mathbf{M}\ddot{\mathbf{x}} + \mathbf{C}\dot{\mathbf{x}} + \mathbf{K}\mathbf{x} = \mathbf{F} \qquad (7.62)$$

An expansion of the response in terms of the mode shapes of the corresponding undamped system does not uncouple the differential equations. Following the discussion of Section 6.3, a state space formulation of Equation 7.62 is used. Defining $\mathbf{y}_1 = \mathbf{x}$ and $\mathbf{y}_2 = \dot{\mathbf{x}}$ leads to

$$\begin{bmatrix} \dot{\mathbf{y}}_1 \\ - \\ \dot{\mathbf{y}}_2 \end{bmatrix} = \begin{bmatrix} \mathbf{0} & | & \mathbf{I} \\ - & | & - \\ -\mathbf{M}^{-1}\mathbf{K} & | & -\mathbf{M}^{-1}\mathbf{C} \end{bmatrix} \begin{bmatrix} \mathbf{y}_1 \\ - \\ \mathbf{y}_2 \end{bmatrix} + \begin{bmatrix} \mathbf{0} \\ - \\ \mathbf{M}^{-1}\mathbf{F} \end{bmatrix} \qquad (7.63)$$

Defining

$$z = \begin{bmatrix} y_1 \\ - \\ y_2 \end{bmatrix} \quad Q = \begin{bmatrix} 0 & | & I \\ - & | & - \\ -M^{-1}K & | & -M^{-1}C \end{bmatrix} \quad h = \begin{bmatrix} 0 \\ - \\ M^{-1}F \end{bmatrix} \quad (7.64)$$

Equation 7.63 can be written as

$$\dot{z} = Qz + h \quad (7.65)$$

The $2n \times 2n$ matrix Q is not self-adjoint; its adjoint, with respect to the standard inner product on R^{2n}, is its transpose. The free response is assumed as $z = \Phi e^{\gamma t}$. The values of γ are the eigenvalues of Q and the Φs are the corresponding right eigenvectors.

The eigenvalues of Q^T are the conjugates of the eigenvalues of Q; if γ is an eigenvalue of Q, then $\bar{\gamma}$ is an eigenvalue of Q^T. The eigenvectors corresponding to Q^T are determined as $\hat{\Phi}$, also called the left eigenvectors of Q. Let Φ_l be an eigenvector of Q corresponding to an eigenvalue γ_i and let $\hat{\Phi}_i$ be an eigenvalue of Q^T corresponding to its eigenvalue $\bar{\gamma}_i$. The right and left eigenvectors of Q corresponding to eigenvalues that are not complex conjugates satisfy a biorthogonality condition

$$(\Phi_i, \hat{\Phi}_j) = 0 \quad i \neq j \quad (7.66)$$

where the inner product is the standard inner product for R^{2n}. The eigenvectors can be normalized such that

$$(\Phi_i, \hat{\Phi}_i) = 1 \quad (7.67)$$

The expansion theorem for expansion of S, an arbitrary member of R^{2n} in terms of the eigenvectors of Q is

$$S = \sum_{i=1}^{n} (S, \hat{\Phi}_i) \Phi_i \quad (7.68)$$

An expansion for z, the solution of Equation 7.64 in terms of the eigenvectors of Q is assumed as

$$\mathbf{z} = \sum_{i=1}^{2n} c_i \Phi_i \tag{7.69}$$

Substituting Equation 7.69 into Equation 7.64 leads to

$$\sum_{i=1}^{2n} \dot{c}_i \Phi_i = \sum_{i=1}^{2n} c_i \mathbf{Q}\Phi_i + \mathbf{h} \tag{7.70}$$

Noting that $\mathbf{Q}\Phi_i = \gamma_i \Phi_i$ and taking the inner product of Equation 7.70 with $\hat{\Phi}_j$ for an arbitrary j between 1 and $2n$ leads to

$$\sum_{i=1}^{2n} \dot{c}_i (\Phi_i, \hat{\Phi}_j) = \sum_{i=1}^{2n} \gamma_i c_i (\Phi_i, \hat{\Phi}_j) + (\mathbf{h}, \hat{\Phi}_j) \tag{7.71}$$

Use of biorthonormality of the eigenvectors in Equation 7.71 leads to

$$\dot{c}_j = \lambda_j c_j + g_j(t) \tag{7.72}$$

where

$$g_j(t) = (\mathbf{h}, \hat{\Phi}_j) \tag{7.73}$$

The general solution of Equation 7.72 is

$$c_j(t) = c_j(0)e^{-\gamma_j t} + \int_0^t g_j(\tau)e^{-\gamma_j(t-\tau)}d\tau \tag{7.74}$$

Example 7.16. The 30 kg block in the system of Example 6.2 is subject to the step pulse

$$F(t) = 10000[u(t) - u(t - 0.175)] \, \text{N} \tag{a}$$

Use modal analysis to determine the system's forced response.

Solution: Using the state-space formulation of Example 6.2, the differential equations for the forced response are

$$
\begin{bmatrix} \dot{y}_{1,1} \\ \dot{y}_{1,2} \\ \dot{y}_{2,1} \\ \dot{y}_{2,2} \end{bmatrix} = \begin{bmatrix} 0 & 0 & 1 & 0 \\ 0 & 0 & 0 & 1 \\ -15000 & 10000 & -150 & 50 \\ 6666.67 & -10000 & 33.33 & -33.33 \end{bmatrix} \begin{bmatrix} y_{1,1} \\ y_{1,2} \\ y_{2,1} \\ y_{2,2} \end{bmatrix} + \begin{bmatrix} 0 \\ 0 \\ 0 \\ \dfrac{1}{30}F(t) \end{bmatrix} \quad \text{(b)}
$$

The eigenvalues and eigenvectors of the matrix $\mathbf{A}$ are determined in Example 6.6. Determination of the forced response also requires determination of the eigenvectors of the adjoint of $\mathbf{A}$. The adjoint of a matrix is its transpose. The eigenvalues of $\mathbf{A}^T$ are the conjugates of the eigenvalues of $\mathbf{A}$, but since the eigenvalues of $\mathbf{A}$ occur in complex conjugate pairs, the eigenvalues of both matrices are the same. The eigenvalues and eigenvectors of the transpose of $\mathbf{A}$ are

$$
\lambda_1 = -73.54 + 114.4i \qquad \hat{\Phi}_1 = \begin{bmatrix} 0.7678 \\ -0.6339 + 0.0925i \\ 0.0039 - 0.0042i \\ 0.0003 - 0.0037i \end{bmatrix} \quad \text{(c)}
$$

$$
\lambda_2 = -73.54 - 114.4 \qquad \hat{\Phi}_2 = \begin{bmatrix} 0.7678 \\ -0.6339 - 0.0925i \\ 0.0039 + 0.0042i \\ 0.0003 + 0.0037i \end{bmatrix} \quad \text{(d)}
$$

$$
\lambda_3 = -18.13 + 64.63i \qquad \hat{\Phi}_3 = \begin{bmatrix} -0.0113 - 0.559i \\ 0.8290 \\ -0.0032 - 0.0054i \\ -0.0017 - 0.0108i \end{bmatrix} \quad \text{(e)}
$$

$$
\lambda_4 = -18.13 - 64.63 \qquad \hat{\Phi}_4 = \begin{bmatrix} -0.0113 + 0.559i \\ 0.8290 \\ -0.0032 + 0.0054i \\ -0.0017 + 0.0108i \end{bmatrix} \tag{f}
$$

The eigenvectors of Equation c through Equation f satisfy biorthogonality relations with the eigenvectors of **A**, Equation f through Equation j of Example 6.2

$$
(\Phi_i, \hat{\Phi}_j) = 0 \quad i \neq j \tag{g}
$$

The eigenvectors are normalized by requiring that

$$
(\Phi_i, \hat{\Phi}_i) = 1 \tag{h}
$$

Let **P** be the matrix of normalized eigenvectors of **A** and $\hat{\mathbf{P}}$ the matrix of normalized eigenvectors of $\mathbf{A}^T$. Using Equation h, these matrices are obtained as

$$
\mathbf{P} = \begin{bmatrix} 0.0228 - 0.0654i & 0.0228 + 0.654i & 0.0626 + 0.0293i & 0.0626 - 0.0293i \\ 0.0142 + 0.0267i & 0.0142 - 0.267i & 0.0550 + 0.0582i & 0.0550 - 0.0582i \\ 5.800 + 7.425i & 5.800 + 7.425i & -3.0305 + 3.512i & -3.0305 - 3.512i \\ -4.101 - 0.339i & -4.103 - 0.331i & -4.761 + 2.499i & -4.761 - 2.499i \end{bmatrix} \tag{i}
$$

$$
\hat{\mathbf{P}} = \begin{bmatrix} 4.858 + 6.221i & 4.858 \quad 6.221i & 1.917 - 3.478i & 1.917 + 3.478i \\ -4.761 - 4.551i & -4.761 + 4.551i & 5.213 - 2.736i & 5.213 - 2.736i \\ 0.0586 + 5.00 \times 10^{-3}i & 0.0586 - 5.00 \times 10^{-3}i & -0.0377 - 0.0236i & -0.0377 + 0.0236i \\ -0.0285 + 0.0260i & -0.0285 - 0.0260i & -0.0459 - 0.0623i & -0.0459 + 0.0623i \end{bmatrix} \tag{j}
$$

The modal equations are those of Equation 7.72, where

$$
g_i(t) = (F(t), \hat{\Phi}_i) \tag{k}
$$

Evaluation of Equation k yields

$$G = \begin{bmatrix} -9.49 + 8.67i \\ -9.49 - 8.67i \\ -15.32 - 20.77i \\ -15.32 + 20.77i \end{bmatrix} [u(t) - u(t - 0.175)] \qquad (l)$$

The modal equations are written as

$$\dot{c}_1 = (-73.54 + 114.4i)c_i + [-9.49 + 8.67i][u(t) - u(t - 0.175)] \qquad (m)$$

$$\ddot{c}_2 = (-73.54 + 114.4i)c_2 + [-9.49 - 8.67i][u(t) - u(t - 0.175)] \qquad (n)$$

$$\ddot{c}_3 = (-18.13 + 64.63i)c_3 + [-15.32 - 20.77i][u(t) - u(t - 0.175)] \qquad (o)$$

$$\ddot{c}_4 = (-18.13 - 64.63i)c_4 + [-15.32 + 20.77i][u(t) - u(t - 0.175)] \qquad (p)$$

Equation m through Equation p are solved using the convolution integral. The results are substituted into Equation 7.69 to solve the original generalized coordinates. After considerable algebra, the responses for the original

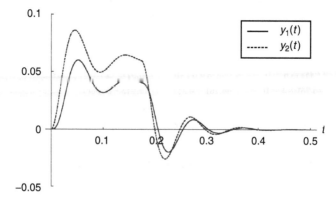

FIGURE 7.24 Forced response of two-degree-of-freedom system with general damping.

generalized coordinates are obtained as

$$y_1(t) = 0.04 - e^{-18.13t}[0.0472 \cos(64.63t) + 0.0241 \sin(64.63t)]$$
$$+ e^{-73.54t}[0.00734 \cos(114.4t) + 0.0183 \sin(114.4t)]$$
$$+ \{-0.04 + e^{-18.13t}[0.862 \cos(64.63t) - 0.915 \sin(64.63t)]$$
$$+ e^{-73.54t}[2441.6 \cos(114.4t) - 4256.3 \sin(114.4t)]\}u(t - 0.175)$$

(q)

$$y_2(t) = 0.06 - e^{-18.13t}[0.613 \cos(64.63t) + 0.00586 \sin(64.63t)]$$
$$+ e^{-73.54t}[0.00130 \cos(114.4t) - 0.00557 \sin(114.4t)]$$
$$+ \{-0.06e^{-18.13t}[0.541 \cos(64.63t) - 1.254 \sin(64.63t)]$$
$$- e^{-74.54t}[2090 \cos(114.4t) - 477.3 \sin(114.4t)]\}u(t - 0.175)$$

(r)

Equation q and Equation r are illustrated in Figure 7.24.

8 Rayleigh–Ritz and Finite-Element Methods

Exact solutions for the natural frequencies, mode shapes, and forced responses do not exist for many continuous systems. Even when they do exist they are often cumbersome to use, often requiring solutions of transcendental equations to determine the natural frequencies and subsequent evaluation of infinite series to evaluate the system response. For these reasons approximate solutions using variational principles are developed and applied.

8.1 FOURIER BEST APPROXIMATION THEOREM

Let $\mathbf{V}$ be a vector space in which a vector $\mathbf{v}$ resides. Let S be a finite-dimensional subspace of $\mathbf{V}$ spanned by $\mathbf{u}_1$, $\mathbf{u}_2$,...,$\mathbf{u}_n$. Let $(\mathbf{u},\mathbf{v})$ be a valid definition of an inner product on $\mathbf{V}$. It is desired to find an approximation to $\mathbf{v}$ from the elements of S. The best approximation is defined as the vector $\mathbf{u}$ in S which minimizes the functional

$$D(\mathbf{u}) = \|\mathbf{v} - \mathbf{u}\| \tag{8.1}$$

where the norm is the inner product generated norm.

Since $\mathbf{u}_1$, $\mathbf{u}_2$,...,$\mathbf{u}_n$ spans S any vector in S may be written as a linear combination of the vectors in this basis

$$\mathbf{u} = \sum_{i=1}^{n} \alpha_i \mathbf{u}_i \tag{8.2}$$

Thus $D(\mathbf{u}) = f(\alpha_1, \alpha_2,...,\alpha_n)$ and is rendered stationary if $\delta D(\mathbf{u}) = 0$. This implies

$$\sum_{i=1}^{n} \frac{\partial D}{\partial \alpha_i} \delta \alpha_i = 0 \tag{8.3}$$

525

Since the variations are independent Equation 8.3 implies

$$\frac{\partial D}{\partial \alpha_i} = 0 \quad i = 1,2,\ldots,n \tag{8.4}$$

Substitution of Equation 8.2 into Equation 8.1 leads to

$$D(\mathbf{u}) = \left\| \mathbf{v} - \sum_{i=1}^{n} \alpha_i \mathbf{u}_i \right\| = \left(\mathbf{v} - \sum_{i=1}^{n} \alpha_i \mathbf{u}_i, \mathbf{v} - \sum_{i=1}^{n} \alpha_i \mathbf{u}_i \right)^{1/2}$$

$$= \left[(\mathbf{v},\mathbf{v}) - \left(\mathbf{v}, \sum_{i=1}^{n} \alpha_i \mathbf{u}_i \right) - \left(\sum_{i=1}^{n} \alpha_i \mathbf{u}_i, \mathbf{v} \right) + \left(\sum_{i=1}^{n} \alpha_i \mathbf{u}_i, \sum_{j=1}^{n} \alpha_j \mathbf{u}_j \right) \right]^{1/2} \tag{8.5}$$

Noting that $\delta\{[D(\mathbf{u})]^2\} = 2D(\mathbf{u})\delta D(\mathbf{u})$, then unless $D(\mathbf{u}) = 0$ for the $\mathbf{u}$ that renders $[D(\mathbf{u})]^2$ stationary, $D(\mathbf{u})$ is stationary if $[D(\mathbf{u})]^2$ is rendered stationary. To this end,

$$[D(\mathbf{u})]^2 = (\mathbf{v},\mathbf{v}) - \left(\mathbf{v}, \sum_{i=1}^{n} \alpha_i \mathbf{u}_i \right) - \left(\sum_{i=1}^{n} \alpha_i \mathbf{u}_i, \mathbf{v} \right)$$

$$+ \left(\sum_{i=1}^{n} \alpha_i \mathbf{u}_i, \sum_{j=1}^{n} \alpha_j \mathbf{u}_j \right)$$

$$= (\mathbf{v},\mathbf{v}) - 2 \sum_{i=1}^{n} \alpha_i (\mathbf{v},\mathbf{u}_i) + \sum_{i=1}^{n} \sum_{j=1}^{n} \alpha_i \alpha_j (\mathbf{u}_i,\mathbf{u}_j) \tag{8.6}$$

$\delta D(\mathbf{u}) = 0$ if $\partial D^2/\partial \alpha_k = 0 \ k = 1,2,\ldots,n$. To this end,

$$\frac{\partial [D(\mathbf{u})]^2}{\partial \alpha_k} = -2 \sum_{i=1}^{n} \frac{\partial \alpha_i}{\partial \alpha_k} (\mathbf{v},\mathbf{u}_i) + \sum_{j=1}^{n} \sum_{i=1}^{n} \frac{\partial}{\partial \alpha_k} (\alpha_i \alpha_j)(\mathbf{u}_i,\mathbf{u}_j) = 0 \tag{8.7}$$

Noting that $\partial \alpha_i/\partial \alpha_k = \delta_{i,k}$ and $\partial/\partial \alpha_k(\alpha_i \alpha_j) = \alpha_i \delta_{j,k} + \alpha_j \delta_{i,k}$ Equation 8.7 simplifies to

$$\sum_{i=1}^{n} \alpha_i (\mathbf{u}_i,\mathbf{u}_k) = (\mathbf{v}, \mathbf{u}_k) \quad k = 1, 2,\ldots,n \tag{8.8}$$

Equation 8.8 represents the kth equation in a set of n equations for n unknowns to solve for α_k, $k = 1,2,\ldots,n$. If the basis vectors are chosen as an orthonormal set then $(\mathbf{u}_i,\mathbf{u}_j) = \delta_{i,j}$, and the only nonzero term in the summation in Equation 8.8 corresponds to $i = k$. The equation collapses to

$$\alpha_k = (\mathbf{v},\mathbf{u}_k) \quad k = 1,2,\ldots,n \tag{8.9}$$

The method described above is called the least squares method and is used to find the best approximation of one vector from a specific subspace from which the vector may not belong.

Application of modal analysis to determine the response of a continuous system due to an external excitation requires that the external force be expanded in a series of eigenvectors of the free response. If S is all of D and the basis is chosen as a set of orthonormal eigenvectors of the stiffness operator and the inner product is the kinetic energy inner product then Equation 8.2 is simply an eigenvector expansion for $\mathbf{u}$. However, the external excitation may not satisfy the homogeneous boundary conditions in which case it is not in the domain of the stiffness operator. The proof of convergence of the eigenvector expansion of Equation 8.2 is beyond the scope of this study. However the Fourier Best Approximation implies that when the series converges the eigenvector expansion is the best approximation to $\mathbf{v}$ from S.

Often approximations are sought for solutions of equations of the form

$$\mathbf{Lu} = \mathbf{f} \tag{8.10}$$

where $\mathbf{L}$ is a linear operator whose domain is S and range is R. Then for a vector $\mathbf{f}$ in R, the solution of Equation 8.10 is the vector $\mathbf{u}$ in S defined such that Equation 8.8 is satisfied. An approximate solution using the least squares method might be the choice of a $\mathbf{w}$ from a subspace W of S such that $\mathbf{Lw}$ provides the best approximation to $\mathbf{f}$ in the least squares sense. To this end, let $\mathbf{w}_1,\mathbf{w}_2,\ldots,\mathbf{w}_n$ be a basis for W. Then any vector in W is written as

$$\mathbf{w} = \sum_{i=1}^{n} \alpha_i \mathbf{w}_i \tag{8.11}$$

The functional to be minimized is

$$D(\mathbf{w}) = \|\mathbf{Lw} - \mathbf{f}\| \tag{8.12}$$

The equations which render the functional stationary are

$$\sum_{i=1}^{n} \alpha_i(\mathbf{Lu}_i,\mathbf{Lu}_k) = (\mathbf{Lu}_k,\mathbf{f}) \tag{8.13}$$

8.2 RAYLEIGH–RITZ METHOD

The least squares method of Section 8.1 used to approximate the solution to
$\mathbf{Lu} = \mathbf{f}$ leads to the best approximation of $\mathbf{f}$, not the best approximation of $\mathbf{u}$.
Suppose $\mathbf{L}$ is self-adjoint and positive definite with respect to a valid inner
product, $(\mathbf{f}, \mathbf{g})$ defined for all $\mathbf{f}$ and $\mathbf{g}$ in the domain of $\mathbf{L}$. In this case an energy
inner product $(\mathbf{f}, \mathbf{g})_E = (\mathbf{Lf}, \mathbf{g})$ may be defined.

As before, let W be a subspace of D, the domain of $\mathbf{L}$. Let $\mathbf{w}_1, \mathbf{w}_2,...,\mathbf{w}_n$ be
a basis for W. Any vector in W has the form of Equation 8.11. Consider the
functional

$$D_E(\mathbf{w}) = \|\mathbf{w} - \mathbf{u}\|_E \tag{8.14}$$

Substitution of the expansion for $\mathbf{w}$ in Equation 8.14 leads to

$$[D_E(\mathbf{w})]^2 = (\mathbf{w} - \mathbf{u}, \mathbf{w} - \mathbf{u})_E = (\mathbf{w}, \mathbf{w})_E - 2(\mathbf{u}, \mathbf{w})_E + (\mathbf{u}, \mathbf{u})_E$$

$$= (\mathbf{Lw}, \mathbf{w})_E - 2(\mathbf{Lu}, \mathbf{w}) + (\mathbf{Lu}, \mathbf{u}) \tag{8.15}$$

Substituting the expansion for $\mathbf{w}$ in terms of the basis functions, noting that
since $\mathbf{u}$ is the true solution $\mathbf{Lu} = \mathbf{f}$ and using properties of inner products
Equation 8.15 becomes

$$[D_E(\mathbf{w})]^2 = \left(\mathbf{L}\left(\sum_{i=1}^{n} \alpha_i \mathbf{w}_i \right), \sum_{j=1}^{n} \alpha_j \mathbf{w}_j \right) - 2\left(\mathbf{f}, \sum_{i=1}^{n} \alpha_i \mathbf{w}_i \right) + (\mathbf{f}, \mathbf{u})$$

$$= \sum_{i=1}^{n} \sum_{j=1}^{n} \alpha_i \alpha_j (\mathbf{Lw}_i, \mathbf{w}_j) - 2 \sum_{i=1}^{n} \alpha_i (\mathbf{f}, \mathbf{w}_i) + (\mathbf{f}, \mathbf{u}) \tag{8.16}$$

$D_E(\mathbf{w})$ is stationary if $[\delta D_E]^2 = 0$, which leads to

$$\frac{\partial D_E}{\partial \alpha_k} = 0 \quad k = 1, 2,...,n \tag{8.17}$$

Application of Equation 8.17 to Equation 8.16 leads to a set of equations
which are summarized by

$$\sum_{i=1}^{n} \alpha_i (\mathbf{Lw}_i, \mathbf{w}_k) = (\mathbf{f}, \mathbf{w}_k) \tag{8.18}$$

or

$$\sum_{i=1}^{n} \alpha_i (\mathbf{w}_i, \mathbf{w}_k)_E = (\mathbf{f}, \mathbf{w}_k) \tag{8.19}$$

If the basis vectors are chosen such that they are orthogonal with respect to the energy inner product then

$$\alpha_k = (\mathbf{f}, \mathbf{w}_k) \tag{8.20}$$

The method developed above in which an energy inner product is used in the least squares formulation to determine an approximation to the solution of $\mathbf{Lu} = \mathbf{f}$ is called the Rayleigh–Ritz method. Since the Rayleigh–Ritz method uses an energy inner product, it can be applied when the operator is self-adjoint and positive definite with respect to a valid inner product.

The Rayleigh–Ritz method provides the best approximation to the solution from the vector space spanned by the chosen basis functions. The method does not speak to how to choose the basis elements. The actual value of the minimum of $D_E(\mathbf{w})$ depends upon the choice of the basis for W.

The question whether the accuracy of the approximation improves as the number of basis elements is increased can be addressed in the following manner. The elements of any basis are, by definition, linearly independent. Because of this and the positive definiteness of $\mathbf{L}$, it can be shown that the matrix represented by Equation 8.19 is nonsingular. Thus the Rayleigh–Ritz approximation exists and is unique. This implies that any choice of a basis for W leads to the same Rayleigh–Ritz approximation. For a chosen basis, the Gram–Schmidt process can be used to determine an orthonormal basis spanning W. Thus in assessing the accuracy of a Rayleigh–Ritz approximation it may be assumed that the basis functions are orthonormal with respect to the energy inner product,

$$(\mathbf{w}_i, \mathbf{w}_j)_E = \delta_{ij} \tag{8.21}$$

Let W_k be a k dimensional subspace of S, the domain of $\mathbf{L}$, with k orthonormal basis vectors satisfying Equation 8.21. Let $\tilde{\mathbf{w}}_k$ be the Rayleigh–Ritz approximation to the solution of $\mathbf{Lu} = \mathbf{f}$ from W_k. From Equation 8.11,

$$\tilde{\mathbf{w}}_k = \sum_{i=1}^{k} (\mathbf{f}, \mathbf{w}_i) \mathbf{w}_i \tag{8.22}$$

Let W_{k+1} be a $(k+1)$th dimensional vector space whose basis elements are those of W_k and $\mathbf{w}_{k+1}$, which is orthonormal with the other basis elements. The Rayleigh–Ritz approximation from W_{k+1} is

$$\tilde{\mathbf{w}}_{k+1} = \tilde{\mathbf{w}}_k + (\mathbf{f}, \mathbf{w}_{k+1}) \mathbf{w}_{k+1} \tag{8.23}$$

Now consider

$$[D_{\mathrm{E}}(\tilde{\mathbf{w}}_{k+1})]^2 = \|\tilde{\mathbf{w}}_{k+1} - \mathbf{u}\|_{\mathrm{E}}^2$$

$$= (\mathbf{L}(\tilde{\mathbf{w}}_{k+1} - \mathbf{u}), \tilde{\mathbf{w}}_{k+1} - \mathbf{u})$$

$$= (\mathbf{L}(\tilde{\mathbf{w}}_k - \mathbf{u}) + (\mathbf{f}, \mathbf{w}_{k+1})\mathbf{L}\mathbf{w}_{k+1}, \tilde{\mathbf{w}}_k - \mathbf{u} + (\mathbf{f}, \mathbf{w}_{k+1})\mathbf{w}_{k+1})$$

$$= (\mathbf{L}(\tilde{\mathbf{w}}_k - \mathbf{u}), \tilde{\mathbf{w}}_k - \mathbf{u}) + (\mathbf{f}, \mathbf{w}_{k+1})(\mathbf{L}(\tilde{\mathbf{w}}_k - \mathbf{u}), \mathbf{w}_{k+1})$$

$$+ (\mathbf{f}, \mathbf{w}_{k+1})(\mathbf{L}\mathbf{w}_{k+1}, \tilde{\mathbf{w}}_k - \mathbf{u}) + (\mathbf{f}, \mathbf{w}_{k+1})^2 \mathbf{L}(\mathbf{w}_{k+1}, \mathbf{w}_{k+1})$$

$$(8.24)$$

L is self-adjoint thus $(\mathbf{L}(\tilde{\mathbf{w}}_k - \mathbf{u}), \mathbf{w}_{k+1}) = (\mathbf{L}\mathbf{w}_{k+1}, \tilde{\mathbf{w}}_k - \mathbf{u})$, $\mathbf{w}_{k+1}$ is normalized with respect to the energy inner product, thus $(\mathbf{L}\mathbf{w}_{k+1}, \mathbf{w}_{k+1}) = 1$, and by definition $(\mathbf{L}(\tilde{\mathbf{w}}_k - \mathbf{u}), \tilde{\mathbf{w}}_k - \mathbf{u}) = [D_{\mathrm{E}}(\tilde{\mathbf{w}}_k)]^2$. Thus Equation 8.24 becomes

$$[D_{\mathrm{E}}(\tilde{\mathbf{w}}_{k+1})]^2 = [D_{\mathrm{E}}(\tilde{\mathbf{w}}_k)]^2 + 2(\mathbf{f}, \mathbf{w}_{k+1})(\mathbf{L}(\tilde{\mathbf{w}}_k - \mathbf{u}), \mathbf{w}_{k+1}) + (\mathbf{f}, \mathbf{w}_{k+1})^2 \quad (8.25)$$

Note that

$$(\mathbf{L}(\tilde{\mathbf{w}}_k - \mathbf{u}), \mathbf{w}_{k+1}) = (\mathbf{L}\tilde{\mathbf{w}}_k, \mathbf{w}_{k+1}) - (\mathbf{L}\mathbf{u}, \mathbf{w}_{k+1})$$

$$= \sum_{i=1}^{k} \alpha_i(\mathbf{L}\mathbf{w}_i, \mathbf{w}_{k+1}) - (\mathbf{f}, \mathbf{w}_{k+1})$$

$$= -(\mathbf{f}, \mathbf{w}_{k+1}) \quad (8.26)$$

The final step in Equation 8.26 is a result of the assumed orthogonality of the basis elements, Equation 8.21. Subsitution of Equation 8.21 into Equation 8.26 leads to

$$[D_{\mathrm{E}}(\tilde{\mathbf{w}}_{k+1})]^2 = [D_{\mathrm{E}}(\tilde{\mathbf{w}}_k)]^2 - (\mathbf{f}, \mathbf{w}_{k+1})^2 \quad (8.27)$$

Equation 8.27 leads to the following theorem:

Theorem 8.1. *Let $\tilde{\mathbf{w}}_k$ be the Rayleigh–Ritz approximation to the solution of $\mathbf{L}\mathbf{u} = \mathbf{f}$ from W_k, a k dimensional subspace of S, the domain of $\mathbf{L}$, and let $\tilde{\mathbf{w}}_{k+1}$ be the Rayleigh–Ritz approximation from W_{k+1}, a subspace of S whose basis is the basis of W_k augmented by one linearly independent vector. Then $\tilde{\mathbf{w}}_{k+1}$ is a better approximation for $\mathbf{u}$ than $\tilde{\mathbf{w}}_k$.*

Theorem 8.1 implies that the Rayleigh–Ritz approximation improves as the number of basis elements used in the approximation increases. While Theorem 8.1

does not guarantee convergence of the sequence $\tilde{\mathbf{w}}_1, \tilde{\mathbf{w}}_2, \ldots, \tilde{\mathbf{w}}_{k+1}, \ldots$, to the true solution, it does guarantee that the accuracy of the approximation improves.

8.3 GALERKIN METHOD

The Rayleigh–Ritz method applies only to liner operators that are self-adjoint and positive definite with respect to a valid inner product. Consider a problem of the form $\mathbf{Lu} = \mathbf{f}$. Let $\mathbf{w}_1, \mathbf{w}_2, \ldots, \mathbf{w}_n$ be a set of trial functions from which an approximation is to $\mathbf{u}$ is assumed as

$$\mathbf{w} = \sum_{i=1}^{n} \beta_i \mathbf{w}_i \qquad (8.28)$$

The error in using Equation 8.28 as an approximation is

$$\mathbf{e} = \mathbf{Lw} - \mathbf{f} \qquad (8.29)$$

If $\mathbf{w}$ is the exact solution, the inner product of the error with any vector is zero. The Galerkin method requires that the error be orthogonal to each element in a set of comparison functions, $\phi_1, \phi_2, \ldots, \phi_k$,

$$(\mathbf{e}, \phi_j) = 0 \qquad (8.30)$$

Substitution of Equation 8.29 into Equation 8.30 leads to

$$(\mathbf{Lw} - \mathbf{f}, \phi_j) = 0 \quad j = 1, 2, \ldots, k$$

$$\left(\mathbf{L} \left(\sum_{i=1}^{n} \beta_i \mathbf{w}_i \right) - \mathbf{f}, \phi_j \right) = 0 \quad j = 1, 2, \ldots, k \qquad (8.31)$$

If $\mathbf{L}$ is linear, then linearity of the operator and properties of inner products applied to Equation 8.31 leads to

$$\sum_{i=1}^{n} \beta_i (\mathbf{Lw}_i, \phi_j) = (\mathbf{f}, \phi_j) \quad j = 1, 2, \ldots, k \qquad (8.32)$$

If the set of comparison functions is chosen to be the set of basis functions, then equations represented by Equation 8.32 is the same as the equations used in the Rayleigh–Ritz method. That is when the operator $\mathbf{L}$ is linear and the comparison functions are the same as the basis functions and the equations for the coefficients are the same as the equations derived using the Rayleigh–Ritz methods. Although the equations are the same, there are significant differences. Many theorems on error analysis for the Rayleigh–Ritz method, such as Theorem 8.1, assume self-adjointness of the operator and thus do not apply to the Galerkin method. In addition, when the operator is self-adjoint and positive definite the matrix whose coefficients are $(\mathbf{L}\mathbf{w}_i, \phi_i)$ is symmetric and positive definite.

If the comparison functions are defined by

$$\phi_i(x) = \delta(x - x_i) \quad i = 1, 2, \ldots, n \tag{8.33}$$

and the inner product is the standard inner product on C[0,1], Equation 8.32 reduces to

$$\sum_{i=1}^{k} \beta_i(\mathbf{L}w_i)\Big|_{x=x_j} = f(x_j) \tag{8.34}$$

The Galerkin method, using the Dirac delta functions as comparison functions, is called the collocation method.

8.4 RAYLEIGH–RITZ METHOD FOR NATURAL FREQUENCIES AND MODE SHAPES

It is shown in Section 5.3 that Rayleigh's Quotient

$$R(\mathbf{w}) = \frac{(\mathbf{w}, \mathbf{w})_K}{(\mathbf{w}, \mathbf{w})_M} \tag{8.35}$$

is stationary if and only if $\mathbf{w}$ is a mode shape of the self-adjoint system whose inertia operator is $\mathbf{M}$ and stiffness operator is $\mathbf{K}$. When $\mathbf{w_i}$ is a mode shape corresponding to a natural frequency ω_i, then

$$R(\mathbf{w}_i) = \omega_i^2 \tag{8.36}$$

An alternate writing of Equation 8.35 is

$$(\mathbf{w}, \mathbf{w})_K - R(\mathbf{w})(\mathbf{w}, \mathbf{w})_M = 0 \tag{8.37}$$

Let $\mathbf{w}_1, \mathbf{w}_2, \ldots, \mathbf{w}_n$ be a set of basis vectors for a n-dimensional subspace of the domain of $\mathbf{K}$. An approximation for the mode shape from this subspace is

$$\mathbf{w} = \sum_{i=1}^{n} \alpha_i \mathbf{w}_i \tag{8.38}$$

It is shown in Section 5.3 that the minimum value of $R(\mathbf{w})$ over all $\mathbf{w}$ in the domain of the stiffness operator is ω_1^2 and occurs when $\mathbf{w}$ is the mode shape vector corresponding to ω_1. Since W is a subspace of the domain of $\mathbf{K}$, the minimum value of $R(\mathbf{w})$ over all vectors in W is larger than ω_1^2. Clearly, the difference, $R(\mathbf{w}) - \omega_1^2$, is minimized when $R(\mathbf{w})$ is a minimum. Thus, the best approximation for the mode shape corresponding to the lowest natural frequency occurs when $R(\mathbf{w})$ is a minimum and this minimum value is the approximation for ω_1^2.

$R(\mathbf{w})$ is minimized over all vectors in W by requiring $\delta R = 0$, which in turn leads to

$$\frac{\partial R}{\partial \alpha_1} = \frac{\partial R}{\partial \alpha_2} = \dots = \frac{\partial R}{\partial \alpha_n} = 0 \tag{8.39}$$

To this end, differentiation of Equation 8.37 with respect to α_k leads to

$$\frac{\partial}{\partial \alpha_k}[(\mathbf{w},\mathbf{w})_K] - R(\mathbf{w}) \frac{\partial}{\partial \alpha_k}[(\mathbf{w},\mathbf{w})_M] - \frac{\partial R}{\partial \alpha_k}(\mathbf{w},\mathbf{w})_M = 0 \tag{8.40}$$

When $\mathbf{w}$ is the minimizing vector, the corresponding value of $R(\mathbf{w})$ is the approximation for the square of the natural frequency. Setting $\partial R/\partial \alpha_k = 0$ in Equation 8.30 leads to

$$\frac{\partial}{\partial \alpha_k}[(\mathbf{w},\mathbf{w})_K] - \omega^2 \frac{\partial}{\partial \alpha_k}[(\mathbf{w},\mathbf{w})_M] = 0 \tag{8.41}$$

It is noted that

$$\begin{aligned}
\frac{\partial}{\partial \alpha_k}[(\mathbf{w},\mathbf{w})_K] &= \frac{\partial}{\partial \alpha_k}\left[\left(\sum_{i=1}^{n}\alpha_i\mathbf{w}_i, \sum_{j=1}^{n}\alpha_j\mathbf{w}_j\right)_K\right] \\
&= \sum_{j=1}^{n}\sum_{i=1}^{n}(\mathbf{w}_i,\mathbf{w}_j)_K\left[\frac{\partial}{\partial \alpha_k}(\alpha_i\alpha_j)\right] \\
&= \sum_{i=1}^{n}\sum_{j=1}^{n}(\mathbf{w}_i,\mathbf{w}_j)_K\left[\alpha_i\frac{\partial \alpha_j}{\partial \alpha_k} + \frac{\partial \alpha_i}{\partial \alpha_k}\right] \\
&= \sum_{i=1}^{n}\sum_{j=1}^{n}(\mathbf{w}_i,\mathbf{w}_j)_K\left[\alpha_i\delta_{j,k} + \alpha_j\delta_{i,k}\right] \\
&= \sum_{i=1}^{n}(\mathbf{w}_i,\mathbf{w}_k)_K\alpha_i + \sum_{j=1}^{n}(\mathbf{w}_k,\mathbf{w}_j)_K\alpha_j
\end{aligned} \tag{8.42}$$

Commutivity of the potential energy inner product is used to recombine the sums in Equation 8.42, leading to

$$\frac{\partial}{\partial \alpha_k}[(\mathbf{w},\mathbf{w})_K] = 2\sum_{i=1}^{n}(\mathbf{w}_i,\mathbf{w}_k)_K \alpha_i \tag{8.43}$$

A similar analysis is used for $\partial/\partial\alpha_k[(\mathbf{w},\mathbf{w})_M]$, which when substituted into Equation 8.41 leads to

$$\sum_{i=1}^{n}(\mathbf{w}_i,\mathbf{w}_k)_K \alpha_i = \omega^2 \sum_{i=1}^{n}(\mathbf{w}_i,\mathbf{w}_k)_M \alpha_i \quad k = 1,2,\dots n \tag{8.44}$$

The system of equations represented by Equation 8.44 can be written in matrix form as

$$\hat{\mathbf{K}}A = \omega^2 \hat{\mathbf{M}}A \tag{8.45}$$

where $\hat{\mathbf{K}}$ is an $n\times n$ matrix whose elements are $\hat{k}_{i,j} = (\mathbf{w}_i, \mathbf{w}_j)_K$, $\hat{\mathbf{M}}$ is an $n\times n$ matrix whose elements are $\hat{m}_{i,j} = (\mathbf{w}_i,\mathbf{w}_j)_M$, and A is a $n\times 1$ column vector whose components are $a_i = \alpha_i$. Equation 8.45 is in the form of a matrix eigenvalue problem. The natural frequency approximations are the square roots of the eigenvalues of $\hat{\mathbf{M}}^{-1}\hat{\mathbf{K}}$. The corresponding eigenvectors are the coefficients in the Rayleigh–Ritz expansion for the mode shapes.

The matrix $\hat{\mathbf{K}}$ is symmetric due to the commutivity of the potential energy inner product. The matrix $\hat{\mathbf{M}}$ is symmetric due to the commutivity of the kinetic energy inner product. Thus both are self-adjoint with respect to the standard inner product on R^n. To avoid confusion between inner products defined on $C^k[0,1]$ with those defined for R^n, the standard inner product on R^n, for this discussion, will be denoted as $(\mathbf{f},\mathbf{g})_{R^n}$. For example, the self-adjointness of the matrix $\hat{\mathbf{M}}$ is described by $(\hat{\mathbf{M}}\mathbf{x},\mathbf{y})_{R^n} - (\mathbf{x},\hat{\mathbf{M}}\mathbf{y})_{R^n}$.

The positive definiteness of $\hat{\mathbf{M}}$ is decided by examining for an arbitrary n-dimensional vector $\mathbf{z}$

$$(\hat{\mathbf{M}}\mathbf{z},\mathbf{z})_{R^n} = \mathbf{z}^T\hat{\mathbf{M}}\mathbf{z} = \sum_{i=1}^{n}\sum_{j=1}^{n}(\mathbf{w}_i,\mathbf{w}_j)_M z_i z_j \tag{8.46}$$

To avoid confusing notation, consider the case where $\hat{\mathbf{M}}$ is a 3×3 matrix. The right-hand side of Equation 8.46 can be expanded such that

$$(\hat{\mathbf{M}}\mathbf{z},\mathbf{z})_{\mathbf{R}^3} = (\mathbf{w}_1, \mathbf{w}_1)_\mathbf{M} z_1^2 + 2(\mathbf{w}_1, \mathbf{w}_2)_\mathbf{M} z_1 z_2 + 2(\mathbf{w}_1, \mathbf{w}_3)_\mathbf{M} z_1 z_3$$

$$+ (\mathbf{w}_2, \mathbf{w}_2)\hat{\mathbf{M}} z_2^2 + 2(\mathbf{w}_2, \mathbf{w}_3)_\mathbf{M} z_2 z_3 + (\mathbf{w}_3, \mathbf{w}_3)_\mathbf{M} z_3^2 \qquad (8.47)$$

The Cauchy–Schwartz inequality, Equation 3.14, states that if $(\mathbf{u},\mathbf{v})$ represents a valid inner product defined for all $\mathbf{u}$ and $\mathbf{v}$ in a vector space V,

$$(\mathbf{u},\mathbf{v})^2 \leq (\mathbf{u},\mathbf{u})(\mathbf{v},\mathbf{v}) \qquad (8.48)$$

The Cauchy–Schwartz inequality implies that

$$(\mathbf{w}_i,\mathbf{w}_j)_\mathbf{M} \leq (\mathbf{w}_i,\mathbf{w}_i)_\mathbf{M}^{1/2} (\mathbf{w}_j,\mathbf{w}_j)_\mathbf{M}^{1/2} \qquad (8.49)$$

Using Equation 8.49 in Equation 8.47 leads to

$$(\hat{\mathbf{M}}\mathbf{z},\mathbf{z})_{\mathbf{R}^3} \geq (\mathbf{w}_1,\mathbf{w}_1)_\mathbf{M} z_1^2 + 2(\mathbf{w}_1,\mathbf{w}_1)_\mathbf{M}^{1/2} (\mathbf{w}_2,\mathbf{w}_2)_\mathbf{M}^{1/2} z_1 z_2$$

$$+ 2(\mathbf{w}_1,\mathbf{w}_1)_\mathbf{M}^{1/2} (\mathbf{w}_3,\mathbf{w}_3)_\mathbf{M}^{1/2} z_1 z_3 + (\mathbf{w}_2,\mathbf{w}_2) z_2^2 + 2(\mathbf{w}_2,\mathbf{w}_2)_\mathbf{M}^{1/2}$$

$$\times (\mathbf{w}_3,\mathbf{w}_3)_\mathbf{M}^{1/2} z_2 z_3 + (\mathbf{w}_3,\mathbf{w}_3)_\mathbf{M} z_3^2$$

$$= [(\mathbf{w}_1,\mathbf{w}_1)_\mathbf{M} z_1 + (\mathbf{w}_2,\mathbf{w}_2)_\mathbf{M} z_2 + (\mathbf{w}_3,\mathbf{w}_3)_\mathbf{M} z_3]^2 \qquad (8.50)$$

Since $\mathbf{M}$ is positive definite, the only vector that leads to $(\hat{\mathbf{M}}z, z)_{\mathbf{R}^3} = 0$ is the zero vector. Thus, $\hat{\mathbf{M}}$ is positive definite with respect to the standard inner product on R^n. The same conclusion is reached about $\hat{\mathbf{K}}$. Since both $\hat{\mathbf{M}}$ and $\hat{\mathbf{K}}$ are positive definite and self-adjoint with respect to the standard inner product on R^n, they can be used to define kinetic and potential energy inner products valid on R^n.

$$(\mathbf{z},\mathbf{z})_{\hat{\mathbf{M}}} = (\hat{\mathbf{M}}z,\mathbf{z})_{\mathbf{R}^n} \qquad (8.51)$$

$$(\mathbf{z},\mathbf{z})_{\hat{\mathbf{K}}} = (\hat{\mathbf{K}}z,\mathbf{z})_{\mathbf{R}^n} \qquad (8.52)$$

The operator $\hat{\mathbf{M}}^{-1}\hat{\mathbf{K}}$ is self-adjoint and positive definite with respect to each of these inner products. Thus, the eigenvalues of $\hat{\mathbf{M}}^{-1}\hat{\mathbf{K}}$ are real and positive, and its eigenvectors corresponding to distinct eigenvalues are mutually orthogonal with respect to both energy inner products.

The above leads to the conclusion that application of the Rayleigh–Ritz method leads to n natural frequency approximations $\omega_1 \leq \omega_2 \leq \omega_3 \ldots \leq \omega_n$. The natural frequency ω_i has a corresponding eigenvector $\mathbf{A}_i$. The eigenvectors satisfy orthogonality conditions

$$(\mathbf{A}_i,\mathbf{A}_j)_{\hat{\mathbf{M}}} = 0 \quad i \neq j \tag{8.53}$$

$$(\mathbf{A}_i,\mathbf{A}_j)_{\hat{\mathbf{K}}} = 0 \quad i \neq j \tag{8.54}$$

The eigenvectrors can be normalized such that

$$(\mathbf{A}_i,\mathbf{A}_i)_{\hat{\mathbf{M}}} = 1 \tag{8.55}$$

$$(\mathbf{A}_i,\mathbf{A}_i)_{\hat{\mathbf{K}}} = \omega_i^2 \tag{8.56}$$

Let the components of the normalized eigenvectors $\mathbf{A}_i$, be described by $(\mathbf{A}_i)_j = \alpha_{i,j}$.

The mode shape approximation, $\tilde{\mathbf{w}}_\ell$, corresponding to the natural frequency approximation $\tilde{\omega}_\ell$, is obtained by substituting into Equation 8.38 leading to

$$\tilde{\mathbf{w}}_\ell = \sum_{i=1}^{n} \alpha_{\ell,i} \mathbf{w}_i \tag{8.57}$$

Consider the kinetic energy inner product

$$(\tilde{\mathbf{w}}_\ell,\tilde{\mathbf{w}}_m)_{\mathbf{M}} = \left(\sum_{i=1}^{n} \alpha_{\ell,i} \mathbf{w}_i, \sum_{j=1}^{n} \alpha_{m,j} \mathbf{w}_j \right)_{\mathbf{M}} = \sum_{i=1}^{n} \sum_{j=1}^{n} \alpha_{\ell,i} \alpha_{m,j} (\mathbf{w}_i,\mathbf{w}_j)_{\mathbf{M}} \tag{8.58}$$

Note that

$$\begin{aligned}
(\mathbf{A}_\ell,\mathbf{A}_m)_{\hat{\mathbf{M}}} &= \mathbf{A}_m^T \hat{\mathbf{M}} \mathbf{A}_\ell \\
&= \sum_{i=1}^{n} \sum_{j=1}^{n} \tilde{m}_{i,j} (\mathbf{A}_\ell)_j (\mathbf{A}_m)_i \\
&= \sum_{i=1}^{n} \sum_{j=1}^{n} (\mathbf{w}_i,\mathbf{w}_j)_{\mathbf{M}} \alpha_{\ell,i} \alpha_{m,i}
\end{aligned} \tag{8.59}$$

Combining Equation 8.58 and Equation 8.59 and using orthonormality of the eigenvectors of $\hat{\mathbf{M}}^{-1}\hat{\mathbf{K}}$ leads to

$$(\tilde{\mathbf{w}}_\ell,\tilde{\mathbf{w}}_m)_{\mathbf{M}} = (\mathbf{A}_\ell,\mathbf{A}_m)_{\hat{\mathbf{M}}} = \delta_{\ell,m} \tag{8.60}$$

A similar analysis is used to show

$$(\tilde{\mathbf{w}}_\ell, \tilde{\mathbf{w}}_m)_K = (\mathbf{A}_\ell, \mathbf{A}_m)_{\hat{K}} = \omega_\ell^2 \delta_{\ell,m} \tag{8.61}$$

Equation 8.61 shows that the mode shape vector approximations are mutually orthogonal with respect to the kinetic energy inner product. Mode shape vectors corresponding to the true natural frequencies are mutually orthogonal with respect to the kinetic energy inner product. Because $\tilde{\mathbf{w}}_1$ is orthogonal to $\tilde{\mathbf{w}}_2$, $\tilde{\mathbf{w}}_1$ is an approximation to the mode shape corresponding to the lowest natural frequency. If the true mode shape vectors $\mathbf{w}_1$ and $\mathbf{w}_2$ are mutually orthogonal, then $\tilde{\mathbf{w}}_2$ must be an approximation to $\mathbf{w}_2$ and then ω_2 must be an approximation to the second lowest natural frequency. This argument continues, and through induction it is shown that application of the Rayleigh–Ritz method with n basis functions to a distributed parameter system leads to approximations for the system's n lowest natural frequencies and their corresponding mode shapes. The mode shape approximations satisfy the orthogonality conditions satisfied by the true mode shape vectors.

Example 8.1. Use the Rayleigh–Ritz method to approximate the natural frequencies and mode shapes for a uniform fixed-fixed beam. Find the Rayleigh–Ritz approximation from the space of polynomials of degree six or less.

Solution: The nondimensional equation for the natural frequencies and mode shapes for a uniform beam is

$$\frac{d^4 w}{dx^4} - \omega^2 w = 0 \tag{a}$$

The boundary conditions for a fixed-fixed beam are

$$w(0) = 0 \tag{b}$$

$$\frac{dw}{dx}(0) = 0 \tag{c}$$

$$w(1) = 0 \tag{d}$$

$$\frac{dw}{dx}(1) = 0 \tag{e}$$

The domain of the stiffness operator, S, is the subspace of the space $C^4[0,1]$ of all functions $f(x)$ such that $f(x)$ satisfies the boundary conditions of Equation b through Equation e.

A basis for the space of polynomials of degree six or less, $P^6[0,1]$, is the set $\{1, x, x^2, x^3, x^4, x^5, x^6\}$. Thus, an arbitrary element of $P^6[0,1]$ is of the form

$$p(x) = c_0 + c_1 x + c_2 x^2 + c_3 x^3 + c_4 x^4 + c_5 x^5 + c_6 x^6 \tag{f}$$

The basis functions for the Rayleigh–Ritz approximation form a basis for $S \cap P^6[0,1]$. The dimension of $P^6[0,1]$ is seven. Selecting the basis so that basis elements also belong to S imposes four constraints on the basis elements. Thus, only three of the coefficients in Equation f may be selected arbitrarily indicating that the dimension of $S \cap P^6[0,1]$ is three. Any three linearly independent polynomials of order six or less, which satisfy the boundary conditions, constitute a basis for $S \cap P^6[0,1]$ and can be used a basis functions for the Rayleigh–Ritz approximation. A method of selecting a basis for such a vector space is described in Example 3.7. Application of this method leads to a basis of

$$w_1(x) = x^6 - 4x^3 + 3x^2 \tag{g}$$

$$w_2(x) = x^5 - 3x^3 + 2x^2 \tag{h}$$

$$w_3(x) = x^4 - 2x^3 + x^2 \tag{i}$$

The basis functions are plotted in Figure 8.1.
The elements of the 3×3 matrix $\hat{\mathbf{M}}$ are $\hat{m}_{i,j} = (w_i(x), w_j(x))_M$. For example,

$$\hat{m}_{1,2} = \int_0^1 (x_i - 4x^3 + 3x^2)(x^5 - 3x^3 + 2x^2) dx = 0.0171 \tag{j}$$

The entire matrix is

$$\hat{\mathbf{M}} = \begin{bmatrix} 0.0293 & 0.0171 & 0.0068 \\ 0.0171 & 0.0100 & 0.0040 \\ 0.0068 & 0.0040 & 0.0016 \end{bmatrix} \tag{k}$$

The elements of the 3×3 matrix $\hat{\mathbf{K}}$ are $\hat{k}_{i,j} = (w_i(x), w_j(x))_K$. For example,

$$\hat{k}_{1,2} = \int_0^1 \left[\frac{d^4}{dx^4}(x^6 - 4x^3 + 3x^2) \right] (x^5 - 3x^3 + 2x^2) dx = 9 \tag{l}$$

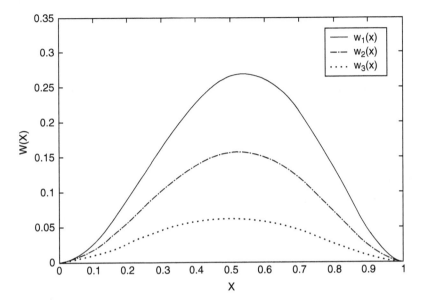

FIGURE 8.1 Basis functions for Example 8.1.

The entire matrix is

$$\hat{\mathbf{K}} = \begin{bmatrix} 16 & 9 & 3.4286 \\ 9 & 5.1429 & 2 \\ 3.4286 & 2 & 0.8 \end{bmatrix} \tag{m}$$

The eigenvalues and eigenvectors of $\hat{\mathbf{M}}^{-1}\hat{\mathbf{K}}$ are obtained as

$$\lambda_1 = 500.6 \quad \mathbf{A}_1 = \begin{bmatrix} 0.7629 \\ -2.2887 \\ 1 \end{bmatrix} \tag{n}$$

$$\lambda_2 = 3960 \quad \mathbf{A}_2 = \begin{bmatrix} 0 \\ -0.4 \\ 1 \end{bmatrix} \tag{o}$$

$$\lambda_3 = 16289.8 \quad \mathbf{A}_3 = \begin{bmatrix} 0.30982 \\ -0.9295 \\ 1 \end{bmatrix} \tag{p}$$

The natural frequency approximations are the square roots of the eigenvalues of $\hat{\mathbf{M}}^{-1}\hat{\mathbf{K}}$ leading to

$$\omega_1 = 22.37, \quad \omega_2 = 62.93 \quad \omega_3 = 127.63 \tag{q}$$

The exact natural frequencies for the first three modes are obtained from Table 5.2 as 22.37, 61.66, and 120.9. Thus, the Rayleigh–Ritz method provides an excellent approximation.

Substitution of the eigenvectors into Equation 8.38 leads to mode shape approximations of

$$\tilde{\mathbf{w}}_1(x) = 0.30982(x^6 - 4x^3 + 3x^2) - 0.9295(x^5 - 3x^3 + 2x^2) + (x^4 - 2x^3 + x^2) \tag{r}$$

$$\tilde{\mathbf{w}}_2(x) = -0.4(x^5 - 3x^3 + 2x^2) + (x^4 - 2x^3 + x^2) \tag{s}$$

$$\tilde{\mathbf{w}}_3(x) = 0.7629(x^6 - 4x^3 + 3x^2) - 2.2887(x^5 - 3x^3 + 2x^2) + (x^4 - 2x^3 + x^2) \tag{t}$$

The mode shapes are normalized such that $(\tilde{w}_i, \tilde{w}_i)_{\mathbf{M}} = 1$. The normalized mode shapes are plotted in Figure 8.2.

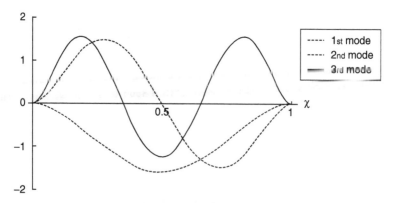

FIGURE 8.2 Rayleigh–Ritz approximation for normalized mode shape vectors corresponding to three lowest natural frequencies of a uniform fixed-fixed beam.

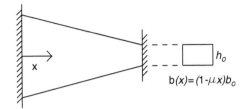

FIGURE 8.3 Fixed-fixed beam with linear taper.

Example 8.2. Use the Rayleigh–Ritz method to approximate the natural frequencies and mode shape vectors for the rectangular beam with a linear taper of Figure 8.3, such that

$$\alpha(x) = 1 - \mu x \tag{a}$$

$$\beta(x) = 1 - \mu x \tag{b}$$

Use the polynomial basis functions of Example 8.1.

Solution: The solution procedure is the same as Example 8.1 except for the definitions of the energy inner products, which become

$$(w_i, w_j)_\mathbf{M} = \int_0^1 w_i(x) w_j(x) \beta(x) \mathrm{d}x \tag{c}$$

$$(w_i, w_j)_\mathrm{K} = \int_0^1 \frac{\mathrm{d}^2}{\mathrm{d}x^2} \left(\alpha(x) \frac{\mathrm{d}^2 w_i}{\mathrm{d}x^2} \right) w_j(x) \mathrm{d}x \tag{d}$$

The first three natural frequencies for varying μ are illustrated in Figure 8.4. The normalized mode shape vectors for $\mu = 0.5$ are illustrated in Figure 8.5.

Example 8.3. Use the Rayleigh–Ritz method to approximate the natural frequencies for the longitudinal motion of a circular fixed-free bar with a linear taper as illustrated in Figure 8.6. Use the normalized mode shapes of the first four modes of a uniform fixed-free bar as basis functions.

Solution: The normalized mode shapes for the first four modes of a uniform fixed-free bar are determined in Example 5.14 as

$$w_1(x) = \sqrt{2}\sin\left(\frac{\pi}{2}x\right) \tag{a}$$

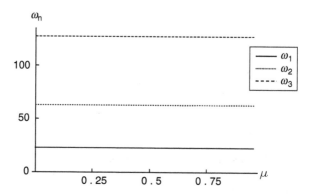

FIGURE 8.4 Rayleigh–Ritz approximation for lowest natural frequencies of a tapered fixed-fixed beam as a function of taper rate.

$$w_2(x) = \sqrt{2}\sin\left(\frac{3\pi}{2}x\right) \tag{b}$$

$$w_3(x) = \sqrt{2}\sin\left(\frac{5\pi}{2}x\right) \tag{c}$$

$$w_4(x) = \sqrt{2}\sin\left(\frac{7\pi}{2}x\right) \tag{d}$$

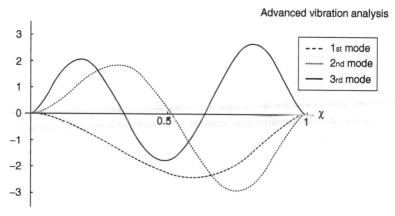

FIGURE 8.5 Mode shapes of non-uniform fixed-fixed beam obtained using Rayleigh–Ritz with polynomial basis.

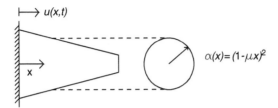

FIGURE 8.6 Fixed-free circular bar with a linear taper of Example 8.3.

$\hat{\mathbf{M}}$ is a 4×4 matrix whose elements are

$$\hat{m}_{i,j} = \int_0^1 \left[\sqrt{2} \sin\left(2(i-1)\frac{\pi}{2}x\right) \right]\left[\sqrt{2}\sin\left(2(j-1)\frac{\pi}{2}x\right) \right](1-\mu x)^2 \, dx \quad (e)$$

$\hat{\mathbf{K}}$ is a 4×4 matrix whose elements are

$$\hat{k}_{i,j} = \int_0^1 \left\{ -\frac{d}{dx}\left([(1-\mu x)^2]\frac{d}{dx}\left[\sqrt{2}\sin\left(2(i-1)\frac{\pi}{2}x\right) \right] \right) \right\}$$
$$\left[\sqrt{2}\sin\left(2(j-1)\frac{\pi}{2}x\right) \right] dx \qquad (f)$$

The natural frequency approximations are the square roots of the eigenvalues of $\hat{\mathbf{M}}^{-1}\hat{\mathbf{K}}$. The four lowest natural frequencies for the system for various values of μ are summarized in Table 8.1.

The exact solution for this problem in terms of Bessel functions as well as a perturbation solution, is determined in Example 5.14. Figure 8.7 provides a comparison between the solutions. The natural frequency approximations using the Rayleigh–Ritz method are almost identical to the exact solution.

Example 8.4. Use the Raylcigh–Ritz method to approximate the natural frequencies and mode shapes for the uniform free-free beam of Figure 8.8. Use polynominal basis functions from the space of polynomials of degree seven or less.

Solution: The boundary conditions for a uniform free-free beam are

$$\frac{d^2 w}{dx^2}(0) = 0 \qquad (a)$$

TABLE 8.1
Natural Frequencies for a Circular Bar with a Linear Taper Calculated Using the Rayleigh–Ritz Method with Polynomial Basis Functions

μ	ω_1	ω_2	ω_3	ω_4
0.1	1.6385	4.7359	7.8682	11.0070
0.2	1.7155	4.7640	7.8859	11.0239
0.3	1.8041	4.8018	7.9086	11.0484
0.4	1.9072	4.8497	7.9387	11.0834
0.5	2.0289	4.9143	7.9805	11.1333
0.6	2.1748	5.0056	8.0421	11.2057
0.7	2.3524	5.1418	8.1406	11.3146
0.8	2.5708	5.3589	8.3159	11.4908
0.9	2.8368	5.7226	8.6743	11.8217

$$\frac{d^3 w}{dx^3}(0) = 0 \qquad\qquad \text{(b)}$$

$$\frac{d^2 w}{dx^2}(1) = 0 \qquad\qquad \text{(c)}$$

$$\frac{d^3 w}{dx^3}(1) = 0 \qquad\qquad \text{(d)}$$

The general form of a vector in $P^7[0,1]$ is

$$p(x) = a_0 + a_1 x + a_2 x^2 + a_3 x^3 + a_4 x^4 + a_5 x^5 + a_6 x^6 + a_7 x^7 \qquad \text{(e)}$$

Basis vectors are sought from the intersection of $P'[0,1]$ with the domain of K which is S, the subspace of $C^4[0,1]$ such that if $f(x)$ is in S then $f(x)$ satisfies the boundary conditions, as shown in Equation a through Equation d. A basis for $P^7[0,1]$ contains eight vectors. A vector in $P^7[0,1] \cap S$ is a polynomial of degree 7 or less that satisfies Equation a) through Equation d). The dimension of this intersection is four.

It is clear that any polynomial of the form $a_1 + a_1 x$ is a member of $P^7[0,1] \cap S$. Thus, two basis functions are a constant and a linear function. It is convenient to choose

Advanced vibration analysis

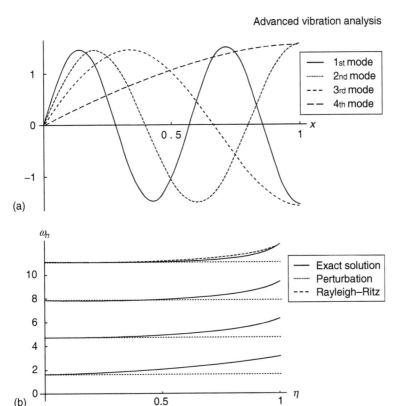

FIGURE 8.7 (a) Mode shapes calculated using Rayleigh–Ritz method for non-uniform bar with ε=0,1. (b) Comparison of natural frequencies of non-uniform bar as calculated from the exact solution method, perturbation method and Rayleigh–Ritz method using mode shapes of uniform bar as basis functions.

$$w_1(x) = 1 \tag{f}$$

$$w_2(x) = x - 0.5 \tag{g}$$

The second function is chosen because it is orthogonal to $w_1(x)$ and is more computationally efficient than choosing x as the second basis function as it represents a rigid-body rotation about the mass center of the beam.

Two other basis members are obtained by applying the boundary conditions to Equation e. Two linearly independent polynomials satisfying all boundary conditions are

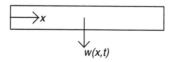

FIGURE 8.8 Free-free beam has two rigid-body modes corresponding to a natural frequency of zero.

$$w_3(x) = \frac{10}{21}x^7 - \frac{4}{3}x^6 + x^5 \tag{h}$$

$$w_4(x) = \frac{4}{7}x^7 - \frac{6}{5}x^6 + x^4 \tag{i}$$

The matrix $\hat{\mathbf{M}}$ is constructed using kinetic energy inner products is

$$\hat{\mathbf{M}} = \begin{bmatrix} 1 & 0 & 0.0357 & 0.1 \\ 0 & 0.08333 & 0.01124 & 0.02302 \\ 0.0357 & 0.01124 & 0.003113 & 0.008406 \\ 0.1 & 0.03016 & 0.008406 & 0.02275 \end{bmatrix} \tag{j}$$

The matrix $\hat{\mathbf{K}}$ constructed using potential energy scalar products is

$$\hat{\mathbf{K}} = \begin{bmatrix} 0 & 0 & 0 & 0 \\ 0 & 0 & 0 & 0 \\ 0 & 0 & 0.1732 & 0.3983 \\ 0 & 0 & 0.3983 & 0.9531 \end{bmatrix} \tag{k}$$

The natural frequencies, approximated as the square roots of the eigenvalues of $\hat{\mathbf{M}}^{-1}\hat{\mathbf{K}}$, and the corresponding mode shape vectors are

$$\omega_1 = 0 \quad \mathbf{A}_1 = \begin{bmatrix} 1 \\ 0 \\ 0 \\ 0 \end{bmatrix} \tag{l}$$

$$\omega_2 = 0 \quad \mathbf{A}_2 = \begin{bmatrix} 0 \\ 1 \\ 0 \\ 0 \end{bmatrix} \tag{m}$$

$$\omega_3 = 22.38 \quad \mathbf{A}_3 = \begin{bmatrix} -0.03626 \\ -0.1269 \\ -0.7615 \\ 0.6346 \end{bmatrix} \tag{n}$$

$$\omega_4 = 61.78 \quad \mathbf{A}_4 = \begin{bmatrix} 0.005494 \\ 0.01465 \\ 0.9229 \\ -0.3846 \end{bmatrix} \tag{o}$$

Exact values of the four lowest natural frequencies are obtained from Table 5.2 as $\omega = 0$, 0, 22.37, and 61.66. Thus the Rayleigh–Ritz method leads to excellent natural frequency approximations.

The normalized mode shape vectors are

$$\tilde{w}_1 = 1 \tag{p}$$

$$\tilde{w}_2 = \frac{2}{\sqrt{3}}(x - 0.5) \tag{q}$$

$$\tilde{w}_3 = 73.753\left[-0.03626 - 0.1266(x - 0.5) - 0.7615\left(\frac{4}{7}x^7 - \frac{6}{5}x^6 + x^4\right)\right.$$
$$\left. + 0.6346\left(\frac{10}{21}x^7 - \frac{4}{3}x^6 + x^5\right)\right] \tag{r}$$

$$\tilde{w}_4 = 1114.44\left[0.005494 + 0.01465(x - 0.5) + 0.9229\left(\frac{4}{7}x^7 - \frac{6}{5}x^6 + x^4\right)\right.$$
$$\left. - 0.3846\left(\frac{10}{21}x^7 - \frac{4}{3}x^6 + x^5\right)\right] \tag{s}$$

The normalized mode shapes are plotted in Figure 8.9.

Example 8.5. Use a Rayleigh–Ritz method to approximate the natural frequencies and mode shapes of the rectangular membrane of Figure 8.10.

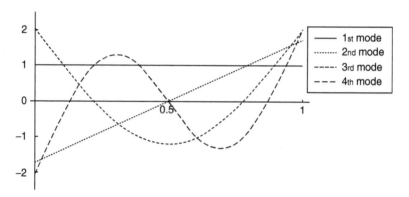

FIGURE 8.9 Mode shapes for free-free beam obtained using Rayleigh–Ritz method with polynomial basis functions.

The membrane is simply supported. Assume $\eta = 2$. (a) Use the following polynomials as basis functions:

$$w_1(x,y) = x(x-1)\left(\frac{y}{\eta}\right)\left(\frac{y}{\eta}-1\right) \tag{a}$$

$$w_2(x,y) = x(x-1)\left(\frac{y}{\eta}\right)^2\left[\left(\frac{y}{\eta}\right)^2-1\right] \tag{b}$$

$$w_3(x,y) = x^2(x^2-1)\left(\frac{y}{\eta}\right)\left(\frac{y}{\eta}-1\right) \tag{c}$$

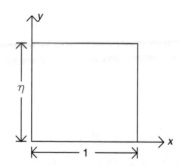

FIGURE 8.10 Rectangular membrane of Example 8.5.

$$w_4(x,y) = x^2(x^2 - 1)\left(\frac{y}{\eta}\right)^2\left[\left(\frac{y}{\eta}\right)^2 - 1\right] \tag{d}$$

(b) Use the following trigonometric basis functions:

$$w_1(x,y) = \sin(\pi x)\sin\left(\pi \frac{y}{\eta}\right) \tag{e}$$

$$w_2(x,y) = \sin(\pi x)\sin\left(2\pi \frac{y}{\eta}\right) \tag{f}$$

$$w_3(x,y) = \sin(2\pi x)\sin\left(\pi \frac{y}{\eta}\right) \tag{g}$$

$$w_4(x,y) = \sin(2\pi x)\sin\left(2\pi \frac{y}{\eta}\right) \tag{h}$$

Solution: The kinetic and potential energy inner products for the rectangular membrane are

$$(f(x,y),g(x,y))_M = \int_0^{\eta}\int_0^1 f(x,y)g(x,y)dx\,dy \tag{i}$$

$$(f(x,y),g(x,y))_K = \int_0^{\eta}\int_0^1 \left[\frac{\partial^2 f}{\partial x^2} + \frac{\partial^2 f}{\partial y^2}\right]g(x,y)dx\,dy \tag{j}$$

The elements of the discrete mass and stiffness matrices are

$$\hat{m}_{i,j} = (w_i,w_j)_M \tag{k}$$

$$\hat{k}_{i,j} = (w_i,w_j)_K \tag{l}$$

(a) Using the polynomial basis functions, the discrete stiffness and mass matrices are

$$\hat{K} = \begin{bmatrix} 0.02778 & 0.0219 & 0.02214 & 0.0175 \\ 0.0219 & 0.02392 & 0.0175 & 0.0190 \\ 0.02214 & 0.0175 & 0.03217 & 0.02534 \\ 0.0175 & 0.0190 & 0.02534 & 0.02661 \end{bmatrix} \tag{m}$$

$$\hat{M} = \begin{bmatrix} 0.002222 & 0.00175 & 0.00175 & 0.001372 \\ 0.00175 & 0.001693 & 0.001372 & 0.001330 \\ 0.00175 & 0.001372 & 0.001693 & 0.001330 \\ 0.001372 & 0.001330 & 0.001330 & 0.001290 \end{bmatrix} \quad \text{(n)}$$

The natural frequency approximations are

$$\omega_{1,1} = 3.5348 \quad \omega_{1,2} = 4.5468 \quad \omega_{2,1} = 6.7238$$

$$\omega_{2,2} = 7.3067 \quad \text{(o)}$$

The exact natural frequencies determined in Equation 5.242 are

$$\omega_{1,1} = 3.5125 \quad \omega_{1,2} = 4.4429 \quad \omega_{2,1} = 6.476$$

$$\omega_{2,2} = 7.0248 \quad \text{(p)}$$

The mode shape approximations are

$$\tilde{w}_{2,2} = 0.6105 w_1(x,y) - 0.7813 w_2(x,y) - 0.7813 w_3(x,y)$$
$$+ w_4(x,y) \quad \text{(q)}$$

$$\tilde{w}_{2,1} = 26.48 w_1(x,y) - 0.7813 w_2(x,y) - 33.8853 w_3(x,y)$$
$$+ w_4(x,y) \quad \text{(r)}$$

$$\tilde{w}_{1,2} = 26.48 w_1(x,y) - 0.7813 w_2(x,y) - 0.7813 w_3(x,y)$$
$$+ w_4(x,y) \quad \text{(s)}$$

$$\tilde{w}_{1,1} = 1148.28 w_1(x,y) - 33.8853 w_2(x,y) - 33.8853 w_3(x,y)$$
$$+ w_4(x,y) \quad \text{(t)}$$

Several mode shape vectors are plotted in Figure 8.11.
(b) The same procedure is applied using the trigonometric basis functions leads to

$$\omega_{1,1} = 3.5124 \quad \omega_{1,2} = 4.5690 \quad \omega_{2,1} = 6.4766$$

$$\omega_{2,2} = 7.1052 \quad \text{(u)}$$

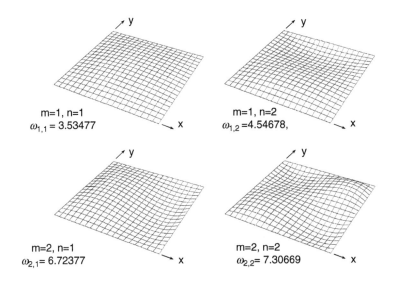

FIGURE 8.11 Mode shapes of rectangular membrane ($\eta=2$) obtained from Rayleigh–Ritz method.

The natural frequencies obtained using the trigonometric basis functions are slightly more accurate.

8.5 RAYLEIGH–RITZ METHODS FOR FORCED RESPONSE

The forced response of a self-adjoint continuous system is obtained by solving

$$\mathbf{K}\mathbf{w} + \mathbf{M}\frac{\partial^2 \mathbf{w}}{\partial t^2} = \mathbf{f} \tag{8.62}$$

A Rayleigh–Ritz approximation of the forced response from a subspace, W, of the domain of $\mathbf{K}$ is assumed as

$$\mathbf{w} = \sum_{i=1}^{n} c_i(t)\mathbf{w}_i \tag{8.63}$$

where $\mathbf{w}_1, \mathbf{w}_2, \dots, \mathbf{w}_n$ are a basis for W and $c_i(t)$ are to be determined from application of the method.

Substitution of Equation 8.63 into Equation 8.62 leads to

$$\mathbf{K}\left(\sum_{i=1}^{n}c_i(t)\mathbf{w}_i\right) + \mathbf{M}\left(\sum_{i=1}^{n}\ddot{c}_i(t)\mathbf{w}_i\right) = \mathbf{f}$$

$$\sum_{i=1}^{n}c_i(t)\mathbf{K}\mathbf{w}_i + \sum_{i=1}^{n}\ddot{c}_i(t)\mathbf{M}\mathbf{w}_i = \mathbf{f} \tag{8.64}$$

Taking the standard inner product of Equation 8.64 with $\mathbf{w}_k$ for any $k = 1,2,\ldots,n$ leads to

$$\sum_{i=1}^{n}c_i(t)(\mathbf{K}\mathbf{w}_i,\mathbf{w}_k) + \sum_{i=1}^{n}\ddot{c}_i(t)(\mathbf{M}\mathbf{w}_i,\mathbf{w}_k) = (\mathbf{f},\mathbf{w}_k)$$

$$\sum_{i=1}^{n}c_i(t)(\mathbf{w}_i,\mathbf{w}_k)_\mathrm{K} + \sum_{i=1}^{n}\ddot{c}_i(t)(\mathbf{w}_i,\mathbf{w}_k)_\mathrm{M} = (\mathbf{f},\mathbf{w}_k) \quad k = 1,2,\ldots,n \tag{8.65}$$

Using the notation of Section 8.4, the equations represented by Equation 8.65 are summarized in a matrix form as

$$\hat{\mathbf{M}}\ddot{\mathbf{c}} + \hat{\mathbf{K}}\mathbf{c} = \hat{\mathbf{f}} \tag{8.66}$$

where $\mathbf{c} = [c_1\, c_2 \ldots c_n]^T$ and the components of $\hat{\mathbf{f}}$ are $\hat{f}_i = (\mathbf{f},\mathbf{w}_i)$.

Equation 8.66 is a set of coupled differential equations whose solution leads to the coefficients in the Rayleigh–Ritz expansion of the approximate solution. The discrete mass and stiffness matrices of Equation 8.66 are symmetric and positive definite (Section 8.4). Thus, modal analysis for discrete self-adjoint systems may be used to determine a solution. The natural frequency approximations are the square roots of the eigenvalues of $\hat{\mathbf{M}}^{-1}\hat{\mathbf{K}}$ such that $\omega_1 < \omega_2 < \ldots \omega_n$ with corresponding normalized eigenvectors $\mathbf{A}_1,\mathbf{A}_2,\ldots,\mathbf{A}_n$. The discrete modal analysis is of the form

$$\mathbf{c} = \sum_{i=1}^{n}p_i(t)\mathbf{A}_i \tag{8.67}$$

Substitution of Equation 8.67 into Equation 8.64 leads to

$$\sum_{i=1}^{n}\ddot{p}_i\hat{\mathbf{M}}\mathbf{A}_i + \sum_{i=1}^{n}p_i(t)\hat{\mathbf{K}}\mathbf{A}_i = \hat{\mathbf{f}} \tag{8.68}$$

Taking the standard inner product of Equation 8.68 with $\mathbf{A}_k$ leads to

$$\sum_{i=1}^{n}\ddot{p}_i(\hat{\mathbf{M}}\mathbf{A}_i,\mathbf{A}_k)_{R^n} + \sum_{i=1}^{n}p_i(t)(\hat{\mathbf{K}}\mathbf{A}_i,\mathbf{A}_k)_{R^n} = (\mathbf{f},\mathbf{A}_k)_{R^n}$$

$$\sum_{i=1}^{n}\ddot{p}_i(\mathbf{A}_i,\mathbf{A}_k)_{\hat{\mathbf{M}}} + \sum_{i=1}^{n}p_i(t)(\mathbf{A}_i,\mathbf{A}_k)_{\hat{\mathbf{K}}} = (\mathbf{f},\mathbf{A}_k)_{R^n} \qquad (8.69)$$

$$\ddot{p}_k(t) + \omega_k^2 p_k(t) = (\mathbf{f},\mathbf{A}_k)_{R^n} = g_k(t)$$

The convolution integral solution of Equation 8.69 is

$$p_k(t) = \frac{1}{\omega_k}\int_0^t g_k(\tau)\sin[\omega_k(t-\tau)]d\tau \qquad (8.70)$$

Example 8.6. Use the Rayleigh–Ritz method to determine the steady-state response of the cantilever beam of Figure 8.12. Use polynomial basis functions of

$$w_1(x) = x^2 - \frac{2}{5}x^5 + \frac{1}{5}x^6 \qquad (a)$$

$$w_2(x) = x^3 - \frac{9}{10}x^5 + \frac{2}{5}x^6 \qquad (b)$$

$$w_3(x) = x^4 - \frac{6}{5}x^5 + \frac{2}{5}x^6 \qquad (c)$$

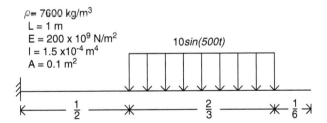

$\rho = 7600$ kg/m³
$L = 1$ m
$E = 200 \times 10^9$ N/m²
$I = 1.5 \times 10^{-4}$ m⁴
$A = 0.1$ m²

$10\sin(500t)$

FIGURE 8.12 System of Example 8.6.

Solution: The problem is formulated dimensionally so

$$(f,g)_M = \int_0^L \rho A f(x) g(x) dx \tag{d}$$

$$(f,g)_K = \int_0^L EI \frac{d^4 f}{dx^4} g dx \tag{e}$$

Using the inner products of Equation d and Equation e and the basis functions of Equation a through Equation c the discrete mass and stiffness matrices are

$$\hat{\mathbf{M}} = \begin{bmatrix} 113.04 & 64.38 & 24.27 \\ 64.38 & 37.09 & 14.08 \\ 24.27 & 14.08 & 5.36 \end{bmatrix} \tag{f}$$

$$\hat{\mathbf{K}} = 10^7 \begin{bmatrix} 5.83 & 3.81 & 1.54 \\ 3.81 & 3.26 & 1.46 \\ 1.54 & 1.46 & 68.6 \end{bmatrix} \tag{g}$$

The natural frequency and mode shape approximations are determined as

$$\omega_1 = 698.6 \text{ r/s} \quad \mathbf{A}_1 = \begin{bmatrix} -0.911 \\ 0.411 \\ 0.0234 \end{bmatrix} \tag{h}$$

$$\omega_2 = 4.38 \times 10^3 \text{ r/s} \quad \mathbf{A}_2 = \begin{bmatrix} -0.510 \\ 0.727 \\ 0.459 \end{bmatrix} \tag{i}$$

$$\omega_3 = 1.32 \times 10^5 \text{r/s} \quad \mathbf{A}_3 = \begin{bmatrix} 0.107 \\ -0.509 \\ 0.854 \end{bmatrix} \tag{j}$$

The Rayleigh–Ritz approximations for the mode shape vectors are

$$\Phi_1 = -0.911 w_1(x) + 0.411 w_2(x) + 0.0234 w_3(x) \tag{k}$$

$$\Phi_2 = -0.510w_1(x) + 0.727w_2(x) + 0.459w_3(x) \tag{l}$$

$$\Phi_3 = 0.107w_1(x) - 0.509w_2(x) + 0.854w_3(x) \tag{m}$$

Recall that the mode shape vectors obtained using Rayleigh–Ritz are orthogonal with respect to the kinetic energy and potential energy inner products. They can be normalized by requiring $(\Phi_i, \Phi_i)_M = 1$. The resulting normalized mode shape vectors are

$$\hat{w}_1(x) = -0.001188x^2(4.804 - 3.841x + x^2)(22.319 + 7.779x + x^2) \tag{n}$$

$$\hat{w}_2(x) = 0.575(-0.7837 + x)x^2(0.846 + x)(2.067 - 2.751x + x^2) \tag{o}$$

$$\hat{w}_3(x) = 6.503(-0.8616 + x)(-0.481 + x)x^2(1.614 - 2.482x + x^2) \tag{p}$$

The force along the span of the beam is

$$F(x,t) = 10 \sin(500t)\left[u\left(x - \frac{1}{2}\right) - u\left(x - \frac{5}{6}\right)\right] \tag{q}$$

The $g_i(t)$ of Equation 8.69 are calculated as

$$g_1 = \int_0^1 F(x,t)\hat{w}_1(x)dx = 10 \sin(500t) \int_{1/2}^{5/6} \hat{w}_1(x)dx = -0.133 \sin(500t) \tag{r}$$

$$g_2 = 10 \sin(500t) \int_{1/2}^{5/6} \hat{w}_2(x)dx = -0.0880 \sin(500t) \tag{s}$$

$$g_3 = 10 \sin(500t) \int_{1/2}^{5/6} \hat{w}_3(x)dx = -0.102 \sin(500t) \tag{t}$$

The modal equations are

$$\ddot{p}_1 + (698.6)^2 p_1 = -0.133 \sin(500t) \tag{u}$$

$$\ddot{p}_2 + (4.38 \times 10^3)^2 p_2 = -0.0080 \sin(500t) \tag{v}$$

$$\ddot{p}_3 + (1.32 \times 10^5)^2 p_3 = -0.102 \sin(500t) \tag{w}$$

The steady-state solutions of Equation u through Equation w are

$$p_1(t) = -5.58 \times 10^{-7}\sin(500t) \tag{x}$$

$$p_2(t) = -4.65 \times 10^{-7}\sin(500t) \tag{y}$$

$$p_3(t) = -5.90 \times 10^{-10}\sin(500t) \tag{z}$$

The steady-state response of the system is

$$w(x,t) = p_1(t)\hat{w}_1(x) + p_2(t)\hat{w}_2(x) + p_3(t)\hat{w}_3(x) \tag{aa}$$

The time dependent response of the midspan and end of the beam are illustrated in Figure 8.13

8.6 ADMISSIBLE FUNCTIONS

The basis functions for use in the Rayleigh–Ritz method are chosen from a finite dimensional subspace of $C^k[0,1]$ where k is the order of the differential equation. In addition, all functions in the subspace satisfy all boundary conditions for the problem. These requirements severely restrict the choices of basis functions for application of the Rayleigh–Ritz method to free and forced vibrations of continuous systems.

Consider the nonuniform bar of Figure 8.14. The bar has a discrete spring at the end. Consider the potential energy inner product

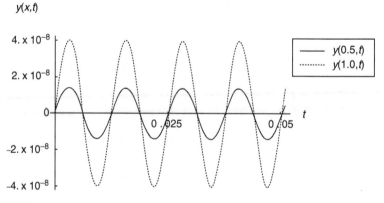

FIGURE 8.13 Rayleigh–Ritz approximation for steady-state response of system of Example 8.6.

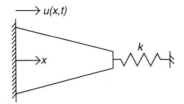

FIGURE 8.14 Non-uniform bar with end spring.

$$(f,g)_K = -\int_0^1 \frac{d}{dx}\left(\alpha(x)\frac{df}{dx}\right)g(x)dx \tag{8.71}$$

Application of integration by parts to the right-hand side of Equation 8.71 leads to

$$(f,g)_K = -\alpha(1)\frac{df}{dx}(1)g(1) + \alpha(0)\frac{df}{dx}(0)g(0) + \int_0^1 \alpha(x)\frac{df}{dx}\frac{dg}{dx}dx \tag{8.72}$$

A similar analysis is used to develop

$$(g,f)_K = -\alpha(1)\frac{dg}{dx}(1)f(1) + \alpha(0)\frac{dg}{dx}(0)f(0) + \int_0^1 \alpha\frac{dg}{dx}\frac{df}{dx}dx \tag{8.73}$$

The boundary conditions for the bar are

$$u(0) = 0 \tag{8.74}$$

$$\alpha(1)\frac{du}{dx}(1) = -\kappa u(1) \tag{8.75}$$

where κ is a nondimensional spring stiffness. The boundary condition in Equation 8.74 is a geometric condition whereas the boundary condition of Equation 8.75 is a natural boundary condition. Assume both $f(x)$ and $g(x)$ satisfy the geometric boundary condition. Equation 8.72 and Equation 8.74 become

$$(f,g)_K = -\alpha(1)\frac{df}{dx}(1)g(1) + \int_0^1 \alpha(x)\frac{df}{dx}\frac{dg}{dx}dx \tag{8.76}$$

$$(g,f)_K = -\alpha(1)\frac{dg}{dx}(1)f(1) + \int_0^1 \alpha\frac{dg}{dx}\frac{df}{dx}dx \qquad (8.77)$$

Now suppose both $f(x)$ and $g(x)$ satisfy the natural boundary condition at $x = 1$, then

$$(f,g)_K = \kappa f(1)g(1) + \int_0^1 \alpha(x)\frac{df}{dx}\frac{dg}{dx}dx \qquad (8.78)$$

$$(g,f)_K = \kappa f(1)g(1) + \int_0^1 \alpha(x)\frac{dg}{dx}\frac{df}{dx}dx \qquad (8.79)$$

If $f(x) = g(x)$, then

$$(f,f)_K = \kappa[f(1)]^2 + \int_0^1 \alpha(x)\left[\frac{df}{dx}\right]^2 dx = 2V \qquad (8.80)$$

where V is the total potential energy of the system.

The above suggests an alternative formulation of the potential energy inner product as

$$(f,g)_K = \int_0^1 \alpha(x)\frac{df}{dx}\frac{dg}{dx}dx + \kappa f(1)g(1) \qquad (8.81)$$

If Equation 8.81 is used then the functions $f(x)$ and $g(x)$ must only be first-order differentiable, whereas use of the inner product of Equation 8.71 requires that the functions be second-order differentiable. In addition, the inner product satisfies $(f,g)_K = (g,f)_K$ independent of whether $f(x)$ and $g(x)$ satisfy the natural boundary condition.

The above suggests that when the potential energy inner product is formulated as Equation 8.81 the basis functions need only to be first-order differentiable and that the basis functions need not satisfy the natural boundary condition. Even after eliminating these restrictions, the evaluation of the inner product of an approximation with itself is equal to twice the potential energy.

If the system has a discrete mass at its end, the kinetic energy inner product is written as

$$(f,g)_{\mathrm{M}} = \int_0^1 \alpha(x)f(x)g(x)\mathrm{d}x + \mu f(1)g(1) \qquad (8.82)$$

Evaluation of the kinetic energy inner product only requires continuity of the basis functions. Since the kinetic energy of the discrete mass is already included in the definition of the inner product, in order for $(f,g)_{\mathrm{M}} = (g,f)_{\mathrm{M}}$ it is not necessary for the functions to satisfy the natural boundary condition at $x = 1$. In addition $(f,f)_{\mathrm{M}} = 2T$ if f is the velocity for the system.

A space of admissible functions for this example is a space of functions that are first-order differentiable and satisfy the geometric boundary conditions. Functions in a space of admissible functions are not required to satisfy the natural boundary conditions.

A space of functions that are twice differentiable and satisfy all boundary conditions is called a space of comparison functions.

In general the basis functions for use in a Rayleigh–Ritz application may be chosen from a space of admissible functions. When admissible functions are used the energy formulation of the potential energy inner product, Equation 8.80, must be used.

The above conclusions may be extended to beam problems by defining a space of admissible functions containing functions that are twice differentiable and satisfy the geometric boundary conditions for a beam. In this case, the potential energy inner product used for a beam with discrete springs at each end is

$$(f,g)_{\mathrm{K}} = \int_0^1 \alpha(x)\frac{\mathrm{d}^2 f}{\mathrm{d}x^2}\frac{\mathrm{d}^2 g}{\mathrm{d}x^2}\mathrm{d}x + \kappa_0 f(0)g(0) + \kappa_1 f(1)g(1) \qquad (8.83)$$

The kinetic and potential energy inner products can be extended to a system with n_m discrete masses located at $x_{1,m}, x_{2,m}, \ldots x_{n_m,m}$ and with n_k discrete springs located at $x_{1,k}, x_{2,k}, \ldots, x_{n_k,k}$ such that

$$(f,g)_{\mathrm{M}} = \int_0^1 \beta(x)f(x)g(x)\mathrm{d}x + \sum_{i=1}^{n_m} \mu_i f(x_{i,m})g(x_{i,m}) \qquad (8.84)$$

$$(f,g)_{\mathrm{K}} = \int_0^1 \alpha(x)\frac{\mathrm{d}^2 f}{\mathrm{d}x^2}\frac{\mathrm{d}^2 g}{\mathrm{d}x^2}\mathrm{d}x + \sum_{i=1}^{n_k} \kappa_i f(x_{i,k})g(x_{i,k}) \qquad (8.85)$$

8.7 ASSUMED MODES METHOD

The energy formulation of the Rayleigh–Ritz method is used to illustrate the connection between the approximation of mode shapes for a continuous system and the discrete system approximation. The normal mode solution is used for the development of Rayleigh's quotient. Consider a Rayleigh–Ritz expansion in which the coefficients in the linear combination of basis functions are functions of time

$$w(x,t) = \sum_{i=1}^{n} \alpha_i(t) w_i(x) \tag{8.86}$$

where the basis functions are selected as admissible functions.

The Rayleigh–Ritz expansion of the form of Equation 8.86 is used in the potential energy functional in the numerator of Equation 8.83 leading to

$$
\begin{aligned}
V &= \frac{1}{2} \int_0^L EA \left\{ \frac{\partial}{\partial x} \left[\sum_{i=1}^{n} \alpha_i w_i(x) \right] \right\}^2 dx + \frac{1}{2} \sum_{\ell=1}^{n_k} k_\ell \left\{ \sum_{i=1}^{n} \alpha_i \left[w_i(x_{k,\ell}) \right] \right\}^2 \\
&= \frac{1}{2} \sum_{i=1}^{n} \sum_{j=1}^{n} \left\{ \alpha_i \alpha_j \left[\int_0^L EA \left(\frac{dw_i}{dx} \right) \left(\frac{dw_j}{dx} \right) dx + \sum_{\ell=1}^{n_k} k_\ell w_i(x_{k,\ell}) w_j(x_{k,\ell}) \right] \right\} \\
&= \frac{1}{2} \sum_{i=1}^{n} \sum_{j=1}^{n} k_{i,j} \alpha_i \alpha_j
\end{aligned}
\tag{8.87}
$$

where

$$k_{i,j} = \left[\int_0^L EA \left(\frac{dw_i}{dx} \right) \left(\frac{dw_j}{dx} \right) dx + \sum_{\ell=1}^{n_k} k_\ell w_i(x_{k,\ell}) w_j(x_{k,\ell}) \right] \tag{8.88}$$

A similar analysis when the Rayleigh–Ritz expansion is used in the kinetic energy functional in the denominator of Equation 8.35 leads to

$$T = \frac{1}{2} \sum_{i=1}^{n} \sum_{j=1}^{n} m_{i,j} \dot{\alpha}_i \dot{\alpha}_j \tag{8.89}$$

where

$$m_{i,j} = \left[\int_0^L \rho A w_i(x) w_j(x) dx + \sum_{\ell=1}^{n_m} m_\ell w_i(x_{k,\ell}) w_j(x_{k,\ell}) \right] \qquad (8.90)$$

Equation 8.87 and Equation 8.89 are equivalent to the quadratic forms of potential and kinetic energy for a linear discrete system whose generalized coordinates are $\alpha_1, \alpha_2, \ldots, \alpha_n$. Application of Lagrange's equations leads to a system of linear equations of the form

$$\mathbf{M\ddot{\alpha}} + \mathbf{K\alpha} = 0 \qquad (8.91)$$

where the elements of the symmetric stiffness and mass matrices are given in Equation 8.88 and Equation 8.90 and α is a column vector of the generalized coordinates.

The normal mode solution is assumed in the form of

$$\alpha = \mathbf{A}e^{i\omega t} \qquad (8.92)$$

where $\mathbf{A}$ is the discrete mode shape vector corresponding to the natural frequency ω. There are n natural frequencies and corresponding mode shape vectors with mode shape orthogonality with respect to a discrete kinetic energy inner product. The natural frequencies are approximations for the first n natural frequencies of the continuous system. The general response for the generalized coordinates is of the form

$$\alpha = \sum_{j=1}^{n} c_j \mathbf{A}_j e^{i\omega_j t} \qquad (8.93)$$

where the c_i's are constants of integration. Substitution of Equation 8.93 into Equation 8.86 leads to

$$w(x,t) = \sum_{i=1}^{n} \left\{ c_i \left[\sum_{j=1}^{n} (\mathbf{A}_i)_j w_j(x) \right] e^{i\omega_i t} \right\} \qquad (8.94)$$

The approximation to the spatial mode shape is

$$\phi_i(x) = \sum_{j=1}^{n} (\mathbf{A}_i)_j w_j(x) \qquad (8.95)$$

When admissible functions are used, the mode shape approximations are not necessarily orthogonal with respect to the kinetic energy inner product of the continuous system. They cannot be orthogonal if the basis functions are not in the domain of the stiffness operator.

The method described above, in which an expansion of the form of Equation 8.86 is assumed and Lagrange's equations are used to derive a discrete set of equations satisfied by the time-dependent coefficients, is called the assumed modes method. The assumed modes method is a variation of the Rayleigh–Ritz method when admissible functions are used as basis functions.

The assumed modes method may be extended to non-conservative systems. Recalling Lagrange's equations for a non-conservative system,

$$\frac{d}{dt}\left(\frac{\partial L}{\partial \dot{\alpha}_i}\right) - \frac{\partial L}{\partial \alpha} - \frac{\partial \mathfrak{I}}{\partial \dot{\alpha}_i} = Q_i \quad i = 1,2,\ldots,n \tag{8.96}$$

where the generalized forces are obtained using the method of virtual work

$$\delta W_{nc} = \sum_{i=1}^{n} Q_i \delta \alpha_i \tag{8.97}$$

and $\mathfrak{I}$ is Rayleigh's dissipation function

$$\mathfrak{I} = \frac{1}{2}\sum_{j=1}^{n}\sum_{i=1}^{n} c_{i,j}\dot{\alpha}_i\dot{\alpha}_j \tag{8.98}$$

which is proportional to the power dissipated by viscous damping forces. Application of the assumed modes method for a non-conservative linear system leads to a discrete set of equations of the form

$$\mathbf{M}\ddot{\alpha} + \mathbf{C}\dot{\alpha} + \mathbf{K}\alpha = \mathbf{F} \tag{8.99}$$

The response of non-conservative systems is described in Chapter 6 and Chapter 7.

Example 8.7. The simply supported beam of Figure 8.15 has a machine fixed to the beam at its midspan as well as a spring attached at the midspan. During operation the machine develops a force of $10\sin(500t)$ in Newtons. Use the assumed modes method to determine the natural frequencies, mode shapes, and the forced response. Use the first three mode shapes of a simply supported beam as basis functions.

Solution: The first three mode shapes of a simply supported beam are

$$w_1(x) = \sin\left(\pi\frac{x}{L}\right) \tag{a}$$

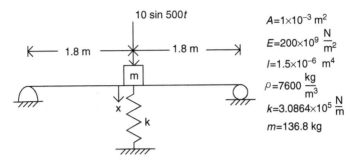

FIGURE 8.15 Simply supported beam with mass and spring attached at midspan.

$$w_2(x) = \sin\left(2\pi\frac{x}{L}\right) \tag{b}$$

$$w_3(x) = \sin\left(3\pi\frac{x}{L}\right) \tag{c}$$

The form of the kinetic energy inner product to apply using the assumed modes method is

$$(f,g)_M = \int_0^L \rho A f(x) g(x) dx + (m) f\left(\frac{L}{2}\right) g\left(\frac{L}{2}\right) \tag{d}$$

The form of the potential energy inner product to apply using the assumed modes method is

$$(f,g)_K = \int_0^L EI \frac{df}{dx}\frac{dg}{dx} dx + (k) f\left(\frac{L}{2}\right) g\left(\frac{L}{2}\right) \tag{e}$$

The discrete mass matrix is a 3×3 matrix whose elements are $m_{i,j} = (w_i,w_j)_M$

$$\mathbf{M} = \begin{bmatrix} 150.5 & 0 & -136.8 \\ 0 & 13.68 & 0 \\ -136.8 & 0 & 150.5 \end{bmatrix} \tag{f}$$

The discrete stiffness matrix is a 3×3 matrix whose elements are $k_{i,j} = (w_i,w_j)_K$

$$\mathbf{K} = 10^7 \begin{bmatrix} 6.218 & 0 & -3.086 \\ 0 & 50.11 & 0 \\ -3.086 & 0 & 256.7 \end{bmatrix} \quad (g)$$

The natural frequency approximations are square roots of the eigenvalues of $\mathbf{M}^{-1}\mathbf{K}$ and the mode shape vectors are the corresponding eigenvectors. The natural frequencies and normalized eigenvectors are

$$\omega_1 = 641.5\frac{r}{s} \quad \mathbf{X}_1 = \begin{bmatrix} 0.0808 \\ 0 \\ -0.000820 \end{bmatrix} \quad (h)$$

$$\omega_2 = 6052\frac{r}{s} \quad \mathbf{X}_1 = \begin{bmatrix} 0 \\ 0.270 \\ 0 \end{bmatrix} \quad (i)$$

$$\omega_3 = 9906\frac{r}{s} \quad \mathbf{X}_3 = \begin{bmatrix} 0.1782 \\ 0 \\ 0.1957 \end{bmatrix} \quad (j)$$

The mathematical form of the force is

$$F(x,t) = 10 \sin(500t)\delta\left(x - \frac{L}{2}\right) \quad (k)$$

The elements of the discrete force vector are

$$F_1 = (F(x,t), w_1(x)) = \int_0^L 10 \sin(500t)\delta\left(x - \frac{L}{2}\right)\sin\left(\pi\frac{x}{L}\right)dx$$

$$= 10 \sin(500t) \sin\left(\frac{\pi}{2}\right) = 10 \sin(500t) \quad (l)$$

Similarly,

$$F_2 = \int_0^L 10 \sin(500t)\delta\left(x - \frac{L}{2}\right)\sin\left(2\pi\frac{x}{L}\right)dx = 0 \quad (m)$$

$$F_3 = \int_0^L 10 \sin(500t)\delta\left(x-\frac{L}{2}\right)\sin\left(3\pi\frac{x}{L}\right)dx = -10 \sin(500t) \qquad \text{(n)}$$

The discrete differential equations are

$$\begin{bmatrix} 150.5 & 0 & -136.8 \\ 0 & 13.68 & 0 \\ -136.8 & 0 & 150.5 \end{bmatrix} \begin{bmatrix} \ddot{\alpha}_1 \\ \ddot{\alpha}_2 \\ \ddot{\alpha}_3 \end{bmatrix} + 10^7 \begin{bmatrix} 6.218 & 0 & -3.086 \\ 0 & 50.11 & 0 \\ -3.086 & 0 & 256.7 \end{bmatrix} \begin{bmatrix} \alpha_1 \\ \alpha_2 \\ \alpha_3 \end{bmatrix}$$

$$= \begin{bmatrix} 10 \\ 0 \\ -10 \end{bmatrix} \sin(500t) \qquad \text{(o)}$$

Modal analysis for discrete systems is used to determine the steady-state response of Equation o. The forces for the modal equations are

$$g_1(t) = (\mathbf{F},\mathbf{X}_1)_{R^3} = \begin{bmatrix} 0.0808 & 0 & -0.000820 \end{bmatrix} \begin{bmatrix} 10 \\ 0 \\ -10 \end{bmatrix} \sin(500t)$$

$$= 0.8159 \sin(500t) \qquad \text{(p)}$$

$$g_2(t) = \begin{bmatrix} 0 & 0.2703 & 0 \end{bmatrix} \begin{bmatrix} 10 \\ 0 \\ -10 \end{bmatrix} \sin(500t) = 0 \qquad \text{(q)}$$

$$g_1(t) = \begin{bmatrix} 0.1783 & 0 & 0.1957 \end{bmatrix} \begin{bmatrix} 10 \\ 0 \\ -10 \end{bmatrix} \sin(500t) = -0.174 \sin(500t) \qquad \text{(r)}$$

The discrete modal equations are

$$\ddot{p}_1 + (641.5)^2 p_1 = 0.8159 \sin(500t) \qquad \text{(s)}$$

$$\ddot{p}_2 + (6052)^2 p_2 = 0 \qquad \text{(t)}$$

$$\ddot{p}_3 + (9906)^2 p_1 = -0.174\sin(500t) \tag{u}$$

The steady-state solutions of Equation s through Equation u are

$$p_1 = \frac{0.81569}{(641.5)^2 - (500)^2}\sin(500t) = 5.05 \times 10^{-6}\sin(500t) \tag{v}$$

$$p_2 = 0 \tag{w}$$

$$p_1 = \frac{-0.1744}{(9906)^2 - (500)^2}\sin(500t) = -1.78 \times 10^{-9}\sin(500t) \tag{x}$$

The solution of the discrete problem is

$$\begin{bmatrix} \alpha_1 \\ \alpha_2 \\ \alpha_3 \end{bmatrix} = \left\{ 5.05 \times 10^{-6} \begin{bmatrix} 0.0808 \\ 0 \\ -0.000820 \end{bmatrix} - 1.78 \times 10^{-9} \begin{bmatrix} 0.1782 \\ 0 \\ 0.1957 \end{bmatrix} \right\} \sin(500t) \tag{y}$$

The Rayleigh–Ritz approximation for the steady-state response is

$$w(x,t) = \left[4.077 \times 10^{-7}\sin(0.8727x) - 4.43 \times 10^{-9}\sin(2.618x)\right]\sin(500t) \tag{z}$$

The time dependent response of the beam at two locations are illustrated in Figure 8.16.

Example 8.8. Use the assumed modes method to approximate the response of the system of Figure 8.17. Choose the basis functions as the mode shapes

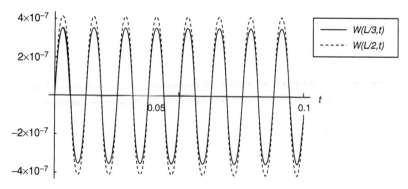

FIGURE 8.16 Assumed mode solution for response of simply supported beam with fixed mass and attached spring at midspan.

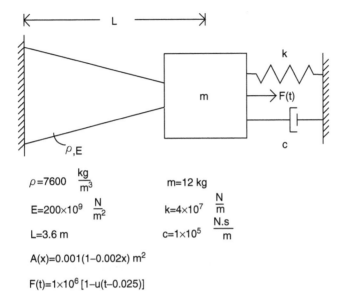

$\rho = 7600 \ \dfrac{kg}{m^3}$ m=12 kg

$E = 200 \times 10^9 \ \dfrac{N}{m^2}$ $k = 4 \times 10^7 \ \dfrac{N}{m}$

L=3.6 m $c = 1 \times 10^5 \ \dfrac{N.s}{m}$

A(x)=0.001(1–0.002x) m²

F(t)=1×10⁶ [1–u(t–0.025)]

FIGURE 8.17 Non-uniform bar with discrete mass-spring-viscous damper system at end.

corresponding to the three lowest modes of a uniform fixed-free bar

$$w_1(x) = \sin\left(\frac{\pi x}{2L}\right) \tag{a}$$

$$w_2(x) = \sin\left(\frac{3\pi x}{2L}\right) \tag{b}$$

$$w_3(x) = \sin\left(\frac{5\pi x}{2L}\right) \tag{c}$$

Solution: An expansion using the basis functions of Equation a through Equation c is assumed as

$$w(x) = \alpha_1(t)w_1(x) + \alpha_2(t)w_2(x) + \alpha_3(t)w_3(x) \tag{d}$$

The kinetic energy of the system is

$$T = \frac{1}{2} \sum_{i=1}^{3} \sum_{j=1}^{3} m_{i,j} \dot{\alpha}_i \dot{\alpha}_j \tag{e}$$

where $m_{i,j}$ are elements of the mass matrix determined by

$$m_{i,j} = (w_i, w_j)_M = \int_0^L \rho A w_i(x) w_j(x) dx + m w_i(L) w_j(L) \tag{f}$$

Evaluation of the mass matrix using Equation f leads to

$$\mathbf{M} = \begin{bmatrix} 25.61 & -11.98 & 12.00 \\ -11.98 & 25.63 & -11.98 \\ 12.0 & -11.98 & 25.63 \end{bmatrix} \tag{g}$$

The potential energy of the system is

$$V = \sum_{i-1}^{3} \sum_{j=1}^{3} k_{i,j} \alpha_i \alpha_j \tag{h}$$

where $k_{i,j}$ are elements of the stiffness matrix calculated by

$$k_{i,j} = (w_i, w_j)_K = \int_0^L EA \frac{dw_i}{dx} \frac{dw_j}{dx} dx + k w_i(L) w_j(L) \tag{i}$$

Evaluation of the stiffness matrix using Equation i leads to

$$\mathbf{K} = 10^7 \begin{bmatrix} 10.84 & -3.97 & 4.01 \\ -3.97 & 65.5 & -3.85 \\ 4.01 & -3.85 & 174.7 \end{bmatrix} \tag{j}$$

The damping force at the end of the bar is

$$F_d = -c \frac{\partial w}{\partial t}(L, t) \tag{k}$$

Use of Equation a through Equation d in Equation k leads to

$$\begin{aligned} F_d &= -c \left[\dot{\alpha}_1 w_1(L) + \dot{\alpha}_2 w_2(L) + \dot{\alpha}_3 w_3(L) \right] \\ &= -c \left[\dot{\alpha}_1 - \dot{\alpha}_2 + \dot{\alpha}_3 \right] \end{aligned} \tag{l}$$

Rayleigh's dissipation function is

$$\begin{aligned} \mathfrak{I} &= \frac{1}{2} c \left(\dot{\alpha}_1 - \dot{\alpha}_2 + \dot{\alpha}_3 \right)^2 \\ &= \frac{1}{2} c \left(\dot{\alpha}_1^2 - 2\dot{\alpha}_1 \dot{\alpha}_2 - 2\dot{\alpha}_1 \dot{\alpha}_3 + \dot{\alpha}_2^2 - 2\dot{\alpha}_2 \dot{\alpha}_3 + \dot{\alpha}_3^2 \right) \end{aligned} \tag{m}$$

The damping matrix is obtained from Equation m as

$$
\mathbf{C} = \begin{bmatrix} c & -c & -c \\ -c & c & -c \\ -c & -c & c \end{bmatrix} = \begin{bmatrix} 100000 & -100000 & -100000 \\ -100000 & 100000 & -100000 \\ -100000 & -100000 & 100000 \end{bmatrix}
\tag{n}
$$

Assuming variations in α_1, α_2, and α_3 the virtual work done by the external force is

$$
\delta W_{\mathrm{nc}} = F(t)\delta w(1,t) = F(t)\left[\delta\alpha_1 - \delta\alpha_2 + \delta\alpha_3\right]
\tag{o}
$$

The force vector is obtained from Equation o as

$$
\mathbf{F} = \begin{bmatrix} 1 \\ -1 \\ 1 \end{bmatrix} F(t)
\tag{p}
$$

The discrete differential equations obtained from the assumed modes model are

$$
\begin{bmatrix} 25.61 & -11.98 & 12.00 \\ -11.98 & 25.63 & -11.98 \\ 12.0 & -11.98 & 25.63 \end{bmatrix} \begin{bmatrix} \ddot{\alpha}_1 \\ \ddot{\alpha}_2 \\ \ddot{\alpha}_3 \end{bmatrix}
$$

$$
+ \begin{bmatrix} 100000 & -100000 & -100000 \\ -100000 & 100000 & -100000 \\ -100000 & -100000 & 100000 \end{bmatrix} \begin{bmatrix} \dot{\alpha}_1 \\ \dot{\alpha}_2 \\ \dot{\alpha}_3 \end{bmatrix}
$$

$$
+ 10^7 \begin{bmatrix} 10.84 & -3.97 & 4.01 \\ -3.97 & 65.5 & -3.85 \\ 4.01 & -3.85 & 174.7 \end{bmatrix} \begin{bmatrix} \alpha_1 \\ \alpha_2 \\ \alpha_3 \end{bmatrix}
$$

$$
= \begin{bmatrix} 1 \\ -1 \\ 1 \end{bmatrix} F(t)
\tag{q}
$$

The forced response of a system governed by Equation q is determined using modal analysis for damped discrete systems as described in Section 4.3.

8.8 FINITE-ELEMENT METHOD

In the broadest sense the finite element method is an application of the Rayleigh–Ritz method or assumed modes method in which piecewise defined basis functions are used. The reduction of restrictions on the selection of basis functions from comparison functions to admissible functions for a second-order system such as a bar or shaft to be only first-order differentiable and those for a beam to be only second-order differentiable. The range of the nondimensional spatial variable, x, is $0 \le x \le 1$. The range is divided into a finite number of subintervals. For simplicity assume the length of each subinterval is the same $\ell = 1/n$ where n is the number of subintervals. A set of $n+1$ basis functions for a second-order problem are defined as shown in Figure 8.18. The mathematical definition of these basis functions is

$$\phi_0(x) = \begin{cases} 1 - \dfrac{x}{\ell} & 0 \le x \le \ell \\[2mm] 0 & \ell \le x \le 1 \end{cases} \tag{8.100}$$

$$\phi_1(x) = \begin{cases} \dfrac{x}{\ell} & 0 \le x \le \ell \\[2mm] 2 - \dfrac{x}{\ell} & \ell \le x \le 2\ell \\[2mm] 0 & 2\ell \le x \le 1 \end{cases} \tag{8.101}$$

$$\vdots$$

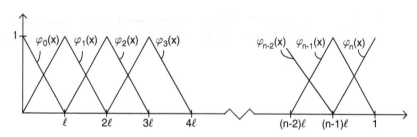

FIGURE 8.18 Basis functions for finite-element model of second-order system.

$$\phi_k(x) = \begin{cases} 0 & 0 \le x \le (k-1)\ell \\ \dfrac{x}{\ell} - (k-1) & (k-1)\ell \le x \le k\ell \\ (k+1) - \dfrac{x}{\ell} & k\ell \le x \le (k+1)\ell \\ 0 & (k+1)\ell \le x \le 1 \end{cases} \tag{8.102}$$

$$\vdots$$

$$\phi_n(x) = \begin{cases} 0 & 0 \le x \le (n-1)\ell \\ \dfrac{x}{\ell} - (n-1) & (n-1)x \le 1 \end{cases} \tag{8.103}$$

The functions defined in Equation 8.100 through Equation 8.103 are summarized using the unit step function as

$$\phi_0(x) = \left(1 - \frac{x}{\ell}\right)[1 - u(x - \ell)]$$

$$\phi_1(x) = \left(\frac{x}{\ell}\right)[u(x) - u(x - \ell)] + \left(2 - \frac{x}{\ell}\right)[u(x - \ell) - u(x - 2\ell)]$$

$$\phi_2(x) = \phi_1(x - \ell)$$

$$\vdots \tag{8.104}$$

$$\phi_k(x) = \phi_1(x - (k-1)\ell)$$

$$\vdots$$

$$\phi_n(x) = \left[\frac{x}{\ell} + (1 - n)\right]u(x - (n-1)\ell)$$

The basis functions of Equation 8.104 can be used for second-order systems for which the admissible functions need only be first-order differentiable. The Rayleigh–Ritz expansion is

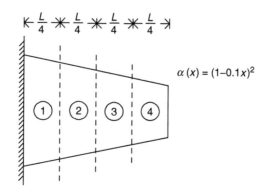

FIGURE 8.19 Four element model is used to model non-uniform fixed-free bar in Example 8.9.

$$w(x) = \sum_{i=0}^{n} a_i \phi_i(x) \tag{8.105}$$

Geometric boundary conditions are satisfied by specifying $a_0 = 0$ if the end at $x = 0$ is fixed and $a_n = 0$ if the end at $x - 1$ is fixed.

Example 8.9. Use the finite-element method with the piecewise defined basis functions to approximate the natural frequencies and mode shapes for the variable area bar of Figure 8.19. Use $n = 4$.

Solution: The basis elements for the fixed-free bar with $n = 4$ are illustrated in Figure 8.20. Note that $\phi_0(x)$ is not included in order to satisfy the geometric boundary condition at $x = 0$. The mathematical forms of the basis functions are

$$\phi_1(x) = 4x[u(x) - u(x - 0.25)] + (2 - 4x)[u(x - 0.25) - u(x - 0.5)] \tag{a}$$

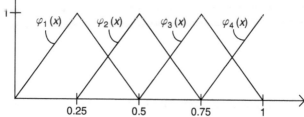

FIGURE 8.20 Basis functions for Example 8.9.

$$\phi_2(x) = 4(x-0.25)[u(x-0.25) - u(x-0.5)] + (3-4x)[u(x-0.5)$$
$$- u(x-0.75)] \tag{b}$$

$$\phi_3(x) = 4(x-0.5)[u(x-0.5) - u(x-0.75)] + (4-4x)[u(x$$
$$= 0.75) - u(x-1)] \tag{c}$$

$$\phi_4(x) = 4(x-0.75)[u(x-0.75) - u(x-1)] \tag{d}$$

The finite element approximation is of the form

$$w(x) = \sum_{i=1}^{4} a_i \phi_i(x) \tag{e}$$

The inner products from which the discrete mass and stiffness matrices are formed are

$$(\phi_i, \phi_j)_M = \int_0^1 (1 - 0.1x)^2 \phi_i(x)\phi_j(x)dx \tag{f}$$

$$(\phi_i, \phi_j)_K = \int_0^1 (1 - 0.1x)^2 \frac{d\phi_i}{dx} \frac{d\phi_j}{dx} dx \tag{g}$$

Application of these inner products to the basis functions of Equation a through Equation d leads to

$$\mathbf{M} = \begin{bmatrix} 0.1584 & 0.03860 & 0 & 0 \\ 0.03860 & 0.1504 & 0.03662 & 0 \\ 0 & 0.03662 & 0.1426 & 0.03469 \\ 0 & 0 & 0.03469 & 0.06844 \end{bmatrix} \tag{h}$$

$$\mathbf{K} = \begin{bmatrix} 7.6068 & -3.7058 & 0 & 0 \\ -3.7058 & 7.2217 & -3.5183 & 0 \\ 0 & -3.5183 & 6.8467 & -3.3308 \\ 0 & 0 & -3.3308 & 3.3308 \end{bmatrix} \tag{i}$$

The natural frequency approximations, the square roots of the eigenvalues of $\mathbf{M}^{-1}\mathbf{K}$, are determined as

$$\omega_1 = 1.6482 \quad \omega_2 = 5.0082 \quad \omega_3 = 9.0637 \quad \omega_4 = 13.0842 \tag{j}$$

The exact natural frequencies as determined in Example 5.5 are

$$\omega_1 = 1.638 \quad \omega_2 = 4.735 \quad \omega_3 = 7.868 \quad \omega_4 = 11.01 \tag{k}$$

The approximate natural frequencies determined using the Rayleigh–Ritz method using the mode shapes corresponding to the four lowest natural frequencies of a uniform bar are

$$\omega_1 = 1.638 \quad \omega_2 = 4,735 \quad \omega_3 = 7.868 \quad \omega_4 = 13.01 \tag{l}$$

The approximations using the Rayleigh–Ritz method are obviously better. However the natural frequencies predicted by the finite element method become better as the number of basis functions increases. The finite-element method has distinct computational advantages over the use of a Rayleigh–Rtiz method using comparison functions. The matrices of Equation h and Equation i are banded matrices, in this case tri-diagonal. Efficient algorithms for calculations of eigenvalues and eigenvectors are available for tri-diagonal matrices.

The mode shape vector approximations are illustrated in Figure 8.21.

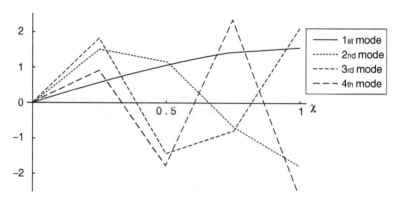

FIGURE 8.21 Finite-element approximations for mode shapes of non-uniform bar with $\varepsilon = 0,1$.

8.9 ASSUMED MODES DEVELOPMENT OF FINITE-ELEMENT METHOD

The finite element method can also be developed from the assumed modes method. When the basis functions of Figure 8.19 are used in an assumed modes expansion for the approximation

$$w(x,t) = \sum_{i=1}^{n} \alpha_i(t)\phi_i(x) \tag{8.106}$$

the displacements at each of the element boundaries, $x_k = k/n$ $k = 0,1,2...,n$ are calculated by

$$w(x_k,t) = \sum_{i=1}^{n} \alpha_i(t)\phi_i(x_k) \tag{8.107}$$

The mathematical definition of the basis functions, Equation 8.104, shows that

$$\phi_i(x_k) = \delta_{i,k} \tag{8.108}$$

which when used in Equation 8.106 leads to

$$w(x_k,t) = \alpha_k(t) \tag{8.109}$$

Equation 8.109 illustrates that the finite-element method is a discrete model for a continuous system. The generalized coordinates used in the discrete model are the displacements at element boundaries. Defining $W_k(t) = w(x_k,t)$ Equation 8.107 is written in terms of the nodal displacements as

$$w(x,t) = \sum_{i=1}^{n} W_i(t)\phi_i(x) \tag{8.110}$$

Equation 8.110 is the assumed modes formulation of the finite-element method for a second-order system. The assumed modes method is used to formulate quadratic forms for the system's kinetic and potential energies from which the mass and stiffness matrices are determined. The method of virtual work is used to determine the generalized force vector. The methods developed for analysis of discrete systems are used to determine the free response and forced response in terms of the displacements at the element boundaries.

Given the discrete approximation for the displacements at element boundaries, the piecewise linear basis functions lead to a spatial variation that is a linear interpolation of displacement between element boundaries.

The assumed modes development of the finite element method leads to a method that is identical to the method developed from application of Rayleigh–Ritz method using the basis functions of Figure 8.18. The assumed modes development is easier to implement as the number of elements increases. In either case the finite element method is a discrete approximation for the continuous system. Using the assumed modes, method the generalized coordinates for the discrete system are the displacements at element boundaries, while using the Rayleigh–Ritz method, the generalized coordinates are coefficients in the expansion.

Using the assumed modes method an "element" is developed. An interpolating function is assumed over each element such that geometric variables are continuous at the element boundaries. The kinetic energy functional is used to develop an element mass matrix; the potential energy functional is used to develop an element stiffness matrix. The method of virtual work is used to develop an element force vector. The element matrices are assembled to form global mass and stiffness matrices and force vectors. The global matrices are used for the discrete model.

A set of global generalized coordinates is selected. A finite element model of a bar uses the displacements at the element boundaries as its global generalized coordinates. The global generalized coordinates for a beam are the displacements and slopes at the element boundaries. A local displacement vector is selected for each element. The local displacement vector is a subset of the global displacement vector and involves displacements at each boundary of the element.

Consider an n element finite element model of a continuous system. Let $\mathbf{W}$ be the global displacement vector. Let $\mathbf{w}_j$ be the displacement vector for element j. The kinetic energy functional is used to determine the kinetic energy for the element in terms of the elements of the local displacement vector and define an element mass matrix such that

$$T_j = \frac{1}{2}\left(\mathbf{m}_j \dot{\mathbf{w}}_j, \dot{\mathbf{w}}_j\right) \tag{8.111}$$

Where the inner product is the standard inner product on R^k where k is the number of elements in $\mathbf{w}_j$. The potential energy functional is used to develop an element stiffness matrix such that

$$V_j = \frac{1}{2}\left(\mathbf{k}_j \mathbf{w}_j, \mathbf{w}_j\right) \tag{8.112}$$

The total kinetic energy of the system is

$$T = \sum_{j=1}^{n} T_j \tag{8.113}$$

which has the quadratic form

$$T = \frac{1}{2}(\mathbf{M}\dot{\mathbf{W}},\dot{\mathbf{W}}) \tag{8.114}$$

where $\mathbf{M}$ is the global mass matrix. The total potential energy has the quadratic form

$$V = \frac{1}{2}(\mathbf{K}\mathbf{W},\mathbf{W})) \tag{8.115}$$

where $\mathbf{K}$ is the global stiffness matrix.

8.10 BAR ELEMENT

The bar of Figure 8.22 is of total length L, is made of a material of mass density ρ and elastic modulus E and has a cross-section area $A(x)$. The longitudinal displacement of the bar is $u(x,t)$. The bar is divided into n finite elements. For simplicity, assume the elements are of equal length, $\ell = L/n$. The global coordinates of the bar are $U_0, U_1, U_2, \ldots, U_n$ defined such that U_k is

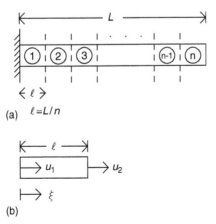

(a) $\ell = L/n$

(b)

FIGURE 8.22 (a) Model of longitudinal motion of bar using n elements. (b) Bar element.

the approximation for $u(k\ell,t)$. The global coordinates are required to be chosen to satisfy all geometric boundary conditions. For example, if the end $x = 0$ is fixed, then $U_0 = 0$.

A bar element is shown in Figure 8.22b. The element is of length ℓ. A local coordinate ξ is defined such that $0 \le \xi \le \ell$. The local coordinates are u_1, the displacement of the end of the element at $\xi = 0$, and u_2, the displacement at $\xi = \ell$. An interpolating function $u(\xi,t)$ is chosen such that $u(0,t) = u_1$ and $u(\ell,t) = u_2$. A linear interpolating function is

$$u(\xi,t) = \frac{1}{\ell}(u_2 - u_1)\xi + u_1 \tag{8.116}$$

The kinetic energy of the element is

$$
\begin{aligned}
T &= \frac{1}{2}\int_0^\ell \rho A \left(\frac{\partial u}{\partial t}\right)^2 d\xi \\
&= \frac{1}{2}\int_0^\ell \rho A \left[\frac{1}{\ell}(\dot{u}_2 - \dot{u}_1)\xi + \dot{u}_1\right]^2 d\xi \\
&= \frac{1}{2}\left\{\left[\int_0^\ell \rho A \left(1 - \frac{\xi}{\ell}\right)^2 d\xi\right]\dot{u}_1^2 + \left[2\int_0^\ell \rho A \left(\frac{\xi}{\ell}\right)\left(1 - \frac{\xi}{\ell}\right)d\xi\right]\dot{u}_1\dot{u}_2 \right. \\
&\qquad \left. + \left[\int_0^\ell \rho A \left(\frac{\xi}{\ell}\right)^2 d\xi\right]\dot{u}_2^2\right\}
\end{aligned}
\tag{8.117}
$$

The components of the 2×2 element mass matrix are

$$m_{1,1} = \int_0^\ell \rho A \left(1 - \frac{\xi}{\ell}\right)^2 d\xi \tag{8.118}$$

$$m_{1,2} = m_{2,1} = \int_0^\ell \rho A \left(\frac{\xi}{\ell}\right)\left(1 - \frac{\xi}{\ell}\right) d\xi \tag{8.119}$$

$$m_{2,2} = \int_0^\ell \rho A \left(\frac{\xi}{\ell}\right)^2 d\xi \tag{8.120}$$

The density and area are constant throughout a uniform element in which case the element mass matrix becomes

$$\mathbf{m} = \frac{\rho A \ell}{6} \begin{bmatrix} 2 & 1 \\ 1 & 2 \end{bmatrix} \tag{8.121}$$

The potential energy in the element is

$$
\begin{aligned}
V &= \frac{1}{2} \int_0^\ell EA \left(\frac{\partial u}{\partial \xi}\right)^2 d\xi \\
&= \frac{1}{2} \int_0^\ell EA \left[\frac{1}{\ell}(u_2 - u_1)\right]^2 d\xi \\
&= \frac{1}{2} \left[\left(\int_0^\ell \frac{EA}{\ell^2} d\xi\right)(u_2^2 - 2u_1 u_2 + u_1^2)\right]
\end{aligned} \tag{8.122}
$$

The components of the element stiffness matrix are determined from Equation 8.122 as

$$k_{1,1} = \int_0^\ell \frac{EA}{\ell^2} d\xi \tag{8.123}$$

$$k_{1,2} = k_{2,1} = -\int_0^\ell \frac{EA}{\ell^2} d\xi \tag{8.124}$$

$$k_{2,2} = \int_0^\ell \frac{EA}{\ell^2} d\xi \tag{8.125}$$

The stiffness matrix for a uniform bar element is

$$\mathbf{k} = \frac{EA}{\ell} \begin{bmatrix} 1 & -1 \\ -1 & 1 \end{bmatrix} \tag{8.126}$$

Determination of the mass and stiffness matrices for non-uniform elements require evaluation of the integrals of Equation 8.118 through Equation 8.120 and Equation 8.123 through Equation 8.125. Before evaluation the cross-sectional area must be expressed in terms of the local coordinates. The transformation between the global coordinate x and the local coordinate ξ for the jth element is

$$\xi = x - (j-1)\ell \tag{8.127}$$

If the bar is subject to an external axial loading in which the load per unit length is described by $f(x, t)$ the virtual work done by the external force is

$$
\begin{aligned}
\delta W &= \int_0^\ell f(\xi,t)\delta u(\xi,t)d\xi \\
&= \left[\int_0^\ell f(\xi,t)\left(1-\frac{\xi}{\ell}\right)d\xi\right]\delta u_1 + \left[\int_0^\ell f(\xi,t)\left(\frac{\xi}{\ell}\right)d\xi\right]\delta u_2
\end{aligned}
\tag{8.128}
$$

The components of the generalized force vector for the element are

$$q_1 = \int_0^\ell f(\xi,t)\left(1-\frac{\xi}{\ell}\right)d\xi \tag{8.129}$$

$$q_2 = \int_0^\ell f(\xi,t)\left(\frac{\xi}{\ell}\right)d\xi \tag{8.130}$$

Evaluation of the integrals of Equation 8.129 and Equation 8.130 requires that the force is written as a function of the local coordinate for the element.

Example 8.10. (a) Use a one-element finite-element model to approximate the lowest natural frequency of a uniform bar that is fixed at $x = 0$ and has a particle of mass M attached at its free end as illustrated in Figure 8.23 (b) Use the finite element model of part (a) to approximate the response of the bar if the particle is subject to a harmonic force $F_0 \sin(\omega t)$.

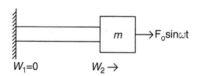

FIGURE 8.23 One element model of fixed-free bar has one degree of freedom.

Solution: (a) The global generalized coordinates for a one-element finite element model of the bar are $W_1(t)$, the displacement of the bar at $x = 0$, and $W_2(t)$, the displacement of the bar at $x = L$. Since the bar is fixed at $x = 0$, $W_1(t) = 0$. Thus, the finite-element model is a one-degree-of-freedom approximation. The total potential energy is

$$V = \frac{1}{2} \left(\frac{EA}{L} \right) W_2^2 \tag{a}$$

The total kinetic energy including the kinetic energy of the particle is

$$T = \frac{1}{2} \left(\frac{\rho A L}{3} \right) \dot{W}_2^2 + \frac{1}{2} M \dot{W}_2^2 \tag{b}$$

The kinetic and potential energies are used to derive the single differential equation as

$$\left(M + \frac{\rho A L}{3} \right) \ddot{W}_2 + \frac{EA}{L} W_2 = 0 \tag{c}$$

Equation c is written in the standard form of a differential equation governing the motion of a one-degree-of-freedom as

$$\ddot{W}_2 + \frac{3EA}{L(3M + \rho A L)} W_2 = 0 \tag{d}$$

The natural frequency is determined from Equation d as

$$\omega_1 = \sqrt{\frac{3EA}{L(3M + \rho A L)}} \tag{e}$$

The natural frequency approximation of Equatione is the same as that obtained when a one-degree-of-freedom approximation is used for the system where the bar is modeled as a spring of stiffness $k = EA/L$ and its inertia effects are approximated by adding one-third of the mass of the bar to the particle.

(b) The virtual work done by the external force is

$$\delta W = F_0 \sin(\omega t) \delta W_2 \tag{f}$$

Thus $Q_2 = F_0 \sin(\omega t)$ and the one-element finite-element approximation is

$$\left(M + \frac{\rho AL}{3}\right)\ddot{W}_2 + \frac{EA}{L} W_2 = F_0 \sin(\omega t) \tag{g}$$

The steady-state response is

$$W_2(t) = \left(\frac{3F_0}{3M + \rho AL}\right) \frac{1}{\omega_n^2 - \omega^2} \sin(\omega t) \tag{h}$$

The local displacement vector for the bar has two components. The global displacement vector for an n element model has $(n+1)$ components. Let $\mathbf{S_j}$ be a matrix of dimensions $2 \times (n+1)$ that serves as a transformation between the global coordinates and the local coordinates for element j

$$\mathbf{w}_j = \mathbf{S}_j \mathbf{W} \tag{8.131}$$

The total kinetic energy of the system is

$$\begin{aligned}
T &= \frac{1}{2}\sum_{j=1}^{n}(\mathbf{m}_j\dot{\mathbf{w}}_j, \dot{\mathbf{w}}_j)_{R^2} \\
&= \frac{1}{2}\sum_{j=1}^{n}(\mathbf{m}_j\mathbf{S}_j\dot{\mathbf{W}}, \mathbf{S}_j\dot{\mathbf{W}})_{R^2}
\end{aligned} \tag{8.132}$$

The inner products of Equation 8.132 are the standard inner product on R^2

$$(\mathbf{m}_j\mathbf{S}_j\dot{\mathbf{W}}, \mathbf{S}_j\dot{\mathbf{W}})_{R^2} = (\mathbf{S}_j\dot{\mathbf{W}})^T\mathbf{m}_j\mathbf{S}_j\dot{\mathbf{W}}_j = \dot{\mathbf{W}}^T\mathbf{S}_j^T\mathbf{m}_j\mathbf{S}_j\dot{\mathbf{W}} \tag{8.133}$$

Substitution of Equation 8.133 in Equation 8.132 leads to

$$\begin{aligned}
T &= \frac{1}{2}\sum_{j=1}^{n}\dot{\mathbf{W}}^T\mathbf{S}_j^T\mathbf{m}_j\mathbf{S}_j\dot{\mathbf{W}} \\
&= \frac{1}{2}\dot{\mathbf{W}}^T\left(\sum_{j=1}^{n}\mathbf{S}_j^T\mathbf{m}_j\mathbf{S}_j\right)\dot{\mathbf{W}} = \frac{1}{2}\dot{\mathbf{W}}^T\mathbf{M}\dot{\mathbf{W}}
\end{aligned} \tag{8.134}$$

where the global mass matrix is

$$\mathbf{M} = \sum_{j=1}^{n}\mathbf{S}_j^T\mathbf{m}_j\mathbf{S}_j \tag{8.135}$$

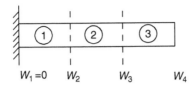

FIGURE 8.24 Three element model of fixed-free bar.

Example 8.11. Develop a three element finite element model for the bar of Figure 8.24.

Solution: The three element model is illustrated in Figure 8.25. The local mass and stiffness matrices are as in Equation 8.121 and Equation 8.126. The global generalized coordinates are W_1, W_2, and W_3

Element 1: $w_1 = W_0 = 0$, $w_2 = W_1$. The mass and stiffness matrices are written in terms of the global coordinates as

$$\mathbf{m}_1 = \frac{\rho A \ell}{6} \begin{bmatrix} 2 & 0 & 0 \\ 0 & 0 & 0 \\ 0 & 0 & 0 \end{bmatrix} \quad \mathbf{k}_1 = \frac{EA}{\ell} \begin{bmatrix} 1 & 0 & 0 \\ 0 & 0 & 0 \\ 0 & 0 & 0 \end{bmatrix} \tag{a}$$

Element 2: $w_1 = W_1$, $w_2 = W_2$. The mass and stiffness matrices written in terms of the global generalized coordinates are

$$\mathbf{m}_2 = \frac{\rho A \ell}{6} \begin{bmatrix} 2 & 1 & 0 \\ 1 & 2 & 0 \\ 0 & 0 & 0 \end{bmatrix} \quad \mathbf{k}_2 = \frac{EA}{\ell} \begin{bmatrix} 1 & -1 & 0 \\ -1 & 1 & 0 \\ 0 & 0 & 0 \end{bmatrix} \tag{b}$$

Element 3: $w_1 = W_2 = 0$, $w_2 = W_3$. The mass and stiffness matrices written in terms of the global generalized coordinates are

$$\mathbf{m}_3 = \frac{\rho A \ell}{6} \begin{bmatrix} 0 & 0 & 0 \\ 0 & 2 & 1 \\ 0 & 1 & 2 \end{bmatrix} \quad \mathbf{k}_3 = \frac{EA}{\ell} \begin{bmatrix} 0 & 0 & 0 \\ 0 & 1 & -1 \\ 0 & 1 & 1 \end{bmatrix} \tag{c}$$

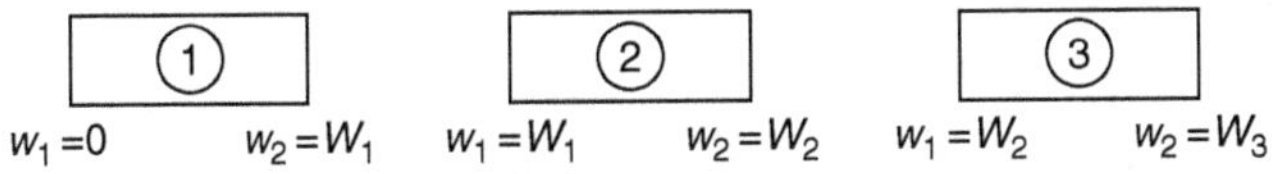

FIGURE 8.25 Relation between local and global coordinates for three element model of bar.

Noting that $\ell = L/3$ The global mass and stiffness matrices are

$$
\mathbf{M} = \frac{\rho A \ell}{6} \left\{ \begin{bmatrix} 2 & 0 & 0 \\ 0 & 0 & 0 \\ 0 & 0 & 0 \end{bmatrix} + \begin{bmatrix} 2 & 1 & 0 \\ 1 & 2 & 0 \\ 0 & 0 & 0 \end{bmatrix} + \begin{bmatrix} 0 & 0 & 0 \\ 0 & 2 & 1 \\ 0 & 1 & 2 \end{bmatrix} \right\}
$$

(d)

$$
= \frac{\rho A L}{18} \begin{bmatrix} 4 & 1 & 0 \\ 1 & 4 & 1 \\ 0 & 1 & 2 \end{bmatrix}
$$

$$
\mathbf{K} = \frac{EA}{\ell} \left\{ \begin{bmatrix} 1 & 0 & 0 \\ 0 & 0 & 0 \\ 0 & 0 & 0 \end{bmatrix} + \begin{bmatrix} 1 & -1 & 0 \\ -1 & 1 & 0 \\ 0 & 0 & 0 \end{bmatrix} + \begin{bmatrix} 0 & 0 & 0 \\ 0 & 1 & -1 \\ 0 & -1 & 1 \end{bmatrix} \right\}
$$

(e)

$$
= \frac{3EA}{L} \begin{bmatrix} 2 & -1 & 0 \\ -1 & 2 & -1 \\ 0 & -1 & 1 \end{bmatrix}
$$

The differential equations, written in terms of the global coordinates, for the three-element model are

$$
\frac{\rho A L}{18} \begin{bmatrix} 4 & 1 & 0 \\ 1 & 4 & 1 \\ 0 & 1 & 2 \end{bmatrix} \begin{bmatrix} \ddot{W}_1 \\ \ddot{W}_2 \\ \ddot{W}_3 \end{bmatrix} + \frac{3EA}{L} \begin{bmatrix} 2 & -1 & 0 \\ -1 & 2 & -1 \\ 0 & -1 & 1 \end{bmatrix} \begin{bmatrix} W_1 \\ W_2 \\ W_3 \end{bmatrix} = \begin{bmatrix} 0 \\ 0 \\ 0 \end{bmatrix}
$$

(f)

The natural frequency approximations obtained as the square roots of the eigenvalues of $\mathbf{M}^{-1}\mathbf{K}$ are

$$
\omega_1 = 1.588 \sqrt{\frac{E}{\rho L^2}} \quad \omega_2 = 5.196 \sqrt{\frac{E}{\rho L^2}} \quad \omega_3 = 9.426 \sqrt{\frac{E}{\rho L^2}}
$$

(g)

8.11 BEAM ELEMENT

A beam is of length L, is made from a material of mass density ρ and the elastic modulus E, has a cross-section area $A(x)$ and area moment of inertia $I(x)$. The beam is broken into n elements of equal length $\ell = L/n$. The global generalized coordinates are the displacements and slopes at the element boundaries.

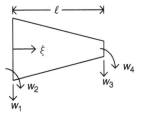

FIGURE 8.26 A beam element has four degrees of freedom.

A beam element is illustrated in Figure 8.26. The local coordinates are

- w_1, the displacement at the left end of the element
- w_2, the slope at the left end of the element
- w_3, the displacement at the right end of the element
- w_4, the slope at the right end of the element.

The local generalized coordinates are illustrated in Figure 8.26. The interpolating function $w(\xi,t)$ is chosen such that:

- $w(0,t) = w_1$
- $\frac{dw}{dx}(0,t) = w_2$
- $w(\ell,t) = w_3$
- $\frac{dw}{dx}(\ell,t) = w_4$

The cubic polynomial that interpolates the above conditions is

$$
w(\xi,t) = \left(1 - 3\frac{\xi^2}{\varrho^2} + 2\frac{\xi^3}{\varrho^3}\right)w_1 + \left(\frac{\xi}{\varrho} - 2\frac{\xi^2}{\varrho^2} + \frac{\xi^3}{\varrho^3}\right)w_2
$$

$$
+ \left(3\frac{\xi^2}{\varrho^2} - 2\frac{\xi^3}{\varrho^3}\right)w_3 + \left(-\frac{\xi^2}{\varrho^2} + \frac{\xi^3}{\varrho^3}\right)w_4 \tag{8.136}
$$

The element mass matrix is obtained by substituting Equation 8.136 into the kinetic energy functional for a beam

$$
T = \frac{1}{2}\int_0^\ell \rho A\left(\frac{\partial w}{\partial t}\right)^2 d\xi \tag{8.137}
$$

The result for a uniform beam is

$$
\mathbf{m} = \frac{\rho A \ell}{420}
\begin{bmatrix}
156 & 22\ell & 54 & -13\ell \\
22\ell & 4\ell^2 & 13\ell & -3\ell^2 \\
54 & 13\ell & 156 & -22\ell \\
-13\ell & -3\ell^2 & -22\ell & 4\ell^2
\end{bmatrix}
\tag{8.138}
$$

The element stiffness matrix is obtained by substituting Equation 8.136 into the potential energy functional for a beam

$$
V = \frac{1}{2} \int_0^\ell EI \left(\frac{\partial w}{\partial \xi} \right)^2 d\xi
\tag{8.139}
$$

The result for a uniform beam is

$$
\mathbf{k} = \frac{EI}{\ell^3}
\begin{bmatrix}
12 & 6\ell & -12 & 6\ell \\
6\ell & 4\ell^2 & -6\ell & 2\ell^2 \\
-12 & -6\ell & 12 & -6\ell \\
6\ell & 2\ell^2 & -6\ell & 4\ell^2
\end{bmatrix}
\tag{8.140}
$$

The method of virtual work is used to determine the components of the local generalized force vector due to a distributed load per unit length $f(\xi,t)$

$$
\delta W = \int_0^\ell f(\xi, y) \delta w(\xi, t) d\xi = \sum_{i=1}^{4} q_i \delta w_i
\tag{8.141}
$$

The results are

$$
q_1 = \int_0^\ell f(\xi,t) \left(1 - 3 \frac{\xi^2}{\ell^2} + 2 \frac{\xi^3}{\ell^3} \right) d\xi
\tag{8.142}
$$

$$
q_2 = \int_0^\ell f(\xi,t) \left(\frac{\xi}{\ell} - 2 \frac{\xi^2}{\ell^2} + \frac{\xi^3}{\ell^3} \right) d\xi
\tag{8.143}
$$

$$
q_3 = \int_0^\ell f(\xi,t) \left(3 \frac{\xi^2}{\ell^2} + 2 \frac{\xi^3}{\ell^3} \right) d\xi
\tag{8.144}
$$

$$q_4 = \int_0^\ell f(\xi,t)\left(-\frac{\xi^2}{\varrho^2} + \frac{\xi^3}{\varrho^3}\right) d\xi \qquad (8.145)$$

Example 8.12. Use a one-element finite-element model to approximate the lowest natural frequency of a uniform fixed-free beam with a spring of stiffness k at its free end. Assume $k = 4EI/L^3$.

Solution: The geometric boundary conditions of a fixed-free beam are that the displacement and slope at $x = 0$ are both zero. Thus the model, illustrated in Figure 8.27, has two degrees of freedom. W_3 is the displacement at $x = L$, and W_4 is the slope at $x = L$. The global mass matrix is obtained by crossing out the first two rows and columns of the element mass matrix. The global stiffness matrix is obtained by crossing out the first two rows and columns of the element stiffness matrix and adding a term to account for the potential energy of the spring $\frac{1}{2}kW_3^2$. The resulting differential equations are

$$\frac{\rho A L}{420}\begin{bmatrix} 156 & -22L \\ -22L & 4L^2 \end{bmatrix}\begin{bmatrix} \ddot{W}_3 \\ \ddot{W}_4 \end{bmatrix} + \frac{EI}{L^3}\begin{bmatrix} 12 + \dfrac{kL^3}{EI} & -6L \\ -6L & 4L^2 \end{bmatrix}\begin{bmatrix} W_3 \\ W_4 \end{bmatrix} = \begin{bmatrix} 0 \\ 0 \end{bmatrix} \qquad (a)$$

The natural frequency approximations are the square roots of the eigenvalues of $\mathbf{M}^{-1}\mathbf{K}$ where for $k = 4EI/L^3$

$$\mathbf{M}^{-1}\mathbf{K} = \left\{\frac{\rho A L}{420}\begin{bmatrix} 156 & -22L \\ -22L & 4L^2 \end{bmatrix}\right\}^{-1}\left\{\frac{EI}{L^3}\begin{bmatrix} 16 & -6L \\ -6L & 4L^2 \end{bmatrix}\right\} \qquad (b)$$

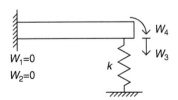

FIGURE 8.27 One element model of fixed-free beam has two-degrees of freedom.

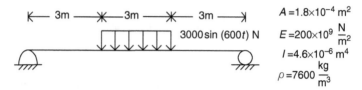

FIGURE 8.28 Simply supported beam of Example 8.12.

The natural frequencies are calculated as

$$\omega_1 = 5.326\sqrt{\frac{EI}{\rho AL^4}} \quad \omega_2 = 35.265\sqrt{\frac{EI}{\rho AL^4}} \tag{c}$$

The mode shape vectors are determined as

$$\begin{bmatrix} W_3 \\ W_4 \end{bmatrix}_1 = \begin{bmatrix} 1 \\ \dfrac{1.21}{L} \end{bmatrix} \quad \begin{bmatrix} W_3 \\ W_4 \end{bmatrix}_2 = \begin{bmatrix} 1 \\ \dfrac{7.54}{L} \end{bmatrix} \tag{d}$$

Example 8.13. Use a three-element finite element model to determine the forced response of the beam of Figure 8.28.

Solution: The three element mode for the beam is illustrated in Figure 8.29. The geometric model results boundary conditions for the simply supported beam are $w(0) = 0$ and $w(L) = 0$ which, in terms of the global coordinate system for a three-element model implies $W_1 = 0$ and $W_7 = 0$. The three-element model result in a six-degree-of-freedom discrete model for the system (Figure 8.30).

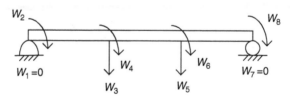

FIGURE 8.29 Definition of global coordinates for beam of Example 8.12.

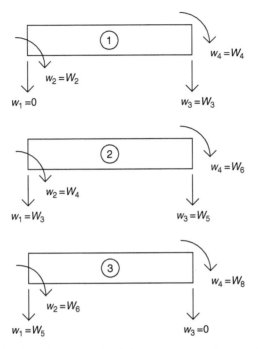

FIGURE 8.30 Relation between local and global coordinates for each element of beam of Example 8.12.

The coefficients multiplying the local mass and stiffness matrices are

$$\rho A \ell / 420 = (7600 \text{ kg/m}^3)(1.8 \times 10^{-4} \text{ m}^2)(3 \text{ m})/420 = 0.009971 \text{ kg} \quad \text{(a)}$$

$$EI/\ell^3 = (200 \times 10^9 \text{ N/m}^2)(4.6 \times 10^{-6} \text{ m}^4)/(3 \text{ m})^3 = 3.41 \times 10^4 \text{ N/m} \quad \text{(b)}$$

The vector of generalized coordinates is $W = [W_2 \ W_3 \ W_4 \ W_5 \ W_6 \ W_8]^T$

Element 1: $w_1 = 0$, $w_2 = W_2$, $w_3 = W_3$, $w_4 = W_4$. The element mass and stiffness matrices are those of Equation 8.131 and Equation 8.133. Its mass and stiffness matrices in terms of the global coordinates is

$$\mathbf{K}_1 = \frac{EI}{\ell^3} \begin{bmatrix} 36 & 18 & 18 & 0 & 0 & 0 \\ -18 & 12 & -18 & 0 & 0 & 0 \\ 18 & -18 & 36 & 0 & 0 & 0 \\ 0 & 0 & 0 & 0 & 0 & 0 \\ 0 & 0 & 0 & 0 & 0 & 0 \\ 0 & 0 & 0 & 0 & 0 & 0 \end{bmatrix} \quad \text{(c)}$$

$$
\mathbf{M}_1 = \frac{\rho A \ell}{420}
\begin{bmatrix}
36 & 39 & -27 & 0 & 0 & 0 \\
39 & 156 & -66 & 0 & 0 & 0 \\
-27 & -66 & 36 & 0 & 0 & 0 \\
0 & 0 & 0 & 0 & 0 & 0 \\
0 & 0 & 0 & 0 & 0 & 0 \\
0 & 0 & 0 & 0 & 0 & 0
\end{bmatrix}
\tag{d}
$$

Element 2: $w_1 = W_3$, $w_2 = W_4$, $w_3 = W_5$, $w_4 = W_6$. The element mass and stiffness matrices written in terms of the global coordinates are

$$
\mathbf{K}_2 = \frac{EI}{\ell^3}
\begin{bmatrix}
0 & 0 & 0 & 0 & 0 & 0 \\
0 & 12 & 18 & -12 & 18 & 0 \\
0 & 18 & 36 & -18 & 18 & 0 \\
0 & -12 & -18 & 12 & -18 & 0 \\
0 & 18 & 18 & -18 & 36 & 0 \\
0 & 0 & 0 & 0 & 0 & 0
\end{bmatrix}
\tag{e}
$$

$$
\mathbf{M}_2 = \frac{\rho A \ell}{420}
\begin{bmatrix}
0 & 0 & 0 & 0 & 0 & 0 \\
0 & 156 & 66 & 54 & -39 & 0 \\
0 & 66 & 36 & 39 & -27 & 0 \\
0 & 54 & 39 & 156 & -66 & 0 \\
0 & -39 & -27 & -66 & 36 & 0 \\
0 & 0 & 0 & 0 & 0 & 0
\end{bmatrix}
\tag{f}
$$

The force is uniform over the entire element. Application of Equation 8.142 through Equation 8.145 leads to

$$
q_1 = Q_3 = 3000 \sin(600t) \int_0^\ell \left(1 - 3\frac{\xi^2}{\ell^2} + 2\frac{\xi^3}{\ell^3}\right) d\xi = 4500 \sin(600t) \tag{g}
$$

$$q_2 = 3000 \sin(600t) \int_0^\ell \left(\frac{\xi}{\ell} - 2\frac{\xi^2}{\ell^2} + \frac{\xi^3}{\ell^3} \right) d\xi = 750 \sin(600t) \qquad \text{(h)}$$

$$q_3 = 3000 \sin(600t) \int_0^\ell f\left(3\frac{\xi^2}{\ell^2} + 2\frac{\xi^3}{\ell^3} \right) d\xi = 4500 \sin(600t) \qquad \text{(i)}$$

$$q_4 = 3000 \sin(600t) \int_0^\ell f\left(-\frac{\xi^2}{\ell^2} + \frac{\xi^3}{\ell^3} \right) d\xi = -750 \sin(600t) \qquad \text{(j)}$$

Element 3: $w_1 = W_5$, $w_2 = W_6$, $w_3 = 0$, $w_4 = W_8$. The element mass and stiffness matrices, written using the global coordinates are

$$\mathbf{K}_3 = \frac{EI}{\ell^3} \begin{bmatrix} 0 & 0 & 0 & 0 & 0 & 0 \\ 0 & 0 & 0 & 0 & 0 & 0 \\ 0 & 0 & 0 & 0 & 0 & 0 \\ 0 & 0 & 0 & 12 & 18 & 18 \\ 0 & 0 & 0 & 18 & 36 & 18 \\ 0 & 0 & 0 & 18 & 18 & 36 \end{bmatrix} \qquad \text{(k)}$$

$$\mathbf{M}_3 = \frac{\rho A \ell}{420} \begin{bmatrix} 0 & 0 & 0 & 0 & 0 & 0 \\ 0 & 0 & 0 & 0 & 0 & 0 \\ 0 & 0 & 0 & 0 & 0 & 0 \\ 0 & 0 & 0 & 156 & 66 & -39 \\ 0 & 0 & 0 & 66 & 36 & -27 \\ 0 & 0 & 0 & -39 & -27 & 36 \end{bmatrix} \qquad \text{(l)}$$

The differential equations written in terms of the global coordinates, are

$$\frac{\rho A \ell}{420} \begin{bmatrix} 36 & 39 & -27 & 0 & 0 & 0 \\ 39 & 312 & 0 & 54 & -39 & 0 \\ -27 & 0 & 72 & 39 & -27 & 0 \\ 0 & 54 & 39 & 312 & 0 & -39 \\ 0 & -39 & -27 & 0 & 72 & -27 \\ 0 & 0 & 0 & -39 & -27 & 36 \end{bmatrix} \begin{bmatrix} \ddot{W}_2 \\ \ddot{W}_3 \\ \ddot{W}_4 \\ \ddot{W}_5 \\ \ddot{W}_6 \\ \ddot{W}_8 \end{bmatrix}$$

$$+ \frac{EI}{\ell^3} \begin{bmatrix} 36 & -18 & 18 & 0 & 0 & 0 \\ -18 & 24 & 0 & -12 & 18 & 0 \\ 18 & 0 & 72 & -18 & 18 & 0 \\ 0 & -12 & -18 & 24 & 0 & 18 \\ 0 & 18 & 18 & 0 & 72 & 18 \\ 0 & 0 & 0 & 18 & 18 & 36 \end{bmatrix} \begin{bmatrix} W_1 \\ W_2 \\ W_3 \\ W_4 \\ W_5 \\ W_6 \end{bmatrix}$$

$$= \begin{bmatrix} 0 \\ 4500 \\ 750 \\ 4500 \\ -750 \\ 0 \end{bmatrix} \sin(600t) \tag{m}$$

The natural frequency approximations determined as the square roots of the eigenvalues of $\mathbf{M}^{-1}\mathbf{K}$, are

$$\omega_1 = 100.0\frac{r}{s} \quad \omega_2 = 404.5\frac{r}{s} \quad \omega_3 = 998.2\frac{r}{s} \quad \omega_4 = 1.86 \times 10^3 \frac{r}{s} \tag{n}$$

$$\omega_5 = 3.32 \times 10^3 \frac{r}{s} \quad \omega_6 = 4.57 \times 10^3 \frac{r}{s}$$

The modal matrix, the matrix whose columns are the normalized eigenvectors is

$$
P = \begin{bmatrix}
-0.1409 & -0.2874 & 0.5203 & 0.8697 & -1.441 & 1.367 \\
-0.3496 & -0.3577 & 0 & -0.3364 & 0.2668 & 0 \\
-0.0705 & -0.1937 & -0.5203 & -0.4349 & -0.7206 & 1.367 \\
-0.3496 & 0.3577 & 0 & 0.3369 & 0.2668 & 0 \\
-0.0705 & 0.1437 & 0.5203 & -0.4349 & 0.7206 & 1.367 \\
0.1409 & -0.2874 & -0.5203 & 0.8697 & 1.441 & 1.367
\end{bmatrix} \quad (o)
$$

The steady-state response is obtained as

$$
\begin{bmatrix}
W_2 \\
W_3 \\
W_4 \\
W_5 \\
W_6 \\
W_8
\end{bmatrix} = \begin{bmatrix}
0.0498 \\
0.0739 \\
-0.00123 \\
-0.00443 \\
0.00114 \\
0
\end{bmatrix} \sin(600t) \quad (p)
$$

9 Exercises

9.1 CHAPTER 1

1.1 Let $\theta(x,t)$ represent the angular deformation of the shaft of Figure 9.1. Derive an expression for the total kinetic energy of the shaft in terms of $\theta(x,t)$.

1.2 Let $\theta(x,t)$ represent the angular deformation of the shaft of Figure 9.1 derive an expression for the total potential energy of the shaft in terms of $\theta(x,t)$.

1.3 Let $\theta(x,t)$ represent the angular deformation of the shaft of Figure 9.2. Derive an expression for the kinetic energy of the shaft including the kinetic energy of the disk.

1.4 Two elastic bars are connected by linear springs in Figure 9.3. The displacement functions in the bars are $u_1(x,t)$ and $u_2(x,t)$ respectively. Derive an expression for the total potential energy in the system in terms of $u_1(x,t)$ and $u_2(x,t)$.

1.5 Derive Equation 1.46 through Equation 1.48 using Equation 1.7 and Equation 1.44.

1.6 Derive Equation 1.51 and Equation 1.52 using Equation 1.50.

1.7 The double pendulum of Figure 9.4 is composed of identical slender bars of length L and mass m. Define $\theta_1(t)$ and $\theta_2(t)$ as angular displacements of the bars from the system's equilibrium position. Determine the kinetic energy of the system at an arbitrary instant in terms of the angular velocities of the bars. Do not assume small angles.

1.8 Repeat Problem 1.7 if the pin support is given a prescribed displacement given by $\mathbf{r}(t) = x(t)\mathbf{i} + y(t)\mathbf{j}$.

1.9 A slender bar of length L and mass m is attached to a spring of stiffness k, as illustrated in Figure 9.5. Define $x(t)$ as the displacement of the end of the spring and $\theta(t)$ as the counter clockwise angular displacement of the bar, both measured from the system's equilibrium position. (a) Determine the kinetic energy of the system at an arbitrary instant and (b) Determine the potential energy of the system at an arbitrary instant. Do not assume small displacements.

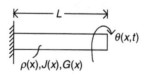

FIGURE 9.1 System for Problems 1.1 and 1.2.

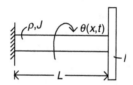

FIGURE 9.2 System for Problem 1.3.

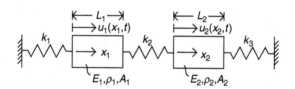

FIGURE 9.3 System for Problem 1.4.

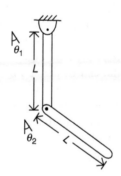

FIGURE 9.4 System for Problems 1.7, 1.8, 2.7 and 2.8.

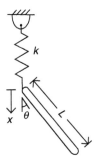

FIGURE 9.5 System for Problems 1.9, 2.9 and 2.10.

1.10 Bars AB and CD of Figure 9.6 are identical slender bars, each of mass m and length L. Define $\theta(t)$ as the clockwise angular displacement, measured from the system's equilibrium position. Assuming small θ determine the kinetic energy of the system at an arbitrary instant.

1.11 The system of Figure 9.6 and Problem 1.10 is designed such that bar AB makes an angle θ_s with the horizontal when the system is in equilibrium. Without assuming small θ determine the kinetic energy of the system and the potential energy of the system at an arbitrary instant.

1.12 Using x_1, x_2, and ϕ as the chosen generalized coordinates for the system of Figure 9.7, determine (a) the kinetic energy at an arbitrary instant and (b) the potential energy at an arbitrary instant.

1.13 Using y, θ, and ϕ as the chosen generalized coordinates for the system of Figure 9.7, determine (a) the kinetic energy at an arbitrary instant and (b) the potential energy at an arbitrary instant.

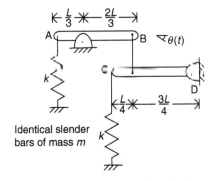

FIGURE 9.6 System for Problems 1.10, 1.11, 1.16 and 1.27.

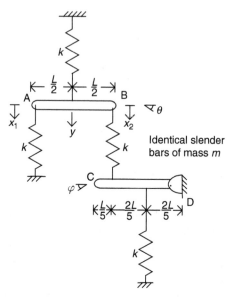

FIGURE 9.7 System for Problems 1.12, 1.13, 2.6, 4.3, 4.4, 4.6 and 5.3.

1.14 Derive the differential equation governing the motion of the system of Figure 9.8 using θ as the chosen generalized coordinate.

1.15 Derive the differential equation governing the motion of the system of Figure 9.9 using x as the chosen generalized coordinate.

1.16 Derive the differential equation governing the motion of the system of Figure 9.6 using θ as the chosen generalized coordinate.

1.17 Derive the differential equation of the system of Figure 9.10 using θ as the chosen generalized coordinate. Assume small θ.

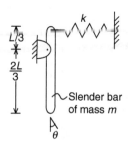

FIGURE 9.8 System for Problem 1.14.

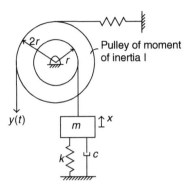

FIGURE 9.9 System for Problems 1.15 and 1.25.

1.18 Consider the system of Figure 9.10 and Problem 1.17 when $m = 2.4$ kg, $L = 2.2$ m, and $k = 4200$ N/m. For what value of c is the system critically damped?

1.19 A 10-kg block is attached to a spring of stiffness 1.8×10^5 N/m in parallel with a viscous damper of damping coefficient 150 N s/m. Determine (a) the natural frequency of the system (b) the damping ratio of the system (c) the period of the damped free response.

1.20 A 20-kg machine rests on a foundation of stiffness 1×10^5 N/m and damping coefficient 400 N s/m. During operation the machine is subject to an impulse of 200 Ns. (a) Determine the response of the machine immediately after application of the impulse (b) Determine the maximum displacement attained by the machine.

1.21 What is the value of the damping ratio of a one-degree-of-freedom system such that 15 percent of the system's energy is dissipated on

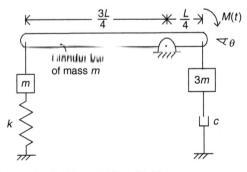

FIGURE 9.10 System for Problems 1.17 and 1.18.

each cycle of motion? What is the ratio of amplitudes on successive cycles of motion for a system with this damping ratio?

1.22 The logarithmic decrement for an underdamped single-degree-of-freedom system is defined as

$$\delta = \ln\left(\frac{x(t)}{x(t + T_\mathrm{d})}\right) \tag{a}$$

Show how knowledge of the logarithmic decrement can be used to determine the system's viscous damping ratio.

1.23 Overshoot occurs in a single-degree-of-freedom system with viscous damping when, after subject to an initial displacement, the system passes through equilibrium before approaching equilibrium. The percent overshoot is defined as the ratio of the maximum displacement after passing through equilibrium to the initial displacement. (a) Derive the relationship between viscous damping ratio and overshoot for an underdamped system. (b) Under what conditions will overshoot occur for a critically damped system? (c) Under what conditions does overshoot occur for an overdamped system?

1.24 A 20-kg block is given an initial displacement of 1.2 mm and released from rest. The block slides on a frictionless surface and is attached to a linear spring of stiffness 1.8×10^4 N/m in parallel with a viscous damper. Determine a mathematical representation of the resulting motion if the viscous damping coefficient is (a) 240 N s/m, (b) 1200 N s/m, (c) 1800 N s/m.

1.25 Determine the steady-state amplitude of block in the system of Figure 9.9 and Problem 1.15 if $y(t) = 5 \sin(100t)$ mm when $r = 10$ cm, $m = 5$ kg, $I = 0.1$ kg m^2, $k = 2.5 \times 10^5$ N/m, and $c = 500$ N s/m.

1.26 Determine the steady-state response of a 100-kg machine attached to a foundation of stiffness 4.5×10^5 N/m and damping coefficient 1000 N s/m when, during operation the machine is subject to a force of 1500 $\sin(35t)$ N.

1.27 Determine the steady-state response of the system of Figure 9.6 and Problem 1.11 when end C of bar CD is subject to a harmonic vertical force of amplitude 125 N and frequency 200 r/s when $L = 80$ cm, $m = 0.6$ kg and $k = 3.2 \times 10^4$ N/m.

1.28 The machine of Figure 1.65(a) has a mass of 10 kg and is attached to the support through a spring of stiffness 4.3×10^5 N/m and damping coefficient 300 N s/m. The base is subject to a harmonic

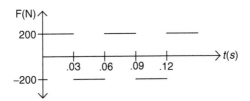

FIGURE 9.11 Periodic input for Problem 1.31.

displacement of amplitude 1.5 mm and frequency of 40 r/s. Determine (a) the steady-state amplitude of the absolute displacement of the machine, (b) the steady-state amplitude of the absolute acceleration of the machine, and (c) the steady-state amplitude of the displacement of the machine relative to the support.

1.29 A 35-kg block is attached to the end of a steel ($\rho = 7600$ kg/m^3, $E = 200 \times 10^9$ N/m^2) cable of diameter 5 cm and length 1.2 m. Approximate the natural frequency of longitudinal vibrations of the block using a one-degree-of-freedom model and (a) neglecting inertia effects of the cable and (b) approximating the inertia effects of the cable.

1.30 A thin disk of mass 1.8 kg and radius 20 cm is attached at the end of a steel ($\rho = 7600$ kg/m^3, $G = 80 \times 10^9$ N/m^2) shaft of radius 8 cm and length 1.5 m. Approximate the natural frequency of torsional oscillations of the disk using a one-degree-of-freedom model and (a) neglecting inertia effects of the disk and (b) approximating the inertial effects of the disk.

1.31 Determine the steady-state response of a single-degree-of-freedom system of mass 20 kg, natural frequency 200 r/s and damping ratio 0.4 when subject to the periodic force of Figure 9.11.

1.32 Determine the steady-state response of a single-degree-of-freedom system of mass 20 kg, natural frequency 200 r/s and damping ratio 0.4 when subject to the periodic force of Figure 9.12.

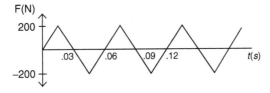

FIGURE 9.12 Periodic input for Problem 1.32.

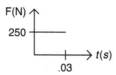

FIGURE 9.13 Pulse input for Problem 1.33.

1.33 Determine the response of a single-degree-of-freedom system of mass 20 kg, natural frequency 200 r/s and damping ratio 0.4 when subject to the step input of Figure 9.13.

1.34 Determine the response of an undamped single-degree-of-freedom system of mass 20 kg and natural frequency 200 r/s when subject to the transient excitation of Figure 9.14.

1.35 The differential equation governing the motion of a single-degree-of-freedom system with a quadratic stiffness is of the form

$$\ddot{x} + 2\zeta\omega_n\dot{x} + \omega_n^2(1 + \varepsilon x^2) = 0 \qquad \text{(a)}$$

Determine and classify the equilibrium points for this system.

1.36 The differential equation governing the motion of a single-degree-of-freedom system with a softening cubic nonlinearity of the form

$$\ddot{x} + 2\zeta\omega_n\dot{x} + \omega_n^2(1 - \varepsilon x^3) = 0 \qquad \text{(a)}$$

Determine and classify the equilibrium points for this system.

9.2 CHAPTER 2

2.1 Let (x_1, y_1) and (x_2, y_2) with $x_2 > x_1$ represent two points in the $x - y$ plane. Let $y(x)$ be an arc connecting the two points, as

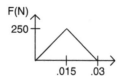

FIGURE 9.14 Pulse input Problem 1.34.

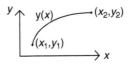

FIGURE 9.15 Planar curve is rotated about x-axis in Problems 2.1 and 2.2.

illustrated in Figure 9.15. Rotation of the curve about the x axis generates a volume of revolution of surface area A. Determine the curve $y(x)$ that generates the volume with the minimum surface area.

2.2 Repeat Problem 2.1, but determining the curve $y(x)$ such that the volume of revolution is minimized.

2.3 Let (x_1, y_1) and (x_2, y_2) with $x_2 > x_1$ represent two points in the $x - y$ plane. It is desired to hang a rope of length L between the two points (It is assumed, of course, that L is greater than the distance between the two points). The principle of minimum potential energy states that the rope will assume a shape that minimizes its potential energy. Use calculus of variations to determine the shape $y(x)$ assumed by the rope.

2.4 What is the equation of a hanging rope of length 10 m which is to be suspended from two points in a horizontal plane separated by 6 m? What is the total potential energy of the rope?

2.5 Fermat's principle of geometrical optics states that the time it takes for a light ray to travel between two fixed points is minimized over all possible paths. (a) Show that Fermat's principle implies that light travels in a straight line in a medium in which its velocity is constant. (b) Use Fermat's principle and the results of part (a) to prove Snell's law which is illustrated in Figure 9.16. A light ray is traveling in a medium with a velocity v_1 when it reaches the boundary between two

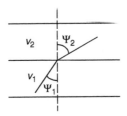

FIGURE 9.16 Refraction of light at interface between two media for Problem 2.5.

media. The light ray makes an angle of ψ_1 with the boundary. If the velocity of light in the second media is v_2 then Snell's law states that the angle of refraction ψ_2 is determined from

$$\sin \psi_2 = \frac{v_2}{v_1} \sin \psi_1 \tag{a}$$

(c) Use Fermat's Principle and calculus of variations to determine the path traveled by light between two points (x_1,y_1) and (x_2,y_2) in an inhomogeneous medium in which the velocity of light is

$$v = cy \tag{b}$$

where c is a constant.

(d) Use Fermat's Principle and calculus of variations to determine the path traveled by light between two points (x_1,y_1) and (x_2,y_2) in an inhomogeneous medium in which the velocity of light is

$$v = cy^{-(1/2)} \tag{c}$$

where c is a constant.

2.6 Use Lagrange's equations to derive the differential equations governing the motion of the system of Figure 9.7 using x_1, x_2, and ϕ as the chosen generalized coordinates. Assume small displacements.

2.7 Use Lagrange's equations to derive the differential equations governing the motion of the system of Figure 9.4 using $\theta_1(t)$ and $\theta_2(t)$ as chosen generalized coordinates. Do not make the small angle assumption.

2.8 Repeat problem 2.7 using the small angle assumption.

2.9 Use Lagrange's equations to derive the differential equations governing the motion of the system of Figure 9.5 using $x(t)$, the displacement of the end of the spring, and $\theta(t)$, the counter-clockwise angular displacement of the bar, both measured from the system's equilibrium position as the chosen generalized coordinates. Do not assume small θ.

2.10 Repeat Problem 2.9 assuming small θ.

2.11 Use Lagrange's equations to derive the equations of motion for the system of Figure 9.17 using x_1, x_2 and θ as the chosen generalized coordinates. Assume small displacements. The rigid body is a slender bar of mass m.

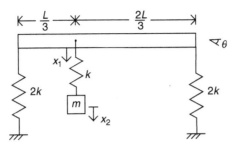

FIGURE 9.17 System for Problems 2.11 and 5.4.

2.12 Use Lagrange's equations to derive the equations of motion for the system of Figure 9.18 using θ and x as the chosen generalized coordinates. The links are slender rods of mass m and 4 m/s. The cart has a mass m.

2.13 Use Lagrange's equations to derive the equations of motion for the system of Figure 9.19 using x and θ as the chosen generalized coordinates. The cart is of mass m.

2.14 Use Lagrange's equations to derive the equations of motion for the system of Figure 9.20 using x_1, x_2 and x_3 as the chosen generalized coordinates.

2.15 Use Lagrange's equations to derive the equations of motion for the system of Figure 9.21 using x_1, x_2 and θ as the chosen generalized coordinates. The bar is of mass m and moment of inertia I. The block is of mass $2m$.

2.16 The arm of the system of Figure 9.22 rotates at constant speed ω. A block of mass m is attached to the arm through a spring. A pendulum of mass m and length ℓ is pinned to the mass, but is constrained to rotate only in the plane of the arm as it rotates. Derive the equations of motion using x, the displacement of the block form the system's equilibrium position and θ the angular displacement of the pendulum in the plane as the chosen generalized coordinates.

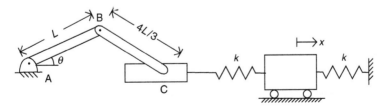

FIGURE 9.18 System for Problem 2.12.

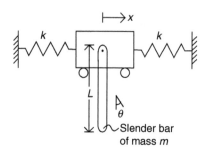

FIGURE 9.19 System for Problems 2.13 and 5.8.

2.17 Derive the differential equations for the system of Figure 9.23 using x and y the coordinates of the end of the bar attached to the spring in the coordinate system rotating with the turntable and θ, the angular displacement of the bar relative to the turntable, as the generalized coordinates. The turntable rotates at a constant speed.

2.18 Use Hamilton's principle to derive the equation of motion governing the torsional oscillations of a circular shaft of length L, made from a material of shear modulus G and mass density ρ, and a radius $r(x)$ where x is measured along the length of the shaft, $0 \le x \le L$. Specify the boundary conditions for a shaft fixed at $x = 0$ and free at $x = L$.

2.19 Determine the appropriate boundary conditions for shaft of Problem 2.18 if it is fixed at $x = 0$ and has a rigid disk of mass moment of inertia I attached at $x = L$.

2.20 Determine the appropriate boundary conditions for the shaft of Problem 2.18 if it is attached to a linear torsional spring of stiffness k_{t0} at $x = 0$ and a linear torsional spring of stiffness k_{t1} at $x = L$.

2.21 Determine the appropriate boundary conditions for the shaft of Problem 2.18 if it is fixed at $x = 0$ and attached to a torsional viscous damper of damping coefficient c_{t1} at $x = 1$.

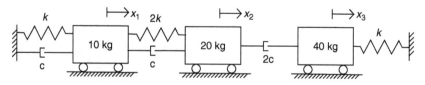

FIGURE 9.20 System for Problems 2.14, 4.1 and 4.5.

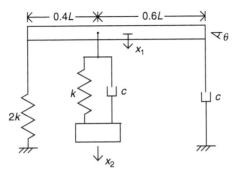

FIGURE 9.21 System for Problems 2.15 and 4.2.

2.22 Determine the differential equation governing the motion of the shaft of Problem 2.18 if it is subject to a distributed torsional load per length of $T(x)$ along the length of the shaft.

2.23 Specify the boundary value problem for the transverse vibrations $w(x,t)$ of a string of length L and mass per unit length $\mu(x)$ which is suspended from a fixed support at $x = 0$ and is free at $x = L$. In its equilibrium position the string is hanging from the support.

2.24 Specify the boundary value problem for the longitudinal vibrations $u(x,t)$ of a uniform bar of length L which is made from a material of elastic modulus E and mass density ρ which is attached to a linear spring of stiffness k at $x = 0$ and has a block of mass m rigidly attached at $x = L$.

2.25 Derive the equation of motion governing an Euler–Bernoulli beam with an axial load $P(x)$ which varies over the span of the beam.

2.26 Derive the equation of motion for the transverse displacement of a pinned-pinned uniform Euler–Bernoulli beam rotating with

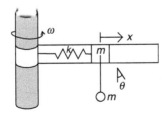

FIGURE 9.22 System for Problem 2.16.

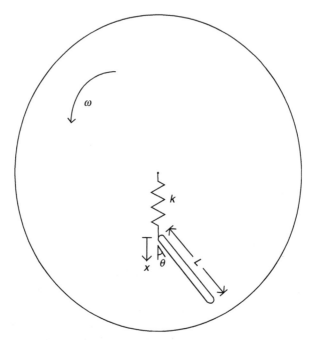

FIGURE 9.23 System for Problems 2.17.

constant angular velocity ω about an axis through the pin supports.

2.27 Specify the boundary value problem for the transverse displacement $w(x,t)$ of a uniform Euler–Bernoulli beam on an elastic foundation of stiffness per length k which is fixed at $x = 0$ and with a rigid particle of mass m attached at $x = L$.

2.28 Specify the boundary value problem for the transverse displacement $w(x,t)$ of a uniform Euler–Bernoulli beam subject to a concentrated load of magnitude $F(t)$ applied at $x = L/3$. The beam is fixed at $x = 0$ and attached to a torsional spring of stiffness k_t at $x = L$.

2.29 Specify the boundary value problem for the transverse displacement $w(x,t)$ of a uniform Euler–Bernoulli beam which is fixed at $x = L$ and has a rigid disk of mass m and mass moment of inertia I attached at $x = 0$.

2.30 Derive the boundary value problem for the transverse displacement of a uniform vertical Euler–Bernoulli beam that is fixed at $x = 0$ but with a tensile load P applied at $x = L$.

2.31 Derive the boundary value problem for the transverse displacement $w(x,t)$ of a uniform Euler–Bernoulli beam which is in a viscous medium which whose effect is to provide a transverse force of the form $c(\partial w/\partial t)$ to act across the span of the beam. The beam is fixed at $x = 0$ and free at $x = L$.

2.32 Derive the boundary value problem for the transverse displacement $w(x,t)$ and the angular rotation due to bending $\psi(x,t)$ of a uniform Timoshenko beam which is fixed at $x = 0$ but has a particle of mass m attached at $x = L$.

2.33 Derive the boundary value problem for the transverse displacement $w(x,t)$ and the angular rotation due to bending $\psi(x,t)$ of a uniform Timoshenko beam which is fixed at $x = L$ but has a linear spring of stiffness k attached at $x = L$.

2.34 Specify the boundary value problem for the transverse displacement $w(x,t)$ and the angular rotation due to bending $\psi(x,t)$ of a uniform Timoshenko beam which is pinned at $x = 0$ and fixed at $x = L$.

2.35 A uniform Timoshenko beam has a discrete mass-spring system attached at $x = L$. Derive the boundary value problem for the transverse displacement $w(x,t)$ and the angular rotation due to bending $\psi(x,t)$ of the Timoshenko beam and the vertical displacement of the mass, $y(t)$.

2.36 Derive the boundary value problem for transverse displacement $w(x,t)$ and the angular rotation due to bending $\psi(x,t)$ of a uniform Timoshenko beam which is subject to a transverse load per unit length $F(x,t)$. The beam is fixed at $x = 0$ and free at $x = L$.

2.37 Derive the boundary value problem for system of Figure 9.24 assuming the beam may be modeled as an Euler–Bernoulli beam.

2.38 Derive the boundary value problem for system of Figure 9.24 assuming the beam may be modeled as a Timoshenko beam.

2.39 Derive the boundary value problem for the system of Figure 9.25 assuming the beam may be modeled as an Euler–Bernoulli beam.

2.40 Specify the boundary value problem for the transverse displacement $w(r,\theta,t)$ for the circular membrane of Figure 9.26.

2.41 Specify the boundary value problem for the transverse displacement $w(x,y,t)$ of a square membrane which is free along two parallel sides and clamped along two parallel sides.

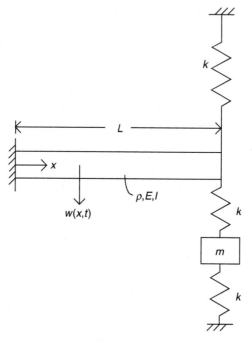

FIGURE 9.24 System for Problems 2.27, 2.38, 7.23 and 7.24.

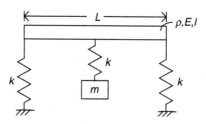

FIGURE 9.25 System for Problems 2.37 and 7.25.

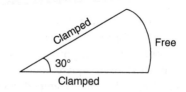

FIGURE 9.26 System for Problems 2.40 and 5.27.

9.3 CHAPTER 3

3.1 Let $P^n[0,1]$ be the vector space consisting of all polynomials of degree n or less defined over the interval $[0,1]$. An element of $P^n[0,1]$ is of the form

$$f(x) = c_0 + c_1 x + c_2 x^2 + \cdots + c_{n-1} x^{n-1} + c_n x^n \qquad \text{(a)}$$

Let S be the set of all polynomials in $P^n[0,1]$ whose coefficients sum to zero, that is

$$c_0 + c_1 + \cdots + c_{n-1} + c_n = 0 \qquad \text{(b)}$$

Is S a subspace of $P^n[0,1]$? Is so, prove it. If not show why not.

3.2 Which of the following is a subspace of R^3? An arbitrary element $\mathbf{u}$ of R^3 is of the form $\mathbf{u} = [u_1 \; u_2 \; u_3]^T$.

a. S is defined as the set of all elements of R^3 such that $u_2 = 0$.

b. S is defined as the set of all elements of R^3 such that $u_1 = 1$.

c. S is defined as the set of all elements of R^3 such that $u_1 + u_3 = 0$.

3.3 Which of the following is a subspace of $C[0,1]$?

a. $P^n[0,1]$, the set of all polynomials of order n or less.

b. S, the set of all functions in $C[0,1]$ such that if $f(x)$ is in S then $f(0.5) = 1$.

c. S, the set of all functions in $C[0,1]$ such that if $f(x)$ is in S then $f(0) = 0$ and $df/dx(1) + 2f(1) = 0$.

d. S, the set of all functions in $C[0,1]$ such that if $f(x)$ is in S then $f(x)$ is of the form of $f(x) = c_1 \sin(\pi x) + c_2 \sin(2\pi x) + c_3 \sin(3\pi x)$.

3.4 The mode shapes for the longitudinal vibrations of a uniform bar fixed at $x = 0$ and free at $x = 1$ are in the vector space D, defined as all elements of $C^2[0,1]$ such that if $f(x)$ is in D then $f(0) = 0$ and $(df/dx)(1) = 0$. $P^5[0,1]$ is the space of all polynomials of degree 5 or less. Determine a basis for $D \cap P^5[0,1]$.

3.5 The mode shapes for the torsional vibrations of a uniform shaft fixed at $x = 0$ and attached to a torsional spring at $x = 0$ are in the vector space D, defined as all elements of $C^2[0,1]$ such that if $f(x)$ is in D then $f(0) = 0$ and $df/dx(1) + 2f(1) = 0$. $P^5[0,1]$ is the space of all polynomials of degree 5 or less. Determine a basis for $D \cap P^5[0,1]$.

3.6 The mode shapes for the transverse vibrations of a uniform Euler–Bernoulli beam fixed at $x = 0$ and free at $x = 1$ are elements of the vector space D, defined as all elements of $C^4[0,1]$ such that if $f(x)$ is in D then $f(0) = 0$, $f'(0) = 0$, $(df/dx)(0) = 0$, $(d^2f/dx^2)(1) = 0$ and $d^3f/dx^3(1) = 0$. $P^6[0,1]$ is the space of all polynomials of degree 6 or less. Determine a basis for $D \cap P^6[0,1]$.

3.7 The mode shapes for the transverse vibrations of a uniform Euler–Bernoulli beam fixed at $x = 0$ and free at $x = 1$ are elements of the vector space D, defined as all elements of $C^4[0,1]$ such that if $f(x)$ is in D then $f(0) = 0$, $f'(0) = 0$, $(df/dx)(0) = 0$, $(d^2f/dx^2)(1) = 0$ and $(d^3f/dx^3)(1) = 0$. $F^k[0,1]$ is defined as all elements of $C^4[0,1]$ of the form

$$f(x) = a_0 + a_1\cos(\pi x) + b_1\sin(\pi x) + a_2\cos(2\pi x) + b_2\sin(2\pi x)$$

$$+ \cdots + a_k\cos(k\pi x) + b_k\sin(k\pi x)$$

Determine a basis for $C^4[0,1] \cap F^3[0,1]$.

3.8 The mode shapes for the transverse vibrations and angular rotation due to bending of a uniform Timoshenko beam fixed at $x = 0$ and pinned at $x = 1$ beam belong to the vector space D which consists of all elements $\mathbf{u}$ of $C^2[0,1] X C^2[0,1]$ which are of the form $\mathbf{u} = \begin{bmatrix} f_1(x) \\ f_2(x) \end{bmatrix}$ such that $f_1(0) = 0$, $f_2(0) = 0$, $f_1(1) = 0$ and $df_2/dx(1) = 0$. Let $V = P^2[0,1] X P^2[0,1]$ be the vector space whose elements are of the form $\mathbf{v} = \begin{bmatrix} p_1(x) \\ p_2(x) \end{bmatrix}$ where $p_1(x)$ and $p_2(x)$ are polynomials of degree 2 or less. Determine a basis for $D \cap V$.

3.9 Determine an orthonormal basis for R^3 with respect to an inner product defined by $(\mathbf{u}, \mathbf{v}) = u_1 v_1 + 2 u_2 v_2 + u_3 v_3$.

3.10 An element of $P^4[0,1]$ is of the form $f(x) = c_0 + c_1 x + c_2 x^2 + c_3 x^3 + c_4 x^4$. Determine an orthonormal basis for $P^4[0,1]$ with respect to the inner product defined by $(f,g) = c_0 d_0 + c_1 d_1 + 2 c_2 d_2 + c_3 d_3 + c_4 d_4$.

3.11 Determine an orthonormal basis for $P^4[0,1]$ with respect to the inner product defined by $(f,g) = \int\limits_0^1 f(x)g(x)(x+1)dx$.

3.12 Orthonormalize the basis determined as solution to Problem 3.4 with respect to the standard inner product on $C^2[0,1]$.

3.13 Orthonormalize the basis determined as a solution to Problem 3.5 with respect to the inner product defined by

$$(f,g) = \int\limits_0^1 f(x)g(x)(1 - 0.5x)dx.$$

3.14 Orthonoramlize the basis determined as a solution to Problem 3.7 with respect to the standard inner product for $C^4[0,1]$.

3.15 Under what conditions is a 3×3 matrix self-adjoint with respect to the inner product defined on R^3 as $(\mathbf{u},\mathbf{v}) = 2u_1v_1 + u_2v_2 + u_3v_3$.

3.16 Let $\mathbf{L}$ be an operator defined by $Lf = -d^2f/dx^2$ such that the domain of L is D, the set of elements of $P^4[0,1]$ which satisfy $f(0) = 0$ and $f(1) = 0$. Is $\mathbf{L}$ self-adjoint with respect to the inner product defined in Problem 3.11?

3.17 Show that the operator $Lf = -(1/x)d^2f/dx^2$ is positive definite and self adjoint with respect to the inner product $(f,g) = \int_1^2 f(x)g(x)x\, dx$ if the domain of $\mathbf{L}$ is the set of functions in $C^2[1,2]$ which satisfy $f(1) = 0$ and $f(2) = 0$.

3.18 Show that the matrix

$$\mathbf{K} = \begin{bmatrix} 2 & -1 \\ -1 & 3 \end{bmatrix}$$

is self adjoint and positive definite with respect to the standard inner product for R^2.

3.19 Consider the matrices

$$\mathbf{M} = \begin{bmatrix} 2 & 0 \\ 0 & 1 \end{bmatrix} \quad \text{and} \quad \mathbf{K} = \begin{bmatrix} 2 & -1 \\ -1 & 2 \end{bmatrix}$$

Since the matrices are symmetric they are self adjoint with respect to the standard inner product for R^2. Is the matrix $\mathbf{M}^{-1}\mathbf{K}$ self-adjoint with respect to the standard inner product for R^2? If not determine a valid inner product for R^2 such that $\mathbf{M}^{-1}\mathbf{K}$ is self-adjoint with respect to this inner product.

3.20 Is the matrix

$$K = \begin{bmatrix} 1 & -1 & 0 \\ -1 & 2 & -1 \\ 0 & -1 & 1 \end{bmatrix}$$

positive definite with respect to the standard inner product for R^3?

3.21 Consider the operator defined by $Lf = -d^2f/dx^2$ on the subspace D of $C^2[0,1]$ such that if $f(x)$ is in D then $(df/dx)(0) = 0$ and $(df/dx)(1) = 0$. (a) Is the operator self adjoint with respect to the standard inner product on D? (b) Is the operator positive definite with respect to the standard inner product on D?

3.22 Show that the operator $Lw = \nabla^4 w$, the biharmonic operator, is positive definite and self adjoint with respect to the inner product $(f(\mathbf{r}), g(\mathbf{r})) = \int_V f(r)g(r)dV$ where the integral is taken over a region in space defined by the domain of L.

9.4 CHAPTER 4

4.1 Use stiffness influence coefficients to derive the stiffness matrix for the system of Figure 9.20.

4.2 Use stiffness influence coefficients to derive the stiffness matrix for the system of Figure 9.21 using x_1, x_2 and θ as generalized coordinates.

4.3 Use stiffness influence coefficients to derive the stiffness matrix for the system of Figure 9.7 using y, θ and ϕ as generalized coordinates.

4.4 Use stiffness influence coefficients to derive the stiffness matrix for the system of Figure 9.7 using x_1, x_2 and ϕ as generalized coordinates.

4.5 Use flexibility influence coefficients to derive the flexibility matrix for the system of Figure 9.20.

4.6 Use flexibility influence coefficients to derive the flexibility matrix for the system of Figure 9.7 using y, θ and ϕ as generalized coordinates.

4.7 Two machines are attached to a beam as illustrated in Figure 9.27. Ignoring the inertia of the beam derive the differential equations for a two-degree-of-freedom model of the vibrations of the system using x_1 and x_2 as generalized coordinates.

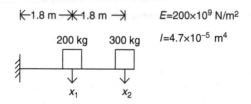

FIGURE 9.27 System for Problems 4.7, 5.5, 5.7, 7.7, 7.8 and 7.11.

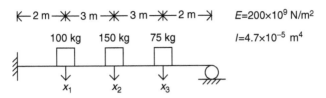

FIGURE 9.28 System for Problems 4.8, 5.6, 6.1 and 7.3.

4.8 Three machines are attached to a fixed-pinned beam as illustrated in Figure 9.28. Ignoring the inertia of the beam derive the differential equations for a two-degree-of-freedom model of the vibrations of the system using x_1, x_2 and x_3 as generalized coordinates.

4.9 Rayleigh's dissipation function for a n-degree-of-freedom linear system can be written as

$$\mathscr{F} = \frac{1}{2} \sum_{i=1}^{n} \sum_{j=1}^{n} c_{ij} \dot{x}_i \dot{x}_j$$

a. Explain why $c_{ij} = c_{ji}$

b. Let $\mathbf{C}$ be the damping matrix generated from Rayleigh's dissipation function. Can an energy inner product be defined as $(\mathbf{u},\mathbf{v})_C = (\mathbf{Cu},\mathbf{v})$?

4.10 The boundary value problem for the torsional oscillations of a shaft fixed at $x = 0$, but with a disk of moment of inertia I attached to a torsional spring of damping coefficient c_t at $x = L$ is

$$\frac{\partial}{\partial x}\left(GJ\frac{\partial\theta}{\partial x}\right) = \rho J \frac{\partial^2\theta}{\partial t^2} \tag{a}$$

$$\theta(0,t) = 0 \tag{b}$$

$$G(L)J(L)\frac{\partial\theta}{\partial x}(L,t) + c_t\frac{\partial\theta}{\partial t} = I\frac{\partial^2\theta}{\partial t^2} \tag{c}$$

Introduce appropriate nondimensional variables to nondimensionalize the boundary value problem defined by Equation a
through Equation c such that the governing equation written in
nondimensional form is

$$\frac{\partial}{\partial x}\left(\alpha(x)\frac{\partial\theta}{\partial x}\right) = \beta(x)\frac{\partial^2\theta}{\partial t^2} \qquad (d)$$

4.11 The governing partial differential equation for an Euler–
Bernoulli beam on an elastic foundation subject to an axial
load P is

$$\frac{\partial^2}{\partial x^2}\left(EI\frac{\partial^2 w}{\partial x^2}\right) - P\frac{\partial^2 w}{\partial x^2} + kw + \rho A\frac{\partial^2 w}{\partial t^2} = 0 \qquad (a)$$

a. Non-dimensionalize Equation a using the non-dimensional
variables of Example 4.4 (b) The non-dimensionalization
leads to identification of a non-dimensional parameter multiplying the w term. Idenitfy this parameter and specify its
physical meaning.

4.12 Determine whether the non-dimensional operator determined in
Problem 7.11 is positive definite and self-adjoint with respect to
the potential energy inner product.

4.13 The non-dimensional partial differential equation governing the
longitudinal oscillations of a variable area bar is

$$\frac{\partial}{\partial x}\left((1-0.1x)\frac{\partial u}{\partial x}\right) = (1-0.1x)\frac{\partial^2 u}{\partial t^2} \qquad (a)$$

a. Specify the appropriate kinetic energy inner product for
this example.

b. Specify the appropriate potential energy inner product for this
example if the bar is fixed at $x = 0$ and free at $x = 1$.

4.14 Formulate the non-dimensional boundary conditions if the end
corresponding to $x = 1$ of a Timoshenko beam is attached to a
spring of stiffness k. Show that the stiffness operator is positive
definite and self adjoint with respect to the inner product defined
in Equation 4.95.

4.15 Consider two uniform Euler–Bernoulli beams connected by an elastic layer. (a) Identify the inertia operator for the system (b) Identify the appropriate kinetic energy inner product for the system (c) Identify the stiffness operator for the system (d) Identify the appropriate potential energy inner product.

4.16 Consider two uniform Timoshenko beams connected by an elastic layer. (a) Identify the inertia operator for the system (b) Identify the appropriate kinetic energy inner product for the system (c) Identify the stiffness operator for the system (d) Identify the appropriate potential energy inner product.

9.5 CHAPTER 5

5.1 Determine the natural frequencies and normalized mode shape vectors for the system of Figure 9.29.

5.2 Determine the natural frequencies and normalized mode shape vectors for the system of Figure 9.30.

5.3 Determine the natural frequencies and normalized mode shape vectors for the system of Figure 9.7 using $m = 5$ kg, $L = 3$ m, and $k = 1000$ N/m.

5.4 Determine the natural frequencies and mode shape vectors for the system of Figure 9.17 using $m = 4$ kg, $L = 3$ m, and $k = 2000$ N/m.

5.5 Determine the natural frequencies and normalized mode shape vectors for the system of Figure 9.27.

5.6 Use matrix iteration to determine the natural frequencies and mode shape vectors for the system of Figure 9.28.

5.7 Use Equation 5.110 to verify Equation 5.111.

5.8 Determine the free response of the system of Figure 9.19 when the cart is given a displacement x_0 from equilibrium while the pendulum is held in the vertical position and then the system released.

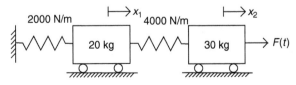

FIGURE 9.29 System for Problems 5.1, 7.1, 7.9 and 7.14.

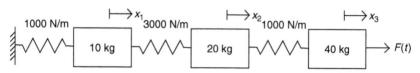

FIGURE 9.30 System for Problems 5.2, 5.10, 5.11 and 7.8.

5.9 Determine the free response of the system of Figure 9.27 when the machine at the end of the beam is subject to an impulse which imparts an initial velocity of 2 m/s to the machine.

5.10 Determine the free response of the system of Figure 9.30 under the initial conditions $x_1(0) = 0$, $x_2(0) = 0$, $x_3(0) = 0.01$ m and $\dot{x}_1(0) = 0, \dot{x}_2(0) = 0, \dot{x}_3(0) = 0$.

5.11 Use the perturbation method of Section 5.4 to predict the changes in the natural frequencies in the system of Figure 9.30 when the 20 kg machine is replaced by a 23 kg machine while the rest of the system remains as is.

5.12 Determine the free response of the system of Figure 9.31 when the 20 kg block is given a displacement of 0.05 m from its equilibrium position while the other blocks are held in their original position and the system then released.

5.13 Determine the natural frequencies for the longitudinal vibrations of a uniform fixed-free bar of length 2 m which is made from a material of elastic modulus 200×10^9 N/m and mass density 7600 kg/m^3.

5.14 Determine the natural frequencies of torsional oscillations of a fixed-free circular shaft of radius 5 cm and length 1.2 m which is made from a material of shear elastic modulus 80×10^9 N/m and mass density 7600 kg/m^3.

5.15 Determine the four lowest natural frequencies and their corresponding normalized mode shapes of longitudinal motion of a uniform circular bar of radius 0.8 cm and length 2.5 which is made from a material of elastic modulus 200×10^9 N/m and mass density 7600 kg/m^3. One end of the bar is fixed while the other is attached to a linear spring of stiffness 2×10^6 N/m.

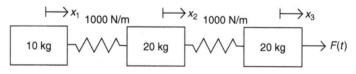

FIGURE 9.31 System for Problems 5.12 and 7.2.

5.16 Determine the four lowest natural frequencies and their corresponding normalized mode shapes of longitudinal motion of a uniform circular bar of radius 0.8 cm and length 2.5 m which is made from a material of elastic modulus 200×10^9 N/m and mass density 7600 kg/m^3. One end of the bar is fixed while the other end is attached to a 5 kg block and a spring of stiffness 2×10^6 N/m.

5.17 Repeat Example 5.14(a) if the radius of the bar varies according to $r(x) = (1 - \mu x)^2$ with $\mu = 0.1$.

5.18 Repeat Example 5.14(b) if the radius of the bar varies according to $r(x) = (1 - \mu x)^2$ with $\mu = 0.1$.

5.19 Determine the characteristic equation for the natural frequencies for the transverse vibrations $w(x,t)$ of a string of length L and mass per unit length $\mu(x)$ which is suspended from a fixed support at $x = 0$ and is free at $x = L$. In its equilibrium position the string is hanging from the support. Determine the four lowest natural frequencies for a uniform string of length 3 m with a mass per unit length of 0.18 kg/m.

5.20 Determine the four lowest natural frequencies of a fixed-pinned uniform Euler–Bernoulli beam of length 3.5 m, cross-section area 3.5×10^{-4} m^2, cross sectional area moment of inertia 7.15×10^{-7} m^4 which is made from a material of elastic modulus 200×10^9 N/m and mass density 7600 kg/m^3.

5.21 Determine the four lowest non-dimensional natural frequencies and their corresponding mode shapes for a uniform Euler–Bernoulli beam pinned at $x = 0$ with an attached spring at $x = 1$ such that $kL^3/(EI) = 2$.

5.22 Determine the four lowest non-dimensional natural frequencies and their corresponding mode shapes for a uniform Euler–Bernoulli beam fixed at $x = 0$ with a particle of mass $m = 0.8\rho AL$ attached at $x = 1$.

5.23 Determine the first five lower and upper natural frequencies of a uniform Timoshenko beam fixed at $x = 0$ and pinned at $x = L$.

5.24 Determine the first five sets of non-dimensional natural frequencies for two identical fixed-free Euler–Bernoulli beams connected by a layer of springs such that $\kappa = 3$.

5.25 Determine the first five sets of non-dimensional natural frequencies for two identical fixed-free Euler–Bernoulli beams connected by springs such that $\kappa_0 = 1$, $\kappa_1 = 2$, and $\kappa_2 = 0$.

5.26 Determine the five lowest natural frequencies and mode shapes for the system of Figure 9.32.

5.27 Determine the natural frequencies and mode shapes for the circular sector membrane of Figure 9.26.

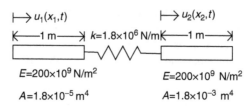

$$\longmapsto u_1(x_1,t) \qquad\qquad \longmapsto u_2(x_2,t)$$

$$\longmapsto\!\!-1\text{ m}\longrightarrow\!\!| \quad k=1.8\times10^6 \text{ N/m} \quad |\!\!\longleftarrow\!\!1\text{ m}\longrightarrow\!\!|$$

E=200×10⁹ N/m²

A=1.8×10⁻⁵ m⁴

E=200×10⁹ N/m²

A=1.8×10⁻³ m⁴

FIGURE 9.32 System for Problem 5.26.

9.6 CHAPTER 6

6.1 It is measured that the modal damping ratio for the lowest mode of the system of Figure 9.28 is 0.35. Assuming the damping is proportional with proportionality to only the stiffness matrix, what are the modal damping ratios for the next two modes.

6.2 The system of Figure 9.33 is proportionally damped. Determine its modal damping ratios.

6.3 The 2-kg block of the system of Figure 9.33 is displaced 1 mm from equilibrium while the 3 kg block is held in its equilibrium position. The system is then released. Determine the resulting free response of the system.

6.4 For what value of c is the system of Figure 9.34 proportionally damped with a lowest modal damping ratio of 0.5? The bar is non-uniform with its mass center at G.

6.5 The 2-kg block of the system of Figure 9.35 is displaced 1 mm from equilibrium while the 3 kg block is held in its equilibrium position. The system is then released. Determine the resulting free response of the system.

6.6 The gyroscopic matrix $\mathbf{G}$ for a n degree-of-freedom system is skew symmetric. (a) Is the matrix self-adjoint with respect to the standard inner product for R^n? (b) Is the gyroscopic matrix positive definite with respect to the standard inner product for R^n? (c) Is it possible to

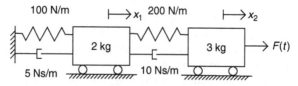

FIGURE 9.33 System for Problems 6.2, 6.3, 7.4, and 7.12.

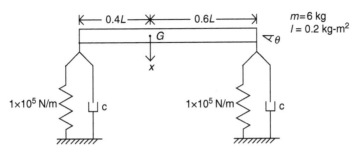

FIGURE 9.34 System for Problem 6.4.

define an inner product for which **G** is self adjoint with respect to that inner product?

6.7 Consider the longitudinal vibrations of the system of Figure 9.36. The uniform elastic bar is fixed at $x = 0$ and connected to a viscous damper at $x = L$. (a) Formulate the boundary value problem for system. (b) Non-dimensionalize the boundary value problem using the non-dimensional variables of Equation 4.36 through Equation 4.38. (c) The system is initially subject to an axial load such that its displacement is $u_0(x) = y_0(x/L)$. The load is removed at $t = 0$, initiating longitudinal motion. Use the method of separation of variables to determine the resulting motion. (d) For what values of c will the resulting motion be umderdamped?

6.8 Two concentric shafts are connected by an elastic layer which is modeled as a layer of torsional springs of stiffness k_t per unit length. Both shafts are fixed at $x = 0$ and free at $x = L$. The shafts are made of the same material of shear modulus G and mass density ρ. The inner shaft is a solid circular shaft of radius r_1. The outer shaft is an annular shaft of inner radius r_1 and outer radius r_2. Define $\theta_1(x,t)$ as the angular displacement of the inner shaft and $\theta_2(x,t)$ as the angular displacement of the outer shaft. The outer shaft rotates in a viscous medium such that as it rotates it is subject to a resisting moment per length of $c(\partial\theta_2/\partial t)$ over its entire length. (a) Formulate the boundary

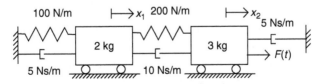

FIGURE 9.35 System for Problems 6.5, 7.5 and 7.13.

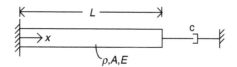

FIGURE 9.36 System for Problem 6.7.

value problem for the free torsional oscillations of the shafts. (b) Non-dimensionalize the boundary value problem using appropriate nondimensional variables (c) Consider the free response when the shafts have initial conditions of $\theta_1(x,0) = \theta_0 x$ and $\theta_2(x,0) = 0$. Use a separation of variables method to determine the free response of each shaft.

6.9 Determine the free response for the transverse oscillations of an uniform Euler–Bernoulli beam fixed at $x = 0$ and attached to a linear spring of stiffness k in parallel with a viscous damper of damping coefficient c at $x = L$.

9.7 CHAPTER 7

7.1 Determine the steady-state response of the system of Figure 9.29 when $F(t) = 20 \sin(10t)$ N.

7.2 Determine the steady-state response of the system of Figure 9.31 when $F(t) = 20 \sin(10t)$ N.

7.3 Determine the steady-state response of the system of Figure 9.28 when, during its operation, the 75-kg machine is subject to a harmonic force of $400 \sin(750t)$ N.

7.4 Determine the steady-state response of the system of Figure 9.33 when $F(t) = 2 \sin(10t)$ N.

7.5 Determine the steady-state response of the system of Figure 9.34 when $F(t) = 20 \sin(10t)$ N.

7.6 During operation a 75-kg machine is subject to a harmonic force of magnitude 500 N at a frequency of 400 r/s. The machine rests on an elastic foundation of stiffness 1.18×10^7 N/m. (a) What is the steady-state amplitude of the machine? (b) What is the required stiffness of an undamped absorber of mass 5 kg such that the steady-state response of the machine is eliminated when the absorber is added? (c) What is the steady-state amplitude of the absorber designed in part (b)? (d) Suppose the frequency of the input varies slightly from 400 rad/s, but the amplitude of the input is constant. What is the range of frequencies near 400 rad/s, such that when the absorber designed in part (b) is added to the machine its steady-state amplitude is less than 1 cm?

7.7 During operation the 300-kg machine of the system of Figure 9.27 is subject to a harmonic force of magnitude 1000 N and frequency 50 rad/s. (a) Design a vibration absorber of mass 50 kg such that when attached to the beam the steady-state vibrations of the 300-kg machine are eliminated? (b) What is the steady-state amplitude of the 200-kg machine when the absorber is attached?

7.8 The 200-kg machine of the system of Figure 9.27 has an unbalanced rotating component which leads to the development of a harmonic force $F(t) = 0.05\omega^2 \sin(\omega t)$ N, where ω is the angular speed of the rotating component in rad/s. The machine operates most frequently at 1000 rad/s. (a) Design a vibration absorber of mass 10 kg such that when attached to the beam, the steady-state amplitude of the 200-kg machine is zero when the machine operates at 1000 rad/s. (b) Develop the frequency response of the system when the absorber is added for both the 200-kg and 300-kg machines.

7.9 Use modal analysis to determine the response of the system of Figure 9.29 when $F(t) = 10(1 - 0.05t)[u(t) - u(t - 20)]$ N.

7.10 Use modal analysis to determine the response of the system of Figure 9.30 when $F(t) = 10(1 - 0.05t)[u(t) - u(t - 20)]$ N.

7.11 Use modal analysis to determine the response of the system of Figure 9.27 when the 300-kg machine is subject to a force $F(t) = 200(1 - 10t)[u(t) - u(t - 0.1)]$ N.

7.12 Use modal analysis to determine the response of the system of Figure 9.33 when $F(t) = 10(1 - 0.05t)[u(t) - u(t - 20)]$ N.

7.13 Use modal analysis to determine the response of the system of Figure 9.35 when $F(t) = 10(1 - 0.05t)[u(t) - u(t - 20)]$ N.

7.14 Use the Laplace transform method to determine the response of the system of Figure 9.29 when $F(t) = 10(1 - 0.05t)[u(t) - u(t - 20)]$ N.

7.15 Determine the steady-state response for the longitudinal motion of a uniform bar fixed at $x = 0$ but subject to a force $F(t) = F_0 \sin(\omega t)$ at $x = 1$. Assume all quantities are non-dimensional.

7.16 Determine the steady state response for the torsional oscillations of a uniform circular shaft fixed at $x = 0$ and with the end at $x = 1$ connected to a torsional spring such that $k_t L/JG = 2$ and subject to a moment $M(t) = 0.5 \sin(2t)$ All quantities are non-dimensional.

7.17 Determine the steady-state response for the transverse vibrations of a uniform fixed-pinned Euler–Bernoulli beam which is subject to a uniform distributed load per length $F(t) = 1.5 \sin(5t)$. All quantities are non-dimensional.

7.18 Determine the steady-state response for the transverse vibrations of a uniform Euler–Bernoulli beam fixed at $x = 0$ and subject to a harmonic force $F(t) = \sin(t)$ at $x = 1$. All quantities are non-dimensional.

7.19 Use the Laplace transform method to determine the response for the longitudinal oscillations of a uniform bar fixed at $x = 0$ but subject to a harmonic force $F(t) = 2 \sin(3t)$ at $x = 1$. All quantities are non-dimensional.

7.20 Use the Laplace transform method to determine the response for the transverse motion of a uniform fixed–fixed Euler–Bernoulli beam subject to a concentrated load of $F(t) = 1.5 \sin(4t)$ applied at $x = 1/3$. All quantities are non-dimensional.

7.21 Determine the response for the longitudinal motion of a uniform bar fixed at $x = 0$ and subject to a force $F(t) = 2e^{-0.5t}$ at $x = 1$. All quantities are non-dimensional.

7.22 Determine the response for the transverse motion of a uniform fixed-free Euler–Bernoulli beam which is subject to a uniform distributed load per length of $F(t) = 2e^{-3t}$ applied from $x = 1/2$ to $x = 2/3$.

7.23 Determine the steady-state response of the system of Figure 9.24 when the block is subject to a force $F(t) = 2.5 \sin(4t)$. Take $kL^3/EI = 0.5$ and $m/\rho AL = 1$. Assume all quantities are non-dimensional.

7.24 Determine the response of the system of Figure 9.24 when the block is subject to a force $F(t) = 2.5e^{-t}$. Take $kL^3/EI = 0.5$ and $m/\rho AL = 1$. Assume all quantities are non-dimensional.

7.25 Determine the response of the system of Figure 9.25 if the mid-span of the beam is subject to a force $F(t) = [u(t) - u(t - 0.5)]$. Take $kL^3/EI = 1$ and $m/\rho AL = 0.5$. Assume all quantities are non-dimensional.

7.26 Determine the response for the transverse oscillations of a rectangular membrane with $\eta = 2$ and the is clamped on all edges when it is subject to a force $F(t) = 1.5 \sin(10t)$ at $x = 0.5$, $y = 0.5$.

7.27 Determine the response for the transverse oscillations of a circular membrane which is clamped around its entire circumference when it is subject to a concentrated load of $F(t) = 10 \sin(10t)$ applied at its geometric center.

9.8 CHAPTER 8

8.1 Use the Rayleigh–Ritz method to approximate the lowest natural frequencies and mode shapes for the longitudinal oscillations of a uniform fixed-free bar. Use a polynomial basis with elements from $P^5[0,1]$.

8.2 Use the Rayleigh–Ritz method to approximate the lowest natural frequencies and mode shapes for the torsional oscillations of a uniform circular shaft fixed at $x = 0$ and connected to a torsional spring at $x = 1$ such that $k_tL/JG = 2$. Use a polynomial basis with elements from $P^5[0,1]$.

8.3 Use the Rayleigh–Ritz method to approximate the lowest natural frequencies and mode shapes of transverse oscillations of a fixed-pinned Euler–Bernoulli beam. Use a polynomial basis with elements from $P^6[0,1]$.

8.4 Use the Rayleigh–Ritz methods to approximate the lowest natural frequencies and mode shapes of a fixed-free Euler–Bernoulli beam with a constant axial load such that $\varepsilon = 0.5$. Use a polynomial basis with elements from $P^6[0,1]$.

8.5 Use the Rayleigh–Ritz method to approximate the lowest natural frequencies and mode shapes of a fixed-free Timoshenko beam. Use polynomial basis functions for both $w(x)$ and $\psi(x)$ which belong to $P^2[0,1]$.

8.6 Use the Rayleigh–Ritz method to approximate the lowest natural frequencies and mode shapes for the longitudinal oscillations of a linearly tapered bar fixed at $x = 0$ and attached to a linear spring of stiffness $k = 2EA_0/L$ at $x = 1$. Assume a taper ratio of 0.1. Use the three lowest mode shapes for the longitudinal motion of a uniform bar attached to the same linear spring as basis functions.

8.7 Consider the system of Figure 8.15 with the left end fixed rather than pinned. Use the assumed modes method to determine the response of the system using the exact mode shapes for the three lowest modes of transverse oscillation of a uniform fixed-pinned beam as basis functions.

8.8 Repeat Problem 8.7 assuming with a fixed-free beam, but using basis functions which only satisfy the geometric boundary conditions.

8.9 Repeat Problem 8.1 using the finite-element method with three elements.

8.10 Repeat Problem 8.2 using the finite-element method with two elements.

8.11 Repeat Problem 8.3 using the finite-element method with three elements.

8.12 Repeat Problem 8.4 using the finite-element method with three elements.

8.13 Repeat Problem 8.6 using the finite-element method with three elements.

References

1. Abramowitz, M and Stegun, I. A., *Handbook of Mathematical Functions*, Dover, Mineola, NY, 1965.
2. Balachandran, B. and Nagrab, E. B., *Vibrations*, Brooks-Cole, Belmont, CA, 2004.
3. Beer, F. W., Johnston, E. R., Clausen, W. R., and Cornwell, P. J., *Vector Mechanics for Engineers-Dynamics*, McGraw-Hill, New York, 2006.
4. Courant, R. and Hilbert, D., *Methods of Mathematical Physics*, Wiley-Interscience, New York, 1962.
5. Den Hartog, J. P., *Mechanical Vibration*, 4th ed., McGraw-Hill, New York, 1956.
6. Franklin, J. N., *Matrix Theory*, Prentice Hall, Englewood Cliffs, NJ, 1968.
7. Ginsberg, J. H., *Mechanical and Structural Vibrations*, Wiley, New York, 2001.
8. Haug, E. and Choi, K. K., *Methods of Engineering Mathematics*, Prentice Hall, Englewood Cliffs, NJ, 1993.
9. Kelly, S. G., Development of normal mode theory for multi-degree-of-freedom systems using an energy inner product, *SIAM Review*, 30, 302–304, 1988.
10. Kelly, S. G., Natural frequency and mode shape perturbations for linear multi-degree-of-freedom systems, *SIAM Review*, 30, 634–638, 1988.
11. Kelly, S. G., *Fundamentals of Mechanical Vibrations*, 2nd ed., McGraw-Hill, New York, 2000.
12. Kreyszig, E., *Advanced Engineering Mathematics*, 5th ed., Wiley, New York, 1983.
13. Langhaar, H. L., *Energy Methods in Applied Mechanics*, Wiley, New York, 1962.
14. Meirovitch, L., *Analytical Methods in Vibrations*, Macmillan, New York, 1967.
15. Meirovitch, L., A new method of the solution of the eigenvalue problem for gyroscopic systems, *AIAA Journal*, 12, 1137–1342, 1974.
16. Meirovitch, L., A modal analysis for the response of linear gyroscopic systems, *Journal of Applied Mechanics*, 12, 446–450, 1975.
17. Meitorvitch, L., *Principles and Techniques of Vibrations*, Prentice Hall, Upper Saddle River, NJ, 1997.
18. Pelesko, J. A. and Bernstein, D. H., *Modeling MEMS and NEMS*, Chapman and Hall/CRC, Boca Raton, Fl, 2003.
19. Prenter, P. M., *Splines and Variational Methods*, Wiley-Ineterscience, New York, 1975.

20. Reddy, J. N., *An Introduction to the Finite Element Method*, McGraw-Hill, New York, 1984.

21. Shames, I. H. and Dym, C. L., *Energy and Finite Element Methods in Structural Mechanics*, McGraw-Hill, New York, 1985.

22. Smith, D. R., *Variational Methods in Optimization*, Prentice Hall, Englewood Cliffs, NJ, 1974.

23. Strang, G. and Fix, G. J., *An Analysis of the Finite Element Method*, Prentice Hall, Englewood Cliffs, NJ, 1973.

24. Taylor, A. E., *Introduction to Functional Analysis*, Wiley, New York, 1957.

25. Thomson, W. T. and Dahileh, M. D., *Theory of Vibrations with Applications*, 5th ed., Prentice Hall, Englewood Cliffs, NJ, 1993.

26. Timoshenko, S. P., On the transverse vibrations of bars of uniform cross section, *Philosophical Magazine*, 6(43), 125–131, 1922.

27. Weinsotck, R., *Calculus of Variations with Applications to Physics and Engineering*, McGraw-Hill, New York, 1952.

28. Yuom, J. and Ru, C. Q., Non coaxial resonance of an isolated multiwalled nanotube, *Physical Review B*, 66, 23340–233402, 2002.

29. Zhang, Y. Y., Wang, C. M., and Tan, V. B. C., Buckling of multiwalled carbon nanotubes using Timoshenko beam theory, *Journal of Engineering Mechanics*, 132, 952–958, 2006.

30. Zienkiewicz, O. C. and Taylor, R. I., *The Finite Element Method*, 4th ed., McGraw-Hill, London, U.K., 1991.

Index

A

Adjoint operators, linear algebra and, 212–219
Admissible functions, Rayleigh–Ritz method and, 556–559
Algebra, linear, 173–224
AM, linear operators and, 240–242
Angular momentum, 8–17
Assumed modes
development, 575–577
method, 560–570

B

Bar element, 577–584
Bars
Hamilton's principle and, 138, 145–150
variational methods and, 138–150
Basis and dimension
linear algebra and, 185–189
vectors, 185–189
Beam element, 584–593
Beams
dual fixed-fixed, 382
dual pinned-pinned, 385
Timoshenko, 262–266, 398–409
Boundary conditions, 139–141
Brachistochrone problem, 93, 97

C

Cauchy sequence, 204
Cauchy–Schwartz inequality, 192
Combined continuous and discrete systems, 409–414
Complete, orthogonal expansions and, 204

Components, vibrating systems and, 17–38
Conservative systems
discrete, 104–112
free vibrations and, 287–435
combined continuous and discrete systems, 409–414
continuous systems, 341–342
discrete systems, 316–326
eigenvalue properties, 292–303
eigenvector properties, 292–303
Euler–Bernoulli beams, 360–375
flexibility matrix, 326–330
free vibrations
free response, 309–316
Green's functions, 430–435
matrix iteration, 330–341
membranes, 414–430
normal mode solution, 287–292
Rayleigh's quotient, 303–306
repeated structures, 375–398
second-order problems, 342–360
solvability conditions, 306–309
Timoshenko beams, 398–409
Continuous and discrete systems, 278–283
combined, 409–414
Continuous systems, 49–56, 136–150
attached inertia elements, 272–278
conservative systems and, 341–342
harmonic excitation, 480–490
Laplace transform method and, 497–501
linear operators and, 242–244
modal analysis and, 504–516

nondimensional boundary
        conditions, 252–253
nondimensionalizations,
        247–248
non-self-adjoint systems and, viscous
        damping, 458–463
partial differential equations,
        242–244
Converge, 204
    convergent sequence, 204
        Cauchy, 204
Coulomb dampers, 32–33

D

Dampers
    Coulomb, 32–33
    Kelvin–Voight model, 34
    viscous, 30–34
Damping ratio, 64
    discrete systems and, 516–523
    general viscous, 517–523
    proportional, 516–517
Degrees of freedom, 1–7
    generalized coordinates, 1–7
Differential equations
    derivation of, 42–48
        Divergence Theorem, 172
        Euler–Bernoulli beams, 150–165
        membranes, 170–172
        Timoshenko beams, 166–169
        variational methods and, 87–172
    linear
        discrete systems and, 125–127
        operators and, discrete systems,
            227
    linearization of, 127–130
    partial, 242–244
    vibration problems and, 227
Discrete and continuous systems,
        278–283
    combined, 409–414
Discrete gyroscopic systems, 452–457
Discrete systems, 316–326
    forced response
        damping, 516

harmonic excitations, 465–480
general damping and,
        non-self-adjoint systems,
        446–452
harmonic excitations
    theory, 465–470
    vibration absorbers, 470–480
Laplace transform method and,
        491–497
linear operators and, 227
matrix eigenvalue problem,
        317–326
proportional damping and,
        non-self-adjoint systems,
        441–446
undamped, 501–504
Divergence Theorem, 172
Dual
    fixed-fixed beams, 382
    pinned-pinned beams, 382

E

Eigenvalue
    problem, matrix, 317–326
    properties, 292–303
    positive definite operators, 297
    self-adjoint operators, 292–297
Eigenvector properties, 292–303
    expansion theorem, 298–301
End conditions, Euler–Bernoulli beams
        and, 367
Energy
    dissipation, 30–34
        viscous dampers, 30–34
    inner products, 222–224
    methods, 87
    sources, external, 34–38
Euler–Bernoulli beams, 150–165
    conservative systems and, 360–375
    end conditions, equations for, 367
    linear operators and, 253–261
Euler–Lagrange equation, 93–99
    brachistochrone problem, 93, 97
Exercises, 595–626

Expansion theorem, eigenvector
    properties and, 298–301
External energy sources, 34–38

F

Finite-element method, 525, 570–574
    assumed modes development,
        575–577
    bar element, 577–584
    beam element, 584–593
Finite-dimensional vector space, 187
Flexibility matrix, 234–240,
        326–330
    natural frequency calculations,
        326–330
Forced response, 465–523
    continuous systems, harmonic
        excitation, 480–490
    discrete systems
        damping, 516
        harmonic excitations, 465–480
    Laplace transform method, 490–501
    Rayleigh–Ritz method and, 551–556
    undamped
        continuous systems, modal
            analysis, 504–516
        discrete systems, modal analysis,
            501–504
Fourier
    best approximation theorem,
        525–527
    series, 80–82
Fredholm alternative, 307
Free response, 309–314
    normal mode solution, 309–316
    principal coordinates, 314–316
Free vibrations, 287–435
    conservative systems
        combined continuous and discrete
            systems, 409–414
        continuous systems, 341–342
        discrete systems, 316–326
        eigenvalue properties, 292–303
        eigenvector properties, 292–303

Euler–Bernoulli beams,
        360–375
    flexibility matrix, 326–330
    free response, 309–316
    Green's functions, 430–435
    matrix iteration, 330–341
    membranes, 414–430
    normal model solution, 287–292
    Rayleigh's quotient, 303–306
    repeated structures, 375–398
    second-order problems,
        342–360
    solvability conditions, 306–309
    Timoshenko beams, 398–409
    linear one-degree-of-freedom systems
        and, 63–70
Frequency squared excitation, 73–75
Functionals, 87–91

G

Galerkin method, 531–532
Gateaux variations, 93
General
    damping, discrete systems and,
        non-self-adjoint systems,
            446–452
    free response, 309–314
    periodic input, Fourier series,
        80–82
    viscous damping, 517–523
Generalized coordinates, 1–7
Gram–Schmidt
    orthonormalization method,
        197–202
    theorem, Hilbert space, 204
Green's functions, conservative systems
    and, 430–435
Gyroscopic systems, 130–136

H

Hamilton's principle, 100–104, 112, 168
    bars, 138, 145–150
    Euler–Bernoulli beams, 150–165
    membranes, 171

shafts, 138, 145–150
strings, 138, 145–150
Harmonic excitations, 70–83
  continuous systems and, 480–490
  discrete systems and, 465–480
  theory, 465–470
  vibration absorbers, 470–480
Hilbert space, 204–206
  Gram–Schmidt theorem and, 204
Hooke's Law, 27–28

I

Inertia elements, 17–22
  continuous systems and, 272–278
Infinite dimensional vector space, 187
  operators and, 209–210
Inner products, 189–193
  Cauchy–Schwartz inequality, 192
  scalar function, 189

K

Kelvin–Voight dampers, 34
Kinetic energy, 8–17, 139–141

L

Lagrange's equation, discrete systems
  conservative, 104–112
  non-conservative, 112–122
Laplace transform method, 82–84,
    490–501
  continuous systems, 497–501
  discrete systems, 491–497
Linear algebra, 173–224
  adjoint operators, 212–219
  basis and dimension, 185–189
  energy inner products, 222–224
  Gram–Schmidt orthonormalization
    method, 197–202
  inner products, 189–193
  linear
    independence, 182–185
    operators, 206–212
  orthogonal expansions, 202–206

positive definite operators, 219–222
three-dimensional space, 174–177
vector spaces, 177–182
vectors, 185–189
  norms, 193–197
Linear discrete systems, 122–130
  differential equations, 125–127
  linearization of, 127–130
  quadratic forms, 122–125
Linear independence, 182–185
  definition of, 183
  linearly dependent, 183
  Wronskian, 184–185
Linear one-degree-of-freedom systems
  damping ratio, 64
  free vibrations, 63–70
Linear operators, 206–212
  continuous systems, attached inertia
    elements, 272–278
  differential equations, discrete
    systems, 227
  vibration problems and, 225–285
    continuous and discrete systems,
      278–283
    continuous systems, 242–244
    Euler–Bernoulli beam, 253–261
    flexibility matrix, 234–240
    M^1K and AM, 240–242
    mass matrix, 233–234
    membranes, 283–285
    multiple deformable bodies,
      266–272
    partial differential equations,
      242–244
    second-order problems,
      245–246, 249–251, 253
    stiffness matrix, 227–233
    Timoshenko beams, 262–266
Linearization, differential equations
    and, 127–130
Linearly dependent, 183

M

M^1K, linear operators and, 240–242
Mass matrix, 227–233

Matrix
    eigenvalue problem, 317–326
    flexibility, 234–240
    iteration, 330–341
    mass, 227–233
    stiffness, 227–233
Membranes, 170–172, 283–285
    conservative systems and, 414–430
    Hamilton's Principle, 171
Modal analysis, undamped
    continuous systems and, 504–516
    discrete systems and, 501–504
Modes
    assumed development, 575–577
    assumed method, 560–570
    shapes, Rayleigh–Ritz method and,
        532–551
Motion input, 75–80
Multiple deformable bodies, linear
        operators and, 266–272

N

Natural frequencies
    calculations, 326–330
    Rayleigh–Ritz method and, 532–551
    Timoshenko beams and, 408
Newton's Second Law, particles, 8–13
Non-conservative discrete systems,
        112–222
    Hamilton's principle, 112
    Rayleigh's dissipation function,
        120–121
Nondimensional boundary conditions
    continuous systems and, 252–253
    Timoshenko beams and, 264–266
Nondimensionalizations, continuous
        systems and, 247–248
Non-self-adjoint systems, 437–463
    continuous systems, viscous damping,
        458–463
    discrete
        gyroscopic systems, 452–457
        systems
            general damping, 446–452
            proportional damping, 441–446

Normal mode solution
    conservative systems and, 287–292
    free response and, 309–316
Norms, 193–197

O

One-degree-of-freedom
    model, second-order problems and,
        348
    systems
        continuous, 49–56
        differential equations, 42–48
        linear, 63–70
        modeling of, 38–56
        qualitative aspects, 56–63
        static spring forces, 39–42
Operators
    adjoint, 212–219
    infinite dimensional vector spaces
        and, 209–210
    linear, 206, 225–285
    self-adjoint, 292–297
Orthogonal expansions, 202–206
    converge, 204
    Hilbert space, 204–206
    vectors, complete, 204
Orthonormalization method,
        Gram–Schmidt, 197–202

P

Partial differential equations, linear
        operators and, 242–244
Particles
    Newton's Second Law and, 8–13
    system of, 9–13
Positive definite operators
    eigenvalue properties and, 297
    linear algebra and, 219–222
Potential energy, 139–141
Principal coordinates, free response and,
        314–316
Proportional damping, 516–517
    discrete systems and, non-self-adjoint
        systems, 441–446

Q

Quadratic forms, linear discrete systems
    and, 122–125
Qualitative aspects,
    one-degree-of-freedom systems
    and, 56–63

R

Rayleigh's
    dissipation function, 120
    quotient, 303–306
Rayleigh–Ritz method, 525, 528–531
    admissible functions, 556–559
    assumed modes method,
        560–570
    forced response, 551–556
    mode shapes, 532–551
    natural frequencies, 532–551
Repeated structures, 375–398
    dual
        fixed-fixed beams, 382
        pinned-pinned beams, 385
        torsional springs, 390
Rigid bodies, 13–17

S

Scalar function, 189
Second-order
    problems, 139–141, 245–246,
        249–251, 253, 342–360
        one-degree-of-freedom model, 348
        Sturm–Liouville, 343
    systems
        boundary conditions, 139–141
        kinetic energy, 139–141
        potential energy, 139–141
Self-adjoint operators,
    eigenvalue properties and, 292–297
    non operators, 437–463
Shafts
    Hamilton's principle and, 138,
        145–150
    variational methods and, 138–150

Single-degree-of-freedom system,
    1–85
    frequency squared excitation,
        73–75
    general periodic input, 80–82
    general theory, 70–73
    harmonic excitation, 70–82
    Laplace transform method,
        82–84
    motion input, 75–80
    transient response, 82–85
    vibrations, 1–85
Solvability conditions, 306–309
    Fredholm alternative, 307
Static spring forces, 39–42
Stiffness
    elements, 22–30
    Hooke's Law, 27–28
    matrix, 227–233
Strings
    Hamilton's principle and, 138,
        145–150
    variational methods and, 138–150
Sturm–Liouville problem, 343
System of particles, 9–13

T

Three-dimensional space, 174–177
Timoshenko beams, 398–409
    Hamilton's Principle, 168
    linear operators and, 262–266
    natural frequencies, 408
    nondimensional boundary
        conditions, 264–266
Torsional springs, 390
Transient response, 82–85

U

Undamped
    continuous systems, modal analysis,
        504–516
    discrete systems, modal analysis,
        501–504

V

Variational methods
  bars, strings and shafts, 138–150
  conservative discrete systems,
    104–112
  continuous systems, 136–150
  differential equation derivation,
    87–172
  energy methods, 87
  Euler–Lagrange equation,
    93–99
  functionals, 87–91
  gyroscopic systems, 130–136
  Hamilton's principle, 100–104
  Lagrange's equation, 104–112
  linear discrete systems, 122–130
  second-order systems, 139–141
  shafts, 138–150
  strings, 138–150
  types, 91–93
Variations, Gateaux, 93
Vector spaces, 177–182
Vectors, 185–189
  basis and dimension, 185–189
  complete, 204
  converge, 204
  finite-dimensional vector space, 187
  infinite dimensional vector space, 187
  norms, 193–197
Vibrating systems
  components, 17–38
    energy dissipation, 30–34
    inertia elements, 17–22
    stiffness elements, 22–30
  compounds, external energy sources,
    34–38
  one-degree-of-freedom systems,
    38–56
  qualitative aspects, 56–63
  single-degree-of-freedom system
    harmonic excitation, 70–82
    transient response, 82–85
Vibration
  absorbers, 470–480

angular momentum, 8–17
conservative systems and,
  287–435
degrees of freedom, 1–7
generalized coordinates, 1–7
kinetic energy, 8–17
linear operators, 225–285
Newton's Second Law, 8–13
problems with, 225–285
  differential equations and, 227
  linear operators, 225–285
    continuous and discrete
      systems, 278–283
    continuous systems, 242–244
    Euler–Bernoulli beam, 253–261
    flexibility matrix, 234–240
    M^1K and AM, 240–242
    mass matrix, 233–234
    membranes, 283–285
    multiple deformable bodies,
      266–272
    partial differential equations,
      242–244
    second-order problems,
      245–253
    stiffness matrix, 227–233
    Timoshenko beams, 262–266
  rigid bodies, 13–17
  single-degree-of-freedom systems
    and, 1–85
  system of particles, 9–13
  vibrating systems, 17–38
Viscous
  dampers, 30–34
    Coulomb, 32–33
    Kelvin–Voight model, 34
  damping, 441–446
  non-self-adjoint systems and,
    458–463

W

Wave equation. *See* Second-order
  problems.
Wronskian, 184–185

Related Titles

Engineering Vibrations
William J. Bottega
ISBN: 0849334209

Mechanical Vibration: Analysis, Uncertainties and Control 2/e
Haym Benaroya
ISBN: 0824753801